Calcined Clays for Sustainable Concrete

RILEM BOOKSERIES

Volume 16

RILEM, The International Union of Laboratories and Experts in Construction Materials, Systems and Structures, founded in 1947, is a non-governmental scientific association whose goal is to contribute to progress in the construction sciences, techniques and industries, essentially by means of the communication it fosters between research and practice. RILEM's focus is on construction materials and their use in building and civil engineering structures, covering all phases of the building process from manufacture to use and recycling of materials. More information on RILEM and its previous publications can be found on www.RILEM.net.

More information about this series at http://www.springer.com/series/8781

Fernando Martirena · Aurélie Favier
Karen Scrivener
Editors

Calcined Clays for Sustainable Concrete

Proceedings of the 2nd International Conference on Calcined Clays for Sustainable Concrete

110 €
T/Werkstoff

Springer

Editors
Fernando Martirena
University Marta Abreu Santa Clara
Santa Clara
Cuba

Karen Scrivener
EPFL/STI/IMX/LMC
Lausanne
Switzerland

Aurélie Favier
EPFL/STI/IMX/LMC
Lausanne
Switzerland

ISSN 2211-0844 ISSN 2211-0852 (electronic)
Calcined Clays for Sustainable Concrete
ISBN 978-94-024-1206-2 ISBN 978-94-024-1207-9 (eBook)
https://doi.org/10.1007/978-94-024-1207-9

Library of Congress Control Number: 2017954894

Printed on acid-free paper

This Springer imprint is published by Springer Nature
The registered company is Springer Science+Business Media B.V.
The registered company address is: Van Godewijckstraat 30, 3311 GX Dordrecht, The Netherlands

Foreword

The 1st International Conference on Calcined Clay for Sustainable Concrete has become a milestone in terms of reviving the use of calcined clays as supplementary cementitious materials (SCMs) in the production of cement worldwide. We have seen an increasing number of publications made by various research groups in different regions, all focused on improving the sustainability of cementitious materials and their application.

Although SCMs like slag and fly ash will continue to be used, their use shall be more limited as the energy generation becomes less carbonized and the industry shifts to more sustainable practices. Calcined clays are proven to abundant in almost every corner of the world, so they become the most promising source of additional SCMs which can make a substantial contribution to lower further the environmental impact of cement and concrete.

This book of proceedings of the **2nd International Conference on Calcined Clay for Sustainable Concrete** presents a snapshot of the intense work carried out by different research groups since 2015. It brings together written versions of the more than 80 presentation and posters delivered at the conference. A broad range of themes are addressed by practitioners and researchers, from cement hydration to concrete durability and environmental impact. This book provides an excellent account of the state of the art on the subject with a truly international perspective.

Topics covered are clay geology, hydration of blended cement, performance and durability, alkali-activated binders, economic and ecological impacts and field applications.

The Editors would like to thank the authors for the outstanding contributions and the sponsors and various other organizations for their contributions and help in making these proceedings and conference a success. The 3rd International Conference is already planned to take place in India in 2019.

Aurélie R. Favier
Fernando Martirena
Karen L. Scrivener

Organization

Organizing Committee

Fernando Martirena	CIDEM, Cuba
Karen Scrivener	EPFL, Switzerland
Aurélie Favier	EPFL, Switzerland
Maude Schneider	EPFL, Switzerland
Adrian Alujas	UCLV, Cuba
Meylin Amador Hernandez	UCLV, Cuba

Scientific Committee

Karen Scrivener	Laboratory of Construction Materials EPFL
Shashank Bishnoi	Indian Institute of Technology, Delhi, India
Anjan Chatterjee	Conmat Technologies PVT. LTD
Jan Elsen	University of Leuven, Belgium
Duncan Herfort	Aalborg Cement Denmark
Juan J. Howland	CUJAE, Cuba
Edgardo F. Irassar	University Central Buenos Aires, Argentina
Maria Juenger	University of Texas at Austin, USA
Harald Justnes	Sintef, Norway
Horst-Michael Ludwig	University of Weimar, Germany
Soumen Maity	TARA, Delhi, India
Fernando Martirena	Universidad central de las Villas, Cuba
John Provis	University of Sheffield, UK
Manu Santhanam	Indian Institute of Technology, Madras, India
Jorgen Skibsted	University Aarhus, Denmark
Zhonghe Shui	Wuhan University of Technology, China
Vanderley John	University of São Paulo, Brazil

Gold Sponsors

thyssenkrupp

IPIAC

Silver and Exhibitor sponsors

gcp
applied technologies

KHD | HUMBOLDT WEDAG

LafargeHolcim

Contents

RILEM Publications

The following list is presenting the global offer of RILEM Publications, sorted by series. Each publication is available in printed version and/or in online version.

RILEM Proceedings (PRO)

PRO 1: Durability of High Performance Concrete (ISBN: 2-912143-03-9; e-ISBN: 2-351580-12-5; e-ISBN: 2351580125); *Ed. H. Sommer*
PRO 2: Chloride Penetration into Concrete (ISBN: 2-912143-00-04; e-ISBN: 2912143454); *Eds. L.-O. Nilsson and J.-P. Ollivier*
PRO 3: Evaluation and Strengthening of Existing Masonry Structures (ISBN: 2-912143-02-0; e-ISBN: 2351580141); *Eds. L. Binda and C. Modena*
PRO 4: Concrete: From Material to Structure (ISBN: 2-912143-04-7; e-ISBN: 2351580206); *Eds. J.-P. Bournazel and Y. Malier*
PRO 5: The Role of Admixtures in High Performance Concrete (ISBN: 2-912143-05-5; e-ISBN: 2351580214); *Eds. J.G. Cabrera and R. Rivera-Villarreal*
PRO 6: High Performance Fiber Reinforced Cement Composites - HPFRCC 3 (ISBN: 2-912143-06-3; e-ISBN: 2351580222); *Eds. H.W. Reinhardt and A.E. Naaman*
PRO 7: 1st International RILEM Symposium on Self-Compacting Concrete (ISBN: 2-912143-09-8; e-ISBN: 2912143721); *Eds. Å. Skarendahl and Ö. Petersson*
PRO 8: International RILEM Symposium on Timber Engineering (ISBN: 2-912143-10-1; e-ISBN: 2351580230); *Ed. L. Boström*
PRO 9: 2nd International RILEM Symposium on Adhesion between Polymers and Concrete ISAP '99 (ISBN: 2-912143-11-X; e-ISBN: 2351580249); *Eds. Y. Ohama and M. Puterman*
PRO 10: 3rd International RILEM Symposium on Durability of Building and Construction Sealants (ISBN: 2-912143-13-6; e-ISBN: 2351580257); *Eds. A.T. Wolf*

PRO 11: 4th International RILEM Conference on Reflective Cracking in Pavements (ISBN: 2-912143-14-4; e-ISBN: 2351580265); *Eds. A.O. Abd El Halim, D.A. Taylor and El H.H. Mohamed*
PRO 12: International RILEM Workshop on Historic Mortars: Characteristics and Tests (ISBN: 2-912143-15-2; e-ISBN: 2351580273); *Eds. P. Bartos, C. Groot and J.J. Hughes*
PRO 13: 2nd International RILEM Symposium on Hydration and Setting (ISBN: 2-912143-16-0; e-ISBN: 2351580281); *Ed. A. Nonat*
PRO 14: Integrated Life-Cycle Design of Materials and Structures - ILCDES 2000 (ISBN: 951-758-408-3; e-ISBN: 235158029X); (ISSN: 0356-9403); *Ed. S. Sarja*
PRO 15: Fifth RILEM Symposium on Fibre-Reinforced Concretes (FRC) - BEFIB'2000 (ISBN: 2-912143-18-7; e-ISBN: 291214373X); *Eds. P. Rossi and G. Chanvillard*
PRO 16: Life Prediction and Management of Concrete Structures (ISBN: 2-912143-19-5; e-ISBN: 2351580303); *Ed. D. Naus*
PRO 17: Shrinkage of Concrete – Shrinkage 2000 (ISBN: 2-912143-20-9; e-ISBN: 2351580311); *Eds. V. Baroghel-Bouny and P.-C. Aïtcin*
PRO 18: Measurement and Interpretation of the On-Site Corrosion Rate (ISBN: 2-912143-21-7; e-ISBN: 235158032X); *Eds. C. Andrade, C. Alonso, J. Fullea, J. Polimon and J. Rodriguez*
PRO 19: Testing and Modelling the Chloride Ingress into Concrete (ISBN: 2-912143-22-5; e-ISBN: 2351580338); *Eds. C. Andrade and J. Kropp*
PRO 20: 1st International RILEM Workshop on Microbial Impacts on Building Materials (CD 02) (e-ISBN 978-2-35158-013-4); *Ed. M. Ribas Silva*
PRO 21: International RILEM Symposium on Connections between Steel and Concrete (ISBN: 2-912143-25-X; e-ISBN: 2351580346); *Ed. R. Eligehausen*
PRO 22: International RILEM Symposium on Joints in Timber Structures (ISBN: 2-912143-28-4; e-ISBN: 2351580354); *Eds. S. Aicher and H.-W. Reinhardt*
PRO 23: International RILEM Conference on Early Age Cracking in Cementitious Systems (ISBN: 2-912143-29-2; e-ISBN: 2351580362); *Eds. K. Kovler and A. Bentur*
PRO 24: 2nd International RILEM Workshop on Frost Resistance of Concrete (ISBN: 2-912143-30-6; e-ISBN: 2351580370); *Eds. M.J. Setzer, R. Auberg and H.-J. Keck*
PRO 25: International RILEM Workshop on Frost Damage in Concrete (ISBN: 2-912143-31-4; e-ISBN: 2351580389); *Eds. D.J. Janssen, M.J. Setzer and M.B. Snyder*
PRO 26: International RILEM Workshop on On-Site Control and Evaluation of Masonry Structures (ISBN: 2-912143-34-9; e-ISBN: 2351580141); *Eds. L. Binda and R.C. de Vekey*
PRO 27: International RILEM Symposium on Building Joint Sealants (CD03; e-ISBN: 235158015X); *Ed. A.T. Wolf*
PRO 28: 6th International RILEM Symposium on Performance Testing and Evaluation of Bituminous Materials - PTEBM'03 (ISBN: 2-912143-35-7; e-ISBN: 978-2-912143-77-8); *Ed. M.N. Partl*

PRO 29: 2nd International RILEM Workshop on Life Prediction and Ageing Management of Concrete Structures (ISBN: 2-912143-36-5; e-ISBN: 2912143780); *Ed. D.J. Naus*

PRO 30: 4th International RILEM Workshop on High Performance Fiber Reinforced Cement Composites - HPFRCC 4 (ISBN: 2-912143-37-3; e-ISBN: 2912143799); *Eds. A.E. Naaman and H.W. Reinhardt*

PRO 31: International RILEM Workshop on Test and Design Methods for Steel Fibre Reinforced Concrete: Background and Experiences (ISBN: 2-912143-38-1; e-ISBN: 2351580168); *Eds. B. Schnütgen and L. Vandewalle*

PRO 32: International Conference on Advances in Concrete and Structures 2 vol. (ISBN (set): 2-912143-41-1; e-ISBN: 2351580176); *Eds. Ying-shu Yuan, Surendra P. Shah and Heng-lin Lü*

PRO 33: 3rd International Symposium on Self-Compacting Concrete (ISBN: 2-912143-42-X; e-ISBN: 2912143713); *Eds. Ó. Wallevik and I. Níelsson*

PRO 34: International RILEM Conference on Microbial Impact on Building Materials (ISBN: 2-912143-43-8; e-ISBN: 2351580184); *Ed. M. Ribas Silva*

PRO 35: International RILEM TC 186-ISA on Internal Sulfate Attack and Delayed Ettringite Formation (ISBN: 2-912143-44-6; e-ISBN: 2912143802); *Eds. K. Scrivener and J. Skalny*

PRO 36: International RILEM Symposium on Concrete Science and Engineering – A Tribute to Arnon Bentur (ISBN: 2-912143-46-2; e-ISBN: 2912143586); *Eds. K. Kovler, J. Marchand, S. Mindess and J. Weiss*

PRO 37: 5th International RILEM Conference on Cracking in Pavements – Mitigation, Risk Assessment and Prevention (ISBN: 2-912143-47-0; e-ISBN: 2912143764); *Eds. C. Petit, I. Al-Qadi and A. Millien*

PRO 38: 3rd International RILEM Workshop on Testing and Modelling the Chloride Ingress into Concrete (ISBN: 2-912143-48-9; e-ISBN: 2912143578); *Eds. C. Andrade and J. Kropp*

PRO 39: 6th International RILEM Symposium on Fibre-Reinforced Concretes - BEFIB 2004 (ISBN: 2-912143-51-9; e-ISBN: 2912143748); *Eds. M. Di Prisco, R. Felicetti and G.A. Plizzari*

PRO 40: International RILEM Conference on the Use of Recycled Materials in Buildings and Structures (ISBN: 2-912143-52-7; e-ISBN: 2912143756); *Eds. E. Vázquez, Ch. F. Hendriks and G.M.T. Janssen*

PRO 41: RILEM International Symposium on Environment-Conscious Materials and Systems for Sustainable Development (ISBN: 2-912143-55-1; e-ISBN: 2912143640); *Eds. N. Kashino and Y. Ohama*

PRO 42: SCC'2005 - China: 1st International Symposium on Design, Performance and Use of Self-Consolidating Concrete (ISBN: 2-912143-61-6; e-ISBN: 2912143624); *Eds. Zhiwu Yu, Caijun Shi, Kamal Henri Khayat and Youjun Xie*

PRO 43: International RILEM Workshop on Bonded Concrete Overlays (e-ISBN: 2-912143-83-7); *Eds. J.L. Granju and J. Silfwerbrand*

PRO 44: 2nd International RILEM Workshop on Microbial Impacts on Building Materials (CD11) (e-ISBN: 2-912143-84-5); *Ed. M. Ribas Silva*

PRO 45: 2nd International Symposium on Nanotechnology in Construction, Bilbao (ISBN: 2-912143-87-X; e-ISBN: 2912143888); *Eds. Peter J.M. Bartos, Yolanda de Miguel and Antonio Porro*

PRO 46: ConcreteLife'06 - International RILEM-JCI Seminar on Concrete Durability and Service Life Planning: Curing, Crack Control, Performance in Harsh Environments (ISBN: 2-912143-89-6; e-ISBN: 291214390X); *Ed. K. Kovler*

PRO 47: International RILEM Workshop on Performance Based Evaluation and Indicators for Concrete Durability (ISBN: 978-2-912143-95-2; e-ISBN: 9782912143969); *Eds. V. Baroghel-Bouny, C. Andrade, R. Torrent and K. Scrivener*

PRO 48: 1st International RILEM Symposium on Advances in Concrete through Science and Engineering (e-ISBN: 2-912143-92-6); *Eds. J. Weiss, K. Kovler, J. Marchand, and S. Mindess*

PRO 49: International RILEM Workshop on High Performance Fiber Reinforced Cementitious Composites in Structural Applications (ISBN: 2-912143-93-4; e-ISBN: 2912143942); *Eds. G. Fischer and V.C. Li*

PRO 50: 1st International RILEM Symposium on Textile Reinforced Concrete (ISBN: 2-912143-97-7; e-ISBN: 2351580087); *Eds. Josef Hegger, Wolfgang Brameshuber and Norbert Will*

PRO 51: 2nd International Symposium on Advances in Concrete through Science and Engineering (ISBN: 2-35158-003-6; e-ISBN: 2-35158-002-8); *Eds. J. Marchand, B. Bissonnette, R. Gagné, M. Jolin and F. Paradis*

PRO 52: Volume Changes of Hardening Concrete: Testing and Mitigation (ISBN: 2-35158-004-4; e-ISBN: 2-35158-005-2); *Eds. O.M. Jensen, P. Lura and K. Kovler*

PRO 53: High Performance Fiber Reinforced Cement Composites - HPFRCC5 (ISBN: 978-2-35158-046-2; e-ISBN: 978-2-35158-089-9); *Eds. H.W. Reinhardt and A.E. Naaman*

PRO 54: 5th International RILEM Symposium on Self-Compacting Concrete (ISBN: 978-2-35158-047-9; e-ISBN: 978-2-35158-088-2); *Eds. G. De Schutter and V. Boel*

PRO 55: International RILEM Symposium Photocatalysis, Environment and Construction Materials (ISBN: 978-2-35158-056-1; e-ISBN: 978-2-35158-057-8); *Eds. P. Baglioni and L. Cassar*

PRO 56: International RILEM Workshop on Integral Service Life Modelling of Concrete Structures (ISBN 978-2-35158-058-5; e-ISBN: 978-2-35158-090-5); *Eds. R.M. Ferreira, J. Gulikers and C. Andrade*

PRO 57: RILEM Workshop on Performance of cement-based materials in aggressive aqueous environments (e-ISBN: 978-2-35158-059-2); *Ed. N. De Belie*

PRO 58: International RILEM Symposium on Concrete Modelling - CONMOD'08 (ISBN: 978-2-35158-060-8; e-ISBN: 978-2-35158-076-9); *Eds. E. Schlangen and G. De Schutter*

PRO 59: International RILEM Conference on Site Assessment of Concrete, Masonry and Timber Structures - SACoMaTiS 2008 (ISBN set: 978-2-35158-061-5; e-ISBN: 978-2-35158-075-2); *Eds. L. Binda, M. di Prisco and R. Felicetti*

PRO 60: Seventh RILEM International Symposium on Fibre Reinforced Concrete: Design and Applications - BEFIB 2008 (ISBN: 978-2-35158-064-6; e-ISBN: 978-2-35158-086-8); *Ed. R. Gettu*

PRO 61: 1st International Conference on Microstructure Related Durability of Cementitious Composites 2 vol., (ISBN: 978-2-35158-065-3; e-ISBN: 978-2-35158-084-4); *Eds. W. Sun, K. van Breugel, C. Miao, G. Ye and H. Chen*

PRO 62: NSF/RILEM Workshop: In-situ Evaluation of Historic Wood and Masonry Structures (e-ISBN: 978-2-35158-068-4); *Eds. B. Kasal, R. Anthony and M. Drdácký*

PRO 63: Concrete in Aggressive Aqueous Environments: Performance, Testing and Modelling, 2 vol., (ISBN: 978-2-35158-071-4; e-ISBN: 978-2-35158-082-0); *Eds. M.G. Alexander and A. Bertron*

PRO 64: Long Term Performance of Cementitious Barriers and Reinforced Concrete in Nuclear Power Plants and Waste Management - NUCPERF 2009 (ISBN: 978-2-35158-072-1; e-ISBN: 978-2-35158-087-5); *Eds. V. L'Hostis, R. Gens, C. Gallé*

PRO 65: Design Performance and Use of Self-consolidating Concrete - SCC'2009 (ISBN: 978-2-35158-073-8; e-ISBN: 978-2-35158-093-6); *Eds. C. Shi, Z. Yu, K.H. Khayat and P. Yan*

PRO 66: 2nd International RILEM Workshop on Concrete Durability and Service Life Planning - ConcreteLife'09 (ISBN: 978-2-35158-074-5; ISBN: 978-2-35158-074-5); *Ed. K. Kovler*

PRO 67: Repairs Mortars for Historic Masonry (e-ISBN: 978-2-35158-083-7); *Ed. C. Groot*

PRO 68: Proceedings of the 3rd International RILEM Symposium on 'Rheology of Cement Suspensions such as Fresh Concrete (ISBN 978-2-35158-091-2; e-ISBN: 978-2-35158-092-9); *Eds. O.H. Wallevik, S. Kubens and S. Oesterheld*

PRO 69: 3rd International PhD Student Workshop on 'Modelling the Durability of Reinforced Concrete (ISBN: 978-2-35158-095-0); *Eds. R.M. Ferreira, J. Gulikers and C. Andrade*

PRO 70: 2nd International Conference on 'Service Life Design for Infrastructure' (ISBN set: 978-2-35158-096-7, e-ISBN: 978-2-35158-097-4); *Ed. K. van Breugel, G. Ye and Y. Yuan*

PRO 71: Advances in Civil Engineering Materials - The 50-year Teaching Anniversary of Prof. Sun Wei' (ISBN: 978-2-35158-098-1; e-ISBN: 978-2-35158-099-8); *Eds. C. Miao, G. Ye, and H. Chen*

PRO 72: First International Conference on 'Advances in Chemically-Activated Materials – CAM'2010' (2010), 264 pp, ISBN: 978-2-35158-101-8; e-ISBN: 978-2-35158-115-5, *Eds. Caijun Shi and Xiaodong Shen*

PRO 73: 2nd International Conference on 'Waste Engineering and Management - ICWEM 2010' (2010), 894 pp, ISBN: 978-2-35158-102-5; e-ISBN: 978-2-35158-103-2, *Eds. J. Zh. Xiao, Y. Zhang, M.S. Cheung and R. Chu*

PRO 74: International RILEM Conference on 'Use of Superabsorsorbent Polymers and Other New Additives in Concrete' (2010) 374 pp., ISBN: 978-2-35158-104-9; e-ISBN: 978-2-35158-105-6; *Eds. O.M. Jensen, M.T. Hasholt, and S. Laustsen*

PRO 75: International Conference on 'Material Science - 2nd ICTRC - Textile Reinforced Concrete - Theme 1' (2010) 436 pp., ISBN: 978-2-35158-106-3; e-ISBN: 978-2-35158-107-0; *Ed. W. Brameshuber*

PRO 76: International Conference on 'Material Science - HetMat - Modelling of Heterogeneous Materials - Theme 2' (2010) 255 pp., ISBN: 978-2-35158-108-7; e-ISBN: 978-2-35158-109-4; *Ed. W. Brameshuber*

PRO 77: International Conference on 'Material Science - AdIPoC - Additions Improving Properties of Concrete - Theme 3' (2010) 459 pp., ISBN: 978-2-35158-110-0; e-ISBN: 978-2-35158-111-7; *Ed. W. Brameshuber*

PRO 78: 2nd Historic Mortars Conference and RILEM TC 203-RHM Final Workshop – HMC2010 (2010) 1416 pp., e-ISBN: 978-2-35158-112-4; *Eds. J. Válek, C. Groot, and J.J. Hughes*

PRO 79: International RILEM Conference on Advances in Construction Materials Through Science and Engineering (2011) 213 pp., ISBN: 978-2-35158-116-2, e-ISBN: 978-2-35158-117-9; *Eds. Christopher Leung and K.T. Wan*

PRO 80: 2nd International RILEM Conference on Concrete Spalling due to Fire Exposure (2011) 453 pp., ISBN: 978-2-35158-118-6, e-ISBN: 978-2-35158-119-3; *Eds. E.A.B. Koenders and F. Dehn*

PRO 81: 2nd International RILEM Conference on Strain Hardening Cementitious Composites (SHCC2-Rio) (2011) 451 pp., ISBN: 978-2-35158-120-9, e-ISBN: 978-2-35158-121-6; *Eds. R.D. Toledo Filho, F.A. Silva, E.A.B. Koenders and E.M.R. Fairbairn*

PRO 82: 2nd International RILEM Conference on Progress of Recycling in the Built Environment (2011) 507 pp., e-ISBN: 978-2-35158-122-3; *Eds. V.M. John, E. Vazquez, S.C. Angulo and C. Ulsen*

PRO 83: 2nd International Conference on Microstructural-related Durability of Cementitious Composites (2012) 250 pp., ISBN: 978-2-35158-129-2; e-ISBN: 978-2-35158-123-0; *Eds. G. Ye, K. van Breugel, W. Sun and C. Miao*

PRO 84: CONSEC13 - Seventh International Conference on Concrete under Severe Conditions – Environment and Loading (2013) 1930 pp., ISBN: 978-2-35158-124-7; e-ISBN: 978-2- 35158-134-6; *Eds. Z.J. Li, W. Sun, C.W. Miao, K. Sakai, O.E. Gjorv and N. Banthia*

PRO 85: RILEM-JCI International Workshop on Crack Control of Mass Concrete and Related issues concerning Early-Age of Concrete Structures – ConCrack 3 – Control of Cracking in Concrete Structures 3 (2012) 237 pp., ISBN: 978-2-35158-125-4; e-ISBN: 978-2-35158-126-1; *Eds. F. Toutlemonde and J.-M. Torrenti*

PRO 86: International Symposium on Life Cycle Assessment and Construction (2012) 414 pp., ISBN: 978-2-35158-127-8, e-ISBN: 978-2-35158-128-5; *Eds. A. Ventura and C. de la Roche*

PRO 87: UHPFRC 2013 – RILEM-fib-AFGC International Symposium on Ultra-High Performance Fibre-Reinforced Concrete (2013), ISBN: 978-2-35158-130-8, e-ISBN: 978-2-35158-131-5; *Ed. F. Toutlemonde*

PRO 88: 8th RILEM International Symposium on Fibre Reinforced Concrete (2012) 344 pp., ISBN: 978-2-35158-132-2, e-ISBN: 978-2-35158-133-9; *Ed. Joaquim A.O. Barros*

PRO 89: RILEM International workshop on performance-based specification and control of concrete durability (2014) 678 pp, ISBN: 978-2-35158-135-3, e-ISBN: 978-2-35158-136-0; *Eds. D. Bjegović, H. Beushausen and M. Serdar*

PRO 90: 7th RILEM International Conference on Self-Compacting Concrete and of the 1st RILEM International Conference on Rheology and Processing of Construction Materials (2013) 396 pp, ISBN: 978-2-35158-137-7, e-ISBN: 978-2-35158-138-4; *Eds. Nicolas Roussel and Hela Bessaies-Bey*

PRO 91: CONMOD 2014 - RILEM International Symposium on Concrete Modelling (2014), ISBN: 978-2-35158-139-1; e-ISBN: 978-2-35158-140-7; *Eds. Kefei Li, Peiyu Yan and Rongwei Yang*

PRO 92: CAM 2014 - 2nd International Conference on advances in chemically-activated materials (2014) 392 pp., ISBN: 978-2-35158-141-4; e-ISBN: 978-2-35158-142-1; *Eds. Caijun Shi and Xiadong Shen*

PRO 93: SCC 2014 - 3rd International Symposium on Design, Performance and Use of Self-Consolidating Concrete (2014) 438 pp., ISBN: 978-2-35158-143-8; e-ISBN: 978-2-35158-144-5; *Eds. Caijun Shi, Zhihua Ou, Kamal H. Khayat*

PRO 94 (online version): HPFRCC-7 - 7th RILEM conference on High performance fiber reinforced cement composites (2015), e-ISBN: 978-2-35158-146-9; *Eds. H.W. Reinhardt, G.J. Parra-Montesinos, H. Garrecht*

PRO 95: International RILEM Conference on Application of superabsorbent polymers and other new admixtures in concrete construction (2014), ISBN: 978-2-35158-147-6; e-ISBN: 978-2-35158-148-3; *Eds. Viktor Mechtcherine, Christof Schroefl*

PRO 96 (online version): XIII DBMC: XIII International Conference on Durability of Building Materials and Components (2015), e-ISBN: 978-2-35158-149-0; *Eds. M. Quattrone, V.M. John*

PRO 97: SHCC3 – 3rd International RILEM Conference on Strain Hardening Cementitious Composites (2014), ISBN: 978-2-35158-150-6; e-ISBN: 978-2-35158-151-3; *Eds. E. Schlangen, M.G. Sierra Beltran, M. Lukovic, G. Ye*

PRO 98: FERRO-11 – 11th International Symposium on Ferrocement and 3rd ICTRC - International Conference on Textile Reinforced Concrete (2015), ISBN: 978-2-35158-152-0; e-ISBN: 978-2-35158-153-7; *Ed. W. Brameshuber*

PRO 99 (online version): ICBBM 2015 - 1st International Conference on Bio-Based Building Materials (2015), e-ISBN: 978-2-35158-154-4; *Eds. S. Amziane, M. Sonebi*

PRO 100: SCC16 - RILEM Self-Consolidating Concrete Conference (2016), ISBN: 978-2-35158-156-8; e-ISBN: 978-2-35158-157-5; *Ed. Kamal H. Kayat*

PRO 101 (online version): III Progress of Recycling in the Built Environment (2015), e-ISBN: 978-2-35158-158-2; *Eds. I. Martins, C. Ulsen and S.C. Angulo*

PRO 102 (online version): RILEM Conference on Microorganisms-Cementitious Materials Interactions (2016), e-ISBN: 978-2-35158-160-5; *Eds. Alexandra Bertron, Henk Jonkers, Virginie Wiktor*

PRO 103 (online version): ACESC'16 - Advances in Civil Engineering and Sustainable Construction (2016), e-ISBN: 978-2-35158-161-2; *Eds. T.Ch. Madhavi, G. Prabhakar, Santhosh Ram and P.M. Rameshwaran*
PRO 104 (online version): SSCS'2015 - Numerical Modeling - Strategies for Sustainable Concrete Structures (2015), e-ISBN: 978-2-35158-162-9
PRO 105: 1st International Conference on UHPC Materials and Structures (2016), ISBN: 978-2-35158-164-3, e-ISBN: 978-2-35158-165-0
PRO 106: AFGC-ACI-fib-RILEM International Conference on Ultra-High-Performance Fibre-Reinforced Concrete – UHPFRC 2017 (2017), ISBN: 978-2-35158-166-7, e-ISBN: 978-2-35158-167-4; *Eds. François Toutlemonde and Jacques Resplendino*
PRO 107 (online version): XIV DBMC – 14th International Conference on Durability of Building Materials and Components (2017), e-ISBN: 978-2-35158-159-9; *Eds. Geert De Schutter, Nele De Belie, Arnold Janssens, Nathan Van Den Bossche*
PRO 108: MSSCE 2016 - Innovation of Teaching in Materials and Structures (2016), ISBN: 978-2-35158-178-0, e-ISBN: 978-2-35158-179-7; *Ed. Per Goltermann*
PRO 109 (2 volumes): MSSCE 2016 - Service Life of Cement-Based Materials and Structures (2016), ISBN Vol. 1: 978-2-35158-170-4, Vol. 2: 978-2-35158-171-4, Set Vol. 1&2: 978-2-35158-172-8, e-ISBN : 978-2-35158-173-5; *Eds. Miguel Azenha, Ivan Gabrijel, Dirk Schlicke, Terje Kanstad and Ole Mejlhede Jensen*
PRO 110: MSSCE 2016 - Historical Masonry (2016), ISBN: 978-2-35158-178-0, e-ISBN: 978-2-35158-179-7; *Eds. Inge Rörig-Dalgaard and Ioannis Ioannou*
PRO 111: MSSCE 2016 - Electrochemistry in Civil Engineering (2016), ISBN: 978-2-35158-176-6, e-ISBN: 978-2-35158-177-3; *Ed. Lisbeth M. Ottosen*
PRO 112: MSSCE 2016 - Moisture in Materials and Structures (2016), ISBN: 978-2-35158-178-0, e-ISBN: 978-2-35158-179-7; *Eds. Kurt Kielsgaard Hansen, Carsten Rode and Lars-Olof Nilsson*
PRO 113: MSSCE 2016 - Concrete with Supplementary Cementitious Materials (2016), ISBN: 978-2-35158-178-0, e-ISBN: 978-2-35158-179-7; *Eds. Ole Mejlhede Jensen, Konstantin Kovler and Nele De Belie*
PRO 114: MSSCE 2016 - Frost Action in Concrete (2016), ISBN: 978-2-35158-182-7, e-ISBN: 978-2-35158-183-4; *Eds. Marianne Tange Hasholt, Katja Fridh and R. Doug Hooton*
PRO 115: MSSCE 2016 - Fresh Concrete (2016), ISBN: 978-2-35158-184-1, e-ISBN: 978-2-35158-185-8; *Eds. Lars N. Thrane, Claus Pade, Oldrich Svec and Nicolas Roussel*
PRO 116: BEFIB 2016 – 9th RILEM International Symposium on Fiber Reinforced Concrete (2016), ISBN: 978-2-35158-187-2, e-ISBN: 978-2-35158-186-5; *Eds. N. Banthia, M. di Prisco and S. Soleimani-Dashtaki*

PRO 117: 3rd International RILEM Conference on Microstructure Related Durability of Cementitious Composites (2016), ISBN: 978-2-35158-188-9, e-ISBN: 978-2-35158-189-6; *Eds. Changwen Miao, Wei Sun, Jiaping Liu, Huisu Chen, Guang Ye and Klaas van Breugel*
PRO 118 (4 volumes): International Conference on Advances in Construction Materials and Systems (2017), ISBN Set: 978-2-35158-190-2, Vol. 1: 978-2-35158-193-3, Vol. 2: 978-2-35158-194-0, Vol. 3: ISBN:978-2-35158-195-7, Vol. 4: ISBN:978-2-35158-196-4, e-ISBN: 978-2-35158-191-9; *Ed. Manu Santhanam*
PRO 119 (online version): ICBBM 2017 - Second International RILEM Conference on Bio-based Building Materials, (2017), e-ISBN: 978-2-35158-192-6; *Ed. Sofiane Amziane*
PRO 120: 2nd International RILEM/COST Conference on Early Age Cracking and Serviceability in Cement-based Materials and Structures (EAC-02), 2017, ISBN: 978-2-35158-197-1, e-ISBN: 978-2-35158-198-8; *Eds. Dimitrios Aggelis and Stéphanie Staquet*

RILEM Reports (REP)

Report 19: Considerations for Use in Managing the Aging of Nuclear Power Plant Concrete Structures (ISBN: 2-912143-07-1); *Ed. D.J. Naus*
Report 20: Engineering and Transport Properties of the Interfacial Transition Zone in Cementitious Composites (ISBN: 2-912143-08-X); *Eds. M.G. Alexander, G. Arliguie, G. Ballivy, A. Bentur and J. Marchand*
Report 21: Durability of Building Sealants (ISBN: 2-912143-12-8); *Ed. A.T. Wolf*
Report 22: Sustainable Raw Materials - Construction and Demolition Waste (ISBN: 2-912143-17-9); *Eds. C.F. Hendriks and H.S. Pietersen*
Report 23: Self-Compacting Concrete state-of-the-art report (ISBN: 2-912143-23-3); *Eds. Å. Skarendahl and Ö. Petersson*
Report 24: Workability and Rheology of Fresh Concrete: Compendium of Tests (ISBN: 2-912143-32-2); *Eds. P.J.M. Bartos, M. Sonebi and A.K. Tamimi*
Report 25: Early Age Cracking in Cementitious Systems (ISBN: 2-912143-33-0); *Ed. A. Bentur*
Report 26: Towards Sustainable Roofing (Joint Committee CIB/RILEM) (CD 07) (e-ISBN 978-2-912143-65-5); *Eds. Thomas W. Hutchinson and Keith Roberts*
Report 27: Condition Assessment of Roofs (Joint Committee CIB/RILEM) (CD 08) (e-ISBN 978-2-912143-66-2); *Ed. CIB W 83/RILEM TC166-RMS*
Report 28: Final report of RILEM TC 167-COM 'Characterisation of Old Mortars with Respect to Their Repair (ISBN: 978-2-912143-56-3); *Eds. C. Groot, G. Ashall and J. Hughes*
Report 29: Pavement Performance Prediction and Evaluation (PPPE): Interlaboratory Tests (e-ISBN: 2-912143-68-3); *Eds. M. Partl and H. Piber*

Report 30: Final Report of RILEM TC 198-URM 'Use of Recycled Materials' (ISBN: 2-912143-82-9; e-ISBN: 2-912143-69-1); *Eds. Ch. F. Hendriks, G.M.T. Janssen and E. Vázquez*

Report 31: Final Report of RILEM TC 185-ATC 'Advanced testing of cement-based materials during setting and hardening' (ISBN: 2-912143-81-0; e-ISBN: 2-912143-70-5); *Eds. H.W. Reinhardt and C.U. Grosse*

Report 32: Probabilistic Assessment of Existing Structures. A JCSS publication (ISBN 2-912143-24-1); *Ed. D. Diamantidis*

Report 33: State-of-the-Art Report of RILEM Technical Committee TC 184-IFE 'Industrial Floors' (ISBN 2-35158-006-0); *Ed. P. Seidler*

Report 34: Report of RILEM Technical Committee TC 147-FMB 'Fracture mechanics applications to anchorage and bond' Tension of Reinforced Concrete Prisms – Round Robin Analysis and Tests on Bond (e-ISBN 2-912143-91-8); *Eds. L. Elfgren and K. Noghabai*

Report 35: Final Report of RILEM Technical Committee TC 188-CSC 'Casting of Self Compacting Concrete' (ISBN 2-35158-001-X; e-ISBN: 2-912143-98-5); *Eds. Å. Skarendahl and P. Billberg*

Report 36: State-of-the-Art Report of RILEM Technical Committee TC 201-TRC 'Textile Reinforced Concrete' (ISBN 2-912143-99-3); *Ed. W. Brameshuber*

Report 37: State-of-the-Art Report of RILEM Technical Committee TC 192-ECM 'Environment-conscious construction materials and systems' (ISBN: 978-2-35158-053-0); *Eds. N. Kashino, D. Van Gemert and K. Imamoto*

Report 38: State-of-the-Art Report of RILEM Technical Committee TC 205-DSC 'Durability of Self-Compacting Concrete' (ISBN: 978-2-35158-048-6); *Eds. G. De Schutter and K. Audenaert*

Report 39: Final Report of RILEM Technical Committee TC 187-SOC 'Experimental determination of the stress-crack opening curve for concrete in tension' (ISBN 978-2-35158-049-3); *Ed. J. Planas*

Report 40: State-of-the-Art Report of RILEM Technical Committee TC 189-NEC 'Non-Destructive Evaluation of the Penetrability and Thickness of the Concrete Cover' (ISBN 978-2-35158-054-7); *Eds. R. Torrent and L. Fernández Luco*

Report 41: State-of-the-Art Report of RILEM Technical Committee TC 196-ICC 'Internal Curing of Concrete' (ISBN 978-2-35158-009-7); *Eds. K. Kovler and O.M. Jensen*

Report 42: 'Acoustic Emission and Related Non-destructive Evaluation Techniques for Crack Detection and Damage Evaluation in Concrete' - Final Report of RILEM Technical Committee 212-ACD (e-ISBN: 978-2-35158-100-1); *Ed. M. Ohtsu*

Report 45: Repair Mortars for Historic Masonry - State-of-the-Art Report of RILEM Technical Committee TC 203-RHM (e-ISBN: 978-2-35158-163-6); *Eds. Paul Maurenbrecher and Caspar Groot*

Calcined Clay-Cement Stabilisation – Physicochemical Attributes and Stabilised Strengths of a-1-a and a-2-6 Soils

O.A. Adekitan[1,2(✉)] and G.M. Ayininuola[1,2]

[1] Department of Civil Engineering, Moshood Abiola Polytechnic, Abeokuta, Nigeria
adekitan.olasunkanmi@mapoly.edu.ng
[2] Department of Civil Engineering, University of Ibadan, Ibadan, Nigeria

Abstract. This work extends the investigations on calcined clay-cement blends to soil stabilisation works as obtained for other pozzolans. Samples of calcined clay (CC) were prepared from a natural kaolinite clay source from south-west Nigeria and were blended with Portland cement (PC) in CC:PC ratios of 0:1 (control), 1:3, 1:1, 3:1 and 1:0 to stabilise samples of lateritic soils from two sources. For each blend, combined stabilisers' percentages of 0, 2.5, 5, 7.5 and 10 of the soils' dry weight were adopted. The physical properties of the soils placed them in the A-1-a and A-2-6 AASHTO classes with fineness moduli (FM) of 4.33 and 2.18 respectively, indicating that the A-2-6 soil is finer, while the corresponding exchangeable cations (ECs) were 1.678 meq/100 gms and 1.738 meq/100 gms; these indicating that the A-2-6 soil has higher capacity for pozzolanic reactions. The 28-day unconfined compressive strength (UCS) tests on the stabilised soils samples show that the 1:1 and 1:3 blends were better stabilisers; this translating to environmental friendly alternatives to cement. The effects of these blends were more pronounced for the A-2-6 soil being the finer and with higher EC of the two soils. The two blends showed enhanced performance on the A-2-6 soil up to 10% binder content with UCS being 120% and 160% of control for 1:1 and 1:3 blends respectively. For the A-1-a soil, the edge over control was limited to 5% and 7% binder contents for the 1:1 and 1:3 blends respectively. The FM and ECs of the parent soil were raised as factors with potentials of affecting the performance of CC:PC blends as soil stabilisers. The paper calls for further work on lateritic soils from other sources to clearly establish the impact of these factors on their strengths when stabilised with CC: PC blends.

1 Introduction

Calcined clay (CC) as a pozzolanic material have been successfully used to replace Portland cement (PC) in concrete works. It is discovered to improve concrete strengths and inhibits porosity with optimum performance occurring at about 10% cement replacement [1–4]. Similarly, it enhances resistances to alkali-silica reaction and chemical attacks (chlorides and sulphates) with optimal performances at 10 to 15% replacement [2, 3]. The use of CC-limestone blend in Portland cement mortars have also

F. Martirena et al. (eds.), *Calcined Clays for Sustainable Concrete*, RILEM Bookseries 16,
https://doi.org/10.1007/978-94-024-1207-9_1

been investigated. The 5%–10%–85% (limestone-CC-ordinary Portland cement) blend reported the highest compressive strengths [5]. The strengths after 1, 7, 28 and 90 days of curing were 105%, 120%, 120% and 93% of control (0% additive) respectively.

Though investigations on the use CC in concretes are still progressing, it becomes necessary to also extend these to soil stabilisation. Pozzolans such as fly ash and silica fume have been investigated in like manner. Moreover, with the intensified global drive at mitigating detrimental CO_2 emissions (including those from cement production), continuous attempt at exploring alternatives to cement becomes necessary. A review of literature on CC showed that there is limited work on its use in soils. It had been considered as a more suitable pozzolan in alkaline activation for deep soft soil improvement [6], used as material for soilcretes in jet grouting and for producing compressed bricks [4]. This paper investigates the behaviour of CC-PC blends in the stabilisation of lateritic soils. It proceeded to discuss these behaviours in the light of the attributes of the parent soils.

1.1 Pozzolan-Cement Soil Stabilisation

The essence of stabilising soils is to increase strength and make them suitable for their intended use. Hydration reaction initiates as cement is added to soil samples in the presence of moisture. Cementitious materials are formed, these fills the soil's voids and provides bonding [7, 8]. As curing progresses, more cementitious materials are formed from further cement hydration and from interaction between the excess lime of earlier cement hydration and the soil's aluminosilicates [8, 9]. The calcium ions of the excess lime interact with the double layer of the soil's particles in a cation exchange reaction leading to dissolution of the soil's double layer and interaction with the soil's aluminosilicates to produce more cementitious materials. This is secondary hydration reaction [8, 9]. The cation exchange capacity of soils measures the exchangeable positive ions held within the double layer of their particles. This determines the amount of cementitious products formed from the secondary hydration reaction, and, by extension, the stabilised strength of the soil.

Stabilisation of soil samples with pozzolan-cement blends was further discussed by [8, 9] who studied the microstructure of fly-ash-cement-soil system. The results showed formation of small clusters instead of the large clusters of soil-cement matrix. The pozzolan's particles disperse the matrix into smaller clusters, increasing the specific surface in the process. This provides the precursor for more hydration products from the initial and secondary hydration reactions, resulting to increased stabilised strength. Pozzolanic dispersions are inferred to be more active in soils of small particle sizes [8, 9].

2 Method

Two burrow sites of lateritic soils were used. Soils of dissimilar physical characteristics were deliberately chosen such that the effect of using CC with PC can be captured across soils of different properties. Disturbed samples were obtained from these sources and labelled A and B respectively.

The soil samples were subjected to index properties tests (particle size and consistency tests) and classified using the AASHTO soil classification system. Tests were in tandem with [10]. Moreover, from the particle size analysis, the fineness modulus (FM) was determined. This measures the fineness of the particles of the lateritic soils. FM is the sum of the cumulative percentages retained on sieves of aperture 150 μm, 300 μm, 600 μm, 1.18 mm, 2.36 mm and 4.75 mm divided by 100. In addition, the lateritic soil samples were subjected to chemical analysis test using X-Ray Fluorescence. The exchangeable cations (ECs) were also determined from the cation exchange capacity test.

Kaolin clay was obtained from a natural deposit in Abeokuta, southwest Nigeria and calcined at 700 °C for 1 h as guided by authors' previous works [11, 12]. The resultant calcined clay (CC) was blended with general purpose Portland Cement (PC) and used as stabilisers on the lateritic soils in total percentages of 2.5, 5.0, 7.5 and 10 of the soil mass. Five blend ratios were established for each total percentage - 0:1, 1:3, 1:1, 3:1 and 1:0 respectively; twenty different CC-PC-soil matrices in all. Soils stabilised with 0:1 blend (PC-only) serve as control. The matrices were tested for their strengths.

The procedures of [10, 13] were adopted for the unconfined compression strength (UCS) test. The test specimens were prepared by remoulding them at their respective optimum moisture contents (earlier obtained from Standard Proctor compaction tests). The remoulded were extruded into cylindrical shapes, 50 mm (diameter) by 100 mm (length), coated with paraffin wax and cured in air at room temperature for 28 days before the test. Tests were repeated two more times and averages were obtained. The unstabilised soil samples (0% binder) were also tested.

3 Results

3.1 Physical Properties of the Lateritic Soils

Lateritic soil sample A was dark-reddish brown sandy clay material. Sample B was light-reddish brown gravelly clay with specks of creamy colourations. Particle size (Table 1) show sample A as fine with over 50% of it passing the 425 μm sieve and sample B gravelly (less than 15% passing 425 μm sieve). Fineness modulus (FM) confirms sample A as finer (FM = 2.18) compared to sample B (FM = 4.33). Moreover, these test results classify the soils as A-2-6 (sample A) and A-1-a (sample B).

Table 1. Physical properties of the lateritic soil samples

Soil sample	%Passing sieve no 10 (2 mm)	%Passing sieve no 40 (425 μm)	%Passing sieve no 200 (75 μm)	Fineness modulus	Liquid limit (%)	Plastic limit (%)	Plasticity index (%)	Group index, GI	AASHTO classification
Sample A	84%	52%	30%	2.18	39. 8	19. 3	20. 5	2	A-2-6
Sample B	36%	12%	2%	4.33	47. 9	41. 6	6. 3	0	A-1-a

3.2 Chemical Properties of the Lateritic Soils

Typical of lateritic soils, the soils predominantly consisted of oxides of aluminium and silicon (Table 2). The A-2-6 soil also has exchangeable cation (EC) capacity of 1.738 meq/100 gm. This was higher than the corresponding EC for the A-1-a soil (1.678 meq/100 gm).

Table 2. Chemical properties of the lateritic soil samples

Chemical constituents	A-2-6 soil	A-1-a soil
SiO_2	53.85	41.26
Al_2O_3	29.01	40.02
Fe_2O_3	6.03	4.32
CaO	4.15	1.55
MgO	0.04	0.00
SO_3	0.10	1.58
K_2O	0.10	0.08
Na_2O	0.07	1.06
P_2O_5	0.04	0.00
LOI	6.66	10.25
Total	100.05	100.12
EC (meq/100 gm)	1.738	1.678

3.3 Unconfined Compression Strength (UCS)

The 28-day UCS were presented in Table 3 and Fig. 1(a, b). The UCS at 0% binder were 0.93 ± 6.85 MPa and 0.94 ± 8.29 MPa for A-2-6 and A-1-a soils respectively. Of the five CC-PC blends, stabilisation with 1:1 and 1:3 blends generally produced strengths that compared favourably with those of control (PC-only or 0:1 blend). For the A-2-6 soil, the 1:3 blend produced the highest strengths. The stabilised strength was 2.31 ± 7.35 MPa at 2.5% stabiliser content and this rose to 5.89 ± 8.38 MPa at 10% stabiliser content. The 1:1 blend produced stabilised strength of 2.97 ± 8.73 MPa at 2.5%, rising to 4.47 ± 4.99 MPa at 10%. The strengths produced using the PC-only (0:1) trailed behind those of the 1:1 and 1:3 soils (Fig. 1a) with 1.11 ± 6.94 MPa at 2.5%, increasing to 3.69 ± 8.06 MPa at 10%. Thus, at the 10% stabiliser content, the strengths achieved with 1:3 and 1:1 blends were 159.6% and 121.1% of the strength of the 0:1 mix.

Table 3. Mean 28-UCS for the stabilised lateritic soils (MPa)

CC:PC ratios	Combined binder%									
	A-2-6					A-1-a				
	0	2.5	5	7.5	10	0	2.5	5	7.5	10
1:0	0.93 ± 6.85	1.52 ± 4.50	1.51 ± 7.59	1.22 ± 7.12	1.12 ± 8.26	0.94 ± 8.29	1.61 ± 4.92	1.73 ± 7.79	1.69 ± 8.18	1.86 ± 7.79
3:1	0.93 ± 6.85	0.97 ± 9.74	1.76 ± 5.89	2.13 ± 8.29	3.36 ± 7.79	0.94 ± 8.29	2.37 ± 9.43	2.62 ± 7.41	3.19 ± 8.16	3.54 ± 8.34
1:1	0.93 ± 6.85	2.97 ± 8.73	3.19 ± 4.03	3.58 ± 9.18	4.47 ± 4.99	0.94 ± 8.29	2.80 ± 7.79	3.43 ± 8.18	3.78 ± 2.05	4.70 ± 5.44
1:3	0.93 ± 6.85	2.31 ± 7.35	4.63 ± 6.13	4.98 ± 7.79	5.89 ± 8.38	0.94 ± 8.29	4.72 ± 7.32	5.22 ± 7.13	6.24 ± 7.76	7.05 ± 9.27
0:1	0.93 ± 6.85	1.11 ± 6.94	1.98 ± 6.94	2.56 ± 9.09	3.69 ± 8.06	0.94 ± 8.29	1.23 ± 6.48	3.36 ± 7.26	6.66 ± 8.16	10.56 ± 6.85

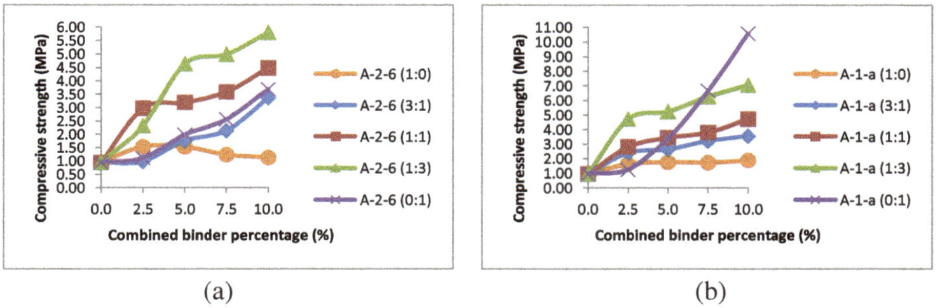

Fig. 1. Compressive strength as function of combined binder percentage.

The stabilised strengths of A-1-a using 1:3 blend were 4.72 ± 7.32 MPa at 2.5% and 7.05 ± 9.27 MPa at 10%. The corresponding strengths for the 1:1 stabilisation were 2.80 ± 7.79 MPa and 4.70 ± 5.44 MPa respectively. Adopting the PC-only blend yields stabilised strengths of 1.23 ± 6.48 MPa at 2.5% and this rose rapidly to 10.56 ± 6.85 MPa at 10%. Thus, unlike the case of the A-2-6 soil, the edges maintained by the application of the 1:1 and 1:3 over PC-only stabiliser were sustained up to about 5% and 7% stabiliser contents respectively (Fig. 1b). Thus, up to these levels, the 1:1 and 1:3 blends were beneficial in both reducing cement use and improving soil strength.

This work presented 1:1 and 1:3 blends as better alternatives to PC-only blend. Their use affords the reduction in the contents of cement for stabilisation by as much as 50% without compromise on the stabilised strengths. More importantly, the reported UCS for these blends (Table 1) met the 1.8 MPa minimum specification for bases of flexible road pavements designed for heavy traffic [14]. These present positive outlooks for environmental health.

The differences in the outcomes of 1:1, 1:3 and 0:1 blends on the two soils is attributed to differences in particle size (fineness modulus) and exchangeable cation (EC) capacity. The A-2-6 soil has lower FM (2.18), indicating that it is finer. As a result, pozzolanic dispersion effects were more pronounced in the face of pozzolan-cement stabilisation [8, 9]. This translates to increased specific surface for formation of more hydration products. Moreover, the EC is higher (1.738 meq/100 gm), ensuring that cation exchange activities and the secondary hydration reactions are more pronounced with this soil. This explains why the pozzolanic edge of the 1:1 and 1:3 blends over PC-only stabiliser was sustained up to 10% stabiliser content for the A-2-6 soil. However, for the A-1-a soil, the edge was not sustained beyond 5% and 7% respectively for the 1:1 and 1:3 soil. The coarser particles (high FM of 4.33) and low EC (1.678 meq/100 gm) did not encourage pozzolanic dispersion and secondary hydration as much as it occurred for the A-2-6 soil.

4 Conclusion

The use of calcined clay (CC) to partly replace general purpose Portland cement (PC) in soil stabilisation was investigated. Typical A-1-a and A-2-6 soils were stabilised and their resultant strengths measured. The study shows that CC:PC blends of 1:1 and 1:3 were better stabilisers than PC-only stabiliser. This implies that, without compromising strength, savings on cement usage by as much as 50% and a reduction of associated CO_2 emission, probably in the same order, is achievable with calcined clay. Moreover, the impact of this pozzolan as blended with cement can be influenced by the physical and chemical attributes of the parent soil. Further works intends to study more soil samples, especially those belonging to other classes, for similarity or otherwise in the observed trends in this work.

References

1. Aboubakar, M.A., Ganjian, E., Pouya, H., Akashi, A.: A study on the effect the addition of thermally treated Libyan natural pozzolan has on the mechanical properties of ordinary Portland cement mortar. Int. J. Sci. Technol. **3**(1), 79–84 (2013)
2. Justice, J.M., Kennison, L.H., Mohr, B.J., Beckwith, S. L., McCormick, L.E., Wiggins, B., Zhang, Z.Z., Kurtis, K. E.: Comparison of two metakaolins and silica fume used as supplementary cementitious material. In: Proceedings of 7th International Symposium on Utilization of High-Strength/High Performance Concrete, Washington DC, June 20–24, 2005
3. Justice, J.M.: Evaluation of metakaolins for use as supplementary cementitious materials. Unpublished M. Sc thesis, School of Material Science and Engineering, Georgia Institute of Technology 2005
4. Kolovos, K.G., Asteris, P.G., Cotsovos, D.M., Badogiannis, E., Tsivilis, S.: Mechanical properties of soilcrete mixtures modified with metakaolin. Constr. Build. Mater. **47**, 1026–1036 (2013)
5. Antoni, M., Rossen, J., Martirena, F., Scrivener, K.: Cement substitution by a combination of metakaolin and limestone. Cem. Conc. Res. **42**, 1579–1589 (2012). doi:10.1016/j.cemconres.2012.09.006
6. Cristelo, N., Glendinning, S., Teixeira, P.A.: Deep soft soil improvement by alkaline activation. Proc. Inst. Civ. Eng. Ground Improv. **164**, 1–10 (2011)
7. Bhattacharja, S., Bhatty, J.: Comparative performance of portland cement and lime in stabilisation of moderate to high plasticity clay soils. Res. Dev. Bull. D Portland Cem. Assoc. **125**, 1–26 (2003)
8. Horpibulsuk, S.: Strength and microstructure of cement stabilised clays. In: Kazminuk, V. (Ed.), Scanning electron microscopy, Croatia: InTech. (2012). ISBN 978-953-51-0092-8, http://www.intechopen.com/books/scanning-electron-microscopy/strength-and-microstructure-of-cementstabilised-clay
9. Horpibulsuk, S., Rachan, R., Chinkulkijniwat, A., Raksachon, Y., Suddeepong, A.: Analysis of strength development in cement-stabilised silty clay from microstructural considerations. Constr. Build. Mater. **24**, 2011–2021 (2013)
10. BS 1377: British standard methods for testing soils in civil engineering, parts 1–9. London: British Standards Institute (2000)

11. Ayininuola, G.M., Adekitan, O.A.: Characterization of Ajebo Kaolinite Clay for Production of Natural Pozzolan. Int. J. Civ. Environ. Struct. Constr. Architectural Engineering **10**(9), 1212–1209 (2016)
12. Adekitan, O.A., Ayininuola, G.M.: Optimizing the thermal treatment of Abeokuta kaolin (south-west Nigeria) for production of natural pozzolan. Afr. J. Sci. Technol. Innov. Dev. **9** (4), 361–365 (2017). in press
13. BS 1924: British standard methods of testing for stabilised soils, British Standards Institute, London (2000)
14. American Association of State Highway and Transportation Officials (AASHTO): AASHTO guide for design of pavement structures, American Association of State Highway and Transportation Officials, Washington DC (2000)

Sulfate Resistance of Cement Mortar Containing Metakaolin

Abdelsalam M. Akasha(✉) and Jamal M. Abdullah

Department of Civil Engineering, Sebha University, Sebha, Libya

Abstract. Many of an extensive researches have been carried out by Libyan Industrial Research Center in natural row materials in south region of Libya [1] and the results found that there are many of material can be used as building material, one of these materials which widespread over large area is Kaolin clay.

From that point of view civil Eng. Dep. at Sebha university-Libya start thinking of using Kaolin clay as a building material and a team work has been formed to carry out a deep research study work on possibility of using the local natural pozzolana in south of Libya as a partially replacement of Portland cement. Previous laboratory study [2] which carried out by the team work proved that there is a possibility to use south Libya natural Kaolin clay as partial replacement of OPC after calcination and milling to get active calcined clay (Metakaolin).

Metakaolin being used very commonly as pozzolanic material or as supplementary cementing materials in mortar and concrete, and has exhibited considerable influence in enhancing the mechanical and durability properties of mortar and concrete as well as utilization of natural calcined clay has widely spread attention in the world to minimize the Portland cement consumption and manufacturing of which being environmentally damaging, in addition to its economic advantage.

Metakaolin when mixed with cement the silica of the pozzolana combines with the free lime released during the hydration forms additional cementitious C-S-H gel. This paper investigates the effect of metakaolin (MK) cement replacement on the resistance of mortar to sulfate attack for two different metakaolin quarries. Four MK replacement levels were considered in the study: 0%, 10%, 15%, and 20% by weight of cement for two quarries. After the specified initial moist curing period, mortar specimens were immersed in 5% sodium sulfate solution for a total period of one year. The degree of sulfate attack was evaluated by measuring expansion of mortar prisms, compressive strength reduction and observation of appearance of deterioration that usually accompanying with sulfate attack. All metakaolins examined in this study led to increases in strength and resistance to sulfate attack in comparing with controls samples or according to standard limits.

Keywords: Metakolin · Sulfate resistance · Expansion · Compressive strength · Visual inspection

F. Martirena et al. (eds.), *Calcined Clays for Sustainable Concrete*, RILEM Bookseries 16,
https://doi.org/10.1007/978-94-024-1207-9_2

1 Experimental Program

1.1 Materials

1.1.1 Sources for Metakaolin

The clay soil (Pozzolana) used in this study was collected from two different locations near to Sebha city (Sebha and Temenhint sites) (Table 1).

Table 1. Chemical composition percentage of calcined clays

Oxides	Samples calcined clay	
	Sebha (A)	Temenhint (B)
SiO2	53.42	70.33
Al2O3	40.84	25.32
Fe2O3	0.975	1.05
Total SiO2, Al2O3, Fe2O3	95.20	96.70
SO3	0.033	0.016
MgO	0.130	0.140
CaO	0.100	0.080
Na_2O	0.220	0.350
K_2O	0.160	0.366
L.O.I	0.880	0.760

The collected stones were generally in dry condition with blocky structure and in order to obtain MK the stones were crushed and calcined at 800 °C for a period of 2.0 h. After the calcination process, the calcined clay stones were cooled then milled to pass on 150 μm sieve.

1.1.2 Cement

Locally available ordinary Portland cement (ASTM Type I).

1.1.3 Standard Sand

The standard used sand is locally siliceous sand.

1.1.4 Sodium Sulfate

Anhydrous sodium sulfate (Na_2So_4) was used.

1.2 Mixture Details

Seven mortar mixtures were prepared and used in this study as shown in Table 2.

Table 2. Mix proportions of blended cement

Sample/Mix	Symbol	Clacined clay MK (%)	Portland cement (%)
Control Sample	CS	0	100
Sebha sample (A)	AI	10	90
	AII	15	85
	AIII	20	80
Temenhint sample (B)	BI	10	90
	BII	15	85
	BIII	20	80

1.2.1 Specimens Preparation

The water-cementitious ratio (w/b) was fixed at 0.49 by mass without plasticizer. For compressive tests, 50.0 mm cubes and for expansion test and visual inspection 25 * 25 * 285 mm prisms were prepared. The cubes and prisms are cured one day in molds by covering with wet cotton. After demolding, specimens were immersed in lime water until day of testing.

1.2.2 Sulfate Solution

According to ASTM C1012 each liter of solution is contained 50.0 g of Na_2SO_4. The solution should be stored at 23.0 ± 2.0 °C. The pH of the solution must be determined before use and the solution is rejected if the pH range is outside 6.0 to 8.0.

2 Test Procedure

2.1 Cubes

The first group of cubes stored in water tank for an initial moist curing period of 3, 7, 28,180 and 365 days. The second group is cured for 28 days in water then immersed in sulfate solution for period of 6 and 12 month. The results of compressive strength are used to determine the Strength Activity Index (SAI) accordance to ASTM C618 as follows:

$$SAI = A/B * 100 \tag{1}$$

Where A is the compressive strength of test mixture, and B is compressive strength of control mixture.

Determination the Compressive Strength Reduction (CSR) for cubes immersed in sulfate solution comparing to that stored in water for the period of 6 and 12 months was done using the next formula:

$$CSR = \frac{\sigma_m - \sigma_s}{\sigma_m} * 100 \tag{2}$$

Where σm is the average compressive strength (in MPa) of three cubes cured in water and σs is the average compressive strength (in MPa) of three cubes immersed in 5% sulfate solution.

2.2 Prisms

The length change measured by using the length comparator in accordance with Specification ASTM C 490. Length change is calculated accordance to ASTM C1012 at any age as follows:

$$\Delta L = \frac{L_x - L_i}{L_g} \times 100 \tag{3}$$

Where ΔL is the change in length at x age, %, Lx is the comparator reading of specimen at x age, Li is the initial comparator reading of specimen at the same time and Lg is the nominal gage length, or 250 mm as applicable.

3 Result and Discussion

3.1 Pozzolanic Activity

Pozzolanic activity is determined by strength activity indexes (SAI) which must be greater than or equal 75% according to ASTM 618-03 as shown on Table 3.

Table 3. Strength activity index

Samples	AI	AII	AIII	BI	BII	BIII
SAI % at 7 days	105.73	93.97	113.77	102.76	87.30	93.26
SAI % at 28 days	104.77	103.15	116.03	116.99	104.27	106.83

3.2 Compressive Strength Reduction CSR

CSR is another reliable measurement to indicate the sulfate attack as shown in Fig. 1.

Loss in strength during the exposure period can be as the result of cracking caused by expansion of ettringite and gypsum formation and/or the loss of C-S-H. Metakaolin suffered less strength loss than the control mortar as shown in Fig. 1. The compressive strength reduction ratio is varying from 2.14% to 24.23% for 6 months period and from 15.86% to 62.03% for 12 months period. The max. value of CSR at both periods was for control sample (CS). The best result is achieved by replacing 20% of cement by MK of sample BIII sample AIII respectively.

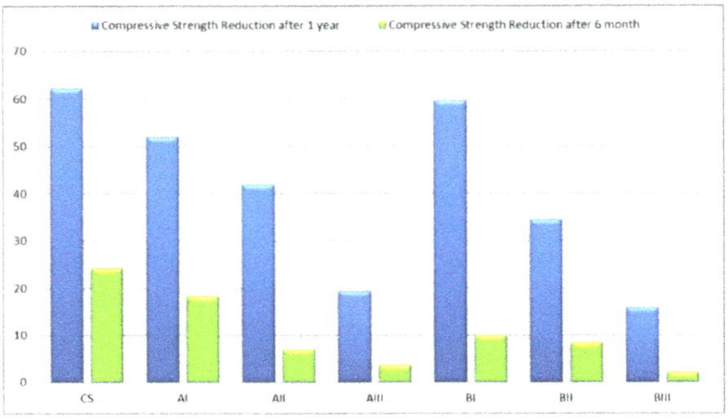

Fig. 1. Compressive Strength Reduction (CSR)

3.3 Expansion Test Results

From the results given in Fig. 2. all MK mortar series (except AI) fall below the standard limit 0.05% and 0.1 for 6 months period of ASTM C 1157 and ASTM C 595. The lowest expansion values were obtained by using 20% of MK of sample AIII and BIII which achieved an expansion of 0.033% and 0.034% respectively which lower than the control sample by 78% and 77% respectively.

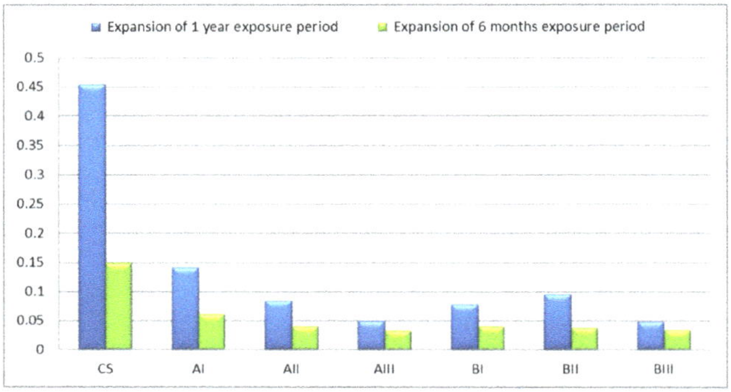

Fig. 2. Expansion of prism mortars after 6 and 12 months

After 12 months of exposure to sulfate solution only CS and AI fail to comply with requirements of ASTM C 1157 of expansion limit 0.1%. The lowest values were achieved also by using 20% of MK of samples (AIII and BIII) which gained an expansion of 0.049% of both samples which lower than the control sample by 89%. The reasons which can be given for the increase in the sulfate resistance in mortar with MK are that the pozzolanic reaction between the MK and calcium hydroxide (CH) released during cement hydration, witch consume part of (CH). Thus the quantity of expansive gypsum formed by the reaction of calcium hydroxide will be less in MK mortar than in plain mortar at specified level. Furthermore, the formation of secondary C-S-H by the pozzolanic reaction, although less dense than the primary C-S-H gel is effective in filling segmenting large capillary pores and increase the impermeability of the mortars [2–8].

3.4 Visual Inspection

A visual inspection of specimens as shown in Fig. 3 revealed the deterioration of the prismatic samples under sulfate attack.

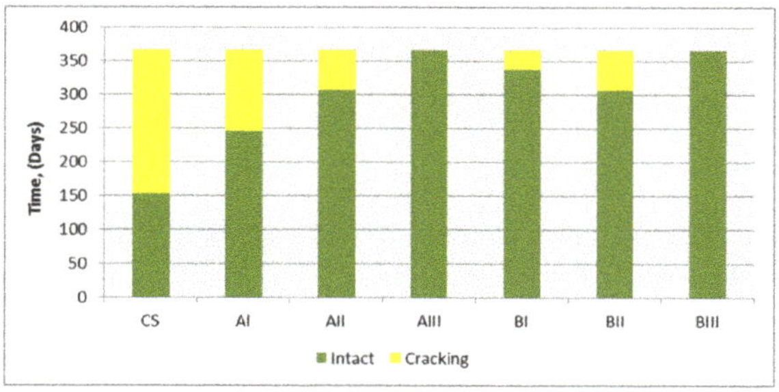

Fig. 3. Visual inspection of prismatic samples of one year period

The presence of MK is beneficial regarding appearance of cracks or any signs of deterioration associated with sulfate attack, since it could delay or avoid formation of ettringite or gypsum which lead to expansion, cracking and deterioration of mortar samples.

Under sulfate attack, using of Sebha and Tenminhint MK with ratio of 20% as partial replacement to cement significantly delayed occurring of any visible cracks and kept the mortar prisms intact up to the 12 months.

4 Conclusions

This study presents the results of the effect of MK cement replacement on the resistance of mortar to sulfate attack. Based on the results obtained from this study, the following conclusions can be drawn:

1. The chemical analysis of all samples collected from two quarries is complied with ASTM C618 and the natural pozzolan (Metakaolin) can be classified as N class accordingly.
2. Metakaolin replacement of cement (10 to 20%) was found effective in improving the resistance of mortar to sulfate attack. The use of optimum value of MK replacement level which 20% for every quarry reduced effectively the sulfate expansion at the end of 6 month by 58–77% and by 69–89% at the end of 12 month in comparing with the control mortar.

3. Most of results of MK mortar expansion results comply with expansion limits of ASTM C 1157 standard performance specification for hydraulic cement and ASTM C595 Standard Specification for blended hydraulic cement.
4. Deteriorations such which usually associated with sulfate attack to mortar sample can be delayed or prevented by using 10 to 20% of metakaolin that produced by fired kaolin clay collected from two studied locations.
5. Metakaolin suffered less strength loss than control mortar. The use of 10–20 MK of two quarries reduces the compressive strength reduction when kept in sulfate solution.
6. For more confirmation of these results, sulfate testing should continue for a period of 18 month or until mortar bars have expanded beyond the length of the comparator.

References

1. Industrrial Research Center: Raw material in SEBHA and WADI SHATTI region. Indus. Res. J. **7**, 126–138 (1998). Tripoli, Libya
2. Alamin, H.M.A.: Use of Local Poaaolanic Material for Concrete Production. Sebha University (2008)
3. American Concrete Institute Reports: ACI C201-2R-01 Guide to Durable Concrete
4. American Concrete Institute: ACI 232.1R-00 Use of Raw or Processed Natural Pozzolans in Concrete
5. Kalny, J.S., Marchand, J., Odler, I.: Sulfate Attack on Concrete. Spon Press, London (2002)
6. Alakhras, N.M.: Durability of metakaolin to sulfate attack. Cem. Concr. Res. **36**, 1727–1734 (2006)
7. Khatib, J.M., Wild, S.: Sulfate resistance of metakaolin concrete. Cem. Concr. Res. **28**(1), 83–92 (1998)
8. Yazici, S., Arel, H.S., Anuk, D.: Influence of metakaolin on the durability and mechanical properties of mortar. Arab. J. Sci. Eng. **39**, 8585–8592 (2014)

Use of R^3 Rapid Screening Test to Determine Reactivity and Chloride Binding Potential of Locally Available Kaolinite Clay

M. Almarshoud(✉), J.L. Saint Rome, and K.A. Riding

Engineering School of Sustainability, Infrastructure and Environment, Civil and Coastal Engineering, University of Florida, 365 Weil Hall, 116580, Gainesville, FL 32611, USA

Abstract. The utilization of calcined clay as a pozzolanic material for concrete has received a lot of attention in recent years because of its sustainability and worldwide availability. The performance of concrete made using a locally sourced high kaolinite clay was examined to determine suitability for use in local transportation projects. For this work, the local clay was tested to determine optimal calcining temperature, limestone blend, and sulfate content using isothermal calorimetry (R^3 method). The clay was calcined at 600 °C, 700 °C, 800 °C, and 850 °C. The effects of material reactivity as measured by calorimetry were compared to the chloride binding capacity.

1 Background

The utilization of calcined clay as a pozzolanic material for concrete took place as a supplementary cementitious materials (SCM) to react with portlandite (CH) that is produced from Portland cement hydration. The consumption of CH has a strong effect on improving concrete strength, sulfate resistance and alkali silica reaction (ASR) [1]. The reactivity of calcined clay is strongly affected by the calcining temperature [2].

2 Aim

The purpose of this paper is to find the correlation between the calcined clay reactivity and chloride binding. The chloride binding process is considered a long process that could take more than 6 to 8 weeks in preparing the paste samples and 2 to 4 weeks for chloride binding. Using the R^3 method to find the calcined clay reactivity and correlate the reactivity to chloride binding reduces the testing process to 12 days. The R^3 method takes between 12 to 14 days depending on the reactivity of the calcined clay and it is considered a rapid indicator for SCM reactivity [3].

F. Martirena et al. (eds.), *Calcined Clays for Sustainable Concrete*, RILEM Bookseries 16,
https://doi.org/10.1007/978-94-024-1207-9_3

3 Method

3.1 Chloride Binding

The chloride binding testing method used is similar to Tang and Nillson chloride binding testing method [4]. The testing was performed on cement paste samples with 15% calcined clay (600, 700, 800 and 850 °C) replacement as SCM after curing for 8 weeks. The paste samples were prepared using a vacuum paste mixer [5] with a water to cement ration of 0.5. The paste was mixed at a mixing speed of 400 rpm in two periods, each period was 90 s with a 10 s rest period in between. During the rest period, the sides of the mixer and the blade were scraped to insure proper mixing. The hydrated cement paste was vertically cut into 3 mm thick disks to avoid bleeding effect and a portlandite solution was used as lubricant to avoid carbonation of the cement paste. After that, about 25 grams of cement paste were placed in 100 ml chloride solution of different concentrations (0.1, 0.3, 0.5, 1.0 and 3.0 M). The bound chloride was calculated using Eq. 1.

$$C_b = \frac{35.45V(c0 - c1)}{W} \tag{1}$$

Where

C_b: bound chloride content (mg/g of cement paste sample).
V: volume of solution (ml).
$c0$, $c1$: initial and final chloride concentration (ml/l).
W: weight of dry sample (g).

3.2 Particle Size Distribution

This test was performed in the cement and calcined clay to analyze the particle size distribution of the cementitious materials using HORIBA LA-950 laser particle size distribution analysis. The samples were diluted in ethanol to avoid hydration.

3.3 Isothermal Calorimetry

Isothermal calorimetry was used to measure the effect of different calcining temperature of clay on cement paste reactivity. The cement paste was prepared using the same method mentioned in Sect. 3.1 and after the mixing was completed, around 15 g of paste were filled in glass ampules and placed in isothermal calorimetry testing unit with a testing temperature of 23 °C and the heat flow was recorded for 7 days.

3.4 R^3 Method

The testing method is well explained in Snellings paper [3]. The calcined clay was mixed with portlandite. The portlandite to SCM ratio was 3 to 1. Gypsum was added depending on the alumina content in the calcined clay to maintain SO_3/Al_2O_3 of the paste to 1. KOH is used to maintain the paste alkalinity at a fixed concentration of 0.5 M. The water to solids ratio was fixed at 0.9. The paste was mixed in a vacuum cement mixer. The mixing process included two periods of mixing at a speed of 400 rpm for 90 s and between the two periods the mixer sides and blade were scraped. After that 11 g of paste were placed in glass ampules then in to the isothermal calorimetry, which was fixed at a temperature of 40 °C for 12 to 14 days.

4 Results

4.1 Chloride Binding

The chloride binding test results are shown in Fig. 1. The results show that using calcined clay as SCM increases chloride binding, especially at high concentrations. The test results show that 800 °C calcining temperature provides the highest chloride binding results.

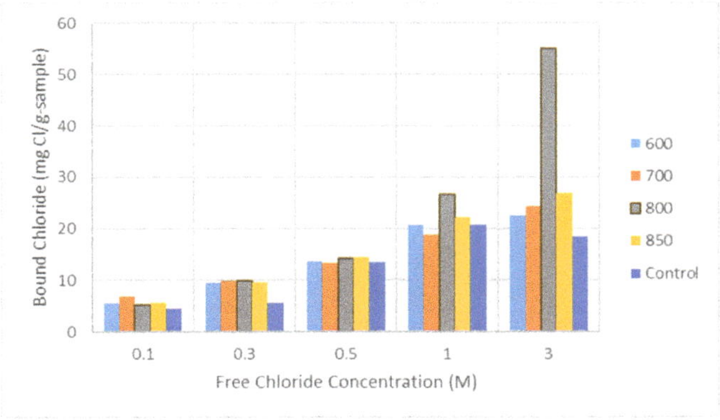

Fig. 1. The chloride binding test results

4.2 Particle Size Distribution

The particle size distribution test shows that the median particle size for the calcined clay was about 17 μm with a standard deviation of 20 μm. The type I/II cement has a median particle size of 3 μm with a standard deviation of 8.5 μm. The particle size distribution is shown in Fig. 2.

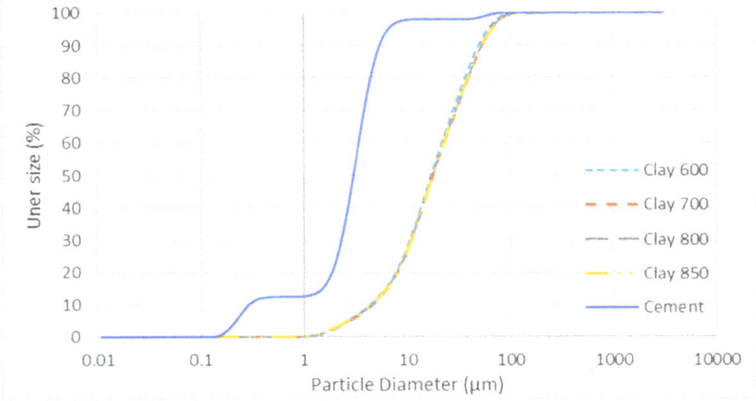

Fig. 2. Particle size distribution for calcined clay and type I/II Portland cement

4.3 Isothermal Calorimetry

The test results showed that using calcined clay increased the paste energy and as the calcining temperature increases, the reactivity of the paste increases. As shown in Fig. 3 calcined clay at 850 °C had slightly the highest energy and the controlled sample had the lowest energy.

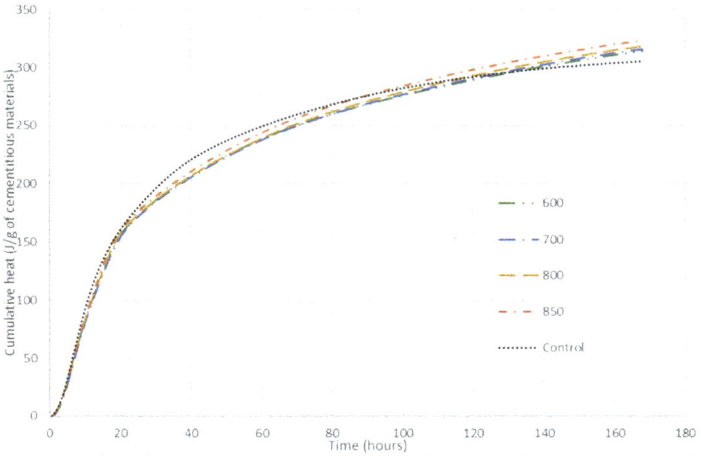

Fig. 3. Isothermal calorimetry test results for calcined clay and controlled cement paste samples

4.4 R^3 Method

The test results could give an indication of the reactivity of the calcined clay in the first three days, depending on the materials reactivity, due to the high temperature of the calorimetry. The results were analyzed using 10 days of isothermal calorimetry, since the change in heat transfer was insignificant.

The R^3 results show that 800 °C calcined clay has the highest heat transfer after that 850 °C, 700 °C and 600 °C in order as shown in Fig. 4.

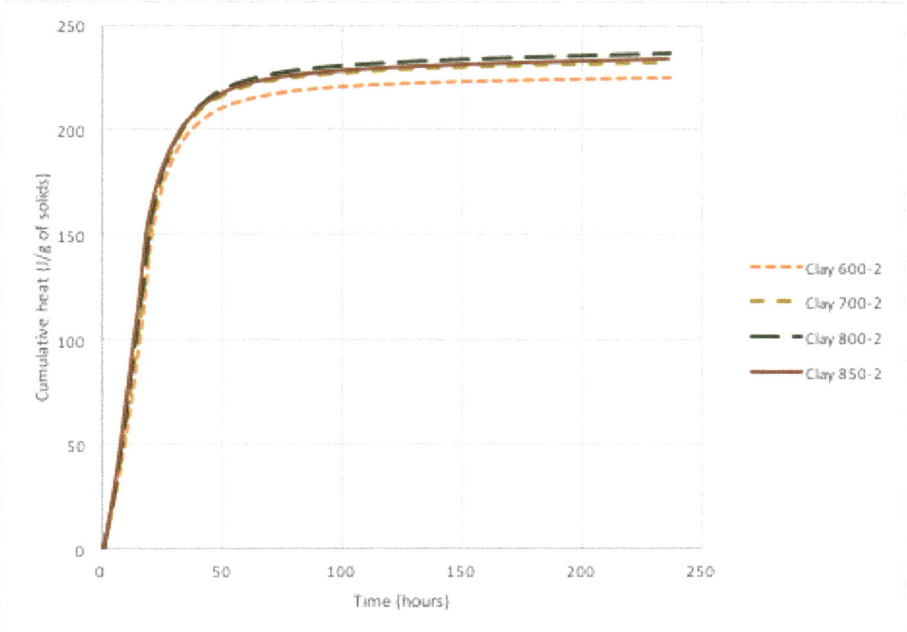

Fig. 4. R^3 isothermal calorimetry results for calcined clay at 600, 700, 800 and 850 °C

5 Conclusion

- The correlation between the R^3 test results and chloride binding was noticed.
- The calcined clay reactivity could be considered as an indicator for chloride binding in case of using calcined clay as SCM.
- The isothermal calorimetry 7-day test results at 23 °C didn't show much correlation with either the chloride binding or the R^3 test results.
- The chloride binding in high free chloride concentrations is strongly noticed to be higher than the controlled sample, especially at 800 °C calcined clay samples.

References

1. Sabir, B., Wild, S., Bai, J.: Metakaolin and calcined clays as pozzolans for concrete: a review. Cement Concr. Compos. **23**, 441–454 (2001)
2. Elimbi, A., Tchakoute, H., Njopwouo, D.: Effects of calcination temperature of kaolinite clays on the properties of geopolymer cements. Constr. Build. Mater. **25**, 2805–2812 (2011)
3. Snellings, R., Scrivener, K.L.: Rapid screening tests for supplementary cementitious materials: past and future. Mater. Struct. **49**, 3265–3279 (2016)
4. Luping, T., Nilsson, L.: Chloride binding capacity and binding isotherms of OPC pastes and mortars. Cem. Concr. Res. **23**, 247–253 (1993)
5. Cao, Y., Zavaterri, P., Youngblood, J., Moon, R., Weiss, J.: The influence of cellulose nanocrystal additions on the performance of cement paste. Cement Concr. Compos. **56**, 73–83 (2015)

Assessment of Cuban Kaolinitic Clays as Source of Supplementary Cementitious Materials to Production of Cement Based on Clinker – Calcined Clay – Limestone

Roger S. Almenares Reyes[1(✉)], Adrián Alujas Díaz[2], Sergio Betancourt Rodríguez[3], Carlos Alberto Leyva Rodríguez[1], and José Fernando Martirena Hernández[4]

[1] Instituto Superior Minero Metalúrgico de Moa, 83330 Moa, Cuba
[2] Centro de Estudios de Química Aplicada, Universidad Central de Las Villas, 54830 Santa Clara, Cuba
[3] Facultad de Construcciones, Universidad Central de Las Villas, 54830 Santa Clara, Cuba
[4] Centro de Investigación y Desarrollo de Estructuras y Materiales, Universidad Central de Las Villas, 54830 Santa Clara, Cuba

Abstract. Four new calcined kaolinitic clays as source of supplementary cementitious materials to production of cement with high level of clinker replacement were assessed in this research. Anhydrous cements were characterized by particle size distribution (PSD), specific surface (BET) and thermogravimetric analysis (TGA). The pastes were assessed by X ray diffraction (XRD), TGA and mercury intrusion porosimetry (MIP). The behavior of blends was too assessed by compressive strength in standard mortars. The specific surface of LC^3 cement depends mainly on the specific surface of calcination product of clay, which depend as well on the mineralogical composition of the raw material and the calcination temperature. Results indicated an agreement with the kaolinite content in the original clay, pozzolanic reactivity and the performances of blended cements. The research showed the potentialities of cuban clay deposits from different geologic origin to be used in the production of ternary blended cements with similar performances to the Portland cement.

1 Introduction

Previous research have demonstrated the potential of thermal activation of clays with kaolinite contents of only 40% to obtain a high reactivity pozzolanic material [1, 2]. However, the simple substitution of clinker for this pozzolanic material is only practical up to 30%, which in many cases as in Cuba, where energy costs have a strong influence on the costs of calcination, this option is not economically viable [3]. Development of a new family of cements named LC^3 (Limestone Calcined Clays Cement), allows to increase the clinker replacement levels and maintains or improve the performance of Portland cement (PC) with high clinker content [4, 5], which compensates for

F. Martirena et al. (eds.), *Calcined Clays for Sustainable Concrete*, RILEM Bookseries 16,
https://doi.org/10.1007/978-94-024-1207-9_4

calcination costs while decreasing environmental impact and costs production with respect to the PC [5, 6].

In Cuba took place the first industrial trial for the calcined clay production to be used as pozzolan in LC^3 cement manufacture [5]. The main limitations for the generalized production of LC^3 in the country is the lack of knowledge about the characteristics and availability of kaolinitic clays with potential to be used as raw material in the production of Supplementary Cementitious Materials (SCMs). Currently, clays resources available are limited to satisfy the characteristics demanded by the ceramic industries, refractories and the white cement production, which require deposits of high kaolinite and low iron content, considerations that are not limiting for the production of SCMs [7]. Reason why the study of clays with high levels of impurities, but its kaolinite content justifies their use as source of SCMs, is a need for the new cementitious system manufacture. The aim of this paper is to assess four new cuban kaolinitic clays as source of SCMs to production of cement based on clinker- calcined clay and limestone.

2 Methods

Four kaolinitic clays (LL, LS, YG, CG) from different deposit of different geological origin were selected. Tables 1 and 2 presents the chemical and mineralogical composition of the composite samples performed by X-ray Fluorescence (XRF) and X-ray Diffraction (XRD), respectively. Clay minerals content in the raw material was determined by Thermogravimetric Analysis (TGA) from the mass loss during clays dehydroxylation, between 350 °C and 850 °C, and reported as equivalent kaolinite (K^E). Clays were calcined at 750 °C and 850 °C. After calcination were grounded in a ball mill. Pozzolanic reactivity of the calcined clays was assessed by R^3 test [8].

Seven LC^3 blended system (LC^3-LL 750, LC^3-LL 850, LC^3-LS 750, LC^3-LS 850, LC^3-YG 750, LC^3-YG 850, LC^3-CG 850) were prepared by mixing of the calcined clays, clinker, pulverized limestone and gypsum, processed previously by separate grinding. The LC^3 systems refers to cements with 45% of substitution of clinker by 30% of calcined clay and 15% of limestone, with a clinker content of 49% and 6% of gypsum. Differences of each blend set in the clay composition and calcination temperature. The chemical composition of the components is presented in Table 1.

Table 1. Chemical composition of the original clays, clinker, gypsum, limestone and PC

	SiO_2	Al_2O_3	Fe_2O_3	CaO	MgO	SO_3	Na_2O	K_2O	TiO_2	P_2O_5	Others	LOI
LL	61,40	18,86	9,61	0,07	0,15	0,02	0,26	0,90	0,62	0,13	0,24	7,80
LS	50,88	25,23	12,58	0,28	0,95	0,02	0,08	0,32	0,98	0,05	0,33	8,39
YG	46,58	20,06	14,41	2,94	0,74	0,04	0,11	0,06	1,12	0,13	0,87	12,74
CG	39,55	31,58	12,68	0,05	0,53	0,13	0,14	0,17	0,52	0,04	0,38	14,37
Clinker	20,81	5,01	4,37	65,70	0,90	0,34	-	-	-	-	-	0,50
Gypsum	8,10	2,03	1,97	30,27	2,81	31,39	-	-	-	-	-	20,91
Limestone	3,02	0,94	0,49	51,49	0,71	-	-	-	-	-	-	41,74
P-35	20,37	4,55	3,08	62,91	1,15	2,49	-	-	-	-	-	3,49

Table 2. Mineralogical composition of the clay samples

	Origin	Clay minerals 1:1	Clay minerals 2:1	Associated minerals	Kaolinite content, %
LL	Hydrothermal	Kaolinite	Muscovite	Quartz, hematite, goethite	42,16
LS	Hydrothermal	Kaolinite, nacrite	Vermiculite, montmorillonite		57,78
YG	Redeposited	Kaolinite, nacrite, halloysite	Vermiculite	Calcite, quartz, goethite, anatase	46,13
CG	Weathering	Kaolinite, halloysite	Kaolinite – montmorillonite	Gibbsite, quartz, hematite, goethite	81,06

Anhydrous cements were characterized by particle size distribution (PSD) and specific surface (BET). A qualitative assessment of the hydration of pastes was carried out by XRD. Bound water, portlandite and calcite content was determined by TGA. Mercury Intrusion Porosimetry (MIP) was used to assess pore structure of LC^3 pastes with calcined clays at 850 °C. Standard mortars were cast in order to determine the strength development of the LC^3 blends according to NC-506 [9]. A Portland cement named P-35 as reference was used.

3 Results

3.1 Reactivity of Calcined Clays

Higher pozzolanic reactivity and lower values of specific surface were obtained for calcined clays at 850 °C (Table 3). Anomalous behavior for YG at 850 °C seems to be related with the rough decrease of the specific surface. In general, the pozzolanic reactivity is in agreement with K^E content.

Table 3. Cumulated heat and specific surface for calcined clays at 750 °C and 850 °C

	LS		LL		YG		CG
Calcination temperature, °C	750	850	750	850	750	850	850
Cumulated heat released at 132 h (J/g)	125,88	137,85	97,00	101,31	118,82	62,43	167,42
Specific surface (m^2/g)	57,08	46,11	19,06	11,61	50,18	5,41	28,86

3.2 Particle Size Distribution and Specific Surface of Anhydrous LC^3 Blends

LC^3 blended system shows similar particles size distribution (Fig. 1). As observed the curve practically are superposed, which it is caused, by the similar preparation conditions of the raw materials. However, more difference, in the specific surface of LC^3 system is observed (Fig. 2a). The specific surface of the LC^3 blends is directly proportional to the specific surface of the calcined clays (Fig. 2b). Decrease of specific surface for samples could not be explained by assuming particle coarsening. Lower specific surface for similar particle size distribution is obtained when grinding clays after calcination. Grinding after calcination have little effect on overall specific surface, as compared with calcination temperature. Therefore, the specific surface of the LC^3 blended system can be controlled by the activation temperature of clays. In addition, decrease of the specific surface from 750 °C to 850 °C seems to be too related with the presence of clay minerals type 2:1 and associate phases. The presence of calcite and vermiculite in YG and vermiculite and montmorillonita in LS sample cause break down at high calcination temperature decreasing specific surface, whereas LL contains more stable minerals (moscovite, quartz) to the analyzed temperatures.

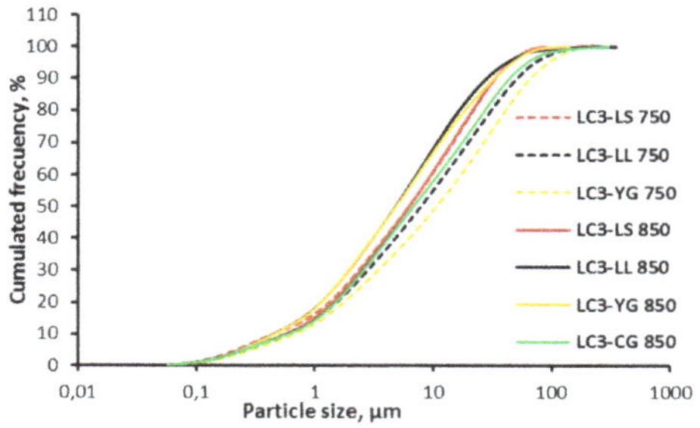

Fig. 1. Particle size distribution of LC^3 blended systems

3.3 Characterization of the LC^3 Pastes

The XRD patterns (Fig. 3) for the LC^3 system show at 2 days the formation of ettringite, hemicarboaluminate (Hc) and a formation growing of portlandite (CH). The presence of Hc is associated at early reaction of the aluminates and calcium carbonate. The peaks assigned to this phase are more intense for LC^3 with calcined clays with higher availability of reactive alumina (higher K^E content). At 28 days decrease the peaks of Hc and monocarboaluminate (Mc) are observed, indicating the advance of the reaction between the aluminates and limestone. Ettringite formation is stabilized at 28 days due to the formation of Hc and Mc allowing a higher sulphate availability. Reduction of the peaks assigned to the CH of 2 to 28 days is a qualitative evidence of the advance of the pozzolanic reaction. Practically imperceptible CH peaks is observed in the paste that contains more reactive calcined clay CG (LC^3–CG 850).

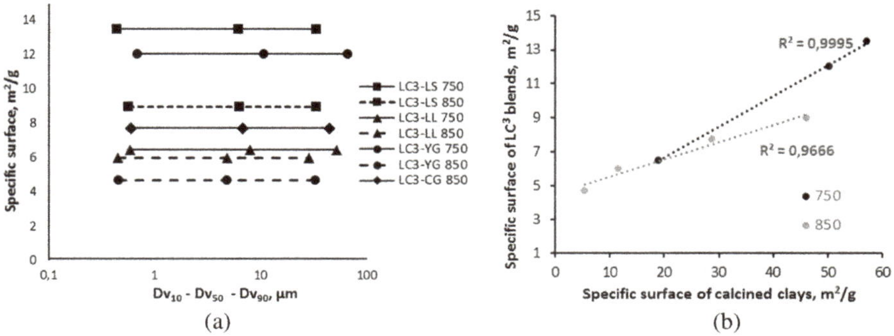

Fig. 2. (a) Specific surface and particles size distribution of the LC^3 blends. (b) Relation between specific surface of the LC^3 blends and the specific surface of the calcined clays

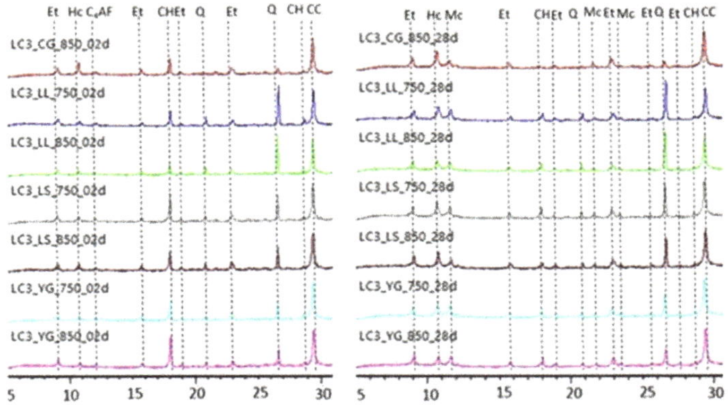

Fig. 3. XRD patterns of the LC^3 paste at 2 days (left) and 28 days (right) of hydration. Et (Ettringite); Hc (Hemicarboaluminate); C_4AF (Ferrite); Mc (Monocarboaluminate); CH (Portlandite); CC (Calcite); Q (Quartz)

For all blended cements, the amount of bound water increases and portlandite content decrease from 2 to 28 days (Fig. 4a). Both are a clear evidence of the pozzolanic reaction in the LC^3 pastes. Decrease of the calcite content (Fig. 4b) from 2 and 28 days show the participation of the calcite as active component in the pozzolanic reaction to form Hc and Mc. Results are in correspondence with the pozzolanic reactivity of the calcined clays determined by the R^3 test (Table 3). In general, the portlandite consumption and bound water at 2 and 28 days are higher in LC^3 pastes containing calcined clays with higher K^E content.

Figure 5 presents the results of pore size distribution of the LC^3 at 2 and 28 days of hydration. In all LC^3 pastes is reduced the total porosity due to the hydration progress.

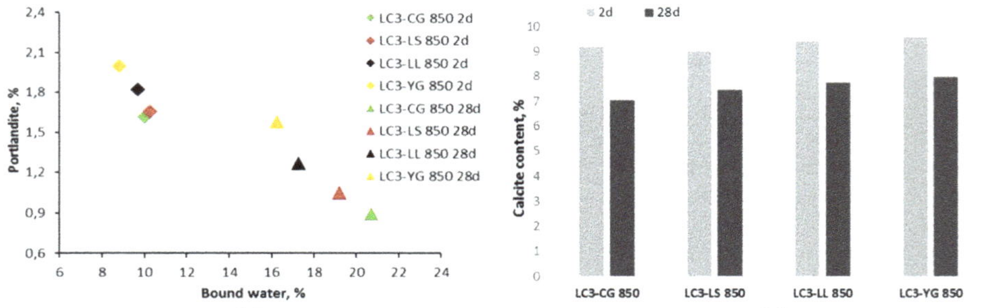

(a) (b)

Fig. 4. (a) Relation between Portlandite content and bound water to 2 (2d) y 28 (28d) days of hydration for LC^3 series with calcined clay to 850 °C. (b) Calcite content at 2 (2d) y 28 (28d) days of hydration

The accumulated porosity seems to be related to the pozzolanic reactivity of the different calcined clays. The reduction of the accumulated porosity and the pore structure refinement are associated to the increase in the volume of hydration products due to the formation of AFm and AFt phases as observed by XRD. Later phases present low density, which contributes very efficiently to the pore structure refinement.

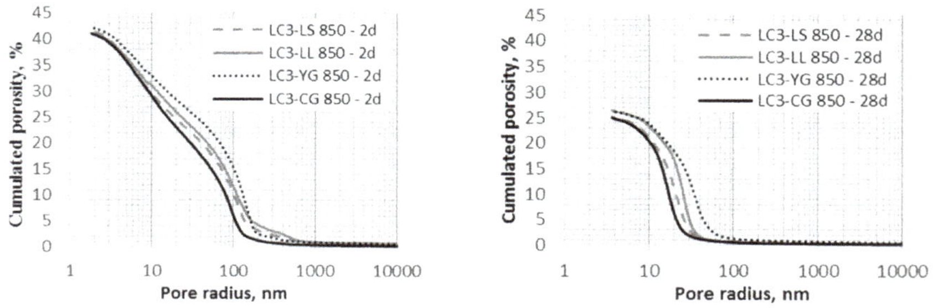

Fig. 5. Cumulated porosity as a function of pore entry radius at 2 and 28 days obtained by MIP

3.4 Performance of the LC^3 Mortars

Compressive strength in standard mortars of the LC^3 blends with calcined clays at 850 °C is presented in Fig. 6. At 3 days, LC^3 blends exhibit lower compressive strength than the reference. However, from 7 days the behavior among the different series begins to be differentiated. The best results are observed for LC^3-LS 850 and LC^3-CG 850, which at 28 days reach and improve, respectively, the reference. Difference between the blended systems is accentuated in relation directly proportional to the calcined clays reactivity.

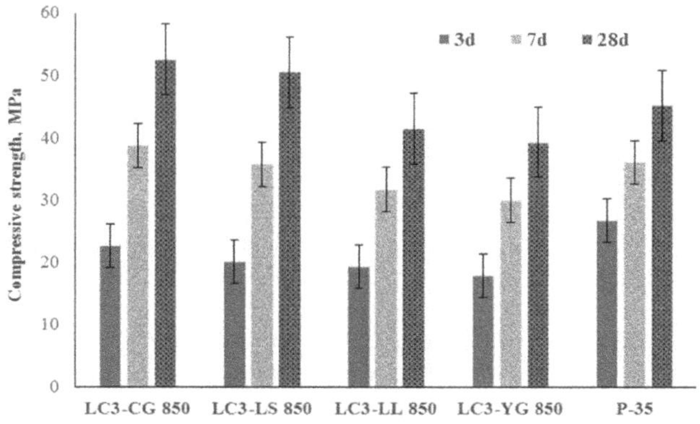

Fig. 6. Compressive strength of the LC^3 mortars at 3, 7 and 28 days

4 Conclusion

Pozzolanic reactivity is consistent with the clay minerals content reported as equivalent kaolinite. For a similar particle size distribution, the specific surface of the LC^3 blended is determined by the specific surface of the calcined clay, which depends on calcination temperature and the mineralogical composition of the original clay. The formation of AFm and AFt phases enhance the reduction of the pore structure. Bound water and portlandite consumption are in agreement with the pozzolanic reactivity of the calcined clays. LC^3 blended systems mortars exhibit compressive strength similar to Portland cement. Results shows the good potential of the clay deposits studied, to be used as a source of MCS in the LC^3 in Cuba.

Acknowledgments. The authors would like to acknowledge to the *Low Carbon Cement Proyect* team from Laboratoires des Matériaux de Construction (LMC) of the École Polytechnique Fédérale de Lausanne (EPFL) in Switzerland, especially to François Avet for their helpful in the experimental work. The authors would like to acknowledge to the Cuban Geological Services for the technical support in clays and limestone deposits sampling.

References

1. Alujas, A., Fernández, R., Quintana, R., Scrivener, K.L., Martirena, F.: Pozzolanic reactivity of low grade kaolinitic clays: influence of calcination temperature and impact of calcination products on OPC hydration. Appl. Clay Sci. **108**, 94–101 (2015). doi:10.1016/j.clay.2015.01.028
2. Alujas, A., Almenares, R.S., Betancourt, S., Leyva, C.: Pozzolanic reactivity of low grade kaolinitic clays: influence of mineralogical composition. In: Scrivener, K., Favier, A. (eds.) Calcined Clays for Sustainable Concrete, pp. 339–345. Springer, Dordrecht (2015). doi:10.1007/978-94-017-9939-3_42
3. Scrivener, K.L.: Options for the future of cement. Indian Concr. J. **88**, 11–21 (2014)

4. Antoni, M., Rossen, J., Martirena, F., Scrivener, K.: Cement substitution by a combination of metakaolin and limestone. Cem. Concr. Res. **42**, 1579–1589 (2012). doi:10.1016/j.cemconres.2012.09.006
5. Vizcaíno, L., Sánchez, S., Pérez, A., Damas, S., Scrivener, K., Martirena, F.: Industrial trial to produce low clinker, low carbon cement. Materiales de Construcción **65**, e045 (2015). doi:10.3989/mc.2015.00614
6. Sánchez Berriel, S., Favier, A., Domínguez, E.R., Sánchez Machado, I.R., Heierli, U., Scrivener, K., et al.: Assessing the environmental and economic potential of Limestone Calcined Clay Cement in Cuba. J. Clean. Prod. **124**, 361–369 (2016). doi:10.1016/j.jclepro.2016.02.125
7. Almenares, R.S., Vizcaíno, L.M., Damas, S., Mathieu, A., Alujas, A., Martirena, F.: Industrial calcination of kaolinitic clays to make reactive pozzolans. Case Stud. Constr. Mater. **6**, 225–232 (2017). doi:10.1016/j.cscm.2017.03.005
8. Avet, F., Snellings, R., Alujas, A., Ben, M., Scrivener, K.: Development of a new rapid, relevant and reliable (R3) test method to evaluate the pozzolanic reactivity of calcined kaolinitic clays. Cem. Concr. Res. **85**, 1–11 (2016). doi:10.1016/j.cemconres.2016.02.015
9. NC 506: 2013, Cemento hidráulico. Método de ensayo. Determinación de la resistencia mecánica, p. 26 (2013). www.nc.cubaindustria.cu

Proposal of a Methodology for the Preliminary Assessment of Kaolinitic Clay Deposits as a Source of SCMs

Adrián Alujas Díaz[1(✉)], Roger S Almenares Reyes[2], Florencio Arcial Carratalá[3], and José F. Martirena Hernández[4]

[1] Centro de Estudios de Química Aplicada, Universidad Central de Las Villas, 54830 Santa Clara, Cuba
[2] Departamento de Metalurgia – Química, Instituto Superior Minero Metalúrgico de Moa, 83330 Moa, Cuba
[3] Empresa Geominera del Centro, 54800 Santa Clara, Cuba
[4] Centro de Investigación y Desarrollo de Estructuras y Materiales, Universidad Central de Las Villas, 54830 Santa Clara, Cuba

Abstract. The growing interest in calcined clays as Supplementary Cementitious Materials (SCMs), from sources other than relatively pure industrial grade clay deposits, demands the development of new tools for the identification and evaluation of the potentialities of a given clay deposit as a source of raw materials for the production of SCMs. In this paper, a methodology for the preliminary assessment of kaolinitic clay deposits as a source of SCMs is presented and discussed. As a case study, the chemical and mineralogical composition of several non industrial grade kaolinitic clay deposits, identified and classified according to the proposed methodology, is related to the pozzolanic reactivity of their calcination products. The developed methodology allows to establish preliminary selection criteria based on the chemical composition of the raw material (% Al_2O_3 > 18,0; Al_2O_3/SiO_2 > 0,3; LOI > 7,0; % CaO < 3,0; % SO_3 < 3,0). Comparison of the potentialities between the different clay deposits is determined from its relative position in a plot that relates the content of Al_2O_3 and the weight loss within the range of temperatures that corresponds to the thermal decomposition of clay minerals, which are, in turn, the main parameters affecting the pozzolanic reactivity of calcination products. The presence of thermally active non clay minerals in the sample may also affects negatively the pozzolanic reactivity. Through the observed results, the developed methodology proves to be a practical and useful tool for the identification and evaluation of kaolinitic clay deposits as a source of pozzolans.

1 Introduction

At a global level, the production of pozzolanic materials from calcined clays focuses primarily on the use of industrial grade kaolin, ignoring the possibilities offered by low grade kaolinitic clay deposits with high degree of impurities, highly abundant in tropical and subtropical areas, where their identified and hypothetical resources exceed billions of tons [1]. This is partly because the current approach for the categorization of kaolinitic

F. Martirena et al. (eds.), *Calcined Clays for Sustainable Concrete*, RILEM Bookseries 16,
https://doi.org/10.1007/978-94-024-1207-9_5

clay deposits and estimation of their reserves responds to criteria derived mainly from the ceramics, paper or pigment industries, where selection parameters are very rigorous as regards to the absence of mineral impurities that may affect color and plasticity [2, 3]. The absence of selection criteria adapted to the evaluation of the potentialities of clay deposits as a source of raw materials for obtaining pozzolanic materials, limits their identification and exploitation on an industrial scale. In this way, the present work is directed towards the development and application of a procedure for the preliminary evaluation of the potentialities of clay deposits as a source of SCM.

2 Analysis and Discussion

Given the relative heterogeneity of chemical and mineralogical composition of clay deposits, it is necessary to identify those essential parameters that influence the behavior of the clays as source of pozzolans. From the results of several research it can be concluded that the most important parameter in the pozzolanic reactivity of calcined clays is their content of clay minerals, more specifically their content of minerals from the kaolinite group, which must be equal to or higher than 40% by weight [4, 5]. A material whose theoretical composition is 40% by weight of minerals of the group of kaolinite and 60% of quartz as an inert material will have a theoretical composition with no less than 15.80% Al_2O_3, an Al_2O_3/SiO_2 ratio of approximately 0.20 and LOI values of no less than 5.58. Based on this calculated values for chemical composition and also in practical experience on the analysis of the different kaolinite clays included in this research (Table 1) and other clay materials reported in previous studies [6, 7], the following parameters could be stablished as empirical selection criteria: $Al_2O_3 \geq 18.00\%$; $Al_2O_3/SiO_2 \geq 0.30$ and LOI $\geq 7.00\%$. In this analysis, it is also necessary to take into account the influence of the accompanying minerals in the pozzolanic reactivity of the final product. For the case of calcite, a negative effect on the pozzolanic reactivity has been reported at calcination temperatures above 800 °C, related to the formation of vitreous phases rich in Ca, Si and Al, of low specific surface area. This phenomenon has been extensively reported in the field of ceramic materials, where it is frequent to add small amounts of calcite to the kaolinite clays to increase the density of the calcination product. Negative effects on the degree of calcination of the final product have also been observed by the authors of this work for samples with relatively high pyrite content. In addition, both the decomposition of sulfides and the decomposition of sulfates during thermal activation generate undesirable SO_3 emissions into the atmosphere, with a negative environmental impact. Therefore, it is proposed to limit the allowed maximum values of SO_3 and CaO to 3.0%. Thus, the chemical composition criteria for accepting or rejecting a clay deposit as source of SCM are as follows: $Al_2O_3 \geq 18.00\%$; $Al_2O_3/SiO_2 \geq 0.30$, LOI $\geq 7.00\%$, $SO_3 \leq 3.0\%$ and CaO $\leq 3.0\%$.

Table 1. Chemical composition of analyzed clays

Sample	SiO_2	Al_2O_3	Fe_2O_3	CaO	MgO	SO_3	Na_2O	K_2O	Others	LOI
MN	57,79	18,71	7,07	1,85	1,80	0,02	2,68	0,65	0,88	8,57
PZ	48,40	29,50	16,50	0,40	1,00	1,10	-	0,70	0,70	8,60
KW	48,00	36,40	0,85	0,14	0,11	0,03	0,02	0,48	0,56	13,41
IW	58,68	19,25	5,04	1,29	2,50	0,17	0,19	6,12	1,05	5,71
MW	63,15	20,09	3,96	1,15	2,27	0,51	2,22	0,54	0,22	5,90
LL	61,40	18,86	9,61	0,07	0,15	0,02	0,26	0,90	0,99	7,80
LS	50,88	25,23	12,58	0,28	0,95	0,02	0,08	0,32	1,26	8,39
YG	46,58	20,06	14,41	2,94	0,74	0,04	0,11	0,06	1,84	12,74
CG	39,55	31,58	12,68	0,05	0,53	0,13	0,14	0,17	0,94	14,37
A	46,96	31,66	5,81	0,14	0,06	0,02	0,02	0,11	0,82	14,44
B	60,33	18,44	6,49	1,03	1,95	0,01	1,64	1,41	1,40	7,38
C	64,34	15,44	9,91	0,08	0,48	0,01	0,15	0,65	1,52	7,49
D	63,95	23,15	1,49	0,07	0,09	0,00	0,02	0,02	0,73	10,53
E	64,61	20,45	4,64	0,15	0,39	0,01	0,03	0,93	0,81	8,03
F	55,75	15,01	5,98	7,09	1,22	0,55	0,58	2,00	1,72	10,31
G	53,78	31,83	8,89	0,91	0,03	0,00	0,00	0,50	1,90	10,84
H	54,74	18,32	4,91	0,41	0,00	0,00	0,00	0,22	1,29	4,66
I	43,25	38,71	1,53	0,10	0,00	0,01	0,12	0,06	3,52	11,70

For the quantification of clay minerals, the weight loss associated with dehydroxylation in the temperature range between 350 and 850 °C [8] is taken as the main criterion. Although some 2:1 clay minerals continue their dehydroxylation process at temperatures above 850 °C, for kaolinite clays the use of calcination temperatures above 850 °C is unusual, since the beginning of the recrystallization processes occurs at temperatures close to 900 °C. On the other hand, limiting the analysis interval to lower temperatures, although allows to quantify most of the contribution of 1.1 clay minerals, discriminates the possible contribution of other clay phases present in the analyzed material and makes the method unsuitable for 2:1 clays. However, even with high accuracy thermal analysis and careful data treatment it is very difficult to separate the individual contribution to the weight loss of each one of the clay minerals presents in the sample. Based on the fact that contributions to weight loss and pozzolanic reactivity is higher for minerals of the kaolinite group, and by assuming that weight loss in the temperature range between 350 and 850 °C are mainly related to kaolinite, the total content of clay minerals could be reported in terms of kaolinite equivalent (KEQ), as calculated by Eq. 1, where 0.1396 is the weight fraction corresponding to hydroxyl groups in kaolinite. Ultimately, the structural disorder is directly proportional to the loss of structural hydroxyls and directly related to the pozzolanic reactivity of the calcined clay.

$$\%KEQ = \frac{m(350^{\circ}C) - m(850^{\circ}C)}{m(200^{\circ}C) * 0.1396} \cdot 100 \tag{1}$$

In the quantification of weight losses, it is also necessary to consider the possible interference caused by the overlapping of the decomposition reactions of associated non-clay minerals. Among the most frequent associated non-clay minerals in clay deposits are carbonates, sulphides, sulphates and hydroxides. These non-clay minerals, although decomposed in a relatively narrow range of temperatures, may have a significant influence on weight losses. However, as in relation to other clay and non-clay minerals, the minerals of the kaolinite group have a relatively high content of Al_2O_3 (~39.50%), it is proposed to correlate the results derived from the quantitative analysis by weight losses with the content of Al_2O_3 in the sample, to correct the possible interferences of the non-clay minerals. This could be expressed in the form of a graph represented in Fig. 1, in which the Al_2O_3 content is related to weight losses. Diagonal line in the graph indicates the linear relationship between both parameters for minerals of the kaolinite group and also indicates the maximum value that the weight losses associated with the clay minerals can take in relation to the content of Al_2O_3. For samples whose composition places them below diagonal line, their content of clay minerals, expressed as %KEQ, is obtained by extrapolating their location to the line, from a line parallel to the axis of the Al_2O_3 content. The location of samples above diagonal line indicates the presence of thermally active non-clay minerals in the range of 350 to 850 °C (carbonates, sulphates, sulphides, hydroxides). In this case, weight losses should be corrected to avoid overestimation of the content of clay minerals.

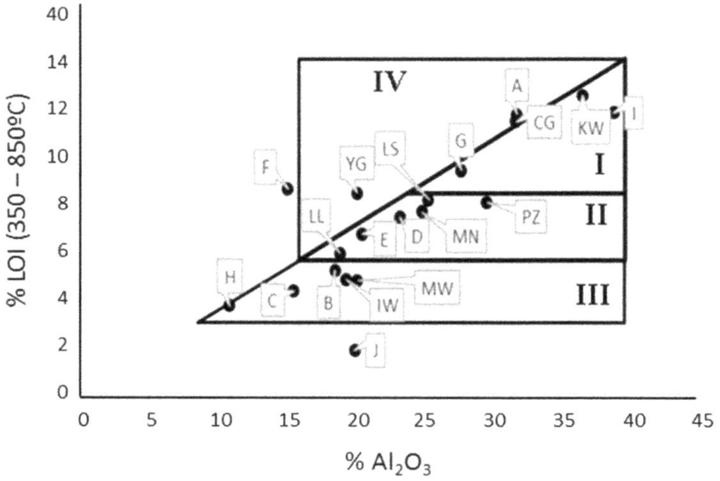

Fig. 1. Representation of several clay samples in a Al2O3 vs LOI plot

Taking into account the characteristics of the sample as a function of the content of clay minerals, the graph represented in Fig. 1 could be divided into four zones, each one with distinctive characteristics. Region I includes those clays with relatively high kaolinitic clay contents (KEQ > 60%), and therefore, excellent potentialities of being used as a source of SCM. These clays may or may not contain relatively high amounts of iron-rich phases; therefore, this region of the diagram includes, but is not limited, to the so-called industrial grade kaolin. Samples in Region II are also representative of kaolinitic clays with good potentialities as a source of SCM, but with lower kaolinite contents (40% < KEQ < 60%). In these zones of the diagram the vast majority of kaolinitic clay deposits yet to be evaluated should be located. Zone III of the diagram refers to those clays with low kaolinite content (KEQ < 40%), but with relatively high content of 2:1 clays. It should be noted that, although this type of material does not generally meet the minimum selection criteria proposed for kaolinitic clays, it could be exploited as an alternative source of SCM in countries where the presence of kaolinitic clay deposits is relatively scarce [9, 10]. Finally, zone IV of the diagram includes clay deposits containing variables amounts of kaolinitic clay minerals, but with an appreciable content of non-clay minerals that decompose in the same temperature range than clay minerals, such as carbonates, sulphates or sulphides. Therefore, in the thermal activation of the clays included in zone IV of the diagram must take into account the occurrence of potentially negative processes and the calcination strategy must be designed accordingly.

As an example of application of the proposed methodology, position of several clay samples are included in Fig. 1, including three high grade samples of Kaolinite (KW), Illite (IW) and Montomorillonite (MW) that were used as reference patterns. As could be observed, for clays with kaolinite content higher than 40% there is an almost linear correlation between content of Al_2O_3 and LOI in the 350–850 °C temperature range, with the noticeable exceptions of samples with relatively high content of CaO (samples F and YG), where LOI are also related to decomposition of carbonate like minerals.

3 Conclusion

Selection criteria based on chemical composition (%Al_2O_3 > 18; Al_2O_3/SiO_2 > 0,3; LOI > 7,0; low % CaO and SO_3), allows a rapid screening of suitable clay deposits on a given area.

Correlation of weight loss with %Al_2O_3 allows a rapid assessment of the potentialities of clay deposits as a source of raw materials for the production of SCMs.

Acknowledgments. The authors would like to acknowledge the financial support of the Low Carbon Cement project, from SDC. They would also like to acknowledge to the Cuban Geological Services for the technical support in clay and limestone deposits sampling.

References

1. Velde, B. (ed.): Origin and Mineralogy of Clays. Clays and the Environment. Springer Science & Business Media, Heidelberg (2013)
2. Chandrasekhar, S., Ramaswamy, S.: Influence of mineral impurities on the properties of kaolin and its thermally treated products. Appl. Clay Sci. **21**, 133–142 (2002). doi:10.1016/S0169-1317(01)00083-7
3. Murray, H.: Industrial Clays Case Study, IIED and WBCSD. MMSD-Indiana University (2002)
4. He, C., Osbaeck, B., Makovicky, E.: Pozzolanic reactions of six principal clay minerals: activation, reactivity assessments and technological effects. Cem. Concr. Res. **25**, 1691–1702 (1995). doi:10.1016/0008-8846(95)00165-4

5. Fernandez, R., Martirena, F., Scrivener, K.L.: The origin of the pozzolanic activity of calcined clay minerals: a comparison between kaolinite, illite and montmorillonite. Cem. Concr. Res. **41**, 113–122 (2011). doi:10.1016/j.cemconres.2010.09.013
6. Fernández, R.: Calcined Clayey Soils as a Potential Replacement for Cement in Developing Countries, École Polytechnique Federale de Lausanne (2009)
7. Alujas, A.: Obtención de un material puzolánico de alta reactividad a partir de la activación térmica de una fracción arcillosa multicomponente, Tesis Doctoral, Universidad Central «Marta Abreu» de las Villas (2010)
8. Földvári, M.: Handbook of Thermogravimetric System of Minerals and Its Use in Geological Practice (2011)
9. Lemma, R., Irassar, E.F., Rahhal, V.: Calcined illitic clays as Portland cement replacements. In: Scrivener, K., Favier, A. (eds.) Calcined Clays for Sustainable Concrete, vol. 10, pp. 269–276. Springer, Dordrecht (2015). doi:10.1007/978-94-017-9939-3_33
10. Lemma, R., Trezza, M., Rahhal, V.F., Irassar, E.F.: Thermal transformation of illitic - chlorite clays and its pozzolanic activity. In: VI Internacional Symposium of Chemestry, Facultad de Química-Farmacia, Universidad Central «Marta Abreu» de Las Villas (UCLV) Cuba, Cayo Santa María, Santa Clara, Cuba (2016)

Hydration Study of Limestone Calcined Clay Cement (LC^3) Using Various Grades of Calcined Kaolinitic Clays

F. Avet(✉) and K. Scrivener

Laboratory of Construction Materials, EPFL, 1015 Lausanne, Switzerland

Abstract. In this study, the influence of the calcined kaolinite content of calcined clay on the hydration of Limestone Calcined Clay Cement (LC^3-50) is investigated. It is shown that with the increase of the calcined kaolinite content of the calcined clay, the clinker hydration degree decreases. The amount of reacted metakaolin increases with the calcined kaolinite content, leading to higher formation of C-A-S-H. However, for the LC^3-50 with the highest-grade calcined clay, the formation of carboaluminate hydrates is limited. For this blend, the porosity is well refined at 3 days of hydration and does not significantly change later on.

1 Introduction

In order to reduce CO_2 emissions of cement production, the use of Supplementary Cementitious Materials (SCMs) is the most promising approach. However, most of common SCMs used nowadays (slag, fly ash) are not able to face the global increase of cement production around the world. Only two SCMs are widely available and can be massively used: calcined kaolinitic clays and limestone. In Limestone Calcined Clay Cements (LC^3), these two SCMs are combined in order to benefit from their synergy [1]. The clinker content is reduced to 50% in LC^3-50 blends. Excellent strengths are obtained for these blends, with an increase of strength with the calcined kaolinite content of the calcined clay [2]. Moreover, higher strengths than plain PC were measured for blends with a calcined kaolinite content $\geq$40%. In this paper, the phase assemblage during hydration is studied in order to understand these strength results. To reach this objective, X-Ray Diffraction (XRD) was combined with Scanning Electron Microscopy (SEM) and a global mass balance to get the whole phase assemblage of the LC^3-50 blends, especially the reaction degree of metakaolin and the amount of C-A-S-H formed. Mercury Intrusion Porosimetry (MIP) was used to characterize the porosity.

2 Materials and Methods

Three clays were used in this study. The kaolinite content was determined by Thermogravimetric Analysis (TGA) from the mass loss over kaolinite dehydroxylation from about 400 °C to 650 °C. In order to take into account the efficiency of calcination, the calcined kaolinite content is introduced as the difference of kaolinite before and after

F. Martirena et al. (eds.), *Calcined Clays for Sustainable Concrete*, RILEM Bookseries 16,
https://doi.org/10.1007/978-94-024-1207-9_6

calcination. It corresponds to the part of kaolinite which is calcined, i.e. the reactive part. The physical characteristics and the XRF chemical composition of the clays used in this study as well as the limestone and cement are shown in Table 1.

Table 1. Physical characteristics and XRF chemical composition of the three calcined clays, the clinker and limestone

Calcined clay	1	2	3	Cement	Limestone
Calcined kaolinite content (%)	95.0	50.3	17.0	–	–
$D_{v,50}$ (μm)	5.1	10.9	5.9	8.4	7.2
BET Specific surface (m^2/g)	9.6	45.7	18.7	0.9	1.8
SiO_2	52.0	44.9	68.4	19.3	0.1
Al_2O_3	43.8	32.3	17.5	5.7	–
Fe_2O_3	0.3	15.4	8.9	3.6	–
CaO	–	1.3	0.6	63.6	55.0
MgO	–	0.8	0.7	1.6	0.2
SO_3	0.1	0.1	–	3.2	–
Na_2O	0.3	0.4	0.1	0.2	0.1
K_2O	0.1	0.2	2.3	1.2	–
TiO_2	1.5	2.4	0.8	0.3	–
P_2O_5	0.2	0.4	0.1	0.2	–
MnO	–	0.1	0.1	0.1	–
Others	0.1	0.2	0.2	0.3	–
LOI (at 950 °C)	1.5	1.7	0.5	0.8	42.6

In addition to reference PC system, LC^3-50 systems were cast, with 30% of calcined clay, 15% of limestone, a clinker content reduced to 50% and 5% of gypsum.

XRD-Rietveld was used to determine the consumption of clinker phases and limestone, and also to quantify the formation of crystalline hydration products, such as AFt, AFm phases and portlandite. The amount of reacted metakaolin was determined by mass balance. Each compound of the reacted phases is redistributed to a series of hydration products. The reacted metakaolin is the free variable of the system. Its correct amount is found when the phase assemblage prediction is similar to what is obtained experimentally.

3 Results

3.1 Phase Assemblage

Figures 1 and 2 show the global phase assemblage of PC and the three LC^3-50 blends containing 17.0%, 50.3% and 95.0% of calcined kaolinite at 3 and 28 days of hydration, respectively. A C-A-S-H density of 2.0 was used to get the volume occupied by this phase [3]. For PC system, monosulfate forms whereas carboaluminate hydrates are present in LC^3-50 blends. At 3 days of hydration, the total volume of solid is higher

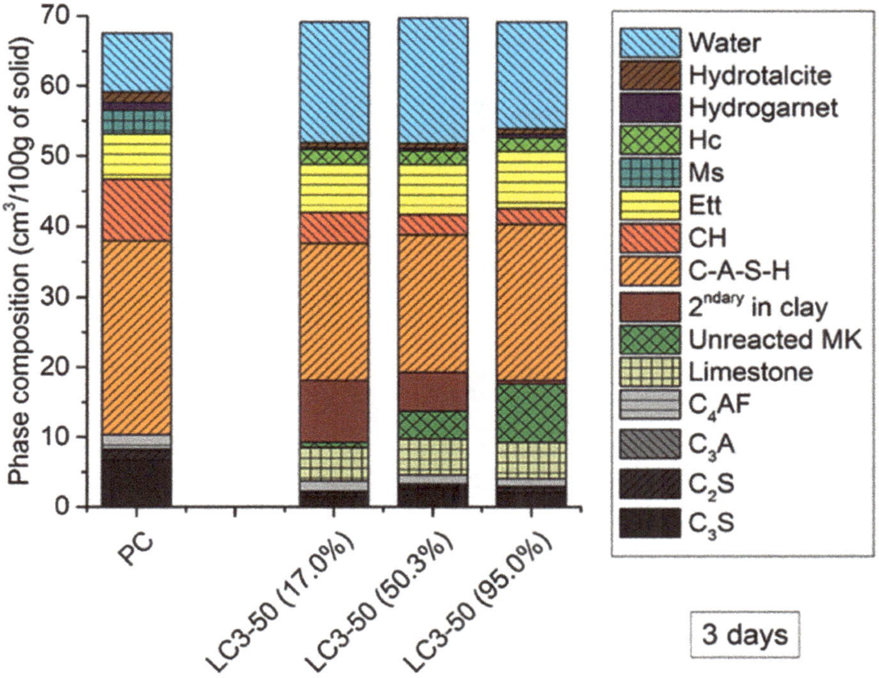

Fig. 1. Phase assemblage at 3 days of hydration for PC and LC3-50 blends with (17.0%), (50.3%) and (95.0%) of calcined kaolinite in the calcined clay.

for plain PC than for LC^3-50 blends. At 28 days of hydration, the difference is much less important.

For LC^3-50 blends at 28 days, Fig. 2 shows that with the increase of the calcined kaolinite content of calcined clay, the amount of unreacted clinker increases, i.e. the degree of hydration of clinker decreases. The portlandite consumption due to the reaction of the metakaolin of the calcined clay also increases with the calcined kaolinite content. This leads to the slightly higher formation of C-A-S-H. The ettringite content is similar for all blends because it depends on the initial sulfate content of the systems, which is constant. The amount of carboaluminate hydrates is expected to increase with the calcined kaolinite content of calcined clay, due to the higher amount of aluminates provided to the system. However, this amount is actually the lowest for the blend with the highest calcined kaolinite content. The total volume of hydrates is similar for both LC^3-50 (50.3%) and (95.0%) blends.

3.2 Porosity

MIP results at 3 days of hydration in Fig. 3 show that there is a significant refinement of pore connectivity for LC^3-50 blend with 95.0% of calcined kaolinite in calcined clay compared with PC. The LC^3-50 (50.3%) shows similar pore refinement to plain PC.

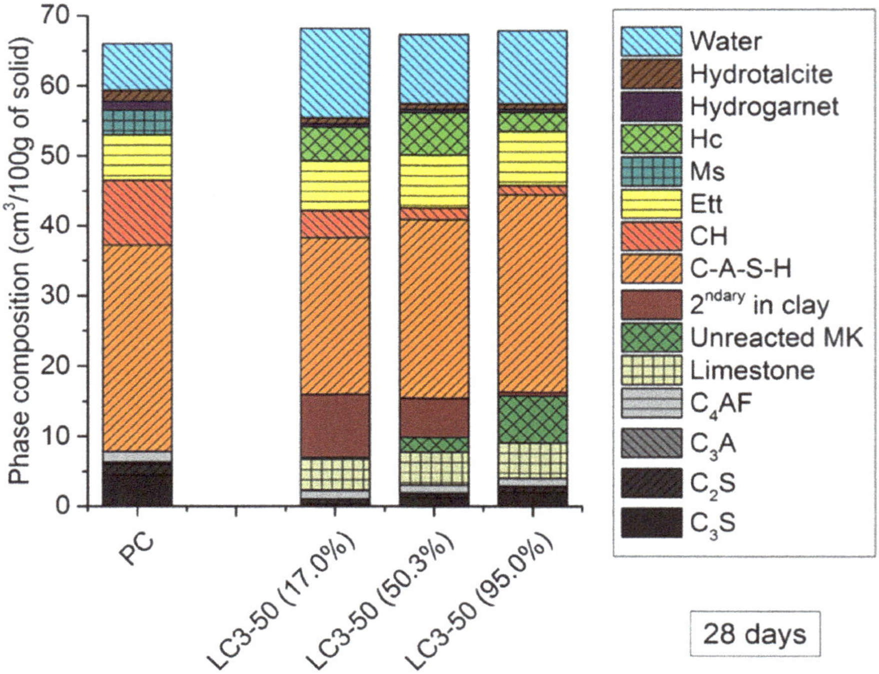

Fig. 2. Phase assemblage at 28 days of hydration for PC and LC3-50 blends with (17.0%), (50.3%) and (95.0%) of calcined kaolinite in the calcined clay

From 3 to 28 days of hydration, the pore connectivity is not further refined for the LC^3-50 with 95.0% of calcined kaolinite in calcined clay, as shown in Fig. 4. However, for the blend containing 50.3% of calcined kaolinite, a significant refinement is observed. These two LC^3-50 blends seem to reach a maximal refinement of pore connectivity, already observed in other studies [4, 5]. The refinement of pore connectivity for LC^3-50 blends compared with PC is even observed for the blend containing only 17.0% of calcined kaolinite.

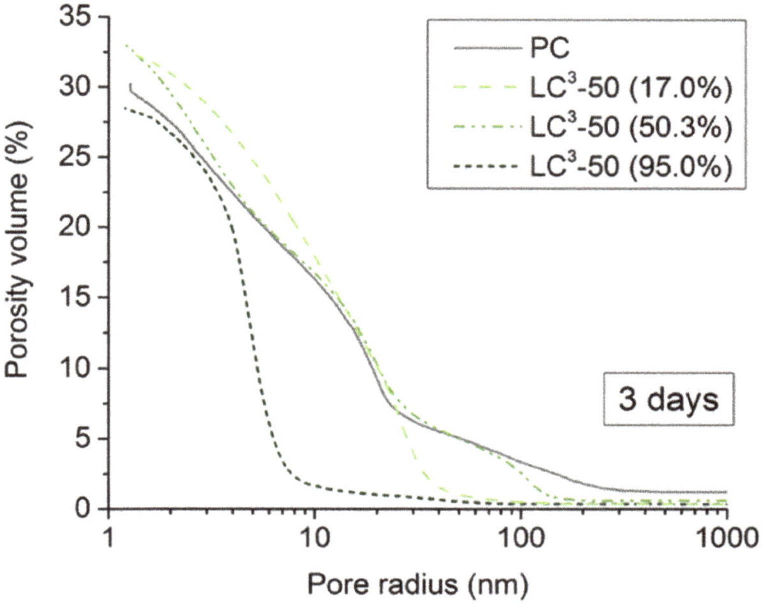

Fig. 3. MIP at 3 days of hydration for PC and LC3-50 blends containing (17.0%), (50.3%) and (95.0%) of calcined kaolinite in the calcined clay

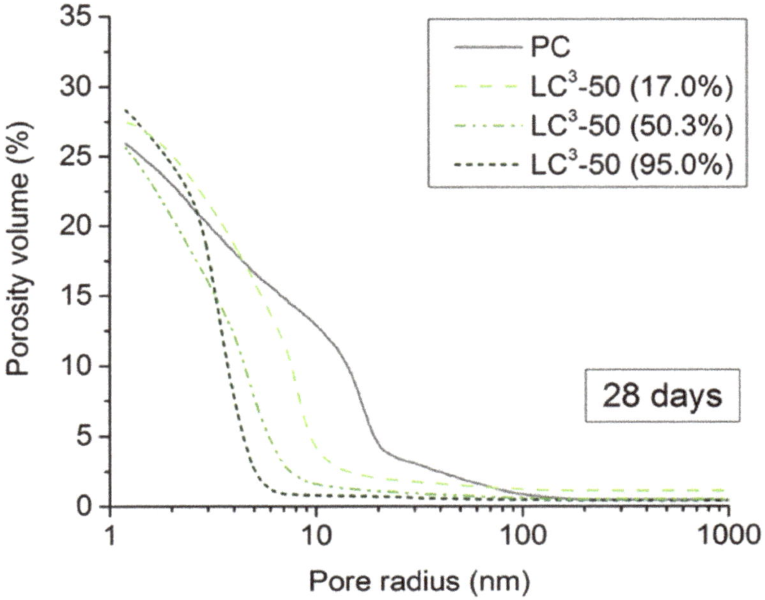

Fig. 4. MIP at 28 days of hydration for PC and LC3-50 blends containing (17.0%), (50.3%) and (95.0%) of calcined kaolinite in the calcined clay

4 Conclusion

The increase of the calcined kaolinite content of calcined clay leads to a decrease of the clinker hydration degree, more metakaolin is consumed, leading to the formation of higher amount of C-A-S-H. Significant refinement of porosity is observed for LC^3-50 systems compared with PC, even for the blend containing only 17.0% of calcined kaolinite in calcined clay.

References

1. Antoni, M., Rossen, J., Martirena, F., Scrivener, K.: Cement substitution by a combination of metakaolin and limestone. Cem. Concr. Res. **42**, 1579–1589 (2012)
2. Avet, F., Snellings, R., Alujas Diaz, A., Ben Haha, M., Scrivener, K.: Development of a new rapid, relevant and reliable (R3) test method to evaluate the pozzolanic reactivity of calcined kaolinitic clays. Cem. Concr. Res. **85**, 1–11 (2016)
3. Muller, A.C.A., Scrivener, K.L., Gajewicz, A.M., McDonald, P.J.: Densification of C-S–H measured by 1H NMR relaxometry. J. Phys. Chem. C **117**, 403–412 (2013)
4. Berodier, E., Scrivener, K.: Evolution of pore structure in blended systems. Cem. Concr. Res. **73**, 25–35 (2015)
5. Bizzozero, J., Gosselin, C., Scrivener, K.L.: Expansion mechanisms in calcium aluminate and sulfoaluminate systems with calcium sulfate. Cem. Concr. Res. **56**, 190–202 (2014)

Reaction Degree of Metakaolin in Limestone Calcined Clay Cement (LC^3)

F. Avet(✉) and K. Scrivener

Laboratory of Construction Materials, EPFL, 1015 Lausanne, Switzerland

Abstract. The amount of reacted metakaolin is determined using two methods: mass balance and thermodynamic modeling. It is found that mass balance is the most promising method. Thermodynamic modeling gives slightly underestimated results at late ages.

1 Introduction

The use of calcined clays allows to reduce the clinker content of cement and leads to a decrease of the CO_2 emissions related to cement production. A more eco-friendly approach is provided by Limestone Calcined Clay Cements (LC^3) which permits a higher clinker substitution [1]. The use of clays with modest kaolinite content is very promising for a massive use in cement due to their wide availability, good performance and low cost [2]. Metakaolin is the reactive phase present in the calcined clay. The determination of the reaction degree of metakaolin in calcined clay blends is the goal of this study. This is a key factor to better understand the influence of the grade of calcined clay, i.e. the metakaolin content of the calcined clay, on the properties of calcined clay blends. Two methods are used in this paper: mass balance and thermodynamic modeling.

2 Materials and Methods

Three calcined clays were used in this study: a fairly pure metakaolin from North America containing 94.2% of metakaolin, a calcined clay from Southeast Asia with 46.5% of metakaolin and a calcined clay from South Asia with only 15.0% of metakaolin. LC^3-50 blends were studied, composed of 50% clinker, 5% gypsum, 30% calcined clay and 15% limestone. The phase composition of the three calcined clays is shown in Table 1.

The XRF of the calcined clays, limestone and cement are shown in Table 2.

For both mass balance and thermodynamic calculations, the amount of phases consumed during hydration (clinker phases, limestone, gypsum) is considered. This phase consumption is then redistributed amongst the hydration products. The metakaolin phase is the free variable of the system. Its reacted amount is varied from 0 to 100% of reaction, and the real amount of reacted metakaolin is obtained when similar results to experimental outputs of phase assemblage are obtained.

F. Martirena et al. (eds.), *Calcined Clays for Sustainable Concrete*, RILEM Bookseries 16,
https://doi.org/10.1007/978-94-024-1207-9_7

Table 1. Phase composition of the three calcined clays

Calcined clay	1	2	3
Origin of clay	North America	South-east Asia	South Asia
Metakaolin	94.2	46.5	15.0
Anatase	1.2	0.6	0.2
Hematite	–	10.0	3.9
Illite	–	–	36.2
Mullite	4.6	1.2	–
Quartz	–	0.8	32.2
Rutile	–	–	1.5
Amorphous	0	40.9	11.0

Table 2. XRF composition of the calcined clays, cement and limestone

Calcined clay	1	2	3	Cement	Limestone
SiO_2	52.0	44.9	68.4	19.3	0.1
Al_2O_3	43.8	32.3	17.5	5.7	–
Fe_2O_3	0.3	15.4	8.9	3.6	–
CaO	–	1.3	0.6	63.6	55.0
MgO	–	0.8	0.7	1.6	0.2
SO_3	0.1	0.1	–	3.2	–
Na_2O	0.3	0.4	0.1	0.2	0.1
K_2O	0.1	0.2	2.3	1.2	–
TiO_2	1.5	2.4	0.8	0.3	–
P_2O_5	0.2	0.4	0.1	0.2	–
MnO	–	0.1	0.1	0.1	–
Others	0.1	0.2	0.2	0.3	–
LOI (at 950 °C)	1.5	1.7	0.5	0.8	42.6

For mass balance, similar to Durdzinski's recent study [3], the consumption of iron, magnesium, silicon, sulfur, aluminum and calcium oxides is used to form Fe-hydrogarnet, hydrotalcite, C-A-S-H, ettringite, monocarboaluminate and portlandite phases, respectively. The aluminium incorporation in C-A-S-H is taken into account by measuring the Al/Si atomic ratio by Scanning Electron Microscopy in Energy Dispersive X-Ray Spectroscopy mode (SEM-EDX).

For thermodynamic modelling, the Gibbs Free Energy Minimization Software (GEMS) is used, with the Cemdata 14 database [4, 5].

3 Results

With the incremental increase of the amount of reacted metakaolin, more aluminum and silicon are provided to the system. This leads to the increase of the consumption of portlandite, and the increase of the formation of C-A-S-H and monocarboaluminate, as shown as example in Fig. 1 for LC^3-50 (46.5%) at 28 days of hydration. The amount of ettringite, hydrotalcite and Fe-hydrogarnet is independent of the amount of reacted metakaolin since its formation is controlled by the sulfur, magnesium and iron availability, respectively. For this system, the amount of reacted metakaolin is determined by comparing the portlandite content obtained experimentally by XRD-Rietveld (dashed line in Fig. 1). The same procedure is used for thermodynamic modelling.

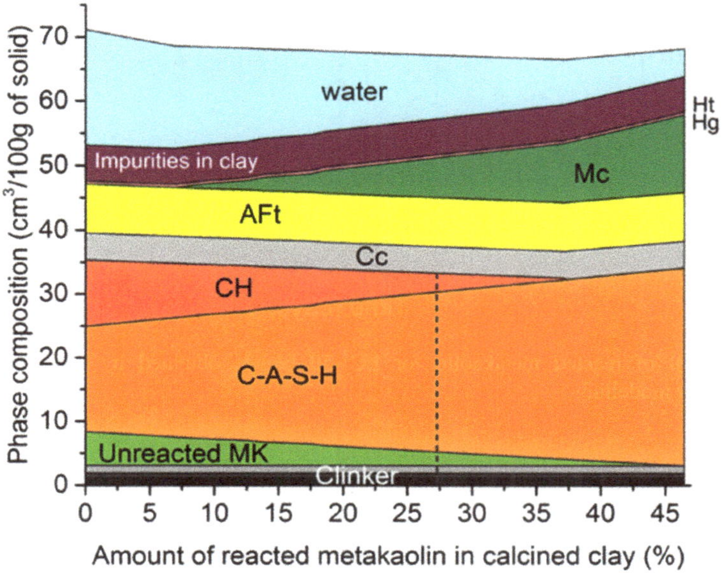

Fig. 1. Phase assemblage obtained by mass balance for LC3-50 (46.5%) at 28 days of hydration

This procedure was then applied to the different calcined clay blends and for different ages. Figure 2 shows the amount of reacted metakaolin in calcined clay for the three different LC^3-50 systems at 1, 3, 7, 28 and 90 days of hydration obtained by mass balance and thermodynamic modelling. The amount of reacted metakaolin increases with time but slows down at late ages.

Both mass balance and thermodynamic modelling results show excellent agreement up to 7 days. At later ages, the amount of reacted metakaolin is slightly lower for thermodynamic modelling. This is explained by the fixed Ca/Si ratio of 1.63 in the thermodynamic software. The values measured for the LC^3-50 blends by SEM-EDX indicate lower Ca/Si ratios, closer to 1.5. If more Ca is considered in the C-A-S-H, less is available to form portlandite. Thus, as observed in Fig. 1, if less portlandite forms, the amount of reacted metakaolin decreases.

Figure 3 shows the reaction degree of metakaolin for these three systems. For the system containing only 15.0% of metakaolin initially in the calcined clay, 100% of

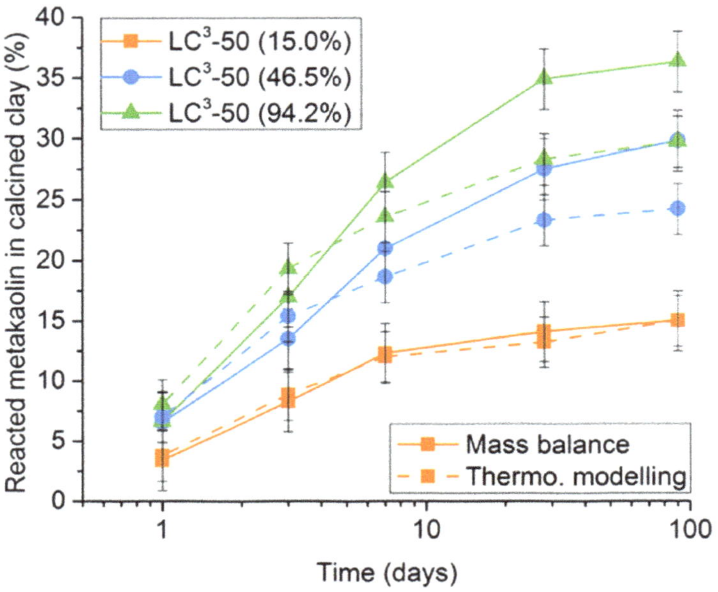

Fig. 2. Amount of reacted metakaolin for LC^3-50 blends obtained by mass balance and thermodynamic modelling

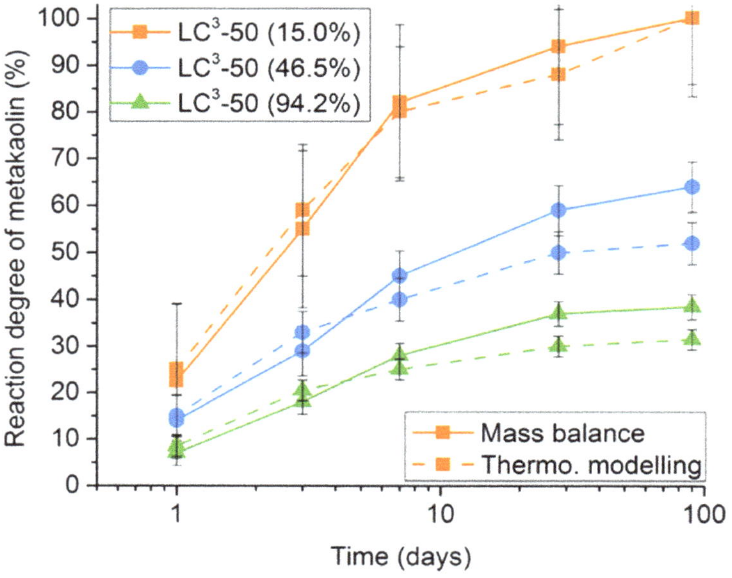

Fig. 3. Reaction degree of metakaolin for LC^3-50 blends obtained by mass balance and thermodynamic modelling

reaction is reached at 90 days of hydration. The reaction degree of metakaolin decreases with the metakaolin content of the calcined clay.

4 Conclusion

Mass balance provides a relatively easy way of determining the amount of reacted metakaolin in calcined clay. It is shown that the amount of reacted metakaolin increases with time and with the metakaolin content of the calcined clay. Thermodynamic modelling shows very similar results to mass balance, except at late ages where thermodynamic modelling slightly underestimates the amount of reacted metakaolin due to the fixed composition of the C-S-H in the model.

References

1. Antoni, M., Rossen, J., Martirena, F., Scrivener, K.: Cement substitution by a combination of metakaolin and limestone. Cem. Concr. Res. **42**, 1579–1589 (2012)
2. Avet, F., Snellings, R., Alujas, A.: Diaz, M. Ben Haha, K. Scrivener, Development of a new rapid, relevant and reliable (R^3) test method to evaluate the pozzolanic reactivity of calcined kaolinitic clays. Cem. Concr. Res. **85**, 1–11 (2016)
3. Durdziński, P.T., Ben Haha, M., Zajac, M., Scrivener, K.L.: Phase assemblage of composite cements. Cem. Concr. Res. **99**, 172–182 (2017)
4. (2014). http://gems.web.psi.ch/
5. Cemdata 14: EMPA (2014). http://www.empa.ch/cemdata

Durability of Steam Cured Pozzolanic Mortars at Atmospheric Pressure

Kübra Ekiz Barış(✉) and Leyla Tanaçan

Architecture Faculty, İstanbul Technical University, İstanbul, Turkey

Abstract. In the present study, the durability properties of Earth of Datça--hydrated lime mortar (ED)M under steam curing conditions at atmospheric pressure was inspected and compared with relatively well-known "horasan" mortars consisting of hydrated lime and fired clay as artificial pozzolan. Resistance of the lime-pozzolan mortars to wetting and drying, salt crystallization, freeze-thaw cycling and ageing by SO_2 action tests was assessed. Tests results showed that both mortars produced under steam curing at atmospheric pressure can be used as load-bearing building bricks. The better behaviour of the ED(M) under ageing tests compared to Fired Clay-hydrated lime mortar (FC)M is in conformity with its efficient physical and mechanical characteristics.

1 Introduction

Pozzolans are silicate or alumino-silicate compositions which by themselves possess no binding capacity, but when finely grinded react with calcium hydroxide and water at ordinary temperature to form calcium silicate hydrates or calcium aluminate hydrates [1]. Pozzolanic mortars can be used for the structures exposed to an excessive amount of moisture and liquid water such as public baths, aqueducts, and bridges due to their strength and durability [2]. Their durability can be affected by various deterioration factors such as wet and dry cycles, rain exposure and related leaching, freezing and thawing cycles, exposure to pollutants, and sun's rays. To what extent mortars are influenced by these deterioration factors depends on the type and properties of the binder and the aggregate, the binder/aggregate ratio, the amount of mixing water, the permeability of mortar, the capillarity coefficient, water intake, the presence of admixtures and the curing conditions [3]. The existence of water is necessary for several deterioration processes [4].

In the previous study, physical and mechanical characteristics of pozzolanic lime binder consisting of hydrated lime and (ED), under steam curing conditions at atmospheric pressure at 70 °C for 6 h was investigated [5]. Test results showed that the binder might be used for the production of materials similar to that of fired clay and calcium silicate bricks. In the present study, durability properties of (ED)M under the same curing conditions was investigated and compared with relatively well-known "horasan" mortars consisting of fired clay and hydrated lime (FC)M.

F. Martirena et al. (eds.), *Calcined Clays for Sustainable Concrete*, RILEM Bookseries 16,
https://doi.org/10.1007/978-94-024-1207-9_8

2 Materials and Methods

Quicklime supplied from its producer was slaked to obtain hydrated lime putty and stored in the moist state by covering it for minimum 15 days. ED obtained from the volcanic rocks of Datça Peninsula [6] was used as the natural pozzolan. FC supplied from a local pottery factory was used as artificial pozzolan. The maximum grain size of the pozzolans was determined as 90 μm. Semi quantitative element analysis (XRF) results of the hydrated lime and pozzolans are given in Table 1. Rılem Cembureau graded standard sand was used as aggregate.

Table 1. Chemical analysis of the Hydrated lime, Earth of Datça (ED), and Fired Clay (FC).

Component	SiO_2	Al_2O_3	Fe_2O_3	MgO	CaO	CO_2	Na_2O	K_2O	TiO_2	P_2O_5	MnO	Cl	Cr_2O_3	SO_3	Total
Lime (%)	1	-	-	1	85	5	-	-	-	-	-	-	-	0.6	92.6
ED (%)	72.07	13.06	1.39	0.44	2.34	-	3.27	4.60	0.20	0.05	0.06	0.10	-	-	99.73
FC (%)	45.21	14.95	14.56	4.52	9.32	-	2.52	2.47	2.12	0.27	0.21	0.05	0.37	0.48	99.68

The pozzolanic activity of (ED)M and (FC)M was examined according to [7]. The mixture proportion of lime:pozzolan:sand is 1:2.5:9, by weight. The mixtures were prepared at water-to-solid ratio of 0.16, by weight. The mixtures were cast in prismatic 40 × 40 × 160 mm moulds. The freshly moulded specimens, pre-cured at 20 ± 2 °C and 95 ± 5% RH for 36 h in the moulds by keeping them inside the polyethylene cover. Then they were removed from the moulds and PE and cured at 20 ± 2 °C and 65 ± 5% RH for 24 h. After pre-curing, the specimens were cured at 70°C under saturated steam in a closed container for 6 h [5]. The specimens were tested for specific density, unit weight [8], capillary water absorption [9], water absorption under

Table 2. Properties of (ED)M and (FC)M before and after durability tests

Properties	Before durability tests									
	Wetting-drying cycles		Salt crystallisation		Freeze-thaw cycles		Ageing by SO_2 action			
ED	FC	ED	FC	ED	FC	ED	FC	ED	FC	
Pozzolanic Act. test (MPa) Compressive str./ Flexural str.	6.12/1.43	4.10/1.03								
Weight variation (%)	-	-	+0.34	+0.54	+4.48	+5.90	−3.59	−9.04	+0.25	+0.46
Unit weight (g/cm^3)	1.81	1.78	1.86	1.84	-	-	-	-	-	-
Capillarity coefficient (cm^2/s)	0.09	0.11	-	-	-	-	-	-	-	-
Water absorption (%)	8.11	9.28	-	-	-	-	-	-	-	-
Apparent porosity (%)	14.71	17.15	13.68	15.43	-	-	-	-	-	-
Porosity (%)	22.98	32.36	-	-	-	-	-	-	-	-
Ultrasound velocity (km/s)	1.88	1.81	2.18	2.27	3.35	3.74	-	-	-	-
Compressive strength (MPa)	10.74	6.10	13.00	9.88	12.78	7.56	-	-	-	-
Flexural strength (MPa)	1.49	1.15	2.28	2.60	1.81	1.50	-	-	-	-
Modulus of elasticity (MPa)	6390	6020	8940	9480	-	-	-	-	-	-
Visual inspection scale	-	-	0	0	0	0	3	3	0	0

Specific density ED: 2.35 g/cm^3; FC: 2.63 g/cm^3; Hydrated lime: 2.15 g/cm^3

atmospheric pressure [10], ultrasound velocity to determine Young's dynamic modulus [11] and flexural and compressive strengths [12]. Mortars' resistance to wetting and drying [13], salt crystallization [14], freeze-thaw cycling [15], and ageing by SO_2 action [16] were assessed to determine the effect of accelerated aging. Following the durability tests, specimens were visually inspected, compared with the reference specimens and their appearance was scored according to the scale given in [15]. The test results are the average of six specimens and given in Table 2.

3 Results and Discussion

3.1 Physical and Mechanical Properties

Results obtained by XRF and pozzolanic activity tests, for both (ED)M and (FC)M, fit well with the requirements of TS 25 [7]. After they were cured at low temperature steam curing their compressive strengths increase 1.75 and 1.45 times and flexural strengths increase 1.04 and 1.12 times respectively for (ED)M and (FC)M than TS 25. This improvement observed also in their physical properties seems low temperature steam curing is the suitable curing condition for them to enhance the formation of CSH gel produced from the pozzolanic reactions [5]. The CSH gel leads to infilling pores, thus, increases pore refinement, reduces total porosity and the amount and connectivity of capillary pores within the mortar [17]. On the other hand (ED)M has more reactivity towards lime regarding the development of successfully hydration reactions; hence, measured compressive and flexural strengths of (ED)M are 1.76 and 1.29 times higher than (FC)M. Parallel to these results, other properties of (ED)M like porosity, apparent porosity, which indicates the existence of exposed pores, and water absorption were respectively found 0.71, 0.86 and 0.87 times lower than (FC)M as expected. Ultrasound velocity, unit weight and capillarity coefficient values of the both mortars are close to each other, which may indicate they have similar pore structure.

3.2 Influence of Wetting-Drying Cycles

After 20 cycles of wetting-drying, the increase of the mass variation in (ED)M and (FC)M was found very low, 0.34% and 0.54%, respectively, and can be considered negligible. All the physical and mechanical properties of both type of the mortars improved after the test. This increment is probably assigned to the fact that there is a pozzolanic reaction enhancement by the application of successive wetting and drying cycles. 70 °C drying at the wetting-drying test is equal to the curing conditions applied in this study and besides; it is well known that pozzolanic mortars prone to water. The ratio of this improvement was 1.34 times higher for (FC)M than (ED)M. The higher CaO content of FC pozzolan may lead to the higher ratio of $Ca(OH)_2$ formation in the solution which may continue its pozzolanic reaction to form CSH gels during the wetting-drying cycles. At the end of the test, there were no flaking, cracking, or any other alteration observed on the specimens which can be classified as '0' according to the numerical scale of visual inspection (Fig. 1).

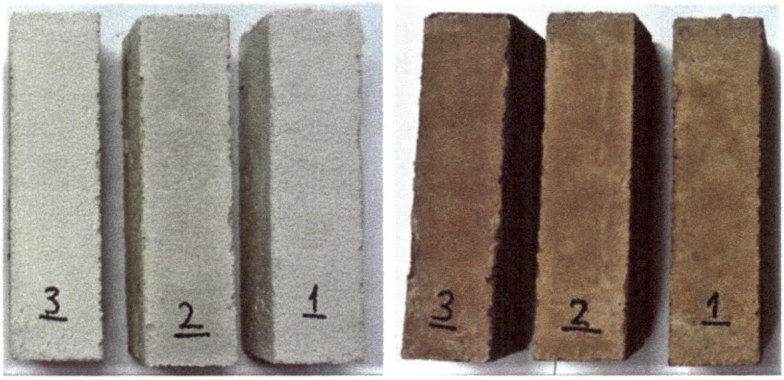

Fig. 1. The specimens after wetting-drying cycles: (ED)M on the left; (FC)M on the right.

3.3 Influence of Salt Crystallization

The salt decay susceptibility is described by the salt uptake and the mass variations for specimens [18]. Mass uptakes of (ED)M and (FC)M gradually increased towards 15th cycles due to the absorption of salt from solution inside the pores accessible to the mortar. Total connected porosity and the microporosity of pores smaller than 5 μm in radius optimally facilitate salt uptake [19]. The quantity of the salt uptake of the (FC)M was 1.32 times higher than that of (ED)M, because of its 1.40 times higher total porosity and 1.14 times higher water absorption ratio than the former one. Proximate improvements were recorded in mechanical properties of the both mortars. High mass uptake of the salts and their subsequent crystallization within the pores of the mortar during the drying stage may partially infill the pores of the material. This increase may also be due to the reaction of Na_2SO_4 with $Ca(OH)_2$ to form gypsum and ettringite which fill in the micro-pores leading to a denser structure [20]. Likewise, ultrasound velocities of the (ED)M and (FC)M increased 1.78 and 2 times respectively. On the other hand, according to the visual inspection of the mortars, neither disintegration into fragments, nor significant surface changes such as hairline cracking or chipping were noted that

Fig. 2. (a) The specimens after salt crystallization test; (b) The section of the mortars:

may occur due to crystal pressure over the pore surface and hence, classified as '0' in the scale (Fig. 2a). After salt crystallization test, relatively darker colour was observed at the exterior side of the sections of (FC)M broken in the flexural tests (Fig. 2b).

Based on this observation, after the test, the specimens were soaked in deionized water for 24 h and were left to dry at constant ambient temperature. While the surface of the (ED)M did not display any visual alteration, efflorescence was seen on surface of the (FC)M due to its higher salt content (Fig. 3).

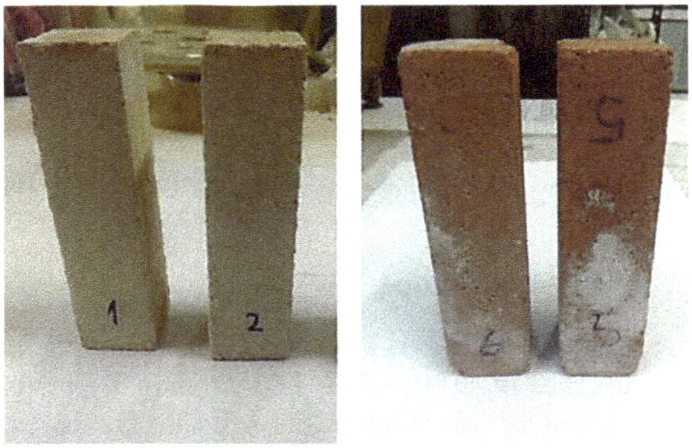

Fig. 3. The specimens after dried at constant ambient temperature: (ED)M on the left; (FC)M on the right.

3.4 Influence of Freeze-thaw

Although saturation coefficients (ST) of both mortars are lower than 80% both mortars loss appreciable amount of material after the freeze-thaw test. Freeze-thaw test of the (FC)M could not be continued further after the 9th cycle because of the formation of the large cracks and dissolution. On the other hand, any substantial differences were noted at the (ED)M after the 9th cycle. The total mass loss of (FC)M and (ED)M was 9.04%

Fig. 4. (ED)M after 20 freeze-thaw cycles (on the left); (FC)M after 9 freeze-thaw cycles (on the right).

(after 9th cycle) and 3.59% (after 20th cycle), respectively. (FC)M was more susceptible to freeze-thaw degradation due to its higher total porosity with small pore sizes, higher water absorption ratio and lower tensile strength. Excessive amount of $Ca(OH)_2$ content stem from the pozzolan type and probably incomplete pozzolanic reaction may accelerate the disintegration of the (FC)M. Visual analysis scale of the (ED)M was determined as '3' after 20 cycles because of the several cracks larger than 0.1 mm width and surface scaling (Fig. 4).

3.5 Influence of Ageing by SO_2 Action

SO_2, NO_x and CO_2 are the atmospherical pollutants and among these SO_2 is the most dangerous one [21]. SO_2 and NO_2 react with water to form HNO_3 and the H_2SO_4 acids. Acidic waters in contact with $CaCO_3$ of the mortars and concretes give place to the formation of calcium nitrates, sulphates, bicarbonates, etc. Following formula shows the conversion of $CaCO_3$ to gypsum with its reaction with sulphuric acid. Gypsum formation is responsible for the development of cracks and spalling because of its expansion during crystallization.

$$H_2O + H_2SO_4 + CaCO_3 \rightarrow CaSO_4, 2H_2O + CO_2 \uparrow \quad (1)$$

After 21 days of exposure to the 5% SO2 concentration, the appearance of the mortars was inspected and in no case were chipping, hairline cracking, scaling or cracks noted (Fig. 5). The increase in mass variation of (ED)M and (FC)M is very low and can be considered negligible. According to the visual analysis scale, '0' could be attained. This may be attributed to the steam curing conditions of the mortars in which pozzolanic reaction is dominant than carbonation reaction.

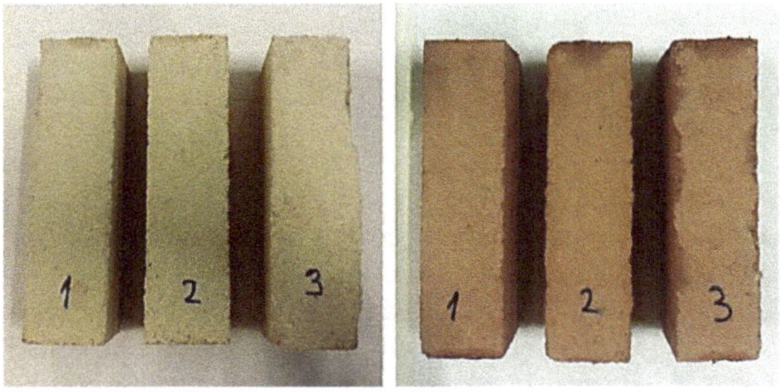

Fig. 5. The specimens after ageing by SO2 action test: (ED)M on the left; (FC)M on the right.

4 Conclusions

- According to the results of the pozzolanic activity, ED seemed to be the more reactive triggering a superior development of the hydraulic compounds.
- Steam curing at atmospheric pressure benefits hydration reactions between lime and both types of pozzolans. A special importance should be given to the curing conditions of the mortars.
- This study reveals the importance of pozzolanic activity and its influence on the mortars durability. The higher the pozzolanic activity the higher the amount of hydrated calcium silicates and aluminates; thus, more resistance to durability tests.
- In the wetting-drying test, wetting in the water and drying at 70 °C may provide additional curing effect on the mortars which result in any mass loss or visual degradation but strength gain.
- Either (ED)M or (FC)M are found resistant to the deterioration of salt crystallization due to their denser structure and strength gains after salt uptake. On the other hand higher porosity of FC mortar keeps higher salt intake which deposits efflorescence on the surface of the mortar carried with water.
- Ageing under freeze-thaw cycles is found as the most aggressive test. On the other hand, (FC)M has lower resistance to freeze-thaw effect due to its lower mechanical strength, higher total porosity with small pore size. Composition of the pozzolan is also an important parameter regarding frost resistance.
- In terms of exposure to SO_2 concentration, any surface level deterioration and changes in volume or expansion on the mortars was observed. This indicates that both mortars may keep their stability against atmospheric pollutants in time.
- Both mortars of (ED)M and (FC)M produced under steam curing at atmospheric pressure can be used as load-bearing building bricks. The better behaviour of (ED) M under ageing tests compared to (FC)M is in conformity with the efficient physical and mechanical characteristics of the (ED)M.

References

1. ASTM C618–15: Standard specification for coal fly ash and raw or calcined natural pozzolan for use in concrete (2015). http://www.astm.org/Standards/C618.htm
2. Navrátilová, E., Rovnaníková, P.: Pozzolanic properties of brick powders and their effect on the properties of modified lime mortars. Constr. Build. Mater. **120**, 530–539 (2016)
3. Izaguirre, A., Lanas, J., Álvarez, J.I.: Ageing of lime mortars with admixtures: durability and strength assessment. Cem. Concr. Res. **40**, 1081–1095 (2010)
4. Rilem, T.C.: 167-COM: Characterisation of old mortars with respect to their repair. Mater. Struct. **37**, 644–648 (2004)
5. Barıs, K.E., Tanaçan, L.: Earth of Datça: development of pozzolanic activity with steam curing. Constr. Build. Mater. **139**, 212–220 (2017)
6. Akgül, E., Tanaçan, L.: Evaluation of the pozzolanic activity of the earth of Datça as a building material. Int. J. Archit. Herit. **5**, 1–26 (2011)
7. TS 25:2008: Tras. Turkish Standard Institution. TSE, Ankara, Turkey

8. TS EN 1015-10:2001: Methods of test for mortar for masonry - Part 10: Determination of dry bulk density of hardened mortar. TSE, Ankara, Turkey
9. TS EN 1925:2000: Natural stone test methods - Determination of water absorption coefficient by capillarity. TSE, Ankara, Turkey
10. TS EN 13755:2009: Natural stone test methods - Determination of water absorption at atmospheric pressure. TSE, Ankara, Turkey
11. TS EN 14579:2006: Natural stone test methods - Determination of sound speed propagation. TSE, Ankara, Turkey
12. TS EN 196-1:2009: Methods of testing cement - Part 1: Determination of strength. TSE, Ankara, Turkey
13. TS EN 14066:2015: Natural stone test methods - Determination of resistance to ageing by thermal shock. TSE, Ankara, Turkey
14. TS EN 12370:2001: Natural stone test methods - Determination of resistance to salt crystallization. TSE, Ankara, Turkey
15. TS EN 12371:2010: Natural stone test methods - Determination of frost resistance. TSE, Ankara, Turkey
16. TS EN 13919:2004: Natural stone test methods - Determination of resistance to ageing by SO_2 action in the presence of humidity. TSE, Ankara, Turkey
17. O'Farrell, M., Wild, S., Sabir, B.B.: Pore size distribution and compressive strength of waste clay brick mortar. Cem. Concr. Compos. **23**, 81–91 (2001)
18. Gulotta, D., Goidanich, S., Tedeschi, C., Toniolo, L.: Commercial NHL-containing mortars for the preservation of historical architecture. Part 2: durability to salt decay. Constr. Build. Mater. **96**, 198–208 (2015)
19. Yu, S., Oguchi, C.T.: Role of pore size distribution in salt uptake, damage, and predicting salt susceptibility of eight types of Japanese building stones. Eng. Geol. **115**, 226–236 (2010)
20. Ghrici, M., Kenai, S., Meziane, E.: Mechanical and durability properties of cement mortar with Algerian natural pozzolana. J. Mater. Sci. **41**, 6965–6972 (2006)
21. Palomo, A., Blanco-Varela, M.T., Martinez-Ramirez, S., Puertas, F., Fortes, C.: Historic Mortars: Characterization and Durability. New Tendencies for Research. http://www.arcchip.cz/w09/w09_palomo.pdf. Accessed 7 Feb 2016

Use of Low-Carbon Cement in the Preparation of Masonry Mortars for Building Restoration

D. Betancourt Cura[1(✉)], Y. Lima Triana[2], and F. Martirena Hernandez[3]

[1] Department of Civil Engineering, Constructions College, Central University "Marta Abreu" of Las Villas (UCLV), Carretera a Camajuanì km 5 1/2, Santa Clara, Villa Clara, Cuba
daniab@uclv.edu.cu

[2] Department of Civil Engineering, Constructions College, Central University "Marta Abreu" of Las Villas (UCLV), Santa Clara, Cuba
yoandil@uclv.edu.cu

[3] Director of CIDEM Constructions College, Central University "Marta Abreu" of Las Villas (UCLV), Santa Clara, Cuba
f.martirena@enet.cu

Abstract. This research work deals with the use of low-carbon cement SIG B45 (SIG B45 is LC^3 cement produced at the Siguaney factory) in the manufacture of masonry mortars according to the dosage established by the NC175:2002 Cuban standard for both types II and III mortars. The experimental part focused on the use of SIG B45 cement, sand from "El Purio" and "Arimao" pit and PP-25 cement as a pattern sample. Tests carried out included flexural-compression strength, water absorption, water permeability, open porosity and bonding. Results attained from tests validate the use of LC^3 cement in masonry mortars.

1 Introduction

Worldwide cement production is responsible for 5% of CO2 emissions to the atmosphere (Martirena Hernandéz 2013; Castillo 2010; Alujas 2010), a fact that favors the current trend to the manufacture of cements by clinker replacement in order to reduce environmental impact, to increase the efficiency of industrial processes and to produce cements resistant to mortars terminations degradation in constructions. (Vizcaino et al. 2015).

Experts from the Materials and Structures Development and Research Center (CIDEM) together with Laussana´s Polytechnical Federal Institute (EPFL) in Switzerland, have carried out multiple research works about the new low-carbon cement type (LC3) since 2009, focused on the manufacture of a cement with up to 60% clinker replacement by calcining clays, producing a puzolanic type binder of quality and performance similar to the PP-25 Cuban cement.

The present work integrates study and assessment of the low-carbon cement (LC3) produced at industrial scale in masonry mortars linked to its comparison with Ordinary Portland cement (CPO) in order to comply with specifications established by the NC175 (2002) Cuban standard Masonry Mortars. Specifications.

F. Martirena et al. (eds.), *Calcined Clays for Sustainable Concrete*, RILEM Bookseries 16,
https://doi.org/10.1007/978-94-024-1207-9_9

2 Method

This research work deals with the use of low-carbon cement (SIG B45) in the manufacture of masonry mortars according to the dosage established by the NC175:2002 Cuban standard for both types II and III mortars.

3 Experimental Design

Properties assessment of masonry mortars with the use of SIG B45, low-carbon cement produced at the Siguaney Factory, in the province of Sancti Spíritus, was begun with a characterization of the raw materials used. Physical-mechanical tests were put into practice with the use of two types of aggregates (natural and artificial) as well as two types of cement (SIG B45 and PP-25), under dosages for types II and III mortars according to the NC175:2002 Cuban standard.

Tests carried out were:

- Compression strength at 7, 14 and 28 days (NC173 2002)
- Water absorption by capillarity at age 28 days (NC171 2002)
- Open porosity
- Bonding (NC172 2002)

4 Results

4.1 Analysis of Prismatic Test Cubes Flexural-Compression Strength

The flexural-compression strength test of masonry mortars was determined at ages 7, 14 and 28 days by using 6 (40 × 40 160) cm prismatic test cubes for each test age.

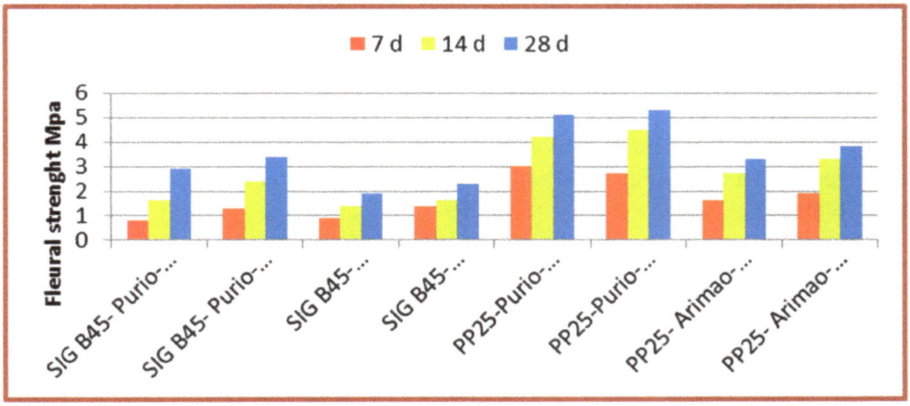

Fig. 1. Flexural strength

Figures 1 and 2 show the results of both strengths for each test as well as compression strength values for types II and III masonry mortars at age 28 days, as demanded by the NC175:2002 Cuban standard.

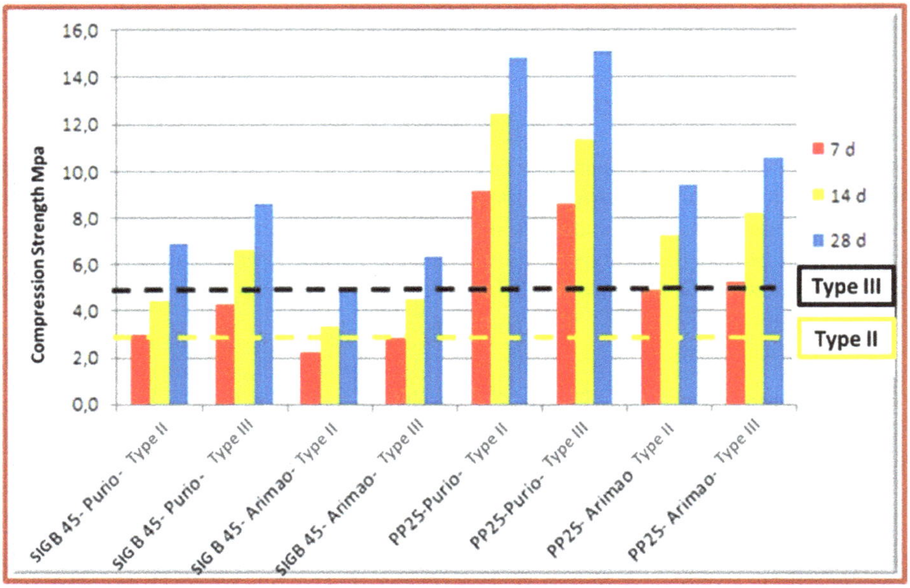

Fig. 2. Compression strength

As observed in Fig. 2, despite presenting lower strength with respect to PP-25, all samples of mortars manufactured with SIG B45 cement comply with specifications established by the Cuban standard for (Type II) mortars 3,5 MPa and for (Type III) mortars 5.2 MPa.

4.2 Analysis of Water Absorption by Capillarity

Figure 3 shows results attained through this test with evident little difference between both cements. SIG B45 cement presents absorption values slightly higher than those of PP-25 cement.

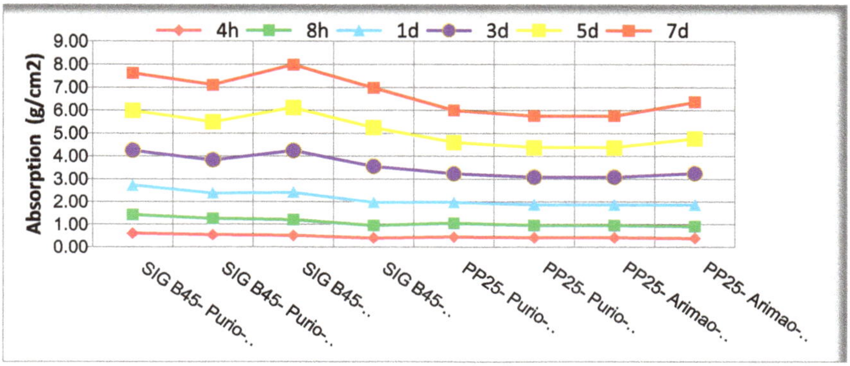

Fig. 3. Results of absorption by capillarity

4.3 Results of Water Absorption by Capillarity in Mortars

Results attained by means of this test show differences among the mortar batches tested when there is a variation of cement type in the manufacture of masonry mortars since absorption values are higher with SIG B45 than the ones with PP-25.

4.4 Analysis of Bond Tests

4.4.1 Bond Test by Direct Traction

Figure 4 shows that bonding for mortars with SIG B45) type II and both sands, on the one hand, and type II mortar with PP-25 and sand from "El Purio" pit, on the other, are the ones that comply with the specified parameters for this type of test.

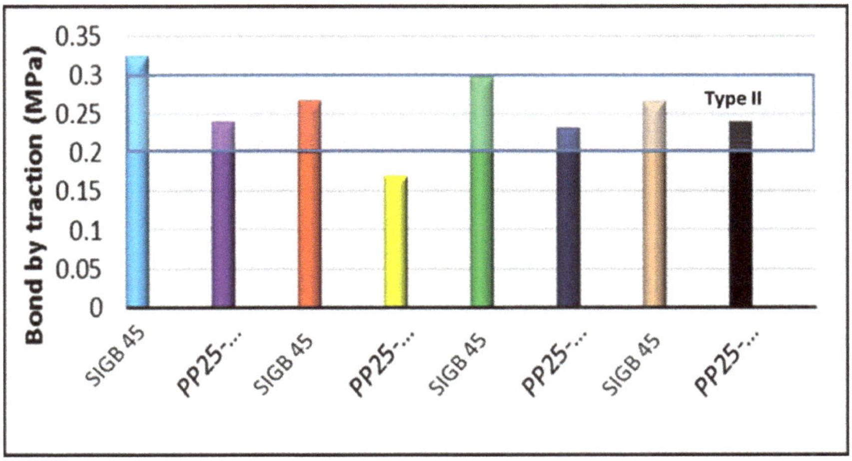

Fig. 4. Results of bond by traction for mortars

It is clear from Fig. 4 that low-carbon cement mortars under the 1:5:1 dosage and the two sands keep the appropriate range representing the NC172 (2002) standard values. Hardened mortar. Determination of bonding strength by traction; however, no mixture gets to comply with the specifications for type III mortars.

4.5 Analysis of the Open Porosity Test

After the cure period (28 days), 48 test cubes of the 8 designs were tested in order to determine the pore % present in each case. Percents of porosity for the samples analyzed were graphically represented in Fig. 5 from the average values obtained.

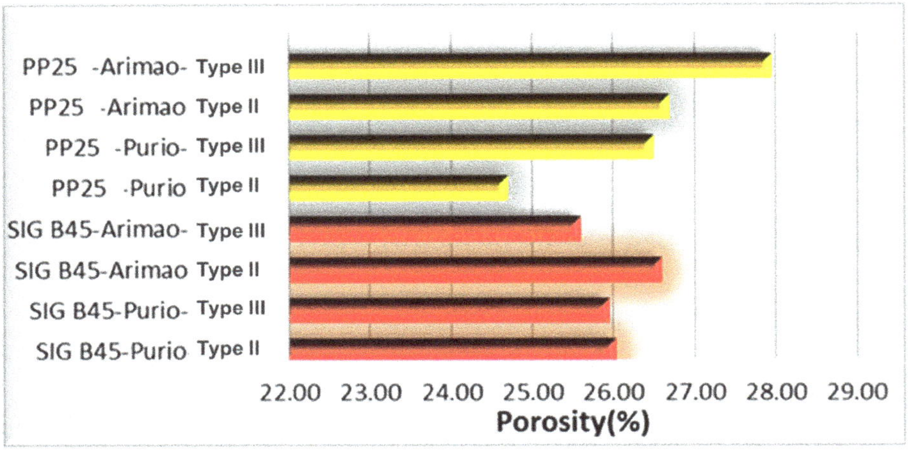

Fig. 5. Results of the porosity test

Porosity per cents reveal significant differences between cements with the two combinations of aggregates and mortars. Mortars manufactured with SIG B45 contain less % of pores compared to PP-25 for type III mortars using "Arimao" sand.

5 Conclusions

- Mechanical strength tests applied to mortars with the use of SIG B45 cement comply with quality specifications established by the NC175:2002 cuban standard, though its results do not reach the PP-25 benchmark.
- Mortars manufactured with SIG B45 cements show higher absorption values by capillarity compared with PP-25 however, the open porosity test made evident a pore per cent decrease with respect to mortars made with PP-25.
- According to the direct traction bond test, mortars made with SIG B45 (produced at Siguaney factory) for type II comply with the requirements established by the NC172 (2002) standard. Hardened mortar. Determination of bond strength by traction.

- Results of the research work implemented from mortars manufactured with SIG B45 show that the use of this type of binder, for the manufacture of types II and III mortars, according to NC175:2002, is feasible.

References

Martirena Hernandéz, J.F.: Project Documentation. Low Carbon Cement (LCC) (2013)
Castillo, R.: High reactivity puzolanas from the mechanical and thermal activation of a low purity kaolinite clay. Tesis en opción al título de Doctor en Ciencias Técnicas Universidad Central "Marta Abreu" de Las Villas (2010)
Alujas, A.: Production of a high reactivity puzolanic material from the activation of a multicomponent argillaceous fraction. Doctorate thesis, Universidad Central "Marta Abreu" de Las Villas (2010)
Vizcaíno-Andrés, L.M., Sánchez-Berriel, S., Damas-Carrera, S., Pérez-Hernández, A., Scrivener, K.L., Martirena-Hernández, J.F.: Industrial trial to produce a low clinker, low carbon cement. Mater. Construcción **65**(317), e045 (2015)
NC175: Masonry mortars. Specifications (2002)
NC173: Hardened mortars. Determination of flexural and compression strength (2002)
NC171: Hardened mortars. Determination of water absorption by capillarity (2002)
NC172: Hardened mortars. Determination of bonding strength by traction (2002)

Quantification of Pore Size Distribution Modification Due to Metakaolin Inclusion in Cement Based Systems

B. Bhattacharjee(✉)

Indian Institute of Technology, Delhi, India

1 Introduction

Hydraulic cement based composites in general and concrete in particular reveal different spatial heterogeneity at different scales. At macro engineering scale, e.g., meter or decimeter scale, the material is considered as isotropic and homogeneous, at a lower scale they are two phase material with aggregate particles dispersed in a matrix of the cement paste. One may also consider a third phase called interfacial transition zone (ITZ) at the interface of aggregate and paste. If one looks at cement paste itself at a finer level, it is heterogeneous at micro level, characterized by solid product of hydration reaction, unhydrated cement, gel and capillary pores. At still finer atomistic level one may encounter further heterogeneity as well, in structure of the material. The macro level engineering properties, such as uni-axial compressive strength of the hardened material, measured on cubic or cylindrical specimens and; often used in structural design is governed by elastic modulus and surface energy of the solid and; also on porosity and mean distribution radius of pores [1]. Permeation related properties are important in the context of long term durability performance of concrete in exposure during service condition. Degradation of concrete material occurs because of chemical action of aggressive agent penetrating in to concrete from external sources leading to deterioration of concrete elements. As all the deterioration mechanisms involve ingress of moisture or some other aggressive agent, permeation properties of concrete is important in the context of durability. Permeability and hydraulic diffusivity of these materials govern the water ingress and hence the durability. These macro level properties of cement based material are again related to porosity, pore inter-connectivity, tortuosity, pore sizes and shape [2, 3]. Thus, pore volume and pore characteristics play a dominant role in governing macro level properties of concrete and cement based materials. Porosity and pore size distribution are common influencing factors for strength and durability. Pore size distribution of these materials can be expressed through the Eq. (1) given below [4]. Equation (1) is reported to represent well the pore size distribution of cement based material [5].

$$V = \frac{p r_{0.5}^{d}}{r_{0.5}^{d} + r^{d}} \tag{1}$$

In Eq. (1) V is the cumulative volume porosity corresponding to equivalent pore radius r, $r_{0.5}$ is median pore radius, p is the total porosity and d is dispersion parameter. The median radius $r_{0.5}$ corresponds to mean distribution radius, which along with

F. Martirena et al. (eds.), *Calcined Clays for Sustainable Concrete*, RILEM Bookseries 16,
https://doi.org/10.1007/978-94-024-1207-9_10

porosity, is related to strength [1]. Thus p, $r_{0.5}$ and d, are the most important micro properties of cement based material controlling macro properties, such as strength and durability performance.

Pozzolanic inclusions in cementitious system, are known to refine the pore structure and same have been reported by many researchers. In past experimental works on effect of meta-kaolin inclusion on pore size distribution (PSD) of cement paste have been also reported [6–9]. Quantification of change in PSD through changes in $r_{0.5}$ and d for fly ash and silica fume inclusions are also reported [4, 10]. However, similar quantification of PSD modification through meta-kaolin inclusion is not reported so far in readily available literature.

In this paper, data available on PSD reported in literature [6–9] is considered and $r_{0.5}$ and d is estimated for various percentage inclusion of meta-kaolin addition in OPC paste, mortar and concrete. Thus, change of $r_{0.5}$ and d due to meta-kaolin inclusion is reported. Implication of these change on strength and durability performance is the discussed to demonstrate the role of micro scale property modification on modification of macro-level engineering property such as strength and durability. The latter is likely to provide better understanding on meta-kaolin inclusion on properties of concrete.

2 Estimation of PSD Parameters from Reported Data

Khatib and wild [7] reported some experimental results on pore size distribution of OPC paste with meta-kaolin inclusion, Sabir et al. [8] in their review reported PSD results on mortar, Poon et al. [9, 10] reported results on pore size distribution of paste and concrete containing meta-kaolin. Thus, pore size distribution data for meta-kaolin incorporated cement paste, mortar and concrete are available in literature. The generalized pore size distribution (PSD) model stated earlier in Eq. (1) is characterized by two parameters, median pore size parameter or mean distribution radius ($r_{0.5}$) and a dispersion coefficient (d). The available data are considered and these parameters of pore size distribution curve are estimated. Cumulative volume or porosity versus pore entry radius data reported in literature is considered. In Eq. (1), at V/P = 0.5, r = $r_{0.5}$, hence from a plot of V versus r, with known value of p obtained from MIP, $r_{0.5}$ can be estimated. For the purpose of estimating d, Eq. (1) is rewritten in the following form given in Eq. (2)

$$\frac{p}{V} - 1 = \left(\frac{r}{r_{0.5}}\right)^d \tag{2}$$

The curve of ln(P/V-1) versus ln($r/r_{0.5}$) is linear with its slope near origin (0,0) is d. The $r_{0.5}$ and d parameters thus estimated completely describe the cumulative pore size distribution curve [11].

Sabir et al. in their review reported results of PSD of mortar with OPC, i.e., mortar A with OPC only and mortar B with 80% OPC, and 20% meta-kaolin, Both with Water to Binder ratio (W/B) 0.4 and aggregate to binder ratio 1 at the age of 100 days. These results were presented in graphical form, from which median porosity can be easily extracted as 50 nm and 10 nm respectively for mortar A and B. The porosity p is 16% and

12% for mortar A and B, respectively. Considering the discrete values extracted from the graphs presented in the paper for V and r and calculating slope at origin for the plot of ln(P/V-1) versus ln(r/r0.5), the values of d are 1.44 and 2.40 respectively, for mortar A and B. Thus, porosity reduced with meta-kaolin inclusion, median pore size also reduced, consequently strength would be increased [1]. Strength is related to $(1 - p)/(r_{0.5})^{½}$ in addition to elastic modulus and surface energy, the relative increase in strength can be obtained from change in $(1 - p)/(r_{0.5})^{½}$ for pore free solid remaining same. The dispersion parameter d increased indicating less dispersion and more uniform pore sizes, with compact PSD, with largest pore size being smaller in case of mortar B, indicating lower permeation characteristics. Similar results are also reported by Poon et al., at W/B equal to 0.3 for paste and concrete containing 0, 5%, 10% and 20% meta-kaolin (MK) in the total binder of cement and meta-kaolin, with total binder content being fixed at 500 kg/m^3. Coarse aggregate and water content of the mix were maintained constant. MIP results at 3, 7, 28 and 90 days are reported. These results and available results on paste with 10% MK and control with no MK at 14 days [6] are analysed in the same manner.

3 Results and Relevant Correlations

The relationships among, age and meta-kaolin content with $r_{0.5}$ and d are presented for paste and concrete with W/B = 0.3. Hence role of meta-kaolin addition on pore size distribution is analysed and quantified. Further possibility of predicting the influence of meta-kaolin content on strength, permeability and hydraulic diffusivity from PSD information is also investigated as described below.

Strength of cement based material in general and concrete in particular is governed by porosity and pore size [1, 10], i.e., p and $r_{0.5}$. Thus, the validity of such relationship for meta-kaolin incorporated concrete is presented and discussed. The charge passed measured for chloride penetration is also correlated with $r_{0.5}$, hence implication of pore size improvement due to MK on permeation quality is also analysed. Since $r_{0.5}$ and d are known, the estimated values of permeability and hydraulic diffusivity of concrete with MK is also presented.

4 Conclusions

Pore size distribution of OPC paste and concrete with MK inclusion is quantified through mean distribution radius and dispersion parameter, which in turn are related to age and MK content.

Relationship between porosity, pore size and strength of paste and concrete with MK inclusion is presented which can help in future strength prediction. Relationship between permeation properties and pore size characteristics for above materials is also proposed.

Methodology presented for quantification of PSD can be used generally for any cement based material and relationship of strength with porosity and pore size distribution can be developed.

References

1. Kumar, R., Bhattacharjee, B.: Porosity Pore size distribution and in-situ strength of concrete. Cem. Concr. Res. **33**, 155–164 (2003)
2. Pradhan, B., Nagesh, M., Bhattacharjee, B.: Prediction of the hydraulic diffusivity from pore size distribution of concrete. Cem. Concr. Res. **35**, 1724–1733 (2005)
3. Kondraivendhan, B., Bhattacharjee, B.: Prediction of strength, permeability and hydraulic diffusivity of OPC paste. ACI Mater. J. **111**, 171–178 (2014)
4. Patil, S.G., Bhattacharjee, B.: Size and volume relationship of pore for construction materials. J. Mater. Civil Eng. **20**, 410–418 (2008). ASCE
5. Chen, Xudong, Shenxin, Wu: Influence of water-to-cement ratio and curing period on pore structure of cement mortar. Constr. Build. Mater. **38**, 804–812 (2013)
6. Khatib, J.M., Wild, S.: Pore size distribution of meta-kaolin paste. Cem. Concr. Res. **26**, 1545–1593 (1996)
7. Sabir, B.B., Wild, S., Bai, J.: Metakaolin and calcined clay as pozzolans for concrete: a review. Cement Concr. Compos. **23**, 441–454 (2001)
8. Poon, C.-S., Lam, L., Kou, S.C., Wong, Y.L., Wong, R.: Rate of pozzolanic reaction of metakaolin in high-performance cement pastes. Cem. Concr. Res. **31**, 1301–1306 (2001)
9. Poon, C.S., Kou, S.C., Lam, L.: Compressive strength, chloride diffusivity and pore structure of high performance meta-kaolin and silica fume concrete. Constr. Build. Mater. **20**, 858–865 (2006)
10. Kondraivendhan, B., Bhattacharjee, B.: Pore size distribution modification of OPC paste through inclusion of fly ash and sand. Mag. Concr. Res. **65**, 673–684 (2013)
11. Bhattacharjee, B.: Influence of pore size distribution on the properties of a stabilized soil cement system. In: Proceedings of Geo-China 2016, Fourth Geo-China International Conference, Shandong, China, ASCE Special Publication, 25–27 July 2016, pp. 53–60 (2016)

Limestone Calcined Clay Cement: The Experience in India This Far

Shashank Bishnoi[1(✉)] and Soumen Maity[2]

[1] Indian Institute of Technology Delhi, New Delhi, India
[2] Technology and Action for Rural Advancement, New Delhi, India

Abstract. Limestone Calcined Clay Cement (LC^3) has been developed on a fast-track for induction in the Indian cement industry for the last four years. This paper presents the key challenges and opportunities, as identified by the Indian team, towards the induction of LC^3 as a mainstream cement. As the development of the cement continues, the challenges towards its introduction continue to change and new opportunities that can be created by its introduction become apparent. While locating the correct raw materials and identifying the correct processing conditions were the challenges identified initially, as the technology become more clear, optimisation of cement characteristics and identification of performance parameters that would help in placing the product in the market remain important challenges for the industry. Development and acceptance of standards that would allow a suitable application of the technology will be important to ensure that the full potential of this cement is realised.

1 Introduction

Limestone Calcined Clay Cements (LC^3) are ternary cements that utilise the synergy of calcined clay and limestone in order to achieve strength development similar to Ordinary Portland Cement (OPC) even at clinker factors as low as 40% to 50% [1]. This cement is being studied and developed in an international collaboration comprised of members from Switzerland, Cuba and India. The main developments on this cement in India are discussed in this article.

2 Research Programme

The main objective of the research programme in India is to support an early uptake of the cement by convincing the industry of the advantages of producing LC^3 and by developing standards that would allow the commercial production of LC^3 in India. A schematic of the main programme in India is shown in Fig. 1.

F. Martirena et al. (eds.), *Calcined Clays for Sustainable Concrete*, RILEM Bookseries 16,
https://doi.org/10.1007/978-94-024-1207-9_11

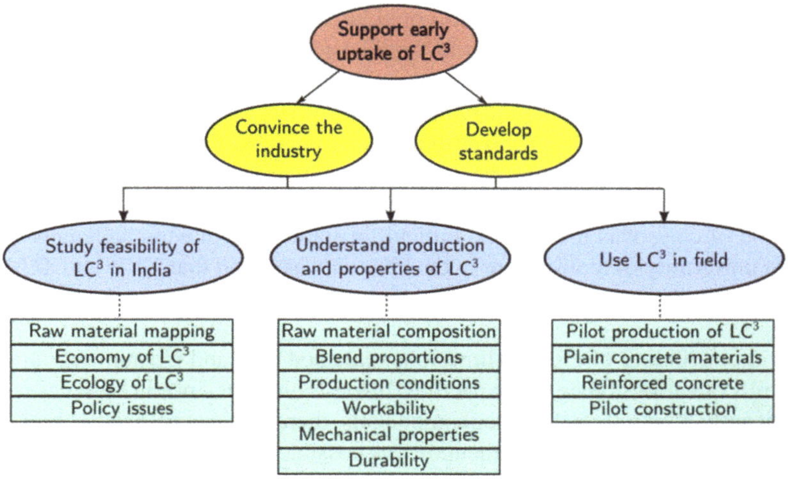

Fig. 1. Schematic of LC^3 research programme in India

At the outset of the programme, it was identified that the following studies would have to be carried out in order to achieve the objectives of the project:

- To study the availability of raw materials suitable for the production of LC^3 in India,
- To study the availability and usability of technology required for the production of LC^3 in India,
- To study the important characteristics of LC^3,
- To study the workability of LC^3,
- To study the hydration characteristics of LC^3,
- To study the mechanical performance of LC^3,
- To study the durability of LC^3,
- To study the economic aspects related to the production of LC^3,
- To study the environmental aspects related to the production of LC^3, including energy, CO_2 emissions and resources and
- To carry out field studies to investigate the usability of LC^3.

3 Characterisation of Raw Materials and Production of LC^3

During the studies, it was found that clays with kaolinite content of 50% to 60% gave satisfactory performance in LC^3. It was also found that limestone with carbonate content of 65% performed well in LC^3 [2, 3]. Large deposits of such limestones were found to be present and unusable for clinker production. The main impurities identified in the limestones were quartz and dolomite. Some clay minerals like kaolinite, muscovite, illite and montmorllionite were also found. Most such limestones were found to be suitable for use in LC^3. Deposits of wastes from the mining of dimensional wastes such as marble were also found to be suitable. It was possible to characterise both the limestone and the clays for suitability using X-Ray Diffraction (XRD) and Thermogravimetric Analysis

(TGA). While the TGA measurements could be used to measure the kaolinite content in the clay and the carbonate content in the limestone, XRD was useful to verify these results and to identify the impurities.

It was found that the calcination of clays could be carried out in existing rotary kilns. While the temperature of the clay was to be kept around 800 °C, the gas in the kilns was fired at 900 °C. It was found that it is important to maintain the temperature in the kiln below 1000 °C to prevent the formation of mullite, which is an unreactive phase. The calcination of clays was found to be difficult in static kilns due to challenges in maintaining a uniformity of temperature in the kilns. It was found that XRD and TGA could also be used for the characterisation of calcined clays and to check their degree of calcination.

The inter-grinding of clinker, limestone, calcined clay and gypsum was attempted in ball mills of various types and capacities. It was found that since in most cases clinker is significantly harder than the other components, there was a tendency to obtain coarse clinker particles in the finished cement. It was possible to achieve much better grinding of the cement in closed circuit ball mills where coarse particles from the mill are sent back using air classifiers. Still, it is felt that separate grinding and dry blending of various components of LC^3 can help in achieving a better overall particle size distribution of the cement. Improper grinding of the cement can lead to a significant reduction in its performance.

4 Properties of LC^3

In most of the productions of LC^3 in India, it was found to have a higher Blaine's fineness than typical OPCs in the market. The Blaine's fineness of well-ground LC^3 was found to be higher than 600 m^2/g, which is almost twice the typical fineness for OPCs. However, there was only a slight increase in the standard consistency, which was found to be of the range of 0.32 to 0.35, against the value of 0.28 to 0.32 for other cements. The setting of LC^3 was found to be slightly faster than for OPC, but it was found to be sensitive to usual admixtures used with cements. The soundness of LC^3 was found to be within acceptable limits despite the use of dolomitic limestones.

5 Design and Strength of LC^3

It was found that it is possible to design LC^3 blends with clinker contents as low as 50% with 28-day strengths that meet the requirements for pozzolanic cements and even 43 grade OPCs. While the chemical composition of clinker, calcined clay and limestone were most important for later strengths, the type and quantity of gypsum had a significant influence on the early strength. The design of LC^3 blends could be verified by using widely available characterisation techniques such as XRD, TGA and isothermal calorimetry. For example, the gypsum content in the cement can be optimised by adjusting the relative position of the peaks in isothermal calorimetry. It was found that gypsum content can have an influence on the early strength of the cement.

6 Durability of LC^3

The pore-structure of LC^3 has been found to be finer than OPC and PPC. For this reason, even when the total porosity in the cement may be slightly higher than the other cements, the permeability is lower. This is partly due to the wider particle size distribution of LC^3 and partly due to the pore refinement from the reactions of the SCMs. This leads to a significant improvement in the durability parameters of LC^3. It has been seen that the surface resistivity of concretes made using LC^3 is significantly higher even than PPC. As a result, the rate of corrosion of steel in LC^3 concretes is lower than in the other cements. Additionally, chloride ingress in the cement is lower due to the finer pores and chloride binding. The rate of carbonation in LC^3 concrete was found to be similar to that in other cements containing slag and fly ash, with similar clinker contents.

7 Economy of LC^3 Production

The economy of the production of LC^3 is a little complex to understand given that it depends on several parameters. While in many countries, the economy of LC^3 is obvious given that it is compared against OPC, in India, since PPC is the major cement and fly ash is widely available, this may not be so. This is because of the additional costs of the calcination of clays. It has been estimated that, if produced using the best available technologies, the cost of calcination of clays should be 30% to 40% lower than the cost of production of clinker [4]. It has been found that the cost of production of LC^3 can be lower than the production of PPC in many scenarios that exist in the country. The limited availability of good quality fly ash and the possibility to better control the properties of calcined clay improves the economy of LC^3 production.

8 Energy and Emissions

It has been found that if others best available technologies are used, CO2 emissions from the production of LC^3 can be 30% lower than from OPC production and as much as 11% lower than PPC production. The energy consumption in the production of LC^3 is also significantly lower than that required for OPC, although it is slightly higher than the energy consumed in the production of PPC.

9 Practical Applications

LC^3 has been used in several practical applications in India. Several hundreds of tonnes of the cement has been produced in three pilot scale productions, one of which has been carried out by a cement company. A two-storey house, including reinforced and plain concrete elements, has been built completely from the first batch of LC^3 produced in India. Autoclaved Aerated Concrete blocks were produced from the second lot of the cement produced and these blocks were used in the construction of a new structure in the Embassy of Switzerland campus in New Delhi. Pavements have been built using the LC^3 produced in the third batch. Plain concrete building materials, including paver blocks, concrete blocks, roofing tiles, etc. have been produced from all batches of the cement.

One of the most important observations from all the practical applications is that LC^3 can be used in a manner similar to the existing cements, requiring no retraining of the workforce. Although there is a slight increase in the water demand of concrete, proper mix designs and the use of admixtures can produce concrete mixes with good workability.

In order to understand the behaviour of LC^3 in various environmental conditions, reinforced concrete columns made using LC^3 have been placed in all major climatic zones in India including temperate, arid, cold and marine conditions. These columns are half buried under soil and half exposed to air and are being monitored on a regular basis.

10 Summary and Conclusions

A large research and development project aiming to support an early uptake of LC^3 is in progress in India. Aspects such as availability of materials, production, cement characteristics, hydration, strength, durability, economy and influence on environment are being studied. Through this study, a deep understanding of the cement is now available. Work on the practical aspects of introducing the cement in the industry are continuing and major implications of this research are expected soon.

References

1. Antoni, M., Rossen, J., Martirena, F., Scrivener, K.: Cement substitution by a combination of metakaolin and limestone. Cem. Concr. Res. **42**, 1579–1589 (2012)
2. Bishnoi, S., Maity, S., Mallik, A., Joseph, S., Krishnan, S.: Pilot scale manufacture of limestone calcined clay cement: the Indian experience. Indian Concr. J. **88**(6), 22–28 (2014)
3. Emmanuel, A., Halder, P., Maity, S., Bishnoi, S.: Second pilot production of limestone calcined clay cement in India: the experience. Indian Concr. J. **90**, 57–64 (2016)
4. Joseph, S., Bishnoi, S., Maity, S.: An economic analysis of the production of limestone calcined clay cement in India. Indian Concr. J. **90**(11), 22–27 (2016)

Pilot Scale Production of Limestone Calcined Clay Cement

Shashank Bishnoi[1(✉)], Soumen Maity[2], Mukesh Kumar[3], S.K. Saxena[3], and S.K. Wali[3]

[1] Indian Institute of Technology Delhi, New Delhi, India
[2] Technology and Action for Rural Advancement, New Delhi, India
[3] JK Lakshmi Cement Ltd., New Delhi, India

Abstract. Limestone Calcined Clay Cement (LC^3) was produced on a pilot scale in a cement grinding unit. Raw clay was obtained from mines in Gujarat. Raw material that was otherwise unusable in applications like paint and paper was found to be suitable for the production. While most of the calcination of clay is being carried out using furnace oil and bio-mass, even petcoke was found to be suitable for application to LC^2 to achieve cost reduction. Quality control was carried out using XRD, TGA and visual observation and the clay was carefully sorted into batches based on the test results. LC^3 and LC^2 were produced by inter-grinding the raw materials and a normal ball mill was found to be adequate for the production process. The test results showed that while Blaine's fineness test may be a good initial measure to control the grinding process, due to differential grindability, laser diffractometry can provide a better assessment of grinding of all components of the cement. It was found that both LC^3 and LC^2 can be produced using widely available technologies with very little modifications. Physical tests on LC^3 and LC^2 were carried out and both were found suitable for general applications and some special applications.

1 Introduction

LC^3 is a new family of ternary cements that is being developed in a large international collaboration [1]. The cement provides several benefits such as improved durability along with lower emissions and a better utilisation of resources. While pilot productions of LC^3 have been carried out earlier in Cuba and India [2, 3], the first worldwide production of LC^3 by a cement company was carried out by JK Lakshmi Cement Ltd. in India. It is expected that both LC^3 and LC^2 can be produced as commercially viable products [4]. A detailed testing and pre-production programme was implemented to ensure a good quality of production. The measures taken, challenges faced and the lessons learnt are discussed in this article.

F. Martirena et al. (eds.), *Calcined Clays for Sustainable Concrete*, RILEM Bookseries 16,
https://doi.org/10.1007/978-94-024-1207-9_12

2 Identification and Preparation of Raw Materials

It was decided to source the clinker, limestone and gypsum from the cement plant of JK Lakshmi at Sirohi in Rajasthan. Large quantities of unused limestone of grade lower than that required for clinker production are available at this plant. The limestone was characterised and it was found to contain approximately 60% of calcite content. The impurities were mainly composed of quartz, kaolinite, muscovite, etc. The clinker was found to contain approximately 50% of alite and XXX% of belite. Imported mineral gypsum was used.

Clays were sampled from several mines in Rajasthan and Gujarat. The approximate kaolinite content was estimated from the TGA measurements. A clay with approximately 60% kaolinite content was selected for the production. Calcination of the clay was carried out in a rotary kiln in Bhuj. The firing in the kiln was carried out using petcoke. Although the kiln was designed for higher temperatures, with some adjustment, it was possible to maintain its temperature at a level that was suitable for the calcination of clay. It was found that an air temperature of 900 °C to 950 °C was required to maintain a solid temperature of 800 °C to 850 °C, which is required for a proper calcination of the clay. Samples of calcined clay were drawn every 10 min initially and tested using XRD and TGA to check if they were properly calcined. Since the samples had to be sent outside the plant for XRD and TGA, loss on ignition measurements between 450 °C and 800 °C were used to approximately assess the calcination. Interestingly, it was found that the plant operators, who were experienced in calcination of clays, were able to assess the quality of calcination visually and that their assessment matched well with laboratory characterisation results.

A large variation in the calcination was observed initially with some samples found to be over calcined and some under calcined. However, after some trials, it was possible to control the temperature in the kiln by controlling the feeding rate of the fuel and the raw material and the rotational speed of the kiln. Around 200 tonnes of clay were calcined using this process.

3 Production of LC^3

All raw materials were packed and transported to the Jhajjar unit of JK Lakshmi Cement Ltd. for the production of LC^3 by inter-grinding in a closed circuit ball mill. The entire process of production was planned in advance and electrical and mechanical maintenance teams were kept on stand-by for adjustments that may be required. The quality control laboratory was set up to measure the retention on 90 μm and 45 μm sieves and to carry out the Blaine's fineness measurement of the cement.

The ball mills were first cleaned and purged with the raw materials to remove residual materials from earlier productions. Blends of LC^3 were prepared by intergrinding 50% clinker, 31% calcined clay, 15% limestone and 4% gypsum. Clinker and gypsum were fed first into the mill and the feeding of the other materials was started once the clinker had been ground to sufficient fineness. The feeding of limestone and calcined clay was then started. It was found that it took approximately 15 min for the product to be stabilised. The stability was assessed from the slightly red colour of the product.

Samples of LC^3 were drawn every 10 min and the fineness of the cement was measured using sieve analysis and Blaine's fineness. The fineness of the cement was consistently found to be over 600 m^2/g. Approximately 6% to 7% of the cement was found to be retained on the 90 μm sieve, while approximately 25% of the material was found to be retained on the 45 μm sieve. Around 200 tonnes of LC^3 and 140 tonnes of LC^2 were produced and packaged separately. Due to the high capacity of the mills, the entire process took less than 4 h. Only minor adjustment of feed and rotational speed of the mill were required during the production process and no major issues were faced. Through this production process it was found that it is possible to produce LC^3 without major changes to existing technologies.

4 Properties of LC^3 Produced

The physical properties of LC^3 were measured. Although the strength of the cement was found to meet the requirements of pozzolanic cements, its strength was lower than that typically observed for the cement. When the particle size distribution of the cement and the composition of various fractions of the cement was analysed, it was found that a significant fraction of the clinker remained in sufficiently ground. This is due to the relatively higher hardness of the clinker compared to the other components. Upon regrinding, acceptable strengths were obtained from the cement. This result shows the importance of controlling that not only the overall fineness, but also the particle size distribution of the cement. Figure 1 shows the influence of further grinding on the particle size distribution of the cement. Figure 2 shows the influence of grinding on the heat of hydration of the cement.

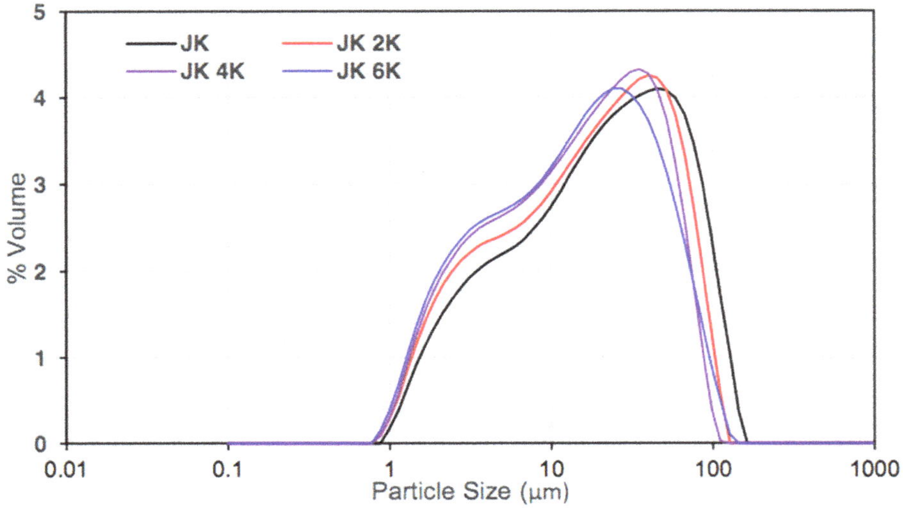

Fig. 1. Influence of additional 2000, 4000 and 6000 revolutions of grinding on particle size distribution of LC^3

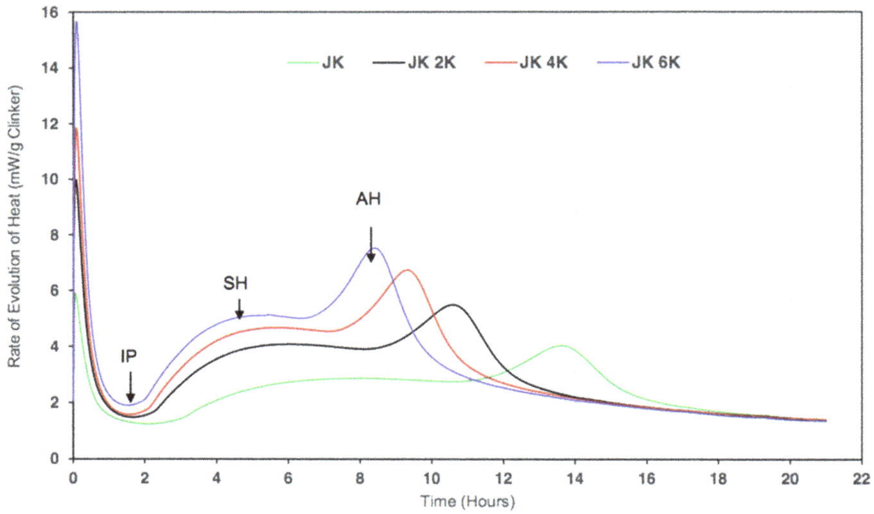

Fig. 2. Influence of additional 2000, 4000 and 6000 revolutions of grinding on isothermal heat of hydration of LC^3

5 Properties of LC^2 Produced

Limestone Calcined Clay Pozzolan, or LC^2, was produced by intergrinding 66% calcined clay and 34% limestone. After the grinding it was found that most of the powder passed through the 45 μm sieve. The Blaine's fineness of the LC^2 produced was more than 1000 m^2/g and was difficult to measure accurately. Due to the similar hardness of calcined clay and limestone, both were found to be properly ground in the LC^2.

6 Applications of LC^2 and LC^3

The LC^3 and LC^2 produced were used in various applications such as in ready mix concrete, paving of a concrete road. It was found that although the water demand was higher, good quality self-compacting concrete could be produced using the cement. In the RMC plants, LC^3 could be used directly in place of cement and LC^2 could be used as a pozzolan to reduce cement content in concrete. It was also found that there was a significant reduction in the sorptivity and RCPT values of concrete upon the replacement of OPC with LC^2 in concrete mixes. Figure 3 shows the influence of replacing up to 20% of OPC with LC^2 on the sorptivity of concrete. Figure 4 shows the influence of replacing up to 15% of OPC with LC^2 on the RCPT values of concrete.

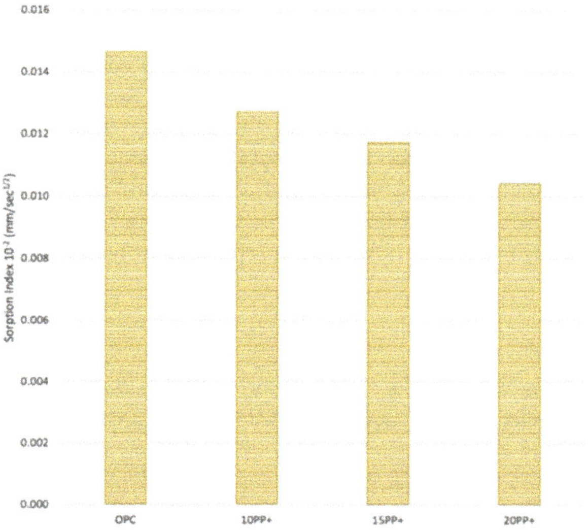

Fig. 3. Influence of replacement of 10%, 15% and 20% of OPC by LC^2 on sorptivity

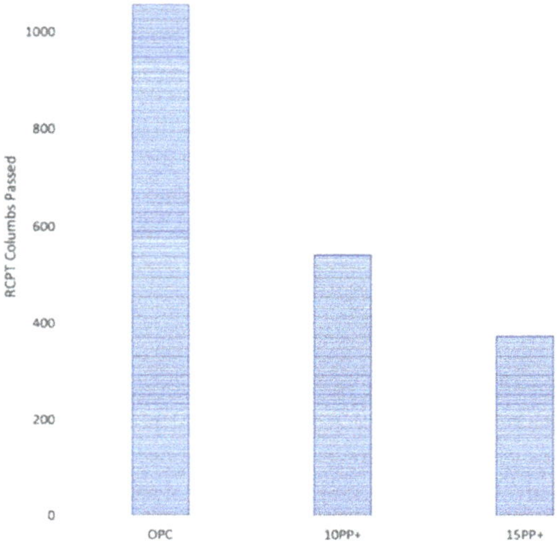

Fig. 4. Influence of replacement of 10% and 15% of OPC by LC^2 on RCPT

7 Conclusions

JK Lakshmi Cement Ltd. carried out the first production of LC^3 and LC^2 by a cement company. It was found that the cement could be easily produced using existing technologies although basic adjustments were required. Petcoke could be used as fuel to control the cost of production and the cement could be inter-ground in a commercially used closed-circuit ball mill. It was found that it is important to control the particle size distribution of the cement. Concretes with high strength and flow were produced using both LC^3 and LC^2 and it was found that there was a significant improvement in the durability indices of concrete upon the partial replacement of OPC with LC^2.

References

1. Antoni, M., Rossen, J., Martirena, F., Scrivener, K.: Cement substitution by a combination of metakaolin and limestone. Cem. Concr. Res. **42**, 1579–1589 (2012)
2. Bishnoi, S., Maity, S., Mallik, A., Joseph, S., Krishnan, S.: Pilot Scale Manufacture of limestone calcined clay cement: the Indian experience. Indian Concr. J. **88**(6), 22–28 (2014)
3. Emmanuel, A., Halder, P., Maity, S., Bishnoi, S.: Second pilot production of limestone calcined clay cement in India: the experience. Indian Concr. J. **90**, 57–64 (2016)
4. Joseph, S., Bishnoi, S., Maity, S.: An economic analysis of the production of limestone calcined clay cement in India. Indian Concr. J. **90**(11), 22–27 (2016)

The Special Case of North-Eastern India for the Production of LC^3

Shashank Bishnoi[1(✉)], Soumen Maity[2], S.P. Pandey[3], and P.K. Tripathy[3]

[1] Indian Institute of Technology Delhi, New Delhi, India
[2] Technology and Action for Rural Advancement, New Delhi, India
[3] Dalmia Cement Bharat Limited, Chennai, India

Abstract. The feasibility of the production of LC^3 and LC^2 in the North-Eastern states of Assam and Meghalaya were studied. Although there are many cement plants in the region, fly ash is not available. The easy availability of kaolinitic clays in the vicinity of the cement plants makes a strong case for the production of LC^3 and LC^2 in the region. Several clays in the area were assessed and suitable clays with more than 50% kaolin and up to 80% kaolin were found to be abundantly available. Trial production using locally available materials was carried out and the physical and chemical characterisation of LC^2 and LC^3 were carried out. The results from these studies are presented in this paper.

1 Introduction

The suitability of limestone calcined clay cement (LC^3) [1] is being studied in a large international collaboration that is investigating various aspects of the cement including its performance [2, 3] and its environmental and economic feasibility [4]. It has been found that LC^3 is most suitable in areas where good quality fly ash is not easily available, but suitable clays are available. This cement provides an interesting opportunity in the North-Eastern region of India where several cement plants are located, but the availability of hydro-electricity prevents the easy availability of fly ash. The transportation of fly ash from eastern India leads to the cost of good fly ash to be at times higher than the cost of production of clinker. This paper presents a study of the clays available in the region and initial results from production of LC^3 using these clays.

2 Availability of Clays

The states of Assam and Meghalaya were focussed upon since most cement plants are located in these regions. Figure 1 shows a map of this region with the locations of clays and cements plants marked on it. The only thermal power plant in the region, with a limited production of fly ash, is shown in the map as well. As can be seen, there is a wide availability of kaolinitic clays in the region.

F. Martirena et al. (eds.), *Calcined Clays for Sustainable Concrete*, RILEM Bookseries 16,
https://doi.org/10.1007/978-94-024-1207-9_13

Fig. 1. Map of the North-Eastern Indian states with the locations of clay marked in yellow and the cement plants marked in green. The only thermal power plant in the region is shown in red.

Samples of clays were collected from two locations in the region. The TGA of two of these clays is shown in Fig. 2. The kaolinite content in the clays was determined to be between 50% and 80% based on the TGA results. The clays were thus found to be suitable for the production of LC^3.

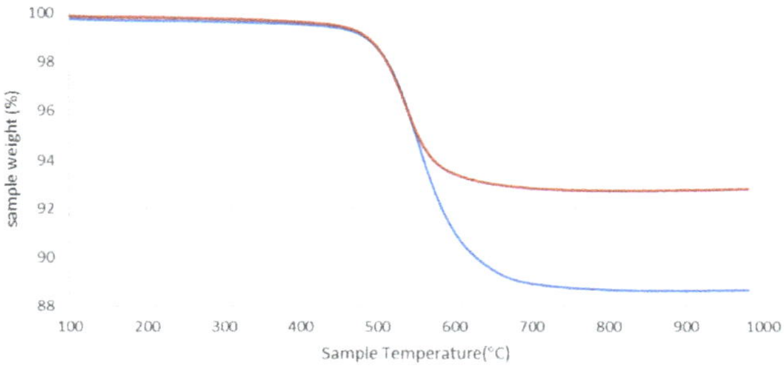

Fig. 2. TGA analysis of clays from North-Eastern India.

The XRD analysis showed that other than kaolinite, quartz was the main component in these clays (Fig. 3). Upon calcination at various temperatures, it was seen that a complete calcination of the clay occurred at 700 °C (Fig. 4).

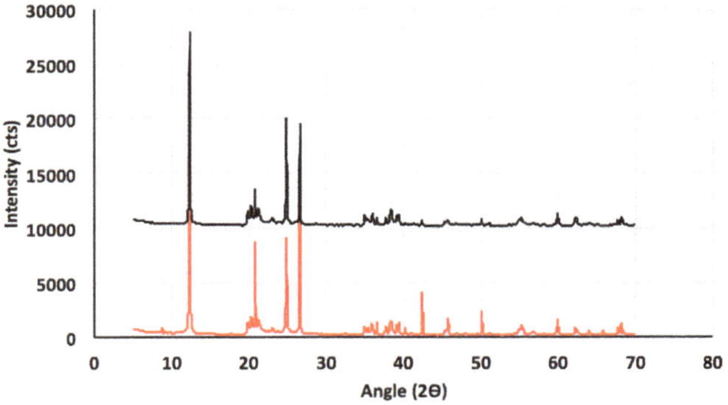

Fig. 3. XRD of raw clays collected from North-Eastern India

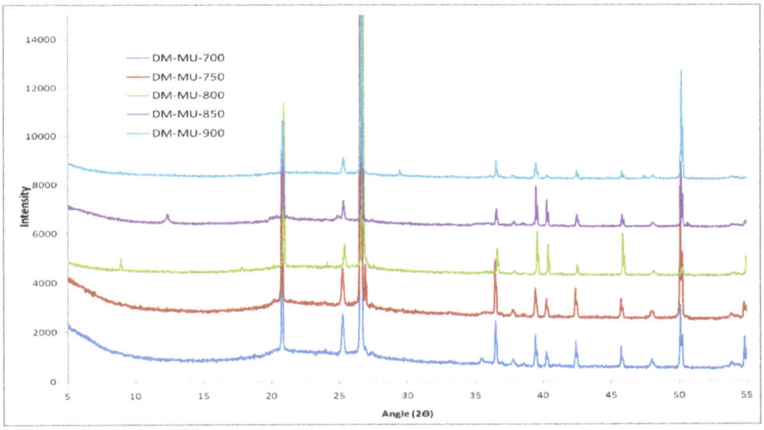

Fig. 4. XRD of clays calcined at various temperatures between 700 °C and 900 °C

The clays were thus found to be suitable for production of LC^3 and LC^2. The lime reactivity of the clays was measured in the lab and was found to be more than 10 MPa for the clay with 50% kaolinite content, again showing the suitability of the clay.

3 Laboratory Trials of LC^3

LC^3 was produced in the laboratory by using clinker, limestone and calcined clays from North-Eastern India. The hydration study of the cements showed that the carbonaluminate and AFt phases were forming as expected. The heat of hydration curves are shown in Fig. 5.

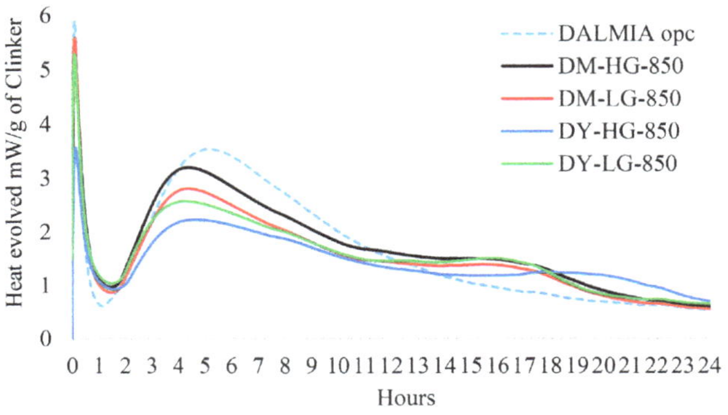

Fig. 5. heat of hydration of various blends of LC3 compared with an OPC

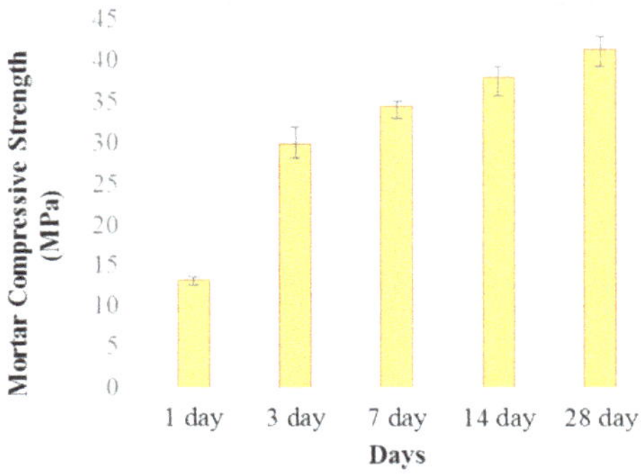

Fig. 6. Strength development of mortars at 0.45 w/c made using LC^3 with materials from North-Eastern India

The compressive strength of the LC^3 produced was measured in the laboratory. The strength at 28 days was found to be of the order of 42 MPa (Fig. 6). Since the production was carried out in the lab and the particle size distribution of the cement wasn't optimised, the strengths are expected to be higher when the cement is produced in a plant. This strength is significantly higher than the requirement of 33 MPa in blended cements.

4 Conclusions

A study of the clays available in North-Eastern India was carried out. The clays were characterised using TGA and XRD and several suitable clays were found in the region. The clays were calcined in a static furnace and 700 °C was found to be sufficient for their calcination. LC^3 blends were prepared using clinker, limestone and clays obtained from North-Eastern India. The strength of the cement produced in the laboratory was found to be suitable and further studies on the cement are planned. LC^3 and LC^2 are expected to bringing significant environmental and economic benefits to cement production due to the limited availability of fly ash in North-Eastern India.

References

1. Antoni, M., Rossen, J., Martirena, F., Scrivener, K.: Cement substitution by a combination of metakaolin and limestone. Cem. Concr. Res. **42**, 1579–1589 (2012)
2. Bishnoi, S., Maity, S., Mallik, A., Joseph, S., Krishnan, S.: Pilot scale manufacture of limestone calcined clay cement: the Indian experience. Indian Concr. J. **88**(6), 22–28 (2014)
3. Emmanuel, A., Halder, P., Maity, S., Bishnoi, S.: Second pilot production of limestone calcined clay cement in India: the experience. Indian Concr. J. **90**, 57–64 (2016)
4. Joseph, S., Bishnoi, S., Maity, S.: An economic analysis of the production of limestone calcined clay cement in India. Indian Concr. J. **90**(11), 22–27 (2016)

Comparative Study of Compressive Creep Behavior of Concrete with Metakaolin or Silica Fume

R. Bucher[1(✉)], H. Cagnon[2], T. Vidal[2], and M. Cyr[2]

[1] Argeco développement, 47500 Fumel, France
[2] Laboratoire Matériaux et Durabilité des Constructions (LMDC), Université de Toulouse, INSA/UPS Génie Civil, 135 Avenue de Rangueil, 31077 Toulouse Cedex 04, France

Abstract. The use of High Performance Concrete (HPC) in specific applications, such as tall buildings or prestressed concrete, causes a creep of the concrete. The durability of the building could be reduced because of a too important deformation of the concrete.

Most HPC are cast with silica fume, therefore data on creep properties exist. According to literature, the use of silica fume allows a decrease in the creep amplitude.

Metakaolin is another cementitious addition which can be used in the case of HPC. However, to the authors' knowledge the work on creep of concrete based on metakaolin is nearly nonexistent.

The aim of this work was to improve the knowledge on creep of concrete based on metakaolin, especially when compared with concrete based on silica fume.

This work allowed measuring the delayed deformation of concrete containing and comparing it with control concrete based on silica fume properties. In order to study the evolution of the materials induced by the creep, compressive strength and Young modulus tests were conducted. Finally, in order to improve the understanding of the creep deformation phenomena, a comparative study of drying kinetic.

1 Introduction

Creep of concrete is an important mechanical characteristic that is always taken into account for tall building or prestressed concrete. Many hypotheses were proposed to explain the deformation under loading, however none actually makes unanimity. Water present in the porous network plays a major role. When the concrete is loaded, water diffuses into the cementitious matrix and changes the distribution of stresses. This phenomenon causes a first phase of deformation [1]. At longer term, the sliding of the C-S-H layers leads to a dimensional variation [2, 3].

The use of supplementary cementitious materials (SCM) in a concrete formulation generally implies a modification of the porous network [4, 5] and a change of the hydrate composition, especially for C-S-H [6]. This microstructural modification could change the macroscopic long term behavior of the concrete. In order to reduce the creep deformation, the most used SCM is silica fume (SF). The significant reduction of the

F. Martirena et al. (eds.), *Calcined Clays for Sustainable Concrete*, RILEM Bookseries 16,
https://doi.org/10.1007/978-94-024-1207-9_14

deformation under loading is explained in several publications dedicated to the creep of concrete with silica fume [7, 8]. According to Jianyong and Yan, the reduction of the pore size could explain lower rate of deformation for concrete with silica fume compared with other ones [9].

Metakaolin (MK) is known as a supplementary cementitious material which allows to increase the concrete durability [10]. However, according to the author's knowledge, only a few articles on the deformation of concrete with metakaolin under loading can be found [11, 12]. For Brooks and Megat Johari [11], the use of metakaolin causes a creep decrease of 50% if the concrete is compared to a reference with cement only. In the case of Ultra High Performance Fiber-Reinforced Concretes (UHPFRC), Nguyen Amanjean [12] showed that a concrete with metakaolin and loaded at 28 days gave lower deformation than a concrete with silica fume.

The goals of this publication were to:

- complete the literature data on the creep of concrete containing supplementary cementitious materials, and particularly metakaolin.
- compare creep deformation of HPC with silica fume to HPC containing flash metakaolin
- compare experimental creep performance of concrete containing metakaolin with creep calculated with the Eurocode 2 [13, 14]

2 Materials and Methods

2.1 Materials and Mixtures

The cement used in this study was a CEM I 52.5 N (clinker > 95%) in accordance with the standard EN 197-1. The characteristic strength, measured on mortar at 28 days, was 62.7 MPa. The silica fume was in accordance with the standard NF EN 13263-1, 2 and the metakaolin to the standard P 18-513. The metakaolin used in this study came from a flash-calcination process [15], i.e. the dehydroxylation (at around 700 °C) of powdered kaolinite clay was made within a few tenths of a second. The purity of the MK was around 50%, the main impurity being quartz. The aggregates were crushed limestone aggregates with two sizes: 0–4 mm and 4–12 mm. The physical and chemical characteristics of cement, silica fume and metakaolin are summarized in Table 1 and the used mixtures in Table 2.

Table 1. Physical and chemical characteristics of cement, silica fume and metakaolin

	CEM I 52.5 N	Silica fume	Metakaolin
SiO_2 (%)	–	96.0	67.10
Al_2O_3 (%)	–	–	26.80
Characteristic strength 28 d (MPa)	62.7		
Specific area (cm^2/g)	3990 (Blaine)	225000 (BET)	165000 (BET)
Specific density (kg/m^3)	3140	2240	2551

Table 2. Mixtures, fresh and hard characteristics of concretes used in this study

	CSF	CMK
Cement (kg/m^3)	450	450
Metakaolin (kg/m^3)	0	50
Silica fume (kg/m^3)	40	0
Efficient water (kg/m^3)	147	143
Sand (kg/m^3)	730	700
Aggregates (kg/m^3)	1060	1070
HRWRA 1 (%)	1.51	1.70
HRWRA 2 (%)	0.00	0.40
W/B	0.300	0.285
Slump t0 (cm)	22.0	24.5
Slump t60 (cm)	18.0	24.0
Compressive strength 90 days (MPa)	115.9 ± 6.4	117.7 ± 6.7
Young Modulus 90 days (GPa)	49.9 ± 0.3	51.4 ± 0.8
Pores accessible to water (%)	9.7 ± 0.3	10.8 ± 0.2
Average pore diameter (nm) (MIP)	26.1	35.6

2.2 Methods

The compressive strength was performed on 4 cylindrical samples of h = 22 cm and diameter = 11 cm, after a 90-day endogenous cure with a loading speed of 0.5 MPa/s. The Young modulus test was performed according to the RILEM recommendation [16] on 3 cylindrical samples of h = 22 cm and diameter = 11 cm after a 90-day endogenous cure. The porosity test was performed [NF P18-459, 2010] on 3 cylinders (h = 5 cm and diameter = 11 cm) taken in one sample (initially h = 22 cm) after a 90-day endogenous cure. The tests of shrinkage were performed on cylindrical samples (h = 22 cm and diameter = 11 cm) in a humidity and temperature controlled room (20 ± 1 °C, 50 ± 5% RH). The measured shrinkage [NF P15-433, 1994] was the total shrinkage (total shrinkage = desiccation shrinkage + autogenous shrinkage). For each mixture, the total creep was measured on 3 cylindrical samples (h = 22 cm and diameter = 11 cm) in a controlled room (20 ± 1 °C, 50 ± 5% RH). The loading was made after a 90-day endogenous cure and fixed to 30% of the 90 days compressive strength. Before creep, samples were stored in autogenous condition at 20 °C. The deformation was measured with 3 gauges per sample, fixed directly on the surface of the cylinder at 120°.

3 Results and Discussion

Table 2 summarizes the mechanical properties of the two studied concretes. At 90 days, it was possible to obtain the same performance whatever the SCM used, the compressive strength being 115.9 ± MPa for concrete with silica fume and 117.7 ± 6.7 MPa for concrete with metakaolin. In order to reach the same strength, a reduction of the W/B ratio was applied to the MK-concrete, but with a small increase in the HRWRA content. The Young Modulus were also similar for the two compositions, even if the MK-concrete was slightly more rigid than the silica fume mixture. The pores accessible to water was 1% higher in the case of the formulation with MK when compared to the mixture with silica fume. This increase of the open porous volume was already observed in a study comparing a reference concrete (only cement with clinker > 95%) with a concrete with MK [17]. Figure 1 shows the drying kinetics of concrete samples dried in a controlled room at 20 °C and 50% RH. For these two formulations, the kinetics was very low compared to ordinary concretes [18], which could be explained by two factors: 1 - the evaporation of the water was slow because of a very dense microstructure; 2 - a significant fraction of the water was consumed by the cement hydration, due to the long autogenous conservation of storage before beginning of desiccation condition and the low W/B used. Nevertheless, the mixture with metakaolin dried faster than the silica fume one. This could mean that the concrete with SF had a finer porous network and/or presented a more important consumption of the water during the hydration.

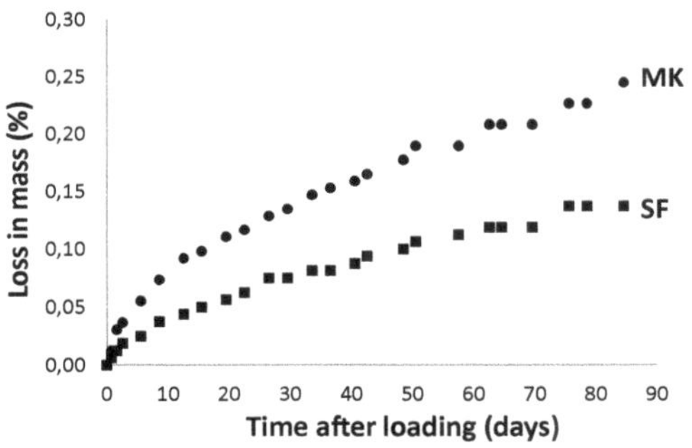

Fig. 1. Loss of mass kinetics for concrete with metakaolin (MK) et silica fume (SF)

Figure 2 shows the delayed strain under loading and the shrinkage strain for the two concretes with t_0 corresponding to the loading time, equal to 90 days after the casting. The creep was calculated by subtracting the shrinkage strain from the total delayed strain under loading.

It can be seen from this figure that the total deformation was higher in the case of the concrete containing metakaolin: after 84 days of loading, deformations reached 485 μm/m for CMK and 282 μm/m for CFS.

This difference of creep could be explained by two hypotheses. On the one hand, the diffusion of the water into the MK-concrete could be facilitated by the higher volume of pores and the greater average pore diameter (Table 2) than the silica fume one. On the other hand, the quantity of HRWRA used in the concrete with MK was higher and Neville showed that a higher HRWRA content could cause an increase of the creep [19].

The design of a concrete structure usually implies the calculation of the creep, based in Europe on "Eurocodes 2 - Bridges" [13] and "Eurocodes 2 - Buildings" [14]. Figure 3 shows the comparison of creep strains determined by experimentation and those calculated with Eurocode 2 (EC2). The EC2 Buildings (EC2 Bu) showed similar results for both MK and SF concretes. In the case of the EC2 Bridges (EC2 Br), for silica fume concrete, the effects of this addition on creep are taken into account, reducing creep strains Nevertheless, the strain of concrete with MK is significantly overestimated by the EC2 Bridges and the EC2 Buildings predictions, meaning that the creep behavior could be improved when MK is used.

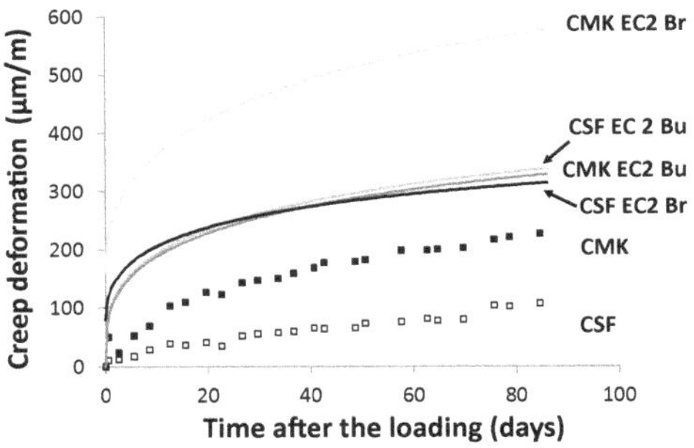

Fig. 3. Comparison of the experimental total creep strains with the values calculated from the Eurocodes 2 Buildings and the Eurocodes 2 Bridges.

4 Conclusions

This study allowed to compare the delayed behavior, shrinkage and creep, of high performance concretes containing two supplementary cementitious materials: silica fume and metakaolin. For concretes of similar compressive strength (116–118 MPa) loaded at 90 days at 30% of their compressive strength, the use of metakaolin led to higher creep deformation than silica fume, probably due to its higher water loss over time induced bay a higher porosity and a less dense and fine porous network. However, the use of metakaolin enabled to obtain concrete with creep strain lower than Eurocode 2 predictions.

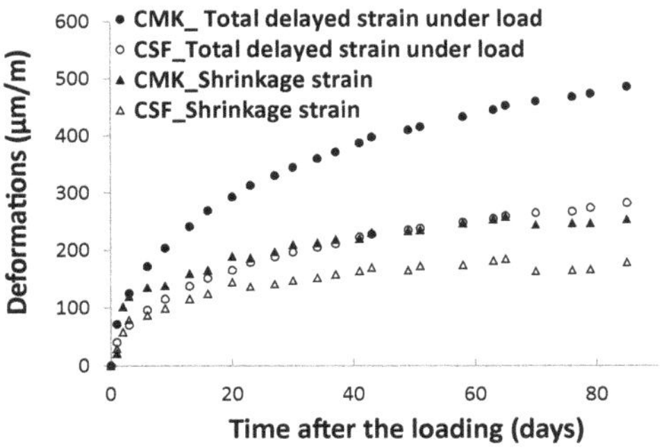

Creep = Total delayed strain under load-Shrinkage strain

Fig. 2. Total delayed strain under load and shrinkage strain of the CMK and CFS mixtures

References

1. Ghosh, R.S.: A hypothesis on mechanism of maturing creep of concrete. Mater. Struct. **6**(2), 23–27 (1973)
2. Bažant, Z.P.: Solidification theory for aging creep. Cem. Concr. Res. **18**(6), 923–932 (1988)
3. Ulm, F.J., Coussy, O., Bažant, Z.P.: The chunnel fire. I; Chemoplastic softening in rapidly heated concrete. J. Eng. Mech. **125**(3), 272–282 (1999)
4. Carles-Gibergues, A.: Les ajouts dans les microbétons. Influence sur l'auréole de transition et sur les propriétés mécaniques. Ph.D. thesis, Université Paul Sabatier, Toulouse (1981)
5. Poon, C.S., Lam, L., Wong, Y.L.: Effects of fly ash and silica fume on interfacial porosity of concrete. J. Mater. Civ. Eng. **11**, 197–205 (1999)
6. Justness, H., Sellevold, E.J., Lundevall, G.: High strength and concrete binders part a: reactivity and composition of cement pastes with and without condensed silica fume. ACI Special Publication, pp. 873–890 (1992)
7. Le Roy, R., Le Maou, F., Torrenti, J.M.: Long term basic creep behavior of high performance concrete: data and modelling. Mater. Struct. **50**(11), 85 (2016)
8. Mazloom, M., Ramezanianpour, A.A., Brooks, J.J.: Effect of silica fume on mechanical properties of high-strength concrete. Cement Concr. Compos. **26**, 347–357 (2004)
9. Jianyong, L., Yan, Y.: A study on creep and drying shrinkage of high performance concrete **31**(8), 1203–1206 (2001)
10. San Nicolas, R., Cyr, M., Escadeillas, G.: Performance-based approach to durability of concrete containing flash-calcined metakaolin as cement replacement. Constr. Build. Mater. **55**, 313–322 (2014)
11. Brooks, J.J., Megat Johari, M.A.: Effect of metakaolin on creep and shrinkage of concrete. Cement Concr. Compos. **23**(6), 495–502 (2001)
12. Nguyen Amanjean, E., Vidal, T.: Low cost ultra-high performance fiber reinforced concrete (UHPFRC) with flash metakaolin. Key Eng. Mater. **629–630**, 55–63 (2015). doi:10.4028/www.scientific.net/KEM.629-630.55
13. EN 1992-2: Design of concrete structures. Part 2: Concrete Bridges - Design and Detailing Rules (2006)
14. EN 1992-1-1: Design of concrete structures. Part 1-1: General rules and rules for building (2005)
15. Salvador, S.: Pozzolanic properties of flash-calcined kaolinite: a comparative study with soak-calcined products. Cem. Concr. Res. **25**, 102–112 (1995)
16. RILEM CPC8: Recommendations RILEM CPC8, Modulus of elasticity of concrete in compression. Mater. Struct. **6**(30), 507–512 (1972)
17. Bucher, R.: Vers une utilisation rationnelle des métakaolins flash: Application aux bétons. Ph.D. thesis, Université Paul Sabatier, Toulouse (2015)
18. Baroghel-Bouny, V., Mainguy, M., Lassabatere, T., Coussy, O.: Characterization and identification of equilibrium and transfer moisture properties for ordinary and high-performance cementitious materials. Cem. Concr. Res. **29**, 1225–1238 (1999)
19. Neville, E.A.M.: Properties of Concrete, 4th edn. Wiley, New York (1996). Final Edition

Effect of Carbonate Minerals and Calcination of Carbonatites and Kamafugites on Their Pozzolanic Performance and Early Age Concrete Properties

A. Buregyeya[1,2(✉)], Y. Ballim[1], S. Nwaubani[1], A.G. Kerali[2], and M. Otieno[1]

[1] School of Civil and Environmental Engineering, University of the Witwatersrand, Johannesburg, South Africa
[2] Department of Civil and Environmental Engineering, Makerere University, Kampala, Uganda

Abstract. Carbonates are primary minerals of the carbonatites and kamafugites sourced from Toro-Ankole geological region of the East African Rift system. Consequently, these materials are silica undersaturated. They are currently utilized as mineral additions in production of Portland pozzolana cements in Uganda. No published work exists to show how their unique composition might affect their pozzolanic performance and other concrete properties. This study investigated the effect of the carbonate minerals in the natural pozzolans and calcination on setting time, standard consistency, workability, soundness, early heat of hydration and strength development of blended cements. Two (2) samples, a carbonatite and a kamafugite sourced from deposits located in the Toro-Ankole geological region of the East African rift system were calcined in a furnace at 825 °C for one hour. The samples were then subjected to XRD analysis for mineralogical composition characterisation and to establish the effect of calcination at 825 °C on mineralogy. Setting time, standard consistency, workability, soundness and strength development were studied following standard procedures for testing blended Portland cements. Calcination led to a gain in compressive strength for both test pozzolans, the kamafugites showing a higher gain in strength than the carbonatite. The higher gain in pozzolanic performance for the kamafugites is likely due to kaolinite, a secondary mineral in the test pozzolan whose pozzolanic reactivity is activated by thermal destabilization. Calcination also led to pacification of the early age properties of cements blended with test carbonatites and kamafugites. The study reveals the carbonate minerals in the test pozzolans as a considered factor in accelerating early hydration of Portland cement. Hydration progression of Portland cement controls the important properties of fresh concrete (workability, setting and paste microstructure), which in turn directly impacts on the strength and durability properties of hardened concrete. Cements blended with carbonate bearing natural pozzolans therefore present interesting perspectives on how paste microstructure composition and durability performance properties might be impacted.

Keywords: Kamafugites · Carbonates · Pozzolans · Hydration · Calcination

F. Martirena et al. (eds.), *Calcined Clays for Sustainable Concrete*, RILEM Bookseries 16,
https://doi.org/10.1007/978-94-024-1207-9_15

1 Introduction

1.1 Background

Portland cement concrete systems present cement as the most important component in the concrete matrix. The hydration reaction between cement phases and mixing water influence workability, setting time, paste microstructure, paste-aggregate interfacial transition zone (ITZ) and consequently strength development, volume change and durability properties of concrete of a specific concrete matrix. The hydration of cement is affected by the presence of supplementary cementitious materials (SCMs) and therefore characterization of different SCMs is important for prediction of concrete performance. Pozzolans represent the majority of SCMs and have been reported to present varied effects on hydration progression of Portland cement [1] mainly on the basis of both their physical and compositional characteristics.

Pozzolans are presented in literature as either natural or artificial depending on the source of the materials and whether the materials have been subjected to any processing leading to activation of constituent compounds to achieve pozzolanic reactivity. Natural pozzolans are mainly of volcanic origin and include tuffs, ash, and lavas but also diatomaceous earth. They require minimal processing that include drying and pulverizing to be utilised as mineral additions in Portland cement. Artificial pozzolans on the other hand include ground granulated blast furnace slag (GGBS), fly ash (FA), metakaolin (MK) and silica fume (SF) and are as a result of industrial processes that are designed to activate the aluminosilicate composition in the raw materials for pozzolanic or latent hydraulic properties. Studies have focused more on application of industrial byproducts (artificial pozzolans) and little has been published on characterization of natural pozzolans as SCMs [2].

Moreover, studies on natural pozzolans tend to focus on zeolitic tuffs [3–10], pumice [5] and ash or trass [5] that are mainly classified as silicate tuffs/lavas/ashes. This paper is part of an experimental study focused on characterizing carbonatites and kamafugites as SCMs. Carbonatites and kamafugites are grouped as materials of volcanic origin whose composition is predominantly carbonate minerals. As a consequence, they are relatively low in silica and alumina composition compared to other volcanic materials [11] that are generally applied as natural pozzolans in Portland cement concrete. Standard methods for testing pozzolanic activity of natural pozzolans fail these materials on the basis of their chemical composition. The materials are however currently applied as natural pozzolans in production of pozzolanic cements in Uganda. There is lack of published works that report on their performance as pozzolanic materials in Portland cement concrete.

This paper reports on the effects of the carbonate minerals in carbonatites and kamafugites on fresh concrete properties, early hydration progression and strength development when the test natural pozzolans are blended with Portland cement.

1.2 Petrology of Volcanic Rocks in Toro-Ankole Geological Region

Kamafugites are ultrapotassic and silica undersaturated zeolitised volcanic tuffs containing feldspathoidal minerals. They appear as either ultramafic or mafic rocks and are defined by the presence of kalsilite, a mineral in the form of $KAlSiO_4$ [12]. They fall within the family of carbonatitic lavas because of their relatively high composition of CO_2 appearing mainly in the form of calcite and dolomite. Their characteristic silica and alumina undersaturation is a result of carbonates and potassium alkalis enrichment [12]. They were first reported by Arthur Holmes and Henry Francis Harwood [13] in the Toro-Ankore geological region of the East Africa Rift system (Fig. 1). More recent publications have reported occurrence of kamafugite deposits in Umbria-Latium region of Italy (central Italy) [14], the Parana Basin region of Brazil and Paraguay [15] and the west Quniling region of China [16]. Olivine is the predominant mineral in the Toro-Ankore kamafugites but also melilitite, Kalsilitite, Leucitite and potassic nephelinite [17–19]. The absence of more regular pozzolanic industrial waste materials in the Toro-Ankole region of Uganda limits access to only carbonatites and kamafugites in production of Portland pozzolana cements.

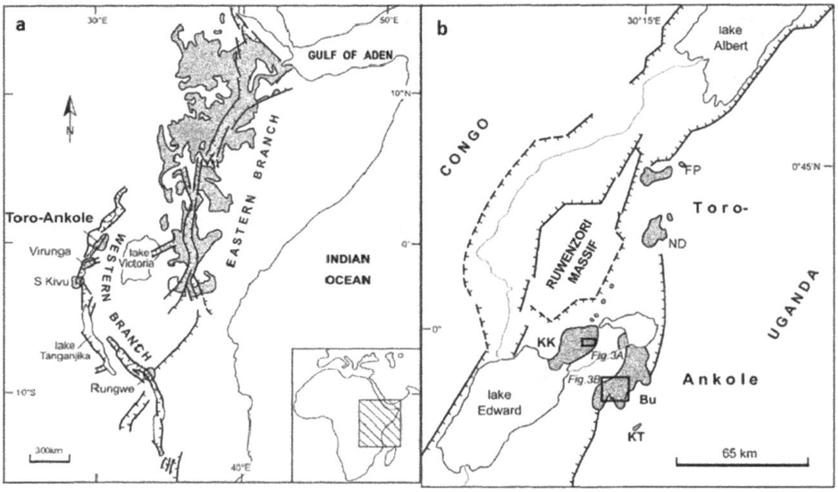

Fig. 1. (a) A sketch map of eastern Africa showing the rift system and occurrence of lava deposits. The Albertine rift or the western branch is composed of Toro-Ankore, Birunga and South Kivu series. The hairy lines represent the main rift faults. (Map adapted)

2 Materials and Methods

All the samples were tested for oxide composition using X-ray fluorescence (XRF) instrumental technique while X-ray Diffraction (XRD) was applied in identification of mineralogical composition for both raw and calcined samples. The test pozzolans milled to conform with the requirements stated in ASTM C618 [20] (maximum of 34% retained on a 45 μ sieve in wet sieving) (Table 1) were then blended with CEM I 52.5R Portland cement at 20% replacement level. The strength activity index (SAI) (Eq. 1) method based on the standard procedure for fly ash provided in BS 3892 [21] and EN 196-1 test protocols presented in literature on pozzolanic tests for

Table 1. Physical properties of test pozzolans

Sample name	Specific surface area (m^2/g)	Retained on 45 μm (%)	d(0.1) (μm)	d(0.5) (μm)	d(0.9) (μm)
Control (CEM I 52.5R)	1.75	2.0	1.422	12.533	32.070
Poz-A	2.18	17.5	0.848	14.695	99.808
Poz-B	2.10	23.9	0.852	21.898	178.719

natural pozzolans [1, 5, 22] was used to characterise strength development of blended samples. Isothermal conduction calorimetry (TAM Air, TA instruments) was used in heat of hydration studies. For standard consistency, setting time and soundness studies, test procedures presented in the European standard EN 196-3 (2009) [23] were followed. A flow table and a standard frustum cone shaped steel mould (EN1015-3, 2007) [24] were used for mortar flow tests at a fixed water content.

$$SAI = \frac{P_{testsample}(MPa)}{P_{controlsample}(MPa)} \times 100\% \tag{1}$$

3 Results and Discussion

3.1 Chemical and Mineralogical Composition

Table 2 presents oxide composition and Loss on Ignition (LOI) results for test samples in raw and calcined state. Both test pozzolans fail the ASTM C618 minimum composition requirement of 70% for the sum of the acidic oxides (Si, Al and Fe) to be used as pozzolanic materials in Portland cement concrete. The test pozzolans have a relatively low silica and alumina content due to carbonates and potassic alkalis enrichment [12].

Table 2. Chemical composition of kamafugites and carbonatites

Sample ID		% oxide composition															
		LOI	SiO_2	Al_2O_3	Fe_2O_3	CaO	MgO	K_2O	Na_2O	P_2O_5	TiO_2	Mn_2O_3	Cr_2O_3	V_2O_5	SO_3	Cl	XRF Total
Poz-A	Raw	13.8	27.32	6.71	11.42	29.24	4.15	1.05	1.06	2.67	1.79	0.48	0.02	0.05	0.1	0.02	99.9
	Calcined	1.01	31.7	7.86	13.15	32.92	4.82	1.25	1.34	2.81	2.17	0.57	0.02	0.06	0.31	0.04	100.0
Poz-B	Raw	13.21	46.91	8.23	7.37	9.78	6.21	4.61	0.2	0.93	2.23	0.2	0.05	0.03	0	0	100.0
	Calcined	1.3	52.95	8.86	8.82	11.33	6.9	5.59	0.26	1.04	2.59	0.23	0.06	0	0.04	0	100.0
Control	CEM I 52.5R	0.09	20.33	4.41	2.53	65.19	1.8	0.48	0.09	0.08	0.41	0.1					95.5

The carbonate minerals in the test carbonatite and kamafugite are also the primary reason for a high LOI value. As a result, their decomposition by calcination leads to an increase in the densities of the different mineral and oxide components of the test pozzolans. It is feasible that this increase in the concentration of mineral and oxide

Table 3. Mineralogical composition of kamafugites and carbonatites (XRD)

Minerals present	Poz-A		Poz-B		Chemical formula
	Raw	Calcined	Raw	Calcined	
Calcite	√	*nd*	√	nd	$(Mg_{0\cdot03}Ca_{0\cdot97})(CO_3)$
Pigeonite	√	√	√	√	$(Ca,Mg,Fe)(Mg,Fe)Si_2O_6$
Quartz	√	√	√	√	SiO_2
Muscovite	√	√	√	√	$KAl_2(AlSi_3O_{10})(F,OH)_2$
Olivine	√	√	√	√	$Na_2(SiO_4)$
Apatite	√	√	√	√	$Ca_{10}(PO_4)_6(OH)_2$
Augite	*x*	√	x	√	$(Ca,Na)(Mg,Fe,Al,Ti)(Si,Al)_2O_6$
Kaolinite	*x*	*x*	√	√	$Al_4(OH)_8(Si_4O_{10})$
Epidote	*x*	*x*	√	√	$Ca_2Al_{2\cdot4}Fe_{0\cdot6}(SiO_4)_3(OH)$

√- Mineral detected; x- Mineral not detected

components of the test pozzolans leads to availability of more reactants per unit of batching and hence a higher relative gain in strength and better performance in the ASTM C618 minimum composition requirement of the sum of the silica, alumina and Iron oxides. The gain in strength, however, does not show any correlation with the relative change in oxide composition after calcination.

XRD mineralogical composition characterisation results are shown in Table 3 and Figs. 2 and 3. The test pozzolans, Poz-A and Poz-B, were effectively decarbonated at 825 °C. A new mineral, Augite, is detected in both calcined test pozzolans possibly due to increase in its concentration to detectable levels by XRD instrumental techniques after the decomposition of the carbonates. Augite is a mineral of the Pyroxene family and is generally associated with low silica igneous rocks. Figures 2 and 3 show calcination to have minimal effect on the structure and composition of the other minerals present in the kamafugites and carbonatite.

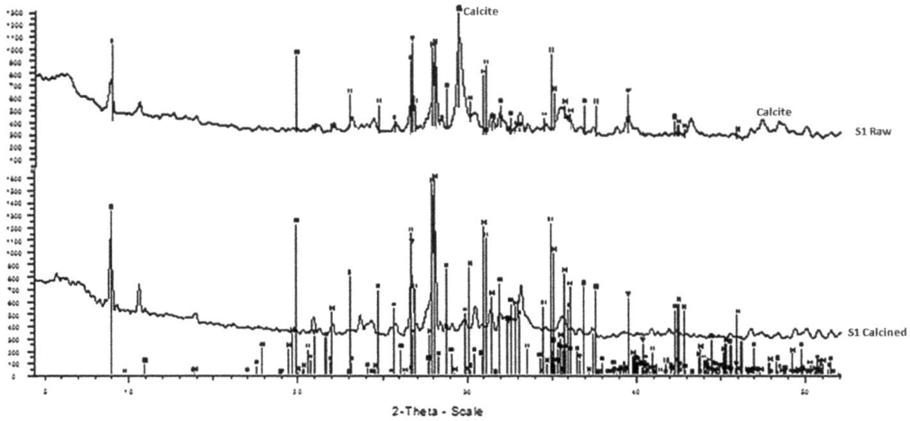

Fig. 2. Mineralogy of Poz-A(S1) in both raw and calcined states

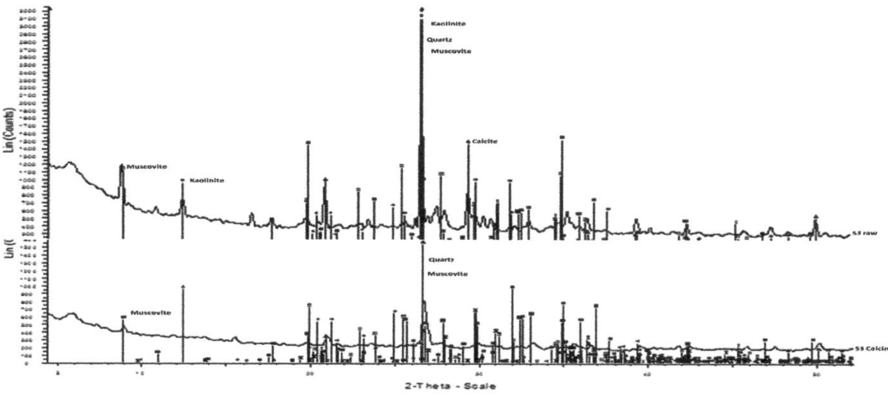

Fig. 3. Mineralogy of Poz-B(S3) in both raw and calcined states

Kaolinite, a secondary mineral formed through alteration of the primary rock minerals [22], was detected in Poz-B, a kamafugite. Habert et al. (2008) [22] reports a higher strength with thermal destabilisation of kaolinite enriched volcanic materials. Kaolinite, a clay mineral, is known to have pozzolanic properties when thermally destabilised. Extensive work on pozzolanic nature of kaolinite and how it influences Portland cement properties is presented in the RILEM book series on Calcined Clays for Sustainable Concrete [25]. Calcination of a kaolinite enriched Poz-B is therefore a considered factor in increasing its compressive strength.

3.2 Effect of Calcination on Fresh Properties of Cement Paste and Mortar

Table 4 presents results on setting time, water demand, slump flow and soundness tests of the test pozzolans in raw and calcined forms. Both samples led to acceleration of setting, increased water demand and reduced slump flow, poz-B showing a relatively higher influence on the fresh properties of Portland cement than Poz-A. On calcination, these effects were pacified significantly. The test pozzolans showed no effect on the soundness property of cement.

Table 4. Properties of fresh cement paste and mortar

Parameters	Control CEM I 52.5R	Poz-A		Poz-B	
		Raw	Calcined	Raw	Calcined
Initial setting (min)	200	180	205	170	190
Final setting (min)	286	246	291	225	269
Water demand (%)	29.2	30.7	29.5	33.1	30.5
Slump flow (%)	100.0	95.3	97.3	87.9	98.0
Soundness (mm)	0	0	1	0	0

3.2.1 Setting Time

The accelerating effect of the test pozzolans is in agreement with the results of a tuff from central Turkey studied by Turanli and others [26] whose alkalis are predominantly potassic. Although Turanli and coworkers did not report on the possible cause of the trend, work by Jawed and Skalny [27] and Odler and Wonnemann [28] show K+ ions to have an accelerating effect on the aluminate phase while Na+ alkali ions have a retarding effect. Considering the ultrapotassic nature of Poz-B, the sample's relative effect on the setting time is higher than that of Poz-A. Calcination completely stopped the set accelerating effect of Poz-A but partially for Poz-B indicating presence of other contributing mechanisms. The ultrapotassic nature of the kamafugite (Poz-B) is advanced as a factor in its superior set accelerating property.

Works on the effect of calcite/limestone on physical properties of Portland cement blended with limestone too report accelerating effect on setting time and a small increase in water demand that is governed by the limestone fineness [28]. Calcite modifies paste microstructure by reacting with the aluminate and sulfate phases to form carboaluminates and AFm respectively [29, 30] with a consequence of accelerating early hydration mechanisms, especially hydration of alite phases [31]. The accelerating phenomena is partly attributed to the "filler effect", which Berodier and Scrivener [32] explain to be driven by the interparticle distance contrary to previous belief that mineral admixtures provide extra surfaces for nucleation sites of hydrates [30]. consequently, calcite gives a higher filler effect compared to other SCMs [32]. The filler effect is the concern of the first few hours of hydration (up to the final setting of cement paste) which is a very important stage of concrete responsible for paste microstructure prognosis and resulting macrostructure properties. Accordingly, the carbonate minerals in the kamafugites and cabonatites are responsible for accelerating the setting time of blended samples.

3.2.2 Water Demand and Slump Flow

The mechanisms that control water demand and slump flow are relatable and depend on both the physical characteristics of the finer particles in test pozzolans and early hydration reactions. Given that both Poz-A and Poz-B have same values of d(0.1) particle size, the difference in water demand and slump flow readings can be attributed to the effect of composition of the test pozzolans on cement early hydration progression. The relationship in trends of the water demand and slump flow results with setting time results further validates the role of the mineralogical composition of the test pozzolans. The accelerating effect of $CaCO_3$ is present in both Poz-A and Poz-B. The high K+ alkali content in Poz-B is considered a factor in causing higher slump and water demand values in relation to Poz-A.

3.2.3 Soundness

On soundness property, the Le Chatelier's soundness tests recorded no expansion for all raw test pozzolans blended with Portland cement at a 20% replacement level. On calcination of the test pozzolans, Poz-A registered a 1 mm expansion possibly due to hydration of calcium oxide made available by the de-carbonation of the calcite minerals in the test pozzolans. Poz-A (a carbonatite) has a relatively higher calcite content that Poz-B (a kamafugites) and consequently more CaO content on calcination.

3.3 Effect of Calcination on Early Heat of Hydration

Figure 4 presents heat of hydration results measured using instrumental techniques (TAM Air calorimeter, TA Instruments) and normalised to unit weight of all the cementitious materials. Heat release in cement hydration is considered as a proxy indicator to the progress of the reactions. Heat release recorded for 24 h show the raw test pozzolans to have a higher heat release than the control sample (CEM I 52.5R). Calcination of the test pozzolans pacified the trend and led to early heat of hydration to be less than that of the control sample. The role of calcite in accelerating early hydration reactions is further revealed through heat of hydration.

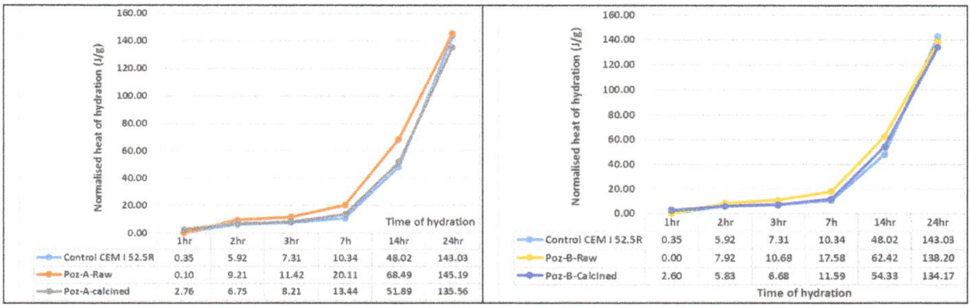

Fig. 4. Heat of hydration progression for the first 24 h

3.4 Effect of Calcination on Strength Development/Pozzolanic Performance

The compressive strength parameter has been used to study the pozzolanic performance of the test natural pozzolans. The results are compared against results of two control samples, CEM I 52.5R and class F Fly Ash. Fly ash is a pozzolanic material that has passed the testing and development stage and therefore a reasonable control in characterisation of pozzolanic properties of natural pozzolans. In fact, Fly Ash and natural pozzolans share the ASTM C 618 [20] characterisation standard. Results presented in Table 5 show the raw natural pozzolans perform better than their calcined counterparts up to 28 days. The hydration reactions in the samples blended with raw natural pozzolans appear to have stopped by 90 days of curing implying depletion of reactants.

Table 5. Compressive strength of mortar prisms

Curing (Days)	Control 1 CEM 1 52.5R	Control 2 class F Fly ash	Poz-A		Poz-B	
			Raw	Calcined	Raw	Calcined
7	49.2	40.0	41.5	40.9	41.2	41.0
28	57.0	54.2	51.1	49.6	56.8	55.0
90	63.4	69.7	53.2	54.4	57.4	58.2
180	65.3	72.1	53.1	54.9	56.6	60.9

The calcined samples continue to gain in strength after 90 days of curing, a trend they share with both FA and CEM I control samples. There is a 3.4% and 7.6% gain in compressive strength with calcination for Poz-A and Poz-B respectively. The gain in compressive strength does not show any correlation with the change in composition of the test natural pozzolans after calcination (Table 2). Poz-B however has a clay mineral, Kaolinite, which exhibits pozzolanic properties when thermally distabilised at temperatures as low as 350 °C [22].

Table 6 presents calculated values of the strength activity index (SAI) based on Eq. 1. All values at all curing times are above the minimum of 80% required for a material to be considered pozzolanic [1, 5, 21, 22]. The calcined samples have higher long-term SAI values in comparison with the raw samples showing a better pozzolanic performance.

Table 6. Strength activity index test for pozzolanicity

Curing (Days)	Control 1 CEM 1 52.5R	Control 2 class F Fly ash	Poz-A		Poz-B	
			Raw	Calcined	Raw	Calcined
7	100	81	84.4	83.3	83.8	83.4
28	100	95	89.6	87.1	99.6	96.4
90	100	110	83.9	85.8	90.6	91.8
180	100	110	81.3	84.1	86.7	93.2

4 Conclusions

The following conclusions have been drawn on the basis of results from the current experimental study:

- Calcination of kamafugites and carbonatites leads to improved rating on the ASTM C618 minimum composition requirement of 70% for the sum of silica, alumina and iron oxides for a volcanic material to be considered pozzolanic enough for application in Portland cement concrete.
- The addition of kamafugites and carbonatites as natural pozzolans in Portland Cement accelerated setting time, early heat of hydration and increased water demand, all early age properties which were eliminated by calcination at 825 °C. It is perceived that the carbon and potassic elements in the test kamafugites and carbonatites are responsible for the hydration accelerating effect observed.
- Calcination resulted in a 3.4% and 7.6% gain in compressive strength for Poz-A and Poz-B respectively at 180 days of hydration. The increase in concentration of available pozzolanic reactants after calcination is advanced as the base reason for the observed increase in strength. The presence of Kaolinite, a secondary mineral that exhibits pozzolanic properties on thermal destabilization, is advanced as the reason for the higher gain in strength observed in Poz-B.
- Calcination had no detectable effect on the soundness property of Poz-B. However, there was an increase in Le Chaterer's expansion for Poz-A though within the range

of maximum permissible limits. The presence of relatively large quantities of CaO as a result of calcination of the carbonatite is considered as the cause for increase in expansion due to CaO hydration.

- Calcination of the carbonatite and kamafugite samples at 825 °C effectively decarbonated the test pozzolans. There was no significant effect on other minerals in the test pozzolans. As a result of the decomposition of calcite, the concentration of Augite, a mineral of the Pyroxene family, was raised to detectable levels.

Based on the above conclusions, cements blended with kamafugites and carbonatites as natural pozzolans therefore present interesting perspectives on how paste microstructure and cement durability properties might be influenced. The conclusions also confirm a need for a different approach in characterizing kamafugites and carbonatites as pozzolans in Portland cement. The calcite enriched materials fail on composition requirements stated in established standard specifications.

Acknowledgements. The authors express their sincere gratitude to the Centre of Product Excellence (CPE) team at Afrisam South Africa Pty (Ltd) who allowed us to use their laboratory facilities both at Rooderport and Vanderbijlpark cement factories to perform most of the experiments in the current study. Also to Walter L. Wirth and his team at PPC Ltd Jupiter laboratories who conducted the X-ray diffraction (XRD) experiments. This study was supported by the National Research Foundation (NRF) of South Africa.

References

1. Donatello, S., Tyrer, M., Cheeseman, C.R.: Comparison of test methods to assess pozzolanic activity. Cem. Concr. Res. **32**(1), 121–127 (2010)
2. Scrivener, K.L., Lothenbach, B., De Belie, N., Gruyaert, E., Skibsted, J., Snellings, R., Vollpracht, A.: TC 238-SCM: hydration and microstructure of concrete with SCMs: State of the art on methods to determine degree of reaction of SCMs. Mater. Struct. **48**, 835–862 (2015)
3. Martinez-Ramirez, S., Blanco-Varela, M.T., Erena, I., Gener, M.: Pozzolanic reactivity of zeolitic rocks from two different Cuban deposits: characterization of reaction products. Appl. Clay Sci. **32**, 40–52 (2006)
4. Caputo, D., Liguori, B., Colella, C.: Some advances in understanding the pozzolanic activity of zeolites: the effect of zeolites structure. Cem. Concr. Compos. **30**, 455–462 (2008)
5. Pourkhorshidi, A.R., Najimi, M., Parhizkar, T., Jafarpour, F., Hillemeier, B.: Applicability of the standard specifications of ASTM C618 for evaluation of natural pozzolans. Cem. Concr. Compos. **32**, 794–800 (2010)
6. Uzal, B., Turanli, L., Yucel, H., Goncuoglu, M.C., Culfaz, A.: Pozzolanic activity of clinoptilolite: a comparative study with silica fume, fly ash and a non-zeolitic natural pozzolan. Cem. Concr. Res. **40**, 398–404 (2010)
7. Luke, K.: The effect of natural zeolites on the composition of cement pore fluids at early ages, 12th International Congress on the Chemistry of Cement (ICCC). Montreal, Canada (2007)
8. Mertens, G., Snellings, R., Van Balen, K., Bicer-Simsir, B., Verlooy, P., Elsen, J.: Pozzolanic reactions of common natural zeolites with lime and parameters affecting their reactivity. Cem. Concr. Res. **39**, 233–240 (2009)

9. Turanli, L., Uzal, B., Bektas, F.: Effect of materials characteristics on the properties of blended cements containing high volumes of natural pozzolans. Cem. Concr. Res. **34**, 2277–2282 (2004)
10. Snellings, R., Mertens, G., Cizer, O., Elsen, J.: Early age hydration and pozzolanic reaction in natural zeolite blended cements: reaction kinetics and products by in situ synchrotron X-ray powder diffraction. Cem. Concr. Res. **40**(12), 1704–1713 (2010)
11. Stoppa, F., Schiazza, M.: An overview of monogenetic carbonatitic magmatism from Uganda, Italy, China and Spain: Volcanologic and geochemical features. J. South Am. Earth Sci. **41**, 140–159 (2013)
12. Tappe, S., Foley, S.F., Pearson, D.G.: The kamafugites of Uganda: a mineralogical and geochemical comparison with their Italian and Brazilian analogues. Periodico di Mineralogia **72**, 51–77 (2003). Special issue: Eurocarb
13. Holmes, A., Harwood, H.F.: Petrology of the Volcanic fields east and south-east of Ruwenzori, Uganda. Q. J. Geol. Soc. Lond. **88**, 370–439 (1932)
14. Stoppa F., Rosatelli G., Vichi G., Principe C.: Extrusive carbonatites and their meaning. In: The case of Italy 32nd International Geological Congress, Florence-Italy (2004)
15. Sgarb, P.B., Gaspari, J.C., Valenca, J.G.: Brazilian kamafugites. Rev. Brasil. Geocienc. **30**, 417–420 (2000)
16. Guo, P., Niu, Y., Yu, X.: A synthesis and new perspective on the petrogenesis of kamafugites from West Qinling, China, in a global context. J. Asian Earth Sci. **79**, 86–96 (2014)
17. Rosenthal, A., Foley, S.F., Pearson, G.D., Nowell, M.G., Tappe, S.: Petrogenesis of strongly alkaline primitive volcanic rocks at the propagating tip of the western branch of the East African Rift. Earth Planet. Sci. Lett. **284**, 236–248 (2009). Elsevier B. V.,
18. Bell, K., Doyle, R.J.: K-Rb relationships in some continental alkali rocks associated with the East African Rift valley system. Geochim. Cosmochim. Acta **35**, 903–915 (1971). Pergamon press
19. Holmes, A.: A suite of volcanic rocks from south-west Uganda containing kalsilite (a polymorph of KAlSiO4). Mineral. Mag. J. Mineral. Soc. **26**(177), 197–217 (1942)
20. ASTM C618. Standard specification for coal fly ash and raw or Calcined natural pozzolan for use in concrete. American society for testing and materials. West Conshohocken, PA, ASTM International (2005)
21. BS 3892-1. Pulverized-fuel ash. Specification for pulverized-fuel ash for use with Portland cement. BSI (1997)
22. Habert, G., Choupay, N., Montel, J.M., Guillaume, D., Escadeillas, G.: Effects of the secondary minerals of the natural pozzolans on their pozzolanic activity. Cem. Concr. Res. **38**, 963975 (2008)
23. DIN EN 196-3. Methods for testing cement- Part 3: Determination of setting times and soundness. European Committee for Standardization (2009)
24. BS EN 1015-3. Methods of test for mortar for masonry; Determination of onsistence of fresh mortar (by flow table). European Committee for Standardization (1999)
25. Scrivener K., Favier A.: Calcined clays for sustainable concrete. In: Proceedings of the 1st International Conference on Calcined Clays for Sustainable Concrete, RELIM Bookseries, ISBN 978-94-017-9939-3. Springer (2015)
26. Turanli, L., Uzal, B., Bektas, F.: Effect of large amounts of natural pozzolan addition on properties of blended cements. Cem. Concr. Res. **35**(6), 1106–1111 (2005)
27. Jawed, I., Skalny, J.: Alkalies in cement: a review, II; Effects of alkalies on hydration and performance of Portland cement. Cem. Concr. Res. **8**, 37–52 (1978)

28. Odler, I., Wonnemann, R.: Effect of alkalies on portland cement hydration I; alkali oxides incorporated in to the crystalline lattice of clinker minerals. Cem. Concr. Res. **13**, 477–482 (1983)
29. Ipavec, A., Gabrovšek, R., Vuk, T., Kaučič, V., Maček, J., Meden, A.: Carboaluminate phases formation during the hydration of calcite-containing Portland cement. J. Am. Ceram. Soc. **94**, 1238–1242 (2011). doi:10.1111/j.1551-2916.2010.04201.x
30. Mohamed, A.R., Elsalamawy, M., Ragab, M.: Modeling the influence of limestone addition on cement hydration. Alex. Eng. J. **54**(1), 1–5 (2015)
31. Matschei, T., Lothenbach, B., Glasser, F.P.: The role of calcium carbonate in cement hydration. Cem. Concr. Res. **37**, 551–558 (2007)
32. Berodier, E., Scrivener, K.: Understanding the Filler Effect on the Nucleation and Growth of C-S-H. J. Am. Ceram. Soc. **97**, 3764–3773 (2014). doi:10.1111/jace.13177

Assessment of the Pozzolanic Reactivity of Calcined Kaolinitic Clays by a Rapid Alkaline Solubility Test

E. Cabrera[1(✉)], R. Almenares[2], and A. Alujas[3]

[1] Faculty of Chemistry, Central University "Marta Abreu" of Las Villas (UCLV), Carretera a Camajuanì km 5 1/2, Santa Clara, Villa Clara, Cuba
ecapolinaire@uclv.edu.cu

[2] Instituto Superior Minero-Metalúrgico de Moa, Moa, Cuba

[3] Center for Research and Development of Structures and Materials (CIDEM), UCLV, Santa Clara, Cuba

Abstract. Given the availability of clay deposits in several geographical areas of the Cuban territory and the use of calcined clays as a source of Supplementary Cementitious Materials (SCM), it is necessary to evaluate the pozzolanic reactivity of its calcination products and because of the complexity in the chemical - mineralogical characteristics of the pozzolans. This work introduces a new rapid method to predict the potential character pozzolanic of these materials by alkaline solubility and presents the possible correlation with isothermal calorimetry in lime-pozzolan pastes following the R3 protocol. The raw materials were characterized by XRF, XRD and TGA.

The clayey materials contain more than 70% of SiO2 + Al2O3 + Fe2O3, which guarantees its pozzolanic character after its thermal activation. The process of the dihydroxylation of the clay minerals happens between 350 °C and 650 °C, and for the deposit of Yaguajay the decomposition of the calcite is between 650 °C and 750 °C. Alkaline solubility test is a viable test to assess the pozzolanic reactivity of calcined kaolinitic clay due a good and directly correlation between the content of soluble aluminum and the total heat released (R = 0.995) and there is not a significant dispersion of the obtained data for each clay deposit (CV% between 0.01 and 2.80). The pozzolanic reactivity of its calcination products determined by isothermal calorimetric follows the sequence CG > LS > LL > YG and matches with Alkaline Solubility test.

1 Introduction

Clay minerals are one of the most employed SCM on the Cuban territory given the availability of clay deposits in several geographical areas of this country. Clays with proven pozzolanic properties, once calcined under specific conditions, represent an attractive source of pozzolans in regions where other resources are not available and helps to reduce CO2 emissions from clinker production [1]. However, clays are often found in the form of deposits in which the clay fraction is composed of a mixture of several mineral phases, without a clear predominance of any of them. Therefore, knowledge and use of raw materials, with varied structural, chemical and mineralogical

F. Martirena et al. (eds.), *Calcined Clays for Sustainable Concrete*, RILEM Bookseries 16,
https://doi.org/10.1007/978-94-024-1207-9_16

characteristics, guarantees a source of pozzolanic materials [2, 3]. Because of this complexity, it is necessary to determinate the pozzolanic character through rapid methods to provide the cement industry with reliable information about the pozzolanic criteria of this clay minerals.

Methods to evaluate the pozzolanic reactivity of kaolinitic clays are classified into two large groups according to the measured property. The direct methods are related to the consumption of portlandite, like the thermogravimetry technique, X-ray diffraction analysis and isothermal calorimetry analysis. Unlike these, the indirect methods are those techniques that detect changes in volume and porosity of the material as well as the solubility of pozzolans related to the quantification of hydration products, therefore conductimetry and mechanical tests have been the main techniques used for these studies [4].

In the case of mechanical strength test, these is a standard method to evaluate the pozzolanic behavior, but this method long testing times are required because depends of the curing time of the samples. Regarding lime consumption uptake tests, they may be interfered by other phenomena no related to pozzolanic reactivity like cation exchange [5]. On the other hand, Isothermal Calorimetry (R3 test) is a rapid, reliable and relevant test and it has a good correlation with compressive strength [6] but, the filler effect must be taken into account and least, this test requires a calorimeter and this equipment is still deficient in the cement factories that it is where the tests of reactivity come true.

For these reasons, methods based on pozzolans alkaline solubility are a viable test to the quickly assess of the pozzolanic reactivity of calcined clays. These methods are based on the first step of the pozzolanic reaction: the chemical attack of OH- and alkali ions released in the hydration of the OPC to the structure that is in a state of high structural disorder within pozzolans, causing rupture of the Si-O and Al-O links and the release of these oxyanions to the solution. That is why pozzolanic reactivity must be directly proportional to the amount of soluble silicon and aluminium species released in an alkaline solution with a pH similar to the hydrated cement pore solution. However, most of the methods employed worldwide make use of too harsh conditions, which do not simulate the environment of the pozzolanic reaction in cement. Those techniques uses in many cases, concentrated acids to dissolve the pozzolans, relatively high test temperatures that could lead to the occurrence of possible collateral reactions, and most require long analysis times in order to arrive at a final decision about its use as SCM [7, 8]. Therefore, the main goal of this paper is to development of a rapid method to assess the pozzolanic reactivity of calcined clays based on its alkaline solubility.

2 Method

Clays used in this study were collected by representative sampling of different clay deposits located at the central and eastern regions in Cuba. The selection was based on its estimated reserves and its distance to cement factories, in order to guarantee future exploitation of these deposits as sources of pozzolanic materials for the cement industry.

- Manifestacion de Yaguajay (YG): located at the central region at the north of Sancti Spiritus city.
- Gaspar Loma Sur (LS) y Gaspar La Loma (LL): located at the central region at **the** southeast of Ciego de Avila city.
- Cayo Guam: located at the oriental region, at the southeast of Moa city, Holguin.

The characterization of the raw material was at the Laboratory of Construction Materials EPFL, Switzerland by X-Ray diffraction, X-Ray fluorescence and Thermogravimetric analysis.

Then, the raw material was thermally activated at 850 °C during 60 min and finally, calcined products were grinding guaranteeing a 90% passed through the sieve of 90 μm.

To evaluate the pozzolanic reactivity of the samples an analytical method was proposed based in the alkaline solubility of the calcined products through Atomic Absorption Spectroscopy. The first one was alkaline solubility. The methodology is mixing 50 ml of a solution of sodium hydroxide (0,5 M) with 0,1 g of calcined clay. Putting the mixture in a double boiler at 50 °C and constant velocity of agitation of 30 revolutions per minutes during 24 h. Next, we took an aliquot of 25 ml of this dissolution and we add 2,5 ml of concentrated Hydrochloric acid in order to decrease the pH below 1. Then, we centrifuge the dissolution at 1500 revolutions per minutes and finally we should filter it to eliminate the remnant solid. Atomic Absorption Spectroscopy technique was used to determine the amount of soluble silicon and aluminum (Fig. 1).

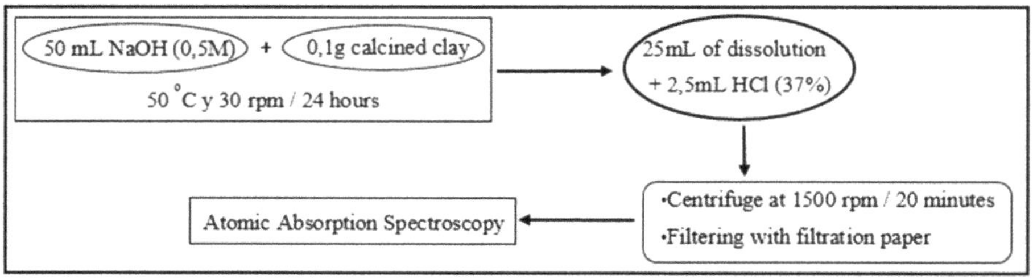

Fig. 1. Methodology proposed to the Alkaline Solubility test.

In order to compare the obtained results was tested the clay deposits by Isothermal Calorimetry using R3 protocol published by Avet, F. in 2015 [6] to evaluate the possible correlation between both methods.

3 Results

The characterization of the raw materials shows that for all materials the total content of aluminium, silicon and iron is greater than 70% which ensures that characterized materials comply with the recommendations of the standards for pozzolanic materials [9]. The Yaguajay deposit presents a relatively high content of calcium oxide, which may be a probable indication of the presence of calcite or some other mineral like this one and the content of equivalent kaolin is in the range between 40% to La Loma deposit (LL) and 80% to Cayo Guam (Table 1).

Table 1. Characterization of the raw materials (XRF, TGA)

Clay	% SiO_2	% Al_2O_3	% Fe_2O_3	% CaO	Al_2O_3/SiO_2	% Keq
MK	51,61	40,44	1,52	0,32	0,78	89,20
CG	39,55	31,58	12,68	0,05	0,80	81,15
LS	50,88	25,23	16,97	0,28	0,50	57,85
LL	61,4	18,86	12,58	0,07	0,30	42,22
YG	46,58	20,06	9,61	2,94	0,43	60,14

Besides, the predominant clay minerals belong to the kaolinite's group, accompanied in most cases by the presence of quartz and mineral phases rich in iron as hematite and goethite. Finally, the presence of calcite as accompanying mineral in the YG deposit corresponds to the calcium oxide content determined by x-rays fluorescence (XRF) (Table 2).

Table 2. Characterization of the raw materials (XRD)

Clay	Origen	1:1 Clay Minerals	2:1 Clay Minerals	Others
CG	Weathering	Halloysite, kaolinite		Quartz and iron oxides and hydroxides (Fe_2O_3 and α-FeO(OH))
LS	Hydrothermal/ weathering	Halloysite, kaolinite	Vermiculite, montmorillonite	
LL	Hydrothermal/ weathering	Kaolinite, nacrite		
YG	Redeposited	Nacrite, kaolinite	Vermiculite	+ Calcite (5,25%)

Results obtained by the Alkaline Solubility test are shown in Fig. 2a, b and c. As it observed, CG presents higher values of soluble alkali, the calcination products of LS and LL clays with intermediate values of solubility and YG with the lowest values due to the presence of calcite as accompanying mineral who can causes formation of glassy phases rich in calcium and aluminium. This phenomenon could decrease the surface area of the pozzolan and also the pozzolanic reactivity of the calcination products.

In order to have an idea of how good it can be this method, we study the possible correlation between this one and R3 test (Fig. 3a, b and c). In the case of the correlation with soluble siliceous and total Siliceous and Aluminium is not good because they have a low correlation coefficient. However, the correlation with soluble Al presents a higher correlation coefficient ($R^2 = 0{,}9953$) because the total accumulated heat comes largely from the reactions of the aluminates. Besides, statistical results for the clay deposits. It is evident that there is not a significant dispersion of the obtained data for each clay deposit because in every case de variation coefficient (CV%) is lower than a 5% (Table 3).

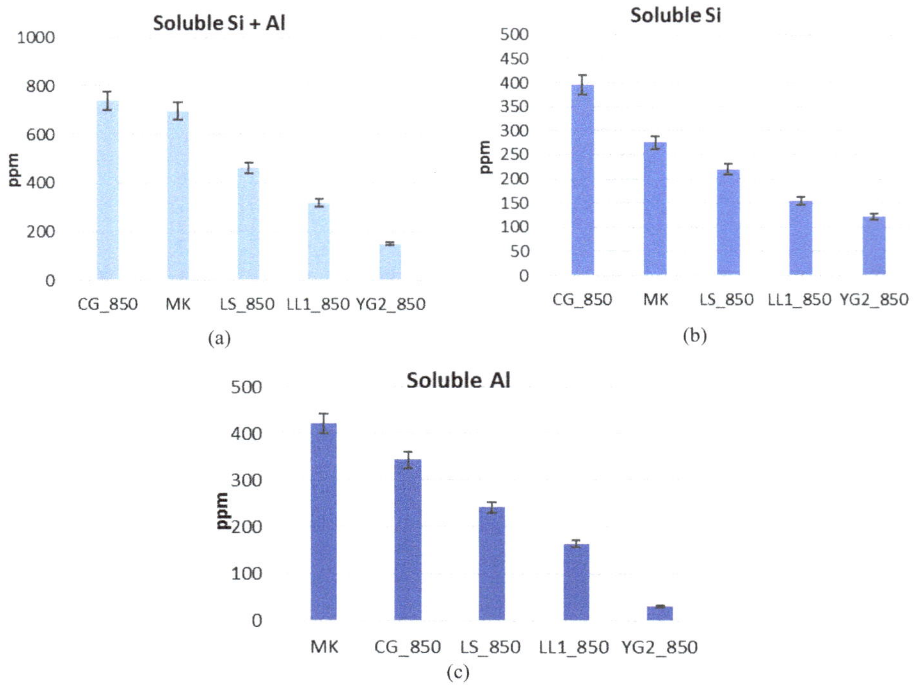

Fig. 2. Soluble Si, Si + Al and Al by alkaline solubility test

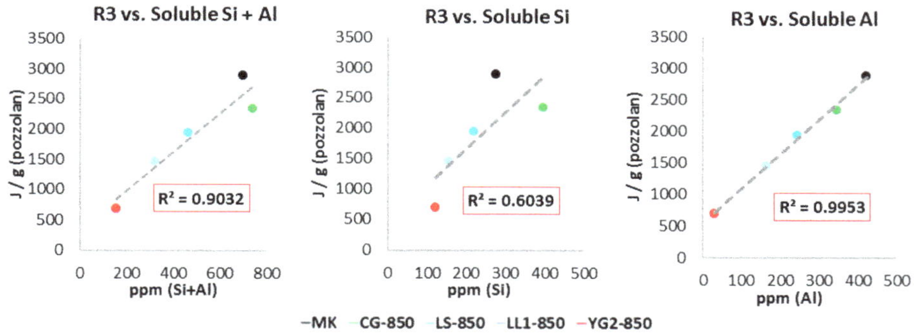

Fig. 3. Correlation between Alkaline solubility test and R3 test

Table 3. Statistical results for each deposit clay for aluminium soluble

Clay	Average (ppm Al)	Standard deviation	CV %
MK	235,43	0,1768	0,08
CG_850	189,13	3,4269	1,81
LS_850	132,23	0,2475	0,19
LL_850	90,50	2,5204	2,78
YG_850	14,75	0,0000	0,00

To study the viability of the proposed method, was decided to increase the number of samples and it is observed that the correlation coefficient is a little lower but is also accepted as a good correlation with aluminium content (Fig. 4).

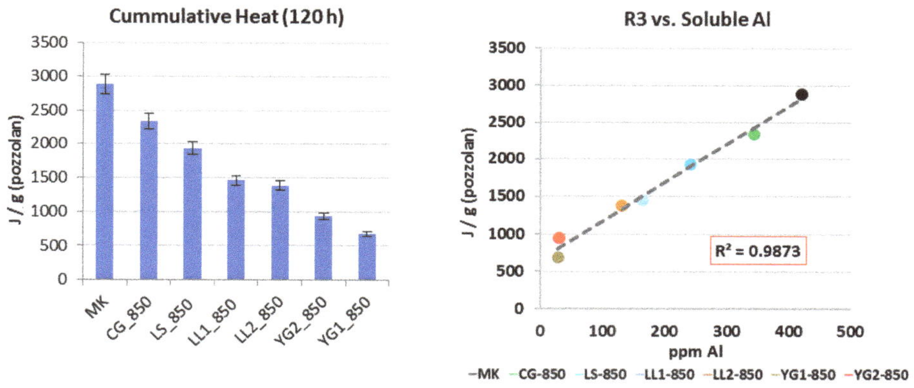

Fig. 4. Results of R3 and its possible correlation with alkaline solubility test increasing the number of samples.

4 Conclusions

Results of characterization of the raw materials shown that all mineral deposits obey with the recommendations of the standards for pozzolanic materials due the total content of SiO, AlO and FeO that is greater than 70%. Also, for kaolinitic clays, there is a direct and good correlation between soluble aluminium and the total heat released by R3 test with a determination coefficient of $R^2 = 0{,}9953$. Also, for this element (Al) there is not a remarkable dispersion of the obtained data for each clay deposit because in every case de variation coefficient (CV%) is lower than a 5%. For all these reasons, alkaline solubility test could be a viable test to evaluate the pozzolanic reactivity of calcined kaolinitic clays.

Acknowledgments. The authors would like to acknowledge the Instituto Superior Minero-Metalurgico of Moa, Cuba, for material support. In addition, the authors would like to thanks to the whole team of the "Low Carbon Cement" Project from CIDEM for all the advisory during this work.

References

1. Castillo, R., et al.: Activación de arcillas de bajo grado a altas temperaturas. Revista Ingeniería de Construcción **25**, 329–352 (2010)
2. Souza, P.S.L., Dal Molin, D.C.C.: Viability of using calcined clays, from industrial by-products, as pozzolans of high reactivity. Cement. Concr. Res. **35**(10), 1993–1998 (2005)
3. Murray, H.H.: Traditional and new applications for kaolin, smectite, and palygorskite: a general overview. Appl. Clay Sci. **17**(5–6), 207–221 (2000)
4. Snellings, R., Mertens, G., Elsen, J.: Supplementary Cementitious Materials. Rev. Mineral. Geochem. **74**, 211–278 (2012)
5. Cabrera, E.: Influencia de las características de los minerales arcillosos en la evaluación de la reactividad puzolánica de sus productos de calcinación mediante métodos químicos basados en el consumo de Ca(OH)2, in Chemistry Department, Central University of Las Villas, p. 43 (2013)
6. Avet, F., et al.: Development of a new rapid, relevant and reliable (R3) testing method to evaluate the pozzolanic reactivity of calcined clays. In: Scrivener, K., Favier, A. (eds.) Calcined Clays for Sustainable Concrete: Proceedings of the 1st International Conference on Calcined Clays for Sustainable Concrete. Springer, Dordrecht, pp. 539–544 (2015)
7. Steopoe, A.: Materiaux Contruct, **492,** 210–212 (1956)
8. Construcción, L.C.d.E.d.M.d., Publicación. El Laboratorio (1947)
9. ASTM International, Standard Specification for Coal Fly Ash and Raw or Calcined Natural Pozzolan for Use in Concrete, ASTM C618-03: West Conshohocken (2003)

Sustainability of Cuban Construction Supply Chain by Means of LC³ Cement: Case Studies in Villa Clara Province

Yudiesky Cancio Díaz[1](✉), Inocencio Raúl Sánchez Machado[1], José Fernando Martirena Hernández[1], and Guillaume Habert[2]

[1] Central University of Las Villas, Santa Clara, Cuba
yudieskycd@uclv.edu.cu
[2] Chair of Sustainable Construction, Swiss Federal Institute of Technology, ETH Zurich, Zurich, Switzerland

Abstract. This paper aims at assessing the implications of using Limestone Calcined Clay Cement (LC^3) in the Cuban construction sector from a sustainability viewpoint. By means of combining Life Cycle Assessment (LCA), economic cost analysis and eco-efficiency approach, two construction techniques have been compared, taking as a functional unit one square meter of usable area. In spite of the inverse correlation between economic efficiency and ecological impact for all confronted techniques, complementary indicators showed some trade-offs at social level. The economic-ecological efficiency of LC^3 potential use is linked to two different sources: the construction method and the cement type itself. Relevant decision-making considerations could support economic policy in the domain of construction industry in Cuba, if taking into consideration the eco-efficiency portfolio provided by this study. Authors conclude that no one construction method is superior *per se* from a sustainability viewpoint, but it rather requires major rethinking beyond economy and environment to embrace social indicators.

1 Introduction

Housing affordability has long been a pressing and challenging issue in Cuba. Alternative solutions need to be rooted under the sustainability umbrella, for at least three reasons: (i) building materials have to be cheaper than usually, (ii) global warming potential should be ameliorated and (iii) social indicators claim for appropriate balance with the economy and the environment. As in most developing countries, housing provision largely relies on concrete. Overall, the production of cement and concrete is estimated to account for around 5–8% of man-made CO_2 emissions [1]. Limestone Calcined Clay Cement (LC^3) is a technological innovation proposed by an international scientific team, led by EPFL-Lausanne, Switzerland. The new product lies on the domain of using supplementary cementitious materials (SCM) to partially replace clinker in the cement content. The cost-effectiveness and environmental advantages of LC^3 have extensively been published in [2–5]. This paper aims at assessing the sustainability of

F. Martirena et al. (eds.), *Calcined Clays for Sustainable Concrete*, RILEM Bookseries 16,
https://doi.org/10.1007/978-94-024-1207-9_17

using LC3 within Cuban construction supply chain, by taking into consideration a case study approach. Two storey-buildings constructed in the city of Santa Clara by employing Grand Panel and Concrete block techniques, respectively, are analysed.

2 Methodology

The used conducted case studies followed a combination of methods. The methodology Life Cycle Assessment (LCA) was employed to determine the environmental impact of LC3 in alternative construction techniques. LCA is well documented in the international standard 14040 and 14044, dated back to 2006 [6]. Figure 1 shows the system boundaries of the LCA conducted in this research (highlighted in dotted lines), which refers to the material phase of buildings under analysis. The functional unit employed for comparability purposes is one squared meter of usable floor area. Two case studies were conducted: (1) first one is a two-storey building with Grand Panel technique and (2) second, two-storey building constructed using traditional concrete block method. The economic dimension of sustainability was covered by means of an economic cost analysis. Some social indicators such as employment have been also discussed within the sustainability approach.

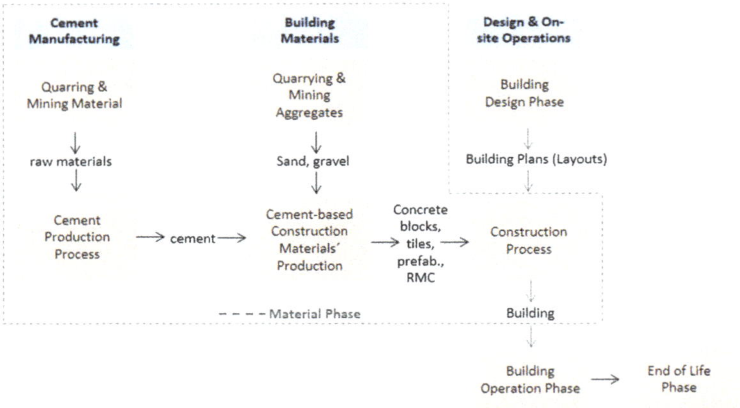

Fig. 1. System boundaries for LCA and economic cost analysis

Table 1 summarizes the data inventory required for LCA as well as for economic assessment. The usable floor area was 168.64 m^2 for Grand Panel building and 161.42 m^2 for concrete blocks building. The column referred to cement entails the amount of cement needed to produce the building materials presented in column 1, whose quantities are shown in the columns labelled "amount".

The environmental impact factor of OPC, PPC and LC3 cements as well as its productions costs in order to determine the impacts at the level of building construction, were taken from previous research of LC3 Project team. Detailed figures are shown in [7], and authors elaborated on its foundations. Economic and ecological impacts are combined into "an eco-efficiency portfolio", as proposed by Schaltegger [8].

Table 1. Data inventory

Material	Unit	Gran Panel building		Concrete blocks building	
		Amount	Cement	Amount	Cement
Ready-Mix Concrete	m^3	24.9	11.95	40.6	19.49
Prefabricated Concrete	m^3	52	20.02	18.46	7.11
Hollow concrete blocks	u	770	1.32	4460	7.67
Mortar	m^3	20.97	6	59.33	16.97
Total	t		39.29		51.23

3 Results

Main results are intended to show the economic and environmental implications of using LC3 in different construction techniques schemes, given the case studies conducted in Santa Clara, Cuba. As can be seen from Fig. 2, the environmental impact of Grand Panel technology is 22% smaller than traditional concrete block method. This green contribution is associated with the shorter amount of material consumed in Grand Panel compared to block's technique. Replacing traditional cements (OPC+PPC) by LC3 leads to lessen the CO_2 emissions along the life cycle of building materials (within system boundaries specified above), in about 28% for Grand Panel buildings and 30% for concrete ones.

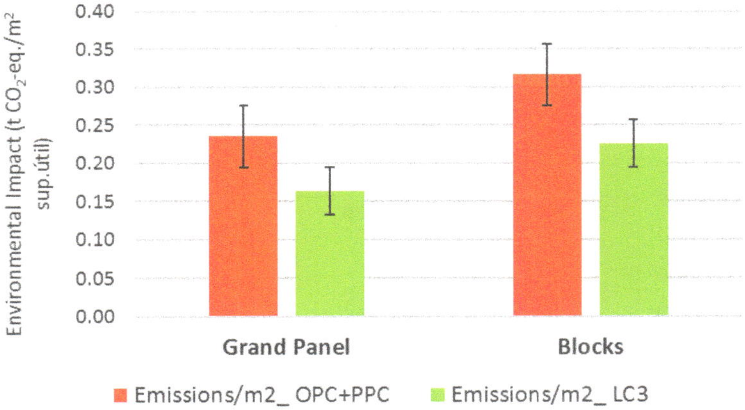

Fig. 2. Environmental impact of cements and construction technologies choices

Figure 3 sheds light on the economic contribution of options under appraisal. Concrete block's technique appears to be 25% less cost-effective than Grand Panel. The introduction of LC^3 in Cuban construction sector might contribute to cost savings of approximately 12% (in concrete blocks houses) and 13% (in Grand Panel ones).

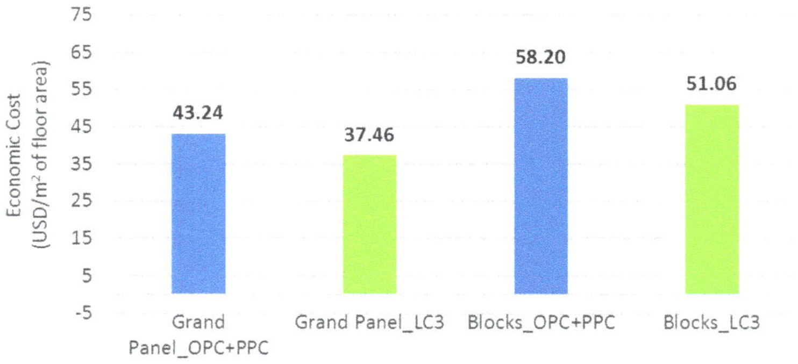

Fig. 3. Economic cost comparison amongst alternative construction techniques and cement types

Figure 4 shows an eco-efficiency portfolio for the conducted case studies. The combination of cost effectiveness and ecological contribution of LC^3 contributes positioning the new cement as an eco-efficient material. It appears to be a positive correlation between economic and environmental achievements for both cement types and construction techniques. This would push some policy decisions in Cuban construction sector if LC^3 is to be implemented at large scale. However, some trade-offs come out when taking into consideration additional indicators. For instance, concrete block construction method is labour-intensive, thus, contributing to employment enhancement in the country. However, Grand Panel, which uses less manpower, is a fast construction process due to the use of prefabricated elements instead of placing the blocks with mortar one by one. These issues provide inputs for policymakers in order to decide which method is more suitable. In authors' viewpoint, despite being proved the eco-efficiency

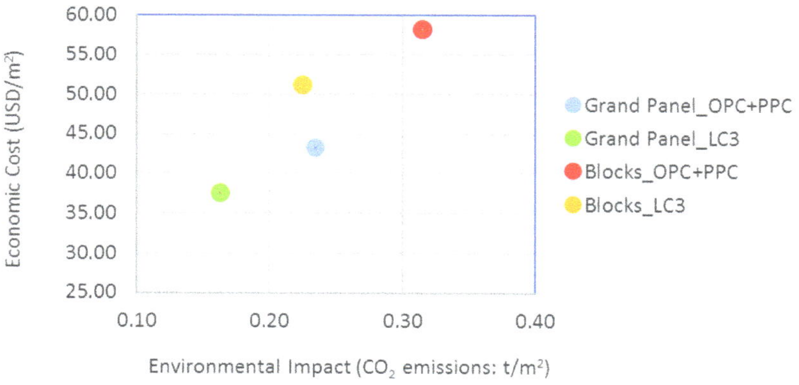

Fig. 4. Eco-efficiency portfolio

profile of Grand Panel versus Blocks, the choices must be tailored in a case-by-case basis, depending on a large number of factors that are underlying.

4 Conclusions

The main purpose of this paper was to assess the implications of using Limestone Calcined Clay Cement (LC^3) in the Cuban construction sector from a sustainability viewpoint. The conducted research has found that implementing housing programs by means of Grand Panel construction technique is rather beneficial than concrete block technique from an economic and environmental viewpoint. However, trade-offs are latent when expanding the system analysis to further consider the social dimension of sustainability. Therefore, decision-making within Cuban construction sector with regards to construction techniques requires a holistic approach which must rely on the specific conditions on a case-by-case basis, rather than concluding that one building method is strictly better than the other one. A sustainability approach must be rooted on an integrative system thinking which should entail discussions beyond the economy and the environment. Moreover, LC^3 has the potentials for affording major sustainability goals even in the wort case scenario.

Acknowledgements. Authors would like to thank the Cuban cement industry and Construction sector for their invaluable support to LC^3 project. We also highly appreciate the Swiss Agency for Development and Cooperation for the financial support to LC^3 project.

References

1. Scrivener, K.: Options for the future of cement. Indian Concr. J. **88**, 11–21 (2014)
2. Vizcaíno, L., Sánchez, S., Martirena, F., Scrivener, K.: Industrial trial to produce a low clinker, low carbon cement. Mater. Construcción. **65** (2015)
3. Sánchez, S., Cancio, Y., Martirena, F., Habert, G.: Assessment of sustainability of low carbon cement in Cuba. Cement pilot production and prospective case. In: Scrivener, K., Favier, A. (eds.) Calcined Clays for Sustainable Concrete, pp. 189–194. Springer, Dordrecht (2015)
4. Cancio, Y., Sánchez, S., Martirena, F., Sánchez, I.R., Scrivener, K., Habert, G.: Economic and ecological assessment of Cuban housing solutions using alternative cement. In: Schlueter, A., Habert, G. (eds.) Sustainable Built Environment Regional Conference, pp. 292–297. ETH Zurich, Zurich (2016)
5. Sánchez, S., Favier, A., Rosa, E.: Assessing the environmental and economic potential of limestone calcined clay cement in Cuba. J. Clean. Prod. **124**, 361–369 (2016)
6. ISO: ISO 14040: Environmental management. Life Cycle Assessment. Principles and Framework, Geneva (2006)
7. Cancio, Y., Sánchez, S., Heierly, U., Favier, A., Sánchez, I.R., Scrivener, K., Martirena, F., Habert, G.: Limestone calcined clay cement as a low carbon solution to meet expanding cement demand in emerging economies. Dev. Eng. **2**, 82–91 (2017)
8. Schaltegger, S.: Accounting for eco-efficiency. Environ. Manag. Pract. **1**, 272–287 (1998)

Degradation of Calcined Clay-Limestone Cementitious Composites Under Sulfate Attack

Cheng Yu[1(✉)], Peng Yuan[2], Xin Yu[1], and Jiaping Liu[2]

[1] State Key Laboratory of High Performance Civil Engineering Materials, Jiangsu Sobute New Materials Co., Ltd., 211103 Nanjing, China
[2] School of Materials Science and Engineering, Southeast University, 211189 Nanjing, China

Abstract. Calcined clay and limestone composite cementitious material system is a newly proposed low-carbon cement, which can effectively reduce energy consumption and carbon emissions of the traditional cement industry without changing the basic mechanical properties of cement-based materials. In this study, the degradation process of mortar samples of limestone and calcined clay cementitious material under sulfate attack was systematically studied by both macroscopic and microscopic analysis. The results show that compared with pure Portland cement, the addition of calcined clay and limestone can significantly reduce the expansion rate, loss of dynamic modulus and mass loss of mortar specimens under sulfate attack. The addition of calcined clay and limestone will refine the pore size distribution of mortar specimens, then inhibiting the diffusion of sulfate and formation of corrosive products, therefore leading to a significant improvement of the sulfate resistance.

1 Introduction

As a mature and effective solution to realize low-carbon development of cement industry, supplementary cementitious materials (SCMs), mainly fly ash and ground blast furnace slag, are extensively applied in cement and concrete to reduce the energy consumption and carbon emissions in the process of cement production and application. However, due to their natural disadvantages like low production, regional disparity distribution, fluctuations in performance and so on, traditional SCMs cannot fully meet the demand of high performance and low carbon emission of cement and concrete materials. Calcined clay and limestone composite cementitious material that proposed by Scrivener has been regarded as a new and potential low-carbon cement solution, owing to its advantages of wide availability of raw materials, low carbon emissions in production, similar production process to Portland cement and so on [1–3].

When clay minerals were calcined in the range of 600–900 °C, due to the dehydroxylation, the lattice structure of kaolin component was damaged to form calcined clay [4, 5]. Metakaolin is the main component of calcined clay which has a high pozzolanic reactivity [6, 7]. Metakaolin can react with $Ca(OH)_2$ to form calcium silicoaluminate hydrate gel (C-A-S-H) in cement paste, which could optimize the pore structure, and then improve the mechanical properties and durability [8, 9]. Limestone is beneficial

F. Martirena et al. (eds.), *Calcined Clays for Sustainable Concrete*, RILEM Bookseries 16,
https://doi.org/10.1007/978-94-024-1207-9_18

to the early strength development of cement when it is added at a low proportion [10, 11]. A suitable amount of limestone plays a role of "filling effect" and "nucleating effect" in accelerating the early stage of hydration [12, 13]. Besides, limestone can react with C_3A to form carbon aluminate hydrate. When calcined clay and limestone were added simultaneously, activated aluminium oxide in clay would react with limestone and calcium hydroxide to form single carbon aluminates and half carbon calcium aluminates [14–16]. It would refine the pore size and the dosage of cement clinker would be decreased without influencing mechanical properties of cement based material.

Some research has shown that calcined clay and limestone composite cementitious material system can reduce the diffusion coefficient of chloride ion of concrete [17]. However, the main hydration product of LC^3 - calcium carboaluminate also has the potential to react with sulfate to form ettringite. A large amount of aluminum phase in calcined clay has been introduced, which could also react with sulfate ions. Therefore, the sulfate resistance of calcined clay-limestone cementitious composites needs to be investigated further. In this study, the expansion rate and dynamic modulus of mortar exposed to sodium sulfate solution were investigated. Corrosion products were analysed using the XRD semi-quantitative method.

2 Materials and Experimental Methods

2.1 Materials

In this study, P·I 42.5 Portland cement (PC) and China ISO Standard Sand are used. Calcined clay (CC) is provided by India Low Carbon Cement Project team. Limestone (LS), Fly ash (FA), slag (SL) are provided by Jiangsu Sobute New Materials Co., Ltd. The chemical composition of cement, calcined clay, limestone, fly ash and slag is shown in Table 1.

Table 1. Chemical composition of the main components given by XRF

Mass fraction/%	SiO_2	Al_2O_3	Fe_2O_3	CaO	MgO	SO_3	Na_2O	K_2O	TiO_2
Portland cement	20.3	5.07	3.10	62.8	3.53	3.54	0.09	1.03	0.237
Calcined clay	51.7	42.1	3.02	0.205	–	0.056	0.113	0.231	2.22
Limestone	1.22	1.18	0.367	96.4	0.121	0.048	–	0.107	–
Fly ash	50.2	28.6	6.21	7.31	1.30	0.969	0.822	1.42	1.36
Slag	30.9	15.9	.0281	41.5	6.86	2.58	0.316	0.335	0.633

2.2 Mix Proportions

In calcined clay and limestone composite cementitious material system, the mass ratio of calcined clay and limestone is fixed as 2:1. The replacement ratio of calcined clay and limestone of Portland cement is 15% (LC15), 30% (LC30) and 45% (LC45). Pure Portland cement (PC) and a traditional SCMs replacement system (SF, 10% fly ash and 20% slag) are also applied as reference group. The water to binder ratio of mortar specimens 0.6. The ratio of cementitious material to sand is fixed to 1:3. Mortar sample with a size of 40 mm × 40 mm × 160 mm were cast and demoulded after 24 h. The compressive strength of mortar was tested After curing for 28d in standard condition. Table 2 gives the details of mix proportions.

Table 2. Mix proportions of mortars

Specimen	W/B	PC/g	FA/g	SL/g	CC/g	LS/g	Water/g	Sand/g	28d Strength/MPa
LC15	0.6	382.5	–	–	45	22.5	270	1350	38.7
LC30		315	–	–	90	45	270	1350	39.7
LC45		247.5	–	–	135	67.5	270	1350	39.1
SF		315	45	90	–	–	270	1350	40.4
P		450	–	–	–	–	270	1350	35.3

2.3 Sulfate Immersion Test

Copper crews were pre-embedded at both ends of mortar specimens used for sulfate immersion test. The original laitance on the surface of specimens were removed to accelerate ion corrosion rate and control the experimental error. Afterwards, specimens were immersed in 5wt% sodium sulfate solution. The volume ratio of solution to specimens was 10 and the temperature was around 20 °C. Sulfate solution was renewed every 30d. The dynamic modulus of mortar specimens at different immersion ages was tested by non-metal ultrasonic detector according to GBT 50082-2009 on the vertical direction of samples. The expansion of specimens was measured by JD18 length measuring instrument and values were accurate to 0.001 mm. The expansion rate of mortars was calculated by the following formula (1) with average value of three specimens.

$$E_T = L_T - L_0/140 \times 100\% \tag{1}$$

In the formula, E_T is the expansion rate of specimens at an immersion age of T (%), L_T the length of specimens at an immersion age of T (mm) and L_0 the initial length of specimens (mm).

After immersed in solution for 90d and 180d, a slice with 2 mm wide was cut from the surface of mortar specimens. Then the slice was grounded into powders and passed through 200 mesh sieve. Phase analysis was performed by Bruker D8 Advance X-Ray Diffractometer. The samples were scanned on a rotating stage between 5 and 20° using a step size of 0.02° with time per step of 0.5 s.

3 Results

3.1 Relative Dynamic Elastic Modulus

The relative dynamic modulus change of mortar specimens is shown in Fig. 1. The relative dynamic modulus of Portland cement mortar has a significant decrease after 90d, which indicates that the internal microstructure specimens have been damaged under sulfate attack. While the relative dynamic modulus of mortar specimens LC samples shows a slight increase after a 180-day exposure. It can be inferred that the addition of calcined clay and limestone refine the pore size distribution of mortars, which could prevent the penetration of sulfate ions into mortars. Therefore, the deterioration that caused by sulfate attack occurs only at the surface of LC Mortars. And the internal structure remains unaltered after a 180-day exposure.

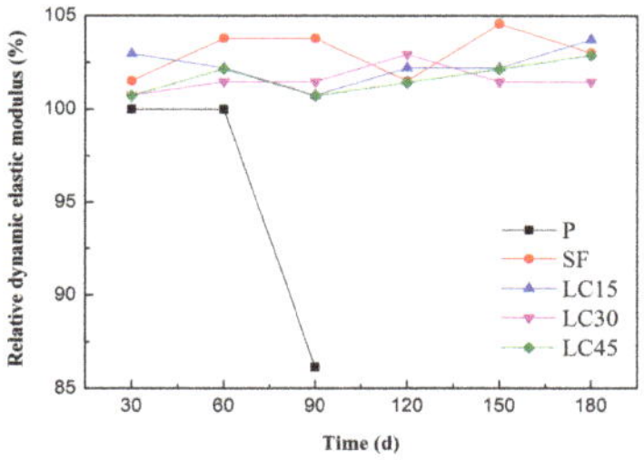

Fig. 1. The relative dynamic modulus of mortars exposed to sulfate solution

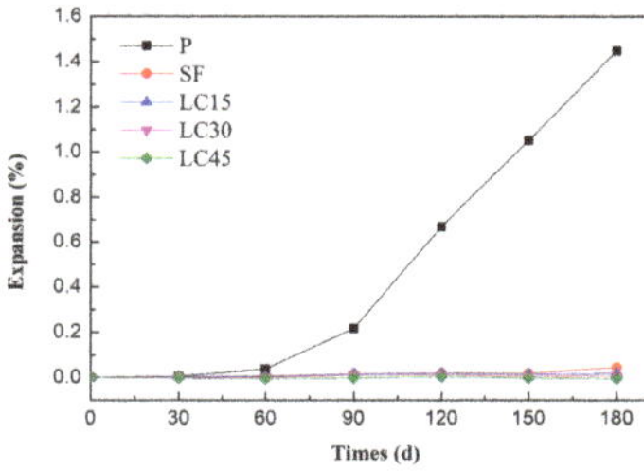

Fig. 2. Expansion of mortars exposed to sulfate solution

3.2 Expansion

Figure 2 shows the expansion rate of mortar specimens under sulfate attack. The expansion rate of P mortar specimen shows a steady increase after a 30-day exposure, and reaches 1.448% at 180d. The SF mortar also starts to expand after 180d. While the addition of calcined clay and limestone significantly reduce the expansion rate of mortars. The expansion rate of LC15, LC30 and LC45 are 0.016%, 0.007% and 0.001% respectively.

3.3 XRD Analysis

The relative amount of main hydration product and corrosion products of mortars can be analyzed by XRD. The characteristic diffraction peak (2θ) that corresponds to monocarboaluminate (Mc), hemicarboaluminate (Hc), portlandite (CH), ettringite (Ettr.) and gypsum (Gyp.) are 11.67°, 10.78°, 18.01°, 9.09° and 11.59° respectively. Figure 3 is the XRD pattern of center part of mortars, which are not affected by the sulfate ions yet, after a 90-day exposure. From the patterns, it is clear that the hydration product of LC mortars is different compared with P and SF mortars. Besides ettringite and protlandite, monocarboaluminate, hemicarboaluminate appears after 90 days. Moreover, the amount Mc and Hc are increased with increasing amount of calcined clay and limestone. Figure 4 shows the XRD pattern of surface part of mortars after a 90-day exposure. Compared with patterns of center part, it is obvious that more ettringite and gypsum have been formed in the surface part of mortar specimens after a 90-day exposure in sulfate solution. The monosulfate and portlandite are reacting with incoming sulfate to form ettringite and gypsum, and the amount of ettringite and gypsum formed in LC mortars is lower than in P and SF samples. This indicates that the addition of calcined clay and limestone could mitigate the forming of corrosion products by sulfate ingress. It is well known that the expansion of mortars under sulfate attack is mainly due to the crystallization of ettringite. LC mortars shows less ettringite forming, thereby leading to a reduced expansion rate and macro degradation. This explains why LC mortars shows lower expansion rate and dynamic modulus loss than P and SF samples.

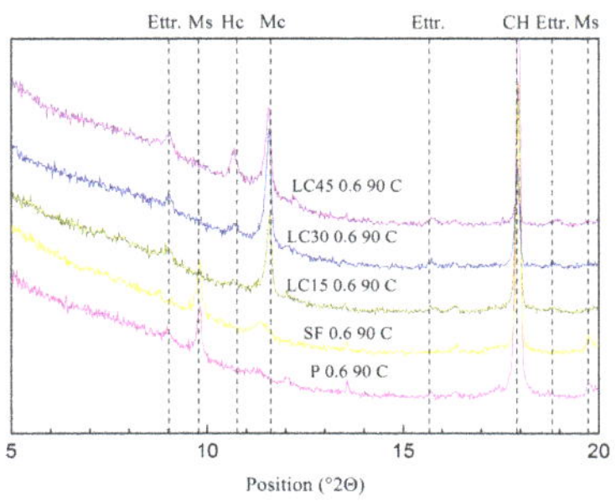

Fig. 3. XRD patterns for the center of the mortars at 90 days

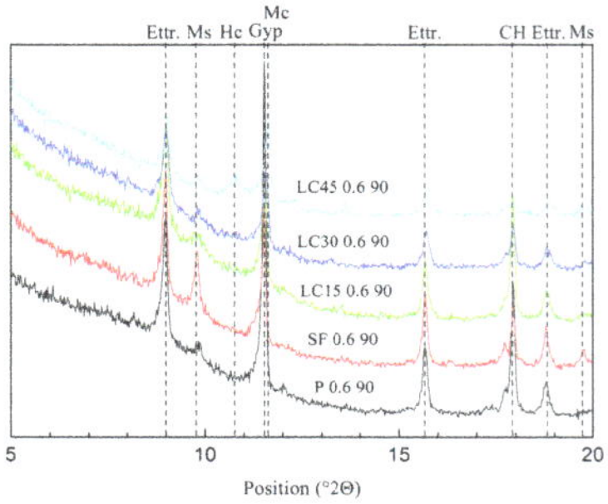

Fig. 4. XRD patterns for the surface of the mortars at 90 days

Figures 5 and 6 show the XRD pattern of center and surface part of mortars after 180d exposure respectively. It can be seen that the amount of ettringite and gypsum increases with the exposure time. After 180 days, nearly all the monosulfate and portlandite have been consumed to form the corrosion product. Ettringite is also forming in the center of P mortars, which indicates that the center has already been affected by the incoming sulfate through cracks. Moreover, in the surface of LC mortars, the amount of Hc and Mc are reduced compare to center samples. This infers that the Hc and Mc can also react with sulfate ions to form ettringite.

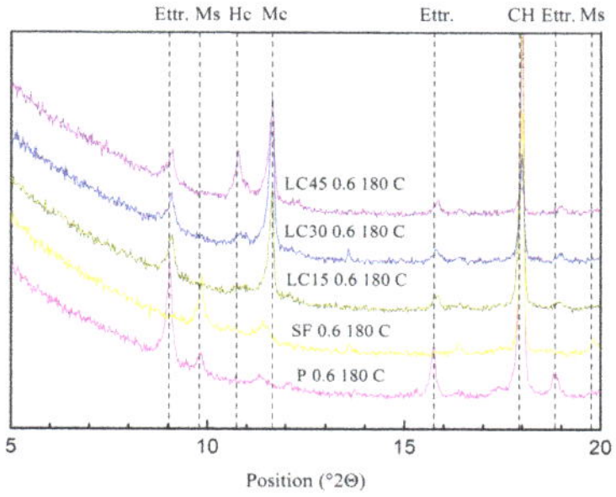

Fig. 5. XRD patterns for the center of mortars after 180 days

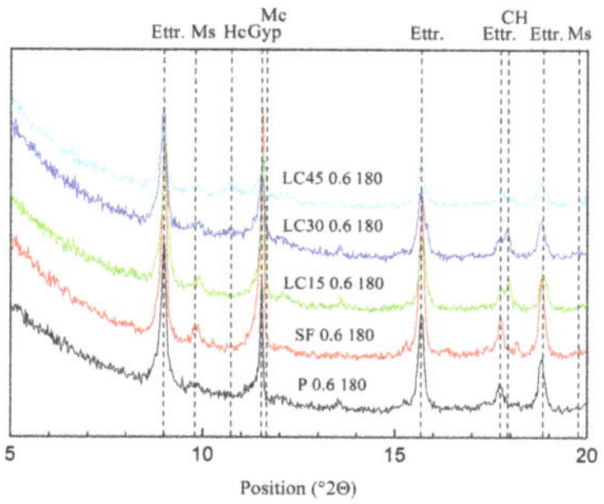

Fig. 6. XRD patterns for the surface of mortars after 180 days

As can been seen from all the patterns, the formation of ettringite and gypsum are significantly reduced by adding calcined clay and limestone. Due to the high pozzolanic reactivity, calcined clay can react with limestone to form mono- and hemi-carboaluminate, leading to a refined pore structure [18]. The optimization of microstructure could strongly mitigate the penetration of corrosive ions, like chloride ion [16] and sulfate ions, which leads to an improved sulfate resistance.

4 Conclusions

1. Partial replacement of calcined clay and limestone to Portland cement can reduce expansion rate and dynamic modulus loss of mortars under sulfate attack for 180 days.
2. Reactive alumina in calcined clay can react with limestone, calcium hydroxide to form mono- and hemi-carboaluminate to get a refined pore structure. Although mono or hemi-carboaluminates can react with sulfate to form ettringite, the improvement of impermeability ensures LC mortars an improved sulfate resistance.

References

1. Love, C.A., Richardson, I.G., Brough, A.R.: Composition and structure of C–S–H in white Portland cement–20% metakaolin pastes hydrated at 25 °C. Cem. Concr. Res. **37**(2), 109–117 (2007)
2. Jiang, G., Rong, Z., Sun, W.: Effects of metakaolin on mechanical properties, pore structure and hydration heat of mortars at 0.17 w/b ratio. Constr. Build. Mater. **93**, 564–572 (2015)
3. Li, Q., Geng, H., Shui, Z., et al.: Effect of metakaolin addition and seawater mixing on the properties and hydration of concrete. Appl. Clay Sci. **115**, 51–60 (2015)
4. Sayanam, R.A., Kalsotra, A.K., Mehta, S.K., et al.: Studies on thermal transformations and pozzolanic activities of clay from Jammu region (India). J. Therm. Anal. Calorim. **35**(1), 99–106 (1989)
5. Ambroise, J., Murat, M., Pera, J.: Investigations on synthetic binders obtained by middle-temperature thermal dissociation of clay minerals. Silic. Indus. **51**(7–8), 99–107 (1986)

6. Sabir, B.B., Wild, S., Bai, J.: Metakaolin and calcined clays as pozzolans for concrete: a review. Cement Concr. Compos. **23**(6), 441–454 (2001)
7. Fernandez, R., Martirena, F., Scrivener, K.L.: The origin of the pozzolanic activity of calcined clay minerals: a comparison between kaolinite, illite and montmorillonite. Cem. Concr. Res. **41**(1), 113–122 (2011)
8. Siddique, R., Klaus, J.: Influence of metakaolin on the properties of mortar and concrete: a review. Appl. Clay Sci. **43**(3), 392–400 (2009)
9. Fŕias, M., Cabrera, J.: Pore size distribution and degree of hydration of metakaolin–cement pastes. Cem. Concr. Res. **30**(4), 561–569 (2000)
10. Tsivilis, S., Chaniotakis, E., Badogiannis, E., et al.: A study on the parameters affecting the properties of Portland limestone cements. Cement Concr. Compos. **21**(2), 107–116 (1999)
11. Voglis, N., Kakali, G., Chaniotakis, E., et al.: Portland-limestone cements. Their properties and hydration compared to those of other composite cements. Cement Concr. Compos. **27**(2), 191–196 (2005)
12. Oey, T., Kumar, A., Bullard, J.W., et al.: The filler effect: the influence of filler content and surface area on cementitious reaction rates. J. Am. Ceram. Soc. **96**(6), 1978–1990 (2013)
13. Lothenbach, B., Saout, G.L., Gallucci, E., et al.: Influence of limestone on the hydration of Portland cements. Cem. Concr. Res. **38**(6), 848–860 (2008)
14. Ingram, K., Poslusny, M., Daugherty, K., et al.: Carboaluminate reactions as influenced by limestone additions (1990)
15. Matschei, T., Lothenbach, B., Glasser, F.P.: The role of calcium carbonate in cement hydration. Cem. Concr. Res. **37**(4), 551–558 (2007)
16. Vance, K., Aguayo, M., Oey, T., et al.: Hydration and strength development in ternary portland cement blends containing limestone and fly ash or metakaolin. Cement Concr. Compos. **39**(5), 93–103 (2013)
17. Antoni, M.: Investigation of cement substitution by blends of calcined clays and limestone. Thèse EPFL n°6339, 2014–2013
18. Antoni, M., Rossen, J., Martirena, F., et al.: Cement substitution by a combination of metakaolin and limestone. Cem. Concr. Res. **42**(12), 1579–1589 (2012)

Sulfate and Alkali-Silica Performance of Blended Cements Containing Illitic Calcined Clays

Gisela Cordoba[1(✉)], Agustin Rossetti[2], Dario Falcone[2], and E.F. Irassar[1]

[1] Facultad de Ingeniería, CIFICEN (UNCPBA-CICPBA-CONICET), B7400JWI Olavarría, Argentina
[2] LEMIT, CIC de la Provincia de Buenos Aires, Av 52 e/121y122, B1900AYB La Plata, Argentina

Abstract. Studies of illitic calcined clays are less developed than that corresponding to kaolinitic clays, but illite is one of the more abundant clayed minerals of the earth's crust, as occurs in the Center of the Buenos Aires Province (Argentina) where the largest cement factories are located. Illite clays develop pozzolanic properties when they are thermally treated at 950 °C, causing dehydroxilation and collapse of structure to form a metastable or amorphous aluminosilicate. Illitic calcined clays don't present a significant water demand and the compressive strength of blended cements attains to the corresponding to portland cement at 90 days. It is characterized as slow pozzolana. Illite incorporates certain proportion of reactive alumina and high proportion of alkalis, modifying the pore structure. From durability point of view, the incorporation of illite can affect the sulfate resistance of portland cements or the alkali-silica reaction (ASR).

The aim of this paper is to study the behavior of two different illite calcined clay blended cements against chemical attack, like sulfate attack and harmful alkali silica reaction, using the test based on the ASTM C 1012 and ASTM C 441, respectively. For sulfate performance, illite calcined clays was blended with a low C_3A in 20% and 40% of weight replacement and a very high C_3A cement (white), using a 30% of weight replacement; while a low (Na_2O_{eq} < 0.5) and high alkali (Na_2O_{eq} = 1.03) cements were used in the ASR-test.

After six months, the low and very high C_3A cements of both illitic clays shows low expansion in sulfate media (<0.05%) for blended cement without water demand. The ASR-expansion results show that illitic calcined clays reduce considerably the expansion of high alkali cements and it is not harmful to low alkali cement, but long test time results will be conclusive.

1 Introduction

Illite is one of the more abundant clayed minerals of the earth's crust coming from the alteration of feldspars and micas of rocks due to the weathering process. Illite clays develop pozzolanic properties when clays are thermally treated at 950 °C [1] causing

F. Martirena et al. (eds.), *Calcined Clays for Sustainable Concrete*, RILEM Bookseries 16,
https://doi.org/10.1007/978-94-024-1207-9_19

dehydroxilation and collapse of structure to form a metastable or amorphous aluminosilicate [2].

Partly substitution of alkali-rich portland cements by calcined clays improves the resistance of the respective concrete against sulfate attack and harmful ASR [3]. It is attributed to the pozzolanic reactivity that consumes the CH to produce C-S-H and C-A-S-H causing a denser microstructure due to the pore and grain size refinements [3, 4]. To controlling ASR, metakaolinite has a very good performance suppressing the expansion. It is attributed to alteration of C-S-H structure, lower pH in pore solution and Al absorbed on the surface of reactive silica that slow down the dissolution rate [3, 5]. High resistance to sulfate attack is normally associated with a segmented pore structure and low level of CH. In addition to the role of CH, the reduction of alumina available to suppress ettringite formation is a main factor to improve the resistance to sulfate attack in Na_2SO_4 solutions [6].

It is largely documented the durability improvements caused by kaolinitic calcined clays used as replacement of PC, but there is limited information about the performance of other type of clays used as supplementary cementitious materials (SCMs). The aim of this paper is to study the behavior of two illitic calcined clay from the province of Buenos Aires in blended cements against sulfate attack and alkali silica reaction.

2 Materials and Methods

Two different illitic clay-stones from quarries near to Olavarria, Buenos Aires Province (Argentine) were reduced to 5 mm particles and fired in a muffle oven by heating at 10.5 °C/min up to 950 °C and maintaining the maximum temperature during 90 min. Then, samples were cooling down into the oven. Calcined clays were grinding in laboratory ball mill to obtain a d_{90} parameter in the PSD curve lower than 10 μm. The chemical composition determined by XRF is reported in Table 1. Both calcined clays meet the chemical requirements for Class N pozzolan (ASTM C 618): S+A+F > 70%; SO_3 < 4% and LOI < 10%. Calcined clays are identified by their color: Red (R) and Orange (O). XRD analysis reveals low intensity peaks of dehydroxilated illite in both clays and the associated minerals are quartz and hematite for R calcined clays and, quartz, hematite, oligoclase and spinel for O-Calcined clay. The density (ASTM C 188), retained on 75 and 45 μm sieves (ASTM D 422 and C 618), the Blaine specific surface (ASTM C 204) and the particle size distribution (PSD) determined using the laser granulometer (Malvern Mastersizer 2000) are reported in Table 2.

Table 1. Chemical composition and loss on ignition of calcined clays and cements, %

Clay/Cement	CaO	SiO_2	Al_2O_3	Fe_2O_3	MgO	SO_3	K_2O	Na_2O	TiO_2	P_2O_5	LOI
R	0.33	66.30	16.28	9.23	1.46	<0.01	5.60	0.08	0.76	0.09	0.6
O	1.13	63.43	18.32	7.89	2.71	0.04	4.29	1.52	0.90	0.06	0.2
SRPC	60.08	20.13	4.19	4.44	0.84	2.47	1.03	0.21	–	–	2.5
WPC	49.90	19.31	5.97	0.30	2.20	3.13	–	–	–	–	6.0
HAPC	63.06	20.74	3.64	1.99	3.63	2.68	1.49	0.05	–	–	2.4
LAPC	61.32	23.53	2.90	2.97	3.50	1.73	0.35	0.14	–	0.14	2.2

Table 2. Physical characteristic of calcined clays and cements

Property/Material		R	O	SRPC	WPC	HAPC	LAPC
Density		2.63	2.65	3.13	2.95	3.07	3.14
Retained on sieve, %	75 μm	0.98	4.02	2.3	1.8	3.20	3.50
	45 μm	4.73	5.68	13.56	12.3	13.79	8.07
Particle size distribution	d_{10}, μm	1.62	1.29	3.22	2.57	3.08	3.29
	d_{50}, μm	8.76	7.34	22.37	12.58	23.99	21.54
	d_{90}, μm	33.65	36.95	58.44	32.77	81.88	60.57
Specific surface Blaine, m^2/kg		552	724	336	432	276	331

For this experimental design, four portland cements were used: A low C_3A portland cement (SRPC); a white portland cement (WPC); a high (HAPC) and a low (LAPC) alkali content portland cements. The chemical and physical properties are also reported in Tables 1 and 2.

The SRPC was used as plain PC to determine the setting time (ASTM C 191) and the amount of water required for normal consistency (ASTM C 187) of blended cements containing 20% and 40% by mass of calcined clays. The mortar flow was assessed according to ASTM C 230. At 2, 7, 28 and 90 days, the compressive strength of blended cement was assessed on three cubes (ASTM C 109).

The **sulfate performance** was studied using the expansion of mortar bars (ASTM C 1012) with two different Portland cements: SRPC (C_3A = 3.8%) and WPC (C3A = 15.31% and C4AF = 0.91%). For SRPC, calcined clays were used as 20 and 40% replacement by weight and for WPC as 30% by weight. The flow of SRPC mortar (w/c = 0.485 and cement-graded sand = 1:2.75) is determined and the water of blended cement is adjusted to obtain the same flow in SRPC-calcined clay mortars. For WPC, constant w/cm was used in blended cements and the mortar flow was adjusted using superplasticizer. Test bars and cubes were molded according to ASTM C 157 and ASTM C 109 procedures, respectively; demolded and they were cured in lime-water at 20 °C. After that mortar compressive strength reached to 20 MPa, test bars were immersed in the Na_2SO_4 solution (0.352 mol/l) at 20 °C with a periodically pH-control. Measurement of length was made up to 6 months and it will be extending up to 18 months. Reported expansion is the average of six specimens. According to ASTM C 1157, to consider the blended cement as sulfate resistant (HS), the expansion should not exceed 0.05% at 6 months or 0.10% at 12 months.

The effectiveness of calcined clays to prevent the expansion due to the **alkali-silica reaction (ASR)** was tested using the ASTM C 441 procedure. Test is based on the expansion developed in mortar bars made with HAPC (1.03% Na_2O_{eq}) and 25% of calcined clay using Pyrex® glass aggregate and stored in container at 38 °C according to ASTM C 227. According to ASTM C 441, the pozzolan effectiveness is measured as the reduction of mortar expansion with calcined clay related to the expansion of mortar bars with HAPC at 14 days. To compare the effectiveness of calcined clay to mitigate the ASR a LAPC (0.37% Na_2O_{eq}) was tested. As suggest the ASTM C 311, the combination of LAPC with calcined clays will not cause an increase of expansion due to their high alkali content (5.65 and 5.29% Na_2O_{eq} for R and O, respectively). For blended cements,

ASTM 1157 standard establishes that the expansion should not exceed the limit of 0.02% at 14 days and 0.06% at 56 days to be considered as Low reactivity with alkali-reactive aggregates (option R).

3 Results

3.1 Setting Time and Water Demand

The time of setting and the water requirement for normal consistency are presented in Table 3. About setting time, one of the illitic calcined clay (N) presents stimulation in the beginning of setting, in both replacement percentages. On the other hand, all blended cements delay the final setting time compared with control cement. Water demand of SRPC was similar to SRPC+20%O and SRPC+40%O pastes, and it was slightly high (~7%) for SRPC+20%R and SRPC+40%R pastes.

Table 3. Setting time and water demand.

Blended cement		SRPC	SRPC +20%R	SRPC +40%R	SRPC +20%O	SRPC +40%O
Setting time (min)	Initial	162	168	182	119	94
	Final	430	490	526	452	497
Water demand (%)		27.0	28.5	29.8	27.3	27.6

3.2 Flow and Compressive Strength

The control mortar (SRPC) has a flow of 80% and the mortars with 20 and 40%R showed a flow comparable with that control with better water retention. Mortars made with 20 and 40%O have a low flow (50%) and the w/cm was increased up to 0.60 to reach the same flow in 20%O-SRPC mortar.

Figure 1 shows the compressive strength for all mortars. At 2 days, the strength of all blended cements is lower than the corresponding to control mortar. For R calcined clays, the 20% of replacement reaches the strength of control at 7 days and the 40% replacement reaches to 87% of the control. On the other hand, mortars with 20 and 40% O do not exceed the 70% of the control strength. At 28 days, both mortars with R calcined clay maintain strength level of control, while SRPC+20%O and SRPC+40%O reach to 95 and 84% of control's strength. At 90 days, mortars with R clays reach to comparable strength (1.05 and 0.94) of SRPC mortar. On contrary, the strength developed is lower for mortars with O calcined clays reaching to 0.70 and 0.82% due to the large water requirements that increase the w/cm. Red calcined clay (R) has a very good pozzolanic activity and mortars develop a similar compressive strength than the control mortar. However, the orange calcined clay (O) presents some incomplete transformation causing the stimulation of Portland cement hydration at early ages. But, the progress of pozzolanic reaction is limited causing low strength at later ages.

3.3 Sulfate Performance

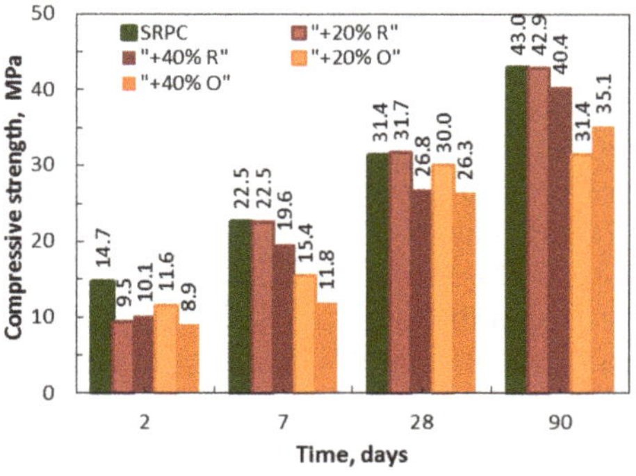

Fig. 1. Compressive strength of plain and blended cements

Expansion of mortar prisms in sulfate solution are shown in Fig. 2. For SRPC (Fig. 2a), it's can be observed that expansion is too low until 28 days and then the expansion increases with uniform rate up to six months attaining to 0.059%. This value is higher than the limit to considerer it as HS cements according to the criterion of ASTM C 1157. It is attributed to the high level of C3S in the SRPC that could cause fails for this type of test under controlled pH [7]. Compared with control mortar, the replacement of SRPC by R calcined clay produces slight increases of initial expansion up to 56 days.

Then, the expansion develops low to null rate of expansion for 20% and 40% R replacements.

At 6 months, both blended cements have lower (0.03% and 0.01%) expansion than limit (0.05%) and it could be considered as high sulfate-resistant (HS). SRPC+20%O shows a worse performance than the corresponding to SRPC and the expansion limit (0.100%) is surpassed after ~150 days instead of the low C_3A-content in this cement. It can be attributed to the increase of w/mc to obtain the flow and consequently the open porosity to sulfates. On the other hand, the SRPC+40%O has lower expansion than the control bars attaining to 0.04% at 6 months.

Figure 2b shows the expansion of mortars bars with WPC. Control mortar has a very quick expansion development after 28 days with exponential rate attaining to failure limit (0.100%) before than 56 days. Conversely, both blended cements present a

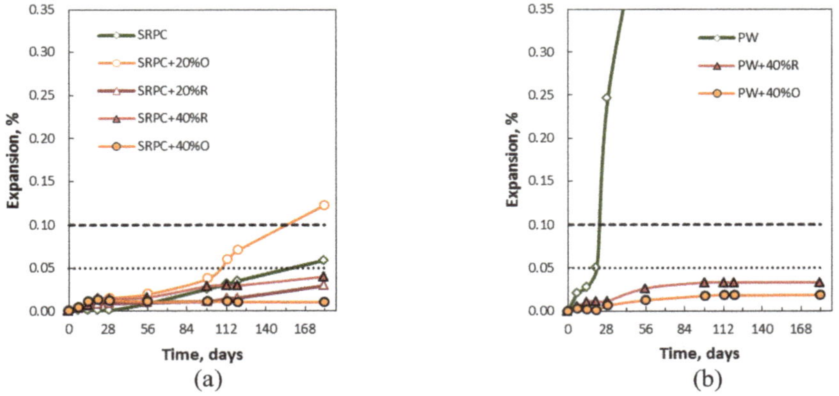

Fig. 2. Sulfate expansion: (a) Low C3A cement (SRPC); B) White Portland cement (WPC)

very low rate of expansions during the six months reported attaining to expansion of 0.018% that is lower than limit to classify as HS.

3.4 ASR Expansion

Figure 3 shows the expansion for mortar bars tested with high and low alkali Portland cements and with blended cements containing 25% of R or O calcined illite clay up to 90 days. For high alkali cement (Fig. 3a), the control bars have 0.221% of expansion at 14 days that is lower than the least value (0.2%) required by the ASTM C 1157. The bars with calcined clays attains to 0.045% and 0.040% for R and O, respectively. For these values, the effectiveness of expansion reduction of illitic calcined clay was 79.6 and 81.8% for R and O, respectively. The expansion of control was increased until 28 days and then it has a slowly rate indicating the consumption of free alkalis in mortar. On the other hand, mortar bars with calcined clays has a very slow increase attaining to 0.057 and 0.048% at 56 days for R and O pozzolans, respectively. It can be attributed that pozzolan do not provide alkalis to the progress of ASR. Although the expansion is not sufficient to consider these blended cements as ASR resistant, because they exceed the expansion limits of 0.02% at 14 days when the slow pozzolanic reaction begins.

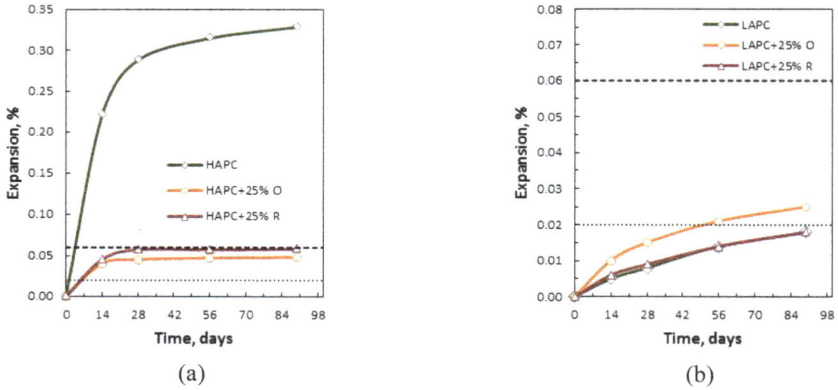

Fig. 3. Expansion due ASR for mortar bars tested with: (a) high alkali PC and (b) low alkali PC

4 Conclusions

Preliminary laboratory test showed that the replacement of illitic calcined clays lead to obtain blended cement with similar final strength than the Portland cement with low water demand.

At six months, illitic clays blended cements with w/c < 0.50, can be considered as high sulfate resistance cements. Sulfate performance of mortars containing 20% of illitic calcined clay was worsened when the water demand increases the w/cm of mortar.

The addition of 25% of illitic calcined clay may decrease the ASR expansion for high alkali Portland cement, but their performance at 14 days is worse than the obtained by low alkali Portland cement. Although, the expansion obtained are not sufficient to consider the blended cement as inhibitor of ASR. Long time performance test with reactive aggregates will be conclusive. The illitic calcined clay can impair the ASR performance of low alkali Portland cement.

References

1. Lemma, R., Irassar, E.F., Rahhal, V.: Calcined illitic clays as portland cement replacements. In: Scrivener, K., Favier, A. (eds.) Calcined Clays for Sustainable Concrete, pp. 269–276 (2015). doi:10.1007/978-94-017-9939-3_33
2. Ramachandran, V.S. (ed.): Concrete Admixtures Handbook: Properties, Science and Technology, 2nd edn. Noyer Publication, Park Ridge (1995)
3. Li, C., Ideker, J.H., Drimalas, T.: The efficacy of calcined clays on mitigating alakli-silica reaction (ASR) in mortar and its influence on microstructure. In: Scrivener, K., Favier, A. (eds.) Calcined Clays for Sustainable Concrete, pp. 211–217 (2015). doi:10.1007/978-94-017-9939-3_26
4. Trümer, A., Ludwig, H.M.: Sulphate and ASR resistance of concrete made with calcined clay blended cements. In: Scrivener, K., Favier, A. (eds.) Calcined Clays for Sustainable Concrete, pp. 3–9 (2015). doi:10.1007/978-94-017-9939-3_1
5. Ramlochan, T., Thomas, M., Gruber, K.A.: The effect of metakaolin on alkali–silica reaction in concrete. Cem. Concr. Res. **30**, 339–344 (2000)
6. Wild, S., Khatib, J.M., O'Farrell, M.: Sulphate resistance of mortar, containing ground brick clay calcined at different temperatures. Cem. Concr. Res. **27**, 697–709 (1997). doi:10.1016/S0008-8846(97)00059-8
7. Gonzalez, M.A., Irassar, E.F.: Ettringite formation in low C_3A portland cement exposed to sodium sulphate solution. Cem. Concr. Res. **27**, 1061–1072 (1997). doi:10.1016/S0008-8846(97)00093-8

Use of Ceramic Waste as a Pozzolanic Addition on Cement

Rayda Crespo Castillo(✉)

Centro de Investigación y Desarrollo de la Construcción (CIDC),
La Habana, Cuba

Abstract. This work studies the real possibility of recycling ceramic waste from the manufacture of structural elements, in order to use it as pozzolanic addition on cement.

The properties of the material where analyzed, the production process and the final product performance were studied to achieve an optimal use of this addition as a pozzolan and also as a fine aggregate in masonry production and waterproofing mortars for the roof cover finishing.

Its technical and economical effectiveness is valued, and its social and ecological impact is also remarked.

Keywords: Ceramic waste · Mortar · Cement · Pozzolan

1 Introduction

In many cases the possibility of using a certain ceramic material to be incorporated as an active addition in the cement has even been suggested, even indications have been made that the addition of a small amount of clay (kaolinite, montmorillonite, illite, and added bentonite clay at a ratio of 1 to 3% of cement) improves plasticity, cohesion, consistency and mechanical strength in mortars and concretes. In previous investigations of minerals containing montmorillonite as a fundamental component, it was concluded that the calcination of the clayey materials was essential for the satisfactory development of the pozzolanic activity, however, the determination of the heat treatment varied with the type of clay minerals present In the composition, for example the hydrated aluminosilicates of the kaolinitic group lost water of hydration between 320 and 500 °C and showed an increase of the pozzolanic activity, although above 900 °C a reduction of the surface area caused a decrease of the pozzolanic activity.

The minerals of the montmorillonite group lost much crystallization water between 150 and 510 °C. Dehydration was completed at 650 °C, but the crystal structure was not destroyed until 870 °C. Consequently, the optimum strengths and properties of the setting time in Portland cement mixtures with calcined montmorillonite were obtained when the calcination temperature was between 650–870 °C. The clays of the type illitica and vermecullitica were not very suitable from the point of view of the development of the resistance.

F. Martirena et al. (eds.), *Calcined Clays for Sustainable Concrete*, RILEM Bookseries 16,
https://doi.org/10.1007/978-94-024-1207-9_20

Establishing controls at each stage of the product burning process will condition the proper cooking of all clay components and will allow observing the structural changes that occur and the possibilities of silica activity in its amorphous state to detect when the disorder occurs in the mineral and the pozzolanic activity is created.

Based on these criteria, this study is carried out on materials for sieving and mortars for seat sills considering the partial replacement of crushed sand with a high modulus of fineness as a component of traditional mortars used in this complementary water-proofing process, by ceramic residues as artificial pozzolana, in this case of the ceramic tile, the results of the characterization of the ceramic wastes will be analyzed, the possibility of being active, and also the results of the resistance that indicate the presence Of a possible pozzolanic activity.

Objectives - To evaluate the possibility of the use of industrial waste of red ceramics in the mixtures of mortar of placement, repels, coatings, and in waterproofing for maintenance and repairs of roofs and walls, benefiting to the grit sands of high modulus of fineness employed in the works of masonry.

2 Method

The following materials were used for the preparation of the seat mortars for the ceramic tile, blocks, bricks and repels.

2.1 Cement

To determine the pozzolanic activity according to the standard C-311-97 a Portland Cement P-35 (350 kg/cm^2) was used from the factory "Mártires de Artemisa", Province Habana, that complies with the established in the Cuban norm.

2.1.1 Aggregates

The Normalized Sand of 4 fractions (coarse, medium, fine, and superfine) was used in the elaboration of the standard mortar.

For the mortar made with ceramic waste, the ceramic tile used in the "24 million" factory, located in Artemisa, were used. The cooking of these ceramic tiles occurs in Hoffman Ovens, at a temperature of 950–980 °C.

The ceramic tile was previously crushed in the Hammer Mill and in the Jaw Mill.

The hammer mill is of the type known as "cross blade", with a delivery capacity of 80 kg/k. A perforated outlet plate with a 4 mm diameter opening was fitted inside.

The Jaw Mill was adjusted with the outlet opening suitable for crushing the material. Once the ceramic residues were analyzed, the one that presented a fineness of the particles was taken and the material was passed by the sieve 325, to be used as artificial pozzolana.

2.1.2 Characterization of the Micronized Tile

For the comparison of the retained, the Micronized Tile was named as Arid 1 and the Victoria Arena as Arid 2 (Tables 1, 2, 3 and Fig. 1).

In the Jaw Mill, 70% of the Arid 1 (Micronized Tile) and 30% of the Arid 2 (Victory Arena) were observed (Tables 4, 5 and Fig. 2).

Table 1. Chemical composition of Micronized Tile

Material	Predominant mineral, %				
	SiO_2	Al_2O_3	Fe_2O_3	CaO	MgO
Rasilla micronizada	79.21	10.62	5.05	0.78	1.41

Table 2. Comparison of the percentages of retained aggregates of the crushed aggregates in the Hammer Mill by standard C-144 "Scaffolding aggregates in brick and block walls"

Sieve, mm	Past in the meshes		Norm C-144		
	Aggregate 1	Aggregate 2	Patron	Calculation	Deviation
9.52	100	100	100	100	0.0
4.76	96.4	92.0	100	95.1	4.9
2.38	85.6	64.4	95.0	79.4	15.6
1.19	60.1	42.7	70.0	55.0	15.0
0.59	47.8	23.5	40.0	40.7	0.7
0.29	38.8	8.2	20.0	29.8	9.8
0.15	29.0	2.7	10.0	21.3	11.3
	71%	29%	Sum of deviation: 57.3		

Table 3. Comparison of retained percentages of crushed aggregates in the Hammer Mill by standard C-897 "Aggregates for mortars of repels and coatings".

Sieve, mm	Past in the meshes		Norm C-897		
	Aggregate 1	Aggregate 2	Patron	Calculation	Deviation
9.52	100	100	100	100	0.0
4.76	96.4	92.0	100	93.6	6.4
2.38	85.6	64.4	90.0	72.5	17.5
1.19	60.1	42.7	60.0	49.4	10.6
0.59	47.8	23.5	35.0	32.8	2.2
0.29	38.8	8.2	10.0	19.9	9.9
0.15	29.0	2.7	0.0	12.8	12.8
	38%	62%	Sum of deviation: 59.4		

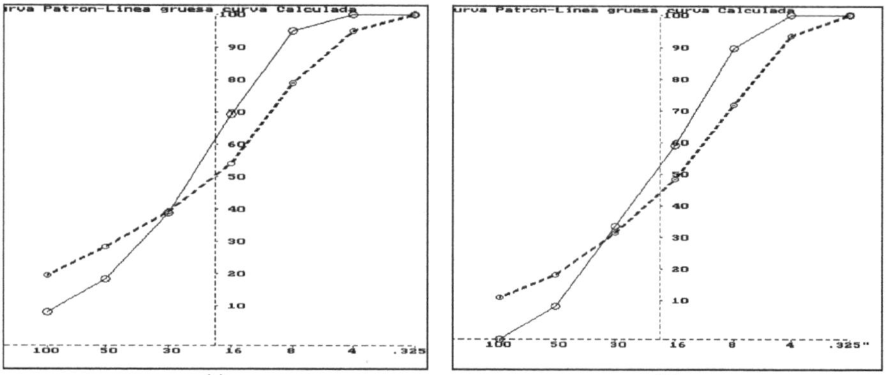

(a) (b)

Fig. 1. Hammer Mill. Granulometric curves (a) According to ASTM Standard C-144 "Arid for seat mortars in walls of blocks and bricks" (b) According to Standard ASTM C-897 897 "Aggregates for mortars of coatings and plasters".

Table 4. Comparison of the retained% of the crushed aggregates in the Jaw Mill by the standard

Sieve, mm	Past in the meshes		Norm C-144		
	Aggregate 1	Aggregate 2	Patron	Calculation	Deviation
9.52	100	100	100	100	0.0
4.76	96.4	92.0	100	92.9	7.1
2.38	85.6	64.4	95.0	68.2	26.8
1.19	60.1	42.7	70.0	48.5	21.5
0.59	47.8	23.5	40.0	35.4	4.6
0.29	38.8	8.2	20.0	27.0	7.0
0.15	29.0	2.7	10.0	19.4	9.4
	70%	30%	Sum of deviation: 76.3		

Table 5. Comparison of the retained percentages of the crushed aggregates in the Jaw Mill by the standard C-897 "Aggregates for mortar of repels and coatings".

Sieve, mm	Past in the meshes		Norm C-897		
	Aggregate 1	Aggregate 2	Patron	Calculation	Deviation
9.52	100	100	100	100	0.0
4.76	96.4	92.0	100	92.4	7.6
2.38	85.6	64.4	90.0	65.9	24.1
1.19	60.1	42.7	60.0	45.0	15.0
0.59	47.8	23.5	35.0	28.3	6.7
0.29	38.8	8.2	10.0	15.8	5.8
0.15	29.0	2.7	0.0	9.4	9.4
	28%	72%	Sum of deviation: 68.5		

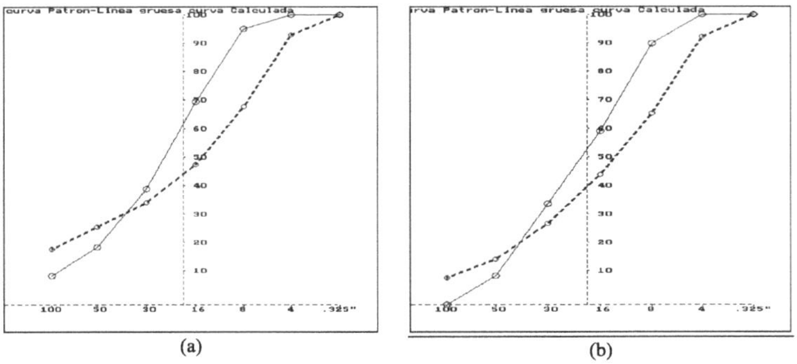

Fig. 2. Jaw Mill. Granulometric curves (a) According to ASTM Standard C-144 "Scaffolding aggregates in brick and block walls" (b) According to ASTM Standard C-897 897 "Aggregates for mortars and coatings".

3 Determination of the Rheological and Physico-Mechanical Properties of the Mortar with the Addition of the Ceramic Waste

3.1 Methods for Determining Pozzolanic Activity

The quality of a pozzolanic material by imparting cementitious properties to a non-hydraulic compound is known as pozzolanic activity. There are several chemical, physical and mechanical methods for its determination and are based on the ability of these materials to absorb lime under standard conditions. For the production of cements, natural and artificial pozzolans are used.

These are siliceous or siliceous aluminous materials which lack cementing properties and hydraulic activity alone but which contain constituents which combine with lime in the presence of water at ordinary temperatures and give rise to permanently soluble and stable compounds which Behave like hydraulic binders. Artificial pozzolans are formed by heating the clays and cooling them quickly, and are used instead of natural pozzolans by spraying bricks, and shingles that are added to coarse materials and sands.

3.2 Tests of the Mortars with the Addition of the Micronized Tile

To prepare the mortar, the standard 4-fraction sands, the Micronized Tile (sieve 325), and the cement were initially weighed. A standard sample was drawn, without addition of the Micronized Tile (P), and a second variant with addition of the same (RM).

The mortars were formed following the test procedure established in the norm NC 54-207-1980, used for the quality control of the cements. The following tables summarize the results achieved (Table 6, 7 and 8):

The average resistance of the Standard Mortar = 39.58 MPa represents more than 78%, and is above the minimum 75% specified by the norm so we conclude that the addition complies in part with the minimum requirements of the pozzolanic activity.

Table 6. Proportions of the standardized mortar according to ASTM C 311 to determine the pozzolanic activity index.

Mortar	Cement, g	Pozzolan, g	Sand, g	A/C
Patron	500	0	1375	0.50
Micronized Tile	400	100	1375	0.50

Table 7. Flowability test (Real water content)

Sample	Water	Fluency	A/C
Patron	250	117.	0.50
Micronized Tile	240	115.	0.48

Table 8. Mechanical strength tests (MPa) of the mortar with addition of the Micronized Tile

Age	Patron mortar	Ceramic tile mortar	% Ceramic tile/Patron
3 days	18.67	17.81	
	18.59	17.34	
	19.22	18.28	
Media	18.83	17.81	94
28 days	41.09	32.19	
	39.37	30.62	
	38.28	30.62	
Media	39.58	31.14	78.7

The average resistance of the mortar with addition of the micronized tile = 31.14 MPa represents 75% and also complies with that specified in the standard.

4 Results

Granulometric properties of the micronized tile: In the grinding of the tiles, a better result of the granulometric characteristics was observed with the use of the hammer mill with respect to that obtained in the grinding mill. A high degree of fineness of the particles was also observed (max. Retained on No. 325 (44 µ) 34% sieve) which satisfy the requirements for pozzolanic activity established by ASTM C 618.

Rheological characteristics of the mortar with the addition of the micronized tile: The results of the fluidity are close to those of the standard mortar made with sand silica, and even the water/cement ratio is below the pattern which can be explained in part by low porosity of the ceramic waste.

Mechanical properties of the mortar with addition of the micronized tile: The performance of the mechanical resistances complies with the requirements of the ASTM C-618 – 1992 standard adopted for the determination of the pozzolanic activity index greater than 75% with the addition of up to 20% micronized tile.

5 Conclusion

It is possible to use ceramic waste in the granulometric benefit of the high modulus fine grinding sands for the production of concrete, avoiding the losses in work by grading the sand for use in masonry mortars.

When using wastes from the national ceramic production, the reservoirs of conventional aggregates are conserved.

The costs of annual waste of crushed tiles are reduced by the partial substitution of high modulus fineness grit sand used in masonry mortar which can reach 35–75% of the volume used per cubic meter of mortar for coating applications or seat of brick walls or blocks.

A reduction in the losses of material resources valued of more than 39.0 MUSD per year is achieved by the granulometric benefit of the crushed sands used in masonry mortar and cement content savings of up to 20%

Expand the studies of the chemical, physical and technological properties of ceramic waste to confirm its introduction in the production of the cements without neglecting the control in the dosages of the same.

References

1. Marta, G.: Application of physicochemical analysis techniques to the study of pozzolanic activity in cement pastes, bulletin N3, CTCM (1984)
2. Masonry mortars for new and old buildings. Master's Thesis, December 1996

Development of the Microstructure in LC^3 Systems and Its Effect on Concrete Properties

Yuvaraj Dhandapani(✉), K. Vignesh, Thangadurai Raja, and Manu Santhanam

Department of Civil Engineering, Indian Institute of Technology Madras, Chennai, India
dyuvarj@gmail.com

Abstract. In this paper, the difference in the microstructure development of limestone calcined clay cement (LC^3) in comparison with the ordinary Portland cement (OPC) and Portland pozzolan cement (represented as FA30) at different water-binder ratios is shown. The results from the studies suggest that a mere adoption of a lower water binder ratio to reduce the capillary pore space for filling by hydrates, ensures only a marginal improvement in the kinetics of microstructural development, as seen from conductivity evolution in OPC and FA30. Highly reactive pozzolans ensure a more rapid drop in conductivity due to low-density hydrates resulting in lower capillary porosity and densification of the hydrate matrix. BSE micrographs also show a more densified binder matrix with LC^3, mainly due to high pozzolanicity of calcined clay resulting in CSH with lower C/S ratio with a hybrid hydrated phase assemblage compared to OPC and FA30. The impact of improved kinetics of LC^3 binder reflects in better durability parameters at an early age in the different concretes made with LC^3 binder. It is seen that the fly ash based systems (FA30) show a marked increase in the concrete resistivity up to an age of 1-year curing. The resultant effect of such microstructural development on the chloride resistance of concretes is also discussed.

1 Introduction

The substitution of cement with calcined clay has shown promising results for compressive strength evolution in cementitious systems [1, 2]. The use of lower grade kaolinite clay along limestone was found to show comparable mechanical properties with respect to plain Portland cement [3]. The durability performance of such binder systems will depend on the development of the physical structure and mineralogical composition of the hydrated cement paste (hcp). In general, the durability mechanisms of cementitious systems can be categorised as transport driven problems and phase transformation driven durability concerns, depending on the predominantly governing mechanism. The problems such as chloride ingress and corrosion involve pore structure and ionic transport as a major governing mechanism. In contrast, in some durability issues such as sulphate attack and carbonation, the mineralogical composition of the hydrated cement paste can have a more dominant effect on the performance. It is known that the amount of sulphoaluminate phases (AFm) and the reserve alkalinity of the binder phase maintained by portlandite control the performance in sulphate rich environment and carbonation

F. Martirena et al. (eds.), *Calcined Clays for Sustainable Concrete*, RILEM Bookseries 16,
https://doi.org/10.1007/978-94-024-1207-9_21

ingress in the concrete systems respectively. It is also widely acknowledged that adopting lower water-binder can ensure better performance due to reduced pore structure [4].

In this study, the effect of the water-cement ratio on the physical structure development of the OPC, FA30 and LC^3 cementitious systems is investigated. The influence of the structure on one durability parameter of concrete, i.e., surface resistivity, is also assessed. Scanning electron microscopy in backscattered electron mode is also conducted to show the densification of the microstructure. The results are supported by compositional analysis of the microstructure by Energy Dispersive X-ray (EDX) analysis and X-Ray Diffraction (XRD). The durability potential of this improved microstructure development seen in LC3 and its effect on the chloride ingress in comparison with OPC and FA30 concretes after an extended period of curing (1 year) is also discussed.

2 Materials and Methods

Plain Portland cement (confirming to IS 12269) and Fly ash (Class-F, confirming to IS 3812-Part-1) were used as the binder component in all cementitious mixtures including cement paste and concrete, and designated as OPC and FA30 (denotes 30% replacement of Portland cement in the binder). The LC^3 from an industry-scale production (details can be found in [2]) was used as a binder in the cementitious systems designated as LC^3. The chemical composition of the materials used in the study can be found elsewhere [5]. The mean d_{50} of the particle size of the Portland cement, fly ash and LC^3 were found to be 18 μm, 23 μm and 14 μm by laser diffraction assessment.

Impedance spectroscopy was carried on the cement paste at three different water-binder ratios (0.3, 0.4 and 0.5). PCE-based superplasticizer was added to 0.3 water-binder ratio mixes to ensure a similar flow as in the systems with 0.4 water-binder (w/b) ratio. The details of the impedance experiments and bulk resistivity assessment are presented in [6]. Scanning Electron microscopy was carried out on a slice of cement paste (w/b ratio: 0.4) hydrated in sealed condition for a period of 28 days. The slices were conditioned for hydration stoppage (in Isopropanol) which was followed by a period of drying. Later, the conditioned slices of the sample were epoxy impregnated (under vacuum), polished with diamond spray (up to 1 micron size) and coated with carbon for observation in the Scanning Electron Microscope in the Back-Scattered Electron (BSE) mode. In addition, the difference in the mineralogical composition of the hydrated phases at 28 days of hydration is evaluated by quantitative X-ray diffraction (XRD) for the three binder systems.

Pore structure assessment by mercury intrusion porosimetry was carried out on a conditioned sample (sample conditioning similar to that for microscopy) using the Pascal 140 and 440 instruments.

The surface resistivity measurement was performed on concrete specimens (details of the concrete mix designs are discussed elsewhere [5]). The measurement was carried out on a cylindrical specimen (100 mm diameter and 200 mm height) using a Wenner four probe resistivity meter (Resipod resistivity meter). A total of 27 measurements on three different circumferential locations of the cylinder was performed.

3 Results and Discussion

3.1 Evolution of Electrical Conductivity in Cement Paste

The conductivity response corresponds to the evolution of the interconnected pore space of the HCP. In general, the electrical response can be a function of interconnected pore space, the ionic strength of the pore fluid and the capacitance of the solid phase formed in the different systems. The changes in the pore space are reported to have a predominant effect on the electrical response [7]. The continuous pore filling process due to the formation of the hydration products occurs during the cement hydration process. The evaluation of the electrical response helps track the kinetics of this change in a continuous non-invasive manner with time. One of the major distinction in the composite cementitious materials is the interaction between the different replacement materials. For instance, in FA30, the interaction between clinker and fly ash depends on portlandite availability, fly ashes reactivity and fineness of fly ashes. Similarly, in the case of LC^3, the interaction between clinker-calcined clay, clinker-limestone and limestone-calcined clay can also be a governing factor for the improved kinetics. The dependence of reaction between the different component can alter the generic tendency of the hydrates formation in both binary and ternary cement composites. The marked difference in the interaction can be due to a physical or chemical factor. The physical factor can be a reduction of the particle space due to an increase in the packing of the binder and a dilution of the reactive component due to partial substitution of the inert component. The chemical part of the interaction is the influence on the degree of hydration, and formation of hydration products which alter the kinetics of microstructural development. However, the basic mechanism of the space filling by the hydration product and underlying reduction in pore space still remains the most important governing factor affecting the development of concrete's properties with the different binder systems. In such a scenario, the effect of water-binder ratio also plays a crucial role in understanding the efficiency of binder systems to produce sustainable binders for a range of concrete mixtures with these low clinker cements.

The evolution of electrical conductivity of the different systems at three water-binder ratios are presented in Fig. 1(a), (b) and (c). In Fig. 1(a), the microstructural evolution in the OPC systems shows a major change occurring at early ages (up to 14 days) beyond which the changes are trivial. The w/b ratio has a predominant effect on the evolution of the microstructure which maintains a considerable difference even at later ages in systems with lower w/b ratio showing lower conductivity than systems higher w/b ratios. The results also show an improved kinetics of the microstructural changes at lower water-binder ratios in all the binder systems. However, it is also found that lower w/b ratio doesn't necessarily invoke progressive reaction at later ages from the anhydrous binder.

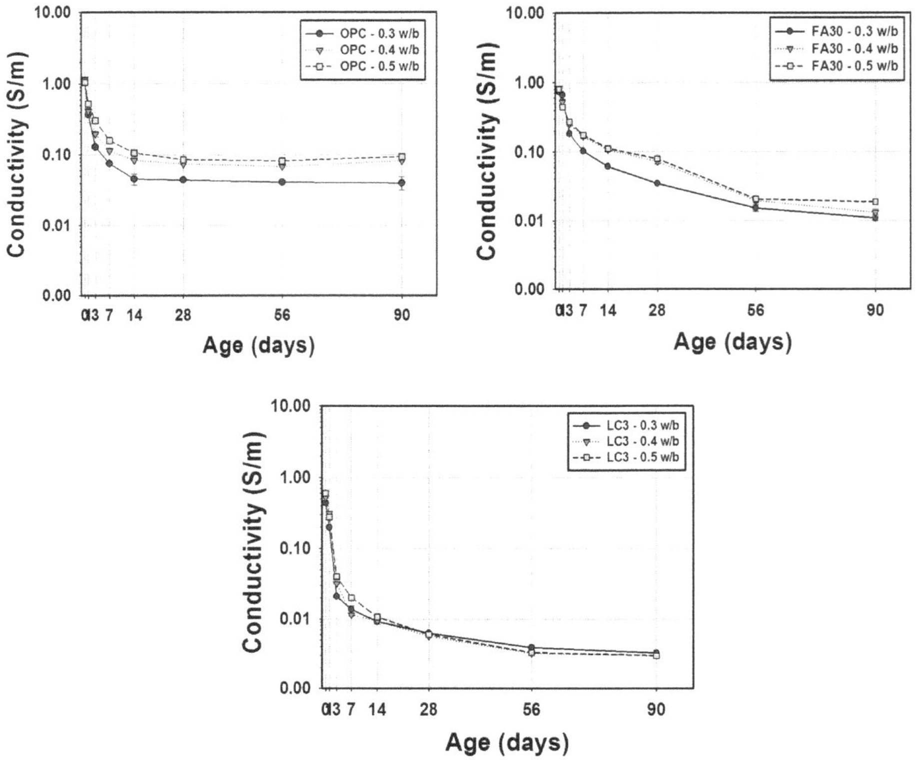

Fig. 1. Evolution of electrical conductivity in the pastes with OPC (a) (top left), FA30 (b) (top right), and LC^3 (c) (bottom) with different water-binder ratio

In contrast to OPC system, for the composite systems with a part of Portland cement replaced with fly ash (FA30), there was a steady sign of microstructural changes post 14 days of curing in contrast to plain Portland cement. This reflects the continuous influence of the fly ash on the microstructure in the composite binders. As seen from Fig. 1(b), the delayed influence of the fly ash on the microstructure led to a continuous drop in the conductivity beyond 14 days. FA30 at lower water-binder ratios had a significantly lower conductivity at all ages, compared to the FA30 systems at higher water ratio (0.4 and 0.5). It is also seen that the FA30 with higher water-binder ratio shows a substantial drop in the conductivity with time. This suggests the effect of continuous and slow dissolution (due to the slow reactivity of the fly ashes) of the fly ash on the microstructural development in the cementitious systems. This also provides insight on the performance of concrete in such binder systems. The traditional plain Portland cement requires a significant reduction of w/b ratio to ensure compliance with required durability parameter. In the case of the composite cement with SCMs, sufficient curing can ensure the durability compliance without significant lowering of the water-binder ratio in the concrete mixtures.

In the case of the composite blend with limestone and calcined clay, the high reactivity of the calcined clay as a pozzolan, ensures that the water-cement ratio makes a minimal difference in the conductivity evolution beyond the early age. It is seen from the results of the FA30, that systems with higher water-binder ratio show prolonged evolution in conductivity beyond 28 days, denoting the space bound kinetics of the binder reactions (in blended systems). The water-cement ratio is one of the major factors controlling the pore space provided to the anhydrous materials to react and thereby, dominates the time-dependent space filling mechanism to limit the effect reaction kinetics in the different binder systems [8]. This difference can be attributed to a reduction in the interparticle spaces due to the finer particles of the limestone and calcined clay, and higher reactivity of the calcined clay. Both these factors led to an acceleration in the development of microstructure even at higher water-binder ratios. The major implication from such development is the improved performance of the concrete mixture at a range of water contents in the concretes.

3.2 Microstructure of LC³ Systems

Backscattered electron images of the three binder systems were obtained for samples sealed cured for 28 days. The evidence of the variation in microstructural characteristics is represented from the micrographs at two different magnification levels. Based on the micrographs for OPC (Fig. 2), the presence of large numbers of pores and presence of portlandite, a loosely packed structure of the plain Portland cement are noticed. The microstructural details of the FA30 (Fig. 3) reveal the surplus amount of partially reacted and/or unreacted fly ash particles. At a much lower magnification, portlandite was not commonly noticed as seen in OPC, and a local densification of the cementitious matrix due to partially reacted fly ash can be identified. In the case of the LC³ cementitious systems, a more compact cementing matrix is seen. In addition, two major differences in the micrographs of the LC³ system are the compactness of the microstructure and densified cement matrix, as shown in Fig. 4.

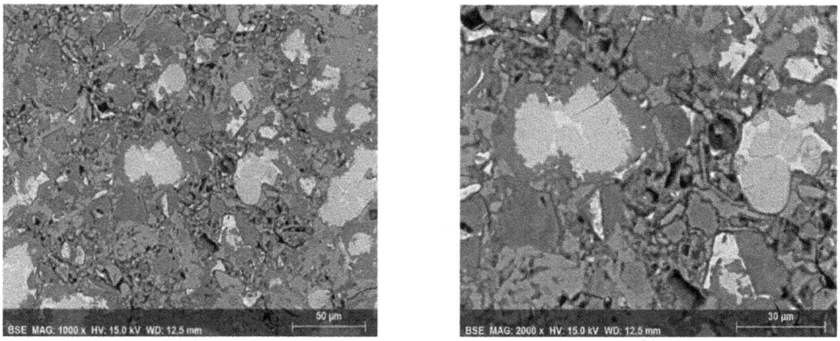

(a) (b)

Fig. 2. Microstructure of OPC at 28 days at two magnifications; (a) 1000 X and (b) 2000 X; EDX shows a C/(S+A) ratio of 1.85 in the CS(A)H matrix

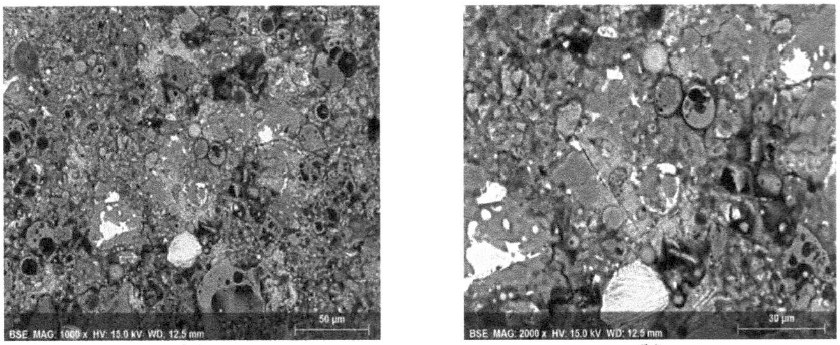

(a) (b)

Fig. 3. Microstructure of FA30 at 28 days at two magnifications; (a) 1000 X and (b) 2000 X; EDX shows a C/(S+A) ratio of 1.48 in the CS(A)H matrix

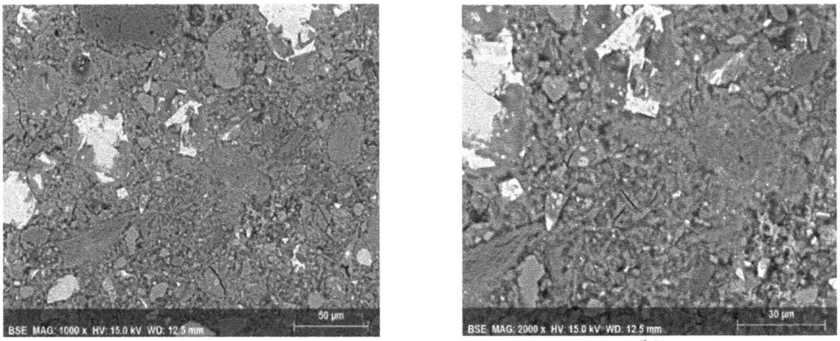

(a) (b)

Fig. 4. Microstructure of LC^3 at 28 days at two magnifications; (a) 1000 X and (b) 2000 X; EDX shows a C/(S+A) ratio of 1.36 in the CS(A)H matrix

The analysis of the chemical composition of the microstructure showed a much lower Ca/Si ratio in the LC^3 in contrast to the FA30 and OPC. This can also be linked to the difference in the kinetics of microstructure formation and thereby, can influence the performance of these binders in concrete systems. The presence of a more compact and dense microstructure in the LC^3 cementitious matrix is mainly due to the improved packing of the LC^3 systems, the higher pozzolanic influence of calcined clays (seen from portlandite consumption) and a more composite hydrate phase assemblage as shown by the XRD results in Fig. 5a. Though kinetics of the microstructural development has a crucial role in the development of the macro-scale properties, the influence of the composite phase assemblage should not be neglected, as it can also influence the chemical interaction involved in the durability concerns. One such effect, the influence of the difference in the hydrated phase assemblage can be an alteration in the chloride binding in the hcp and secondary ettringite formation from monosulphate in the cementitious systems. In addition, a compact capillary pore space indicates that the concrete made with LC³ binder can have much better resistance to transport properties than the conventional cementitious materials investigated in the study; barring the effect of water-cement ratio which leads to changes in the capillary space for hydrates, and induces early densification of microstructure in the finer pores even in the blended systems. From this study, we show that microstructural development in LC³ cementitious systems at higher water-binder ratio is better compared to the fly ash blended FA30 at the lower water-binder ratio. The trends are in agreement with the MIP data presented in Fig. 5b. From the conductivity data, it is seen that it takes 90 days for lower water-binder ratio (0.3) FA30 system to reach a similar conductivity state of an early age (7 days to 14 days) LC³ system with the higher water-binder ratio (0.5). Though a comparison of the absolute conductivity between the binders can be misleading due to the difference in ionic conductivity of the pore solution varying in the different systems, a relative change in conductivity or a normalised conductivity (with pore solution) would also suggest similar trends of an improved kinetics of the microstructural development with the LC³ binder system.

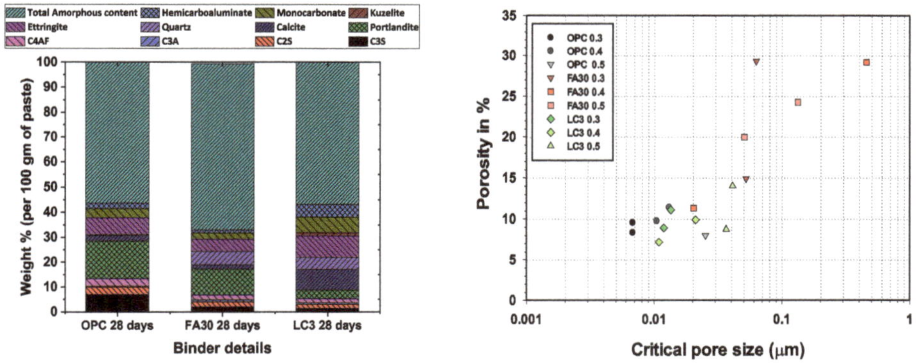

(a) (b)

Fig. 5. XRD phase composition of the hydrated cement paste at 28 days (a) and difference in critical pore size (28 days) with w/b ratio (b) in sealed and cured samples

3.3 Effect on Concrete Properties

Over the years, Wenner four-probe surface resistivity value has been established as an ideal tool assessment of concrete quality due to the ease in the measurement of on-site concrete quality for performance-based specifications [9, 10].

To draw a link between improved conductivity evolution in LC³ cement paste, we performed surface resistivity assessment on different concrete mixes. In equivalent strength grade (M30 and M50) concretes, FA30 mixes demand lower water-binder ratio to achieve similar 28 days strength grade. Figure 6 show the positive connect of the improved kinetics of the LC³ binder to the surface resistivity assessment of concrete mixes. The trends in concrete suggest that the surface resistivity of FA30 at 365 days is similar to the surface resistivity of LC³ concrete made at 28 days. The resistivity values between different concrete in the three classifications follow the order of

LC3 > FA30 > OPC. Based on the order of an increase in the resistivity value between 28 days and 365 days, the systems ranged in the order FA30 > LC^3 > OPC. This shows the impact of extended curing on the performance of fly ash cementitious systems. From the literature connecting the resistivity on the concrete to durability performance, we can suggest that concrete made with LC^3 binder can have higher resistance to transport driven durability problems such as ingress of chloride and propagation of corrosion in a chloride laden environment, and this property would be significantly better than other blended systems at early ages.

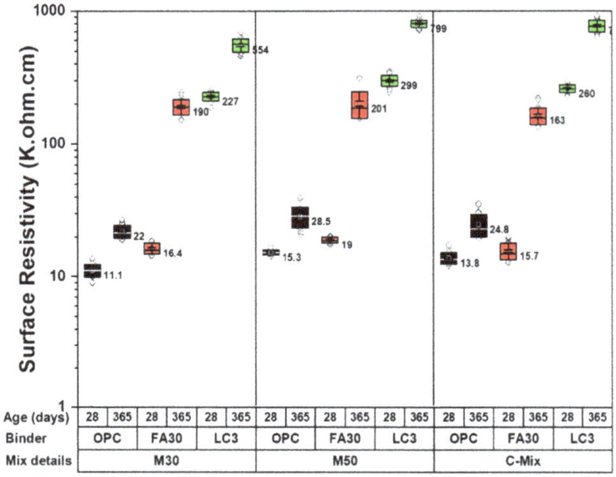

Fig. 6. Surface resistivity of concrete made with OPC, FA30 and LC3 systems at 28 days and 365 days of curing in high humidity conditions

The variation in the physical structure development and chemical composition of microstructure leads to a superior durability performance of LC^3 concretes. This can be due to combined effect of better resistance to transport and higher binding of chlorides in these binder systems (Fig. 7). The chloride profile of the concrete after 1 year of curing shows less than 5 mm penetration, much lower than the penetration depth of 8 mm and 17 mm in FA30 and OPC respectively. Lowered pore sizes, higher pore solution resistivity and increased binding of chloride in the hydrated phases are the major reasons for such performance.

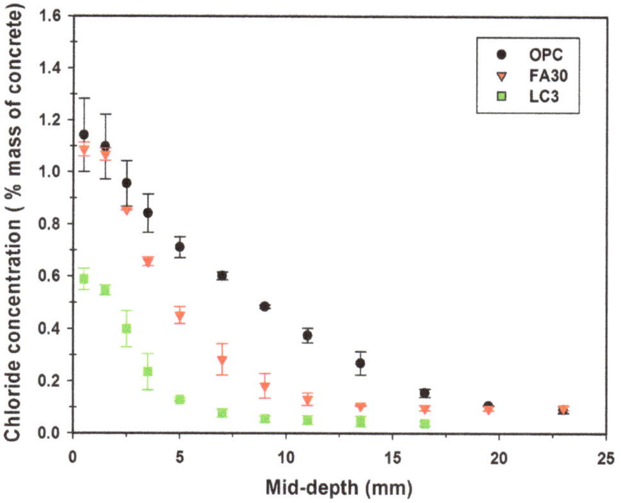

Fig. 7. Chloride profile of concrete specimen subjected to Bulk diffusion (as per ASTM C1556 [11] for 56 days exposure period) for 1-year curing duration; concrete made with a binder content of 360 kg/cu.m and the water-binder ratio of 0.45

4 Conclusion

Based on the results presented in the paper, the following conclusions can be drawn on the effect of the difference in microstructural development on concrete properties:

1. The evolution of microstructure development studied from the changes in conductivity with time shows the positive effect of calcined clay on the LC^3 cementitious matrix. The results show that mere lowering of water-binder ratio ensure only a minor improvement in the kinetics of microstructural development within a particular binder system. A reduction in water-binder ratio doesn't necessarily ensure better durability parameter similar to the improved effectiveness seen in composite systems (FA30 and LC^3) due to better densification of the microstructure with hydration period.
2. In agreement with the improved kinetics, the SEM micrographs showed a denser microstructure in the LC^3 binder matrix along with variation in the composition of the microstructure. It is also found that the effect on the porosity of the binder is also positive.
3. The improved kinetics has a positive impact on the concrete resistivity development of the LC^3 concrete mixes, with a higher resistivity at early curing period. It is also found that the FA30 concrete mixes at equivalent concrete grade or equivalent mix design could attain similar high resistivity only after an extended period of curing.
4. Concrete made with LC^3 has excellent resistance to ingress of chlorides. The chloride profile of 1 year cured concretes showed lower and steeper trend than for the other two cements. The depth of chloride ingress was only 5 mm which was nearly twice

lower than FA30 and thrice lower than OPC, confirming the combined positive effect of the combined physical and chemical variation of the microstructure in the LC^3 binder systems on the performance of concrete.

Acknowledgements. The authors would like to acknowledge the partial financial support Swiss Agency for Development and Cooperation.

References

1. Antoni, M., Rossen, J., Martirena, F., Scrivener, K.: Cement substitution by a combination of metakaolin and limestone. Cem. Concr. Res. **42**, 1579–1589 (2012)
2. Emmanuel, A.C., Haldar, P., Maity, S., Bishnoi, S.: Second pilot production of limestone calcined clay in India: the experience. Indian Concr. J. **90**, 57–64 (2016)
3. Avet, F., Snellings, R., Alujas, A., Scrivener, K.: Development of a new rapid, relevant and reliable (R3) testing method to evaluate the pozzolanic activity of calcined clays. Cem. Concr. Res. **85**, 1–11 (2016). doi:10.1016/j.cemconres.2016.02.015
4. Yu, Z., Ye, G.: The pore structure of cement paste blended with fly ash. Constr. Build. Mater. **45**, 30–35 (2013)
5. Dhandapani, Y., Sakthivel, T., Santhanam, M., Gettu, R., Pillai, R.G.: Mechanical properties and durability performance of concretes with Limestone Calcined Clay Cement (LC^3), Unpublished Results (n.d.)
6. Sanish, K.B., Neithalath, N., Santhanam, M.: Monitoring the evolution of material structure in cement pastes and concretes using electrical property measurements. Constr. Build. Mater. **49**, 288–297 (2013). doi:10.1016/j.conbuildmat.2013.08.038
7. Andrade, C., Blanco, V.M.M., Collazo, A., Keddam, M., Nóvoa, X.R., Takenouti, H.: Cement paste hardening process studied by impedance spectroscopy. Electrochim. Acta **44**, 4313–4318 (1999)
8. Berodier, E., Scrivener, K.: Evolution of pore structure in blended systems. Cem. Concr. Res. **73**, 25–35 (2015)
9. Polder, R.B., Peelen, W.H.A.: Electrical resistivity testing for as-built concrete performance assessment of chloride penetration. In: RILEM International Workshop on Performance-Based Specification and Control of Concrete Durability, pp. 151–157 (2014)
10. ACI, 222R-01, Protection of metals in concrete against corrosion (2001)
11. ASTM C1556-11a, Standard test method for determining the apparent chloride diffusion coefficient of cementitious mixtures by bulk diffusion. ASTM International, pp. 1–7 (2011)

Carbonation of Concrete with Low Carbon Cement LC3 Exposed to Different Environmental Conditions

Ernesto Díaz[1(✉)], Raúl González[2], Dayran Rocha[2], Adrian Alujas[2], and Fernando Martirena[2]

[1] Centro de Investigación y Desarrollo de la Construcción, CIDC, La Habana, Cuba
[2] CIDEM, Universidad Central "Marta Abreu" de las Villas, Santa Clara, Cuba

Abstract. This investigation evaluates the behavior of processes of carbonation in the concrete made with cement low carbon LC3, under different environmental conditions. To determine the carbonated depth, the phenolphthalein technique is used.

Generally the elements made from ternary cements are more sensitive to carbonation than those produced with Portland cements (PC), due to the effect of substitution of clinker with consequent reduction of the alkalinity, so it corresponds to test results conducted with elements manufactured at the Center for Research and Development of Construction (CIDC) in Havana (area of low aggressiveness) (Fig. 2) of 30 MPa characteristic strength, kept in a zone of relative humidity (HR) between 60 and 70%.

On the other hand, when evaluating concrete characteristic design strength of 25 MPa produced in the company of Industrial Production (EPI) Villa Clara, located in Cayo Santa Maria (marine area of high environmental aggressiveness) (Fig. 1), items made with LC3 demonstrate a carbonated depth lower compared to concrete produced with PC, an atypical situation, which may be due to the high levels of porosity in conjunction with the effects of leaching more aggressively in concrete with a higher content of compounds soluble as CO_2.

1 Introduction

Global demand for cement and concrete has increased exponentially over the last twenty years and is the result of a combination of strong ongoing trends, such as population growth and the growing need for infrastructure and housing. The cement industry has undergone significant pressure and has made efforts to improve its production efficiency as well as its environmental impact [1]. The real potential for improvement is based on the replacement of clinker by Complementary Cementitious Materials (SCMs) [2, 3], but the problem is that the availability of these materials is limited and therefore the potential to reduce energy and Emissions through clinker replacement is compromised [4–6].

Kaolinitic clays, however, are widely available in the tropical zone of the planet. Extensive research has demonstrated the reactivity of kaolinic clays when fired at

F. Martirena et al. (eds.), *Calcined Clays for Sustainable Concrete*, RILEM Bookseries 16,
https://doi.org/10.1007/978-94-024-1207-9_22

Fig. 1. Exposure site with high aggressiveness

Fig. 2. Exposure site with low aggressiveness

temperatures between 600–800 °C [7, 8]. It has recently been shown that clays with relatively low kaolinite content (less than 50%) could also be highly reactive [9, 10]. In addition, it has been shown that the synergism between calcined clays and limestone allows for 50% clinker replacement rates without compromising either early and late strength and cement durability [11, 12]. A tertiary cement of clinker-calcined clay up to 50% has been developed and produced at different scales, including a complete industrial test at a cement plant, for use in general-purpose construction applications in Cuba and India [13, 14].

This work presents an evaluation of the durability of the concrete made with the cement produced during the industrial test in Cuba. The tests were performed on concrete specimens placed on the sea shore in a highly aggressive environment and exposed concrete samples in a site of low aggressiveness. Carbonation was one of the main experiments for the evaluation of durability.

2 Description of the Preliminary Tests

Concrete designs were made with 25 MPa for the precast elements placed in conditions of high aggressiveness and 30 MPa for the specimens (150 × 300 mm) exposed to low aggressiveness, using Cuban calcareous aggregates. The M381, M25 and SGR series were made with the new cement, while the M32 and P series were cast with the Cuban Portland Cement, type I with 88% clinker, known as P35. Tables 1 and 2 show the details of the mix design and the mechanical properties of the manufactured concrete.

Table 1. Main characteristics of the elements to evaluate in high aggressiveness site

ID	Target strength	Cement kg/m^3	w/c	Slump (cm)	Date
M381	25 MPa	360	0,47	8	2013-11-02
M-25	25 MPa	360	0,47	12	2014-02-06
M-32 P-35	25 MPa	300	0,40	8	2014-02-10

Table 2. Main characteristics of the elements to evaluate in low aggressiveness site

ID	Target strength	Cement kg/m^3	w/c	Slump (cm)	Date
SGR	30 MPa	404	0,42	17,5	2014-03-14
P	30 MPa	417	0,42	17,8	2014-03-14

The elements were cured for 28 days, while the specimens had different curing regimes (0, 3 and 28 days).

For the preparation of the samples the same procedure was used for all the samples. A traverse was carried out (for the surface originally exposed to the atmosphere) dry cut with 50 mm of thickness (Fig. 3).

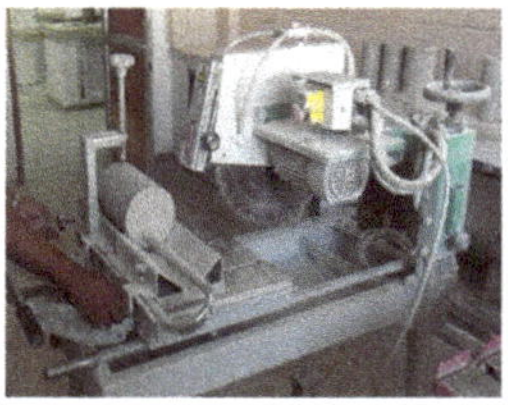

Fig. 3. Preparation of the samples

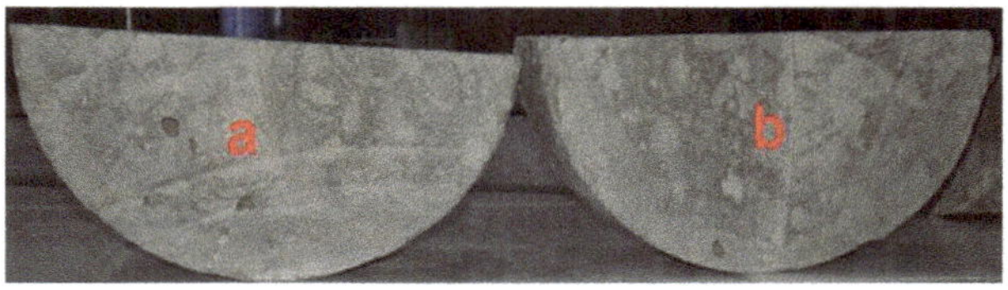

Fig. 4. Cut of the samples

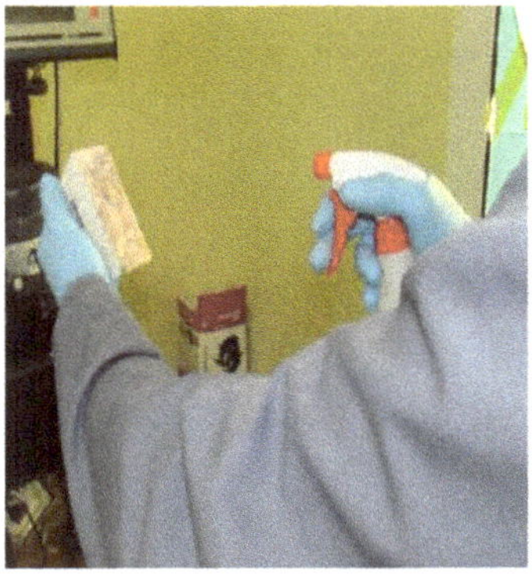

Fig. 5. Application of the phe-nolphthalein

Later, this slice was fractioned in two identified halves as half "a" and half "b" of its corresponding to each one of the samples (Fig. 4).

The phenolphthalein takes a red - purple color for values of superior pH to 9.5 (not carbonated concrete) and becomes colorless in inferior values at 8 (carbonated concrete). For values among 8 and 9.5, the color is between rose and red purple (Fig. 5).

1 or 2 min after the application and before lapsing 15 min, the mensuration of the carbonation depth (the longitude of the area colorless measured from the surface) is made, and the maximum and minimum values, as well as the arithmetic stocking are determined. The procedure should not take more than 20 min. The measures are carried out with a mensuration instrument able to guarantee a precision of 0.5 mm.

3 Test Results and Analysis

The pictures after application of phenolphthalein are show in Fig. 6.

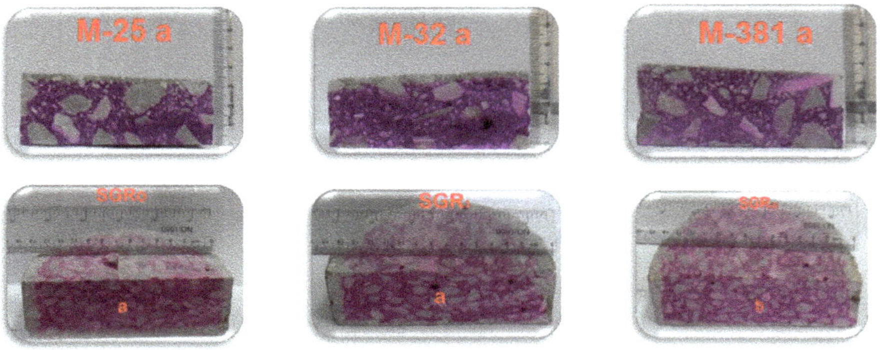

Fig. 6. Concrete samples with phenolphthalein and mensuration instrument.

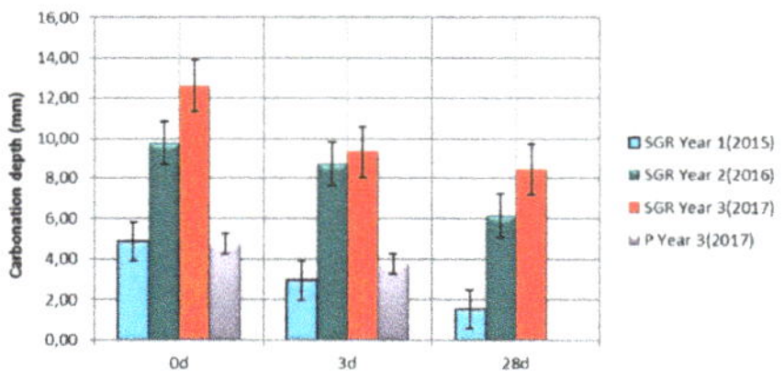

Fig. 7. Low aggressiveness site

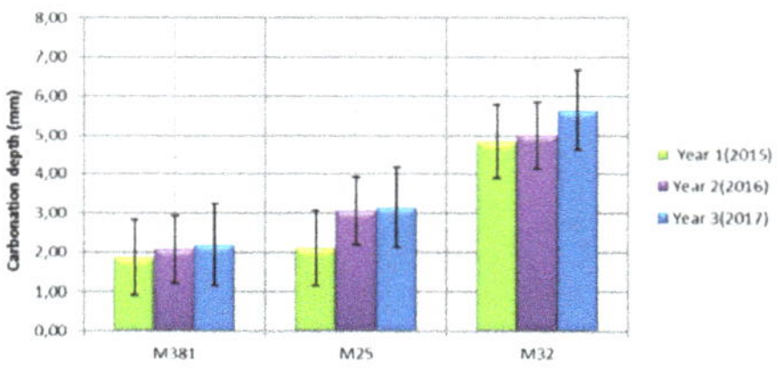

Fig. 8. High aggressiveness site

The results of carbonation obtained are presented in Figs. 7 and 8, comparing values between LC3 with PC.

It can be observed that the concretes manufactured with both types of cements in a marine tropical area, where the effect of the exchange of the tides the relative humidity of the environment is manifested it is generally high, is above 80% during a great part of the year, which provokes the saturation of the pores of the concrete, a factor that facilitates the diffusion of CO_2 inside the concrete structure. This results in a smaller aggressiveness of the phenomenon of the carbonation in the concretes [11], compared to other types of atmospheres.

The atypical behavior of the carbonation can be due to the tide phenomenon. This provokes a greater dissolution of portlandite ($CaOH_2$) in elements that contain PC. This results in a higher porosity and a lower CO_2 absorption capacity. Nevertheless, concretes with cement content of low carbon LC3 are advantageous, because the additions that compose the cement LC3 combined with the products of reaction of the PC, specifically the portlandite are insoluble [13], and hinder the entrance of CO_2 into the concrete.

The located elements the place of low aggressiveness, with intervals of HR among from 60 to 70%, present a favorable condition so that it is triggers the carbonation reactions. This factor, associated to the grade of basicity of the concretes elaborated with LC3, can be the reason of the aggressive behavior of the carbonation front.

In situations contrary to the behavior of the carbonation of the elements elaborated with cement P-35, in CIDC Havana, the effect of the carbonation was insignificant. This is due to a higher alkaline content present in the dosages.

4 Conclusion

The results, that are just preliminary due to the limited amount of data collected so far, highlight the following facts:

The concretes elaborated with LC3 confer a better behavior in areas of exchanges of tides with which (HR 80%) the elements taken place with cement P-35, since the effects of the lixiviation in the latter are more aggressive when possessing a higher $CaOH_2$ content. During dissolution, soluble compounds leave high indexes of pores and permeability, provoking in the periods of drying a bigger diffusion of CO_2 inside the concrete and therefore the acid reacts with more depth.

The elements elaborated with LC3 located in areas of HR stocking are much more susceptible to the effect of the carbonatation even with higher indexes of impermeability and porosity, due to the first layer of alkalinity in the pores of the elements, a primordial factor in the effects of the carbonation.

The results of these tests are encouraging in terms of the potential durability of LC3 concretes and confirm the importance of the execution quality on the performance of full scale elements as well as the need for measurements on site for a realistic durability appraisal.

References

1. CEMBUREAU: World cement production by region Evolution 2001–2012. 2014 (2014)
2. Scrivener, K.L.: Options for the future of cement. Indian Concr. J. **88**, 11–19 (2014)
3. CSI & WBCSD: Development of State of the Art-Techniques in Cement Manufacturing: Trying to Look Ahead (CSI/ECRA-Technology Papers) Final Draft 31 March 2009 (2009)
4. Damtoft, J.S., Lukasik, J., Herfort, D., Sorrentino, D., Gartner, E.M.: Sustainable development and climate change initiatives. Cem. Concr. Res. **38**, 115–127 (2008)
5. Gartner, E.: Industrially interesting approaches to 'low-CO_2' cements. Cem. Concr. Res. **34**, 1489–1498 (2004)
6. Scrivener, K.L., Kirkpatrick, R.J.: Innovation in use and research on cementitious material. Cem. Concr. Res. **38**, 128–136 (2008)
7. Tironi, A., Trezza, M.A., Scian, A.N., Irassar, E.F.: Assessment of pozzolanic activity of different calcined clays. Cem. Concr. Compos. **37**, 319–327 (2013)
8. Sabir, B., Wild, S., Bai, J.: Metakaolin and calcined clays as pozzolans for concrete: a review. Cement Concr. Compos. **23**, 441–454 (2001)

9. Alujas, A., Fernández, R., Quintana, R., Scrivener, K.L., Martirena, F.: Pozzolanic reactivity of low grade kaolinitic clays: influence of calcination temperature and impact of calcination products in OPC hydration. Appl. Clay Sci. (2015)
10. Fernandez, R., Martirena, F., Scrivener, K.L.: The origin of the pozzolanic activity of calcined clay minerals: a comparison between kaolinite, illite and montmorillonite. Cem. Concr. Res. **41**, 113–122 (2011)
11. Antoni, M., Rossen, J., Martirena, F., Scrivener, K.: Cement substitution by a combination of metakaolin and limestone. Cem. Concr. Res. **42**, 1579–1589 (2012)
12. Matschei, T., Lothenbach, B., Glasser, F.P.: The role of calcium carbonate in cement hydration. Cem. Concr. Res. **37**, 551–558 (2007)
13. Vizcaíno-Andrés, L., et al.: Industrial trial to produce a low clinker, low carbon cement. Mater. Construcción **65** (2015)
14. Bishnoi, S., Maity, S., Mallik, A., Joseph, S., Krisnnan, S.: Pilot scale manufacture of limestone calcined clay cement: the Indian experience. India **88**, 22–28 (2014)

Evaluation of Compressive Strength and Microstructure of Cement Pastes Containing Different Qualities of Metakaolin

N. Dumani(✉) and J. Mapiravana

Building Science and Technology, CSIR Built Environment,
Pretoria, South Africa

Abstract. This paper shows the effect of metakaolin with different amorphous phase content on 2 and 7 day compressive strengths and microstructure of MK-blended cement pastes. The cement pastes that contained 0%, 10% and 50% MK were prepared at a constant water/binder ratio of 0.35 and cured at 22 °C for 2 days and 7 days. It was found that cement paste with high-quality MK (high amorphous phase content) resulted in a higher early age compressive strength than pastes with lower amorphous phase contents and this was attributed to high initial reactivity. The lowest amorphous phase content MK/cement blends achieved early age compressive strengths comparable to higher amorphous phase content MK/cement blends. The MK-blended pastes, prepared without a plasticizer, had higher degrees of agglomeration and lower strengths.

1 Introduction

Ordinary Portland cement (OPC) is commonly used as the binder for concrete. However, its production is energy-intensive and releases large amounts of carbon dioxide (CO_2) into the atmosphere [1, 2]. Globally, the production of OPC accounts for 5–7% of total CO_2 emissions [3, 4]. Consequently, less energy-intensive supplementary cementitous materials such as fly ash, ground granulated blast furnace and silica fume are commonly used as cement extenders [5]. However, these materials are subjected to availability problems. The use of metakaolin (MK) as cement extender has also been considered [6, 7]. However, despite the number of studies in recent years on the use of MK as an OPC replacement, in practice its use is still low. The utilisation of MK as an OPC partial replacement greatly improves the mechanical and durability properties of concrete in comparison to OPC alone due to its high pozzolanic reactivity, significantly lower the cement carbon footprint [6, 8, 9].

Kaolinitic clays that are used to produce MK are ubiquitously and abundantly available. In South Africa MK can be produced in large quantities, as the kaolinitic clays occur in widespread deposits available in the country. Locations with deposits of suitable kaolinitic clays for MK production in South Africa include Makana in the Eastern Cape, Hammanskraal outside Pretoria, Zebediela in Limpopo, Northern Cape and Western Cape.

There have been several studies on the compressive strength of concrete containing MK. These studies have shown that partial replacement of OPC with MK can improve

F. Martirena et al. (eds.), *Calcined Clays for Sustainable Concrete*, RILEM Bookseries 16,
https://doi.org/10.1007/978-94-024-1207-9_23

concrete compressive strength. The enhancement in compressive strength of concrete incorporating MK may be attributed to the filler effect, the ability of MK to accelerate cement hydration and pozzolanic reaction of MK [6]. Silica fume exhibits similar performance and therefore, MK has the capability to replace silica fume (SF) as an alternative cement extender [10]. Caldarone et al. [11] produced concretes with 5% and 10% MK by weight of Type 1 cement, with w/c of 0.40, which showed enhanced strengths of up to 365 days. They reported that their MK-OPC concretes exhibited strengths, which were slightly greater than SF-OPC mixtures at the same levels of cement replacement by the pozzolans. Controls had the lowest strength at all ages. Similar results were reported by Wild et al. [6], who tested concretes with 5% to 30% MK, ranging from one to 90 days in age, produced at a w/c of 0.45. They found that 20% replacement with MK was optimal for achieving maximum long-term strength enhancement. Poon et al. [12] also reported that cement pastes containing 5% to 20% MK had higher compressive strengths than the controls at all ages from 3 to 90 days, with the paste containing 10% MK performing the best. Similar results were also reported by Li and Ding [13] where concrete achieved the highest compressive strength with 10% MK. It is clear that MK improves the compressive strength of concrete. Overall, MK has great promise as an SCM.

2 Method

The CSIR has successfully produced MK from kaolinitic clays, initially using an electric furnace under very strict controlled conditions in the laboratory and then semi-industrially using coal-fired vertical shaft kilns (VSKs). High grade kaolinitic clay containing 95% kaolinite was calcined in all cases. Additionally, a low grade kaolinitic clay (red brick clay) was used in the laboratory scale to produce MK. The aim of study was to evaluate the compressive strength of cement pastes containing 10% and 50% MK from the 3 different metakaolins and to examine microstructural differences of cements extended using the metakaolins using scanning electron microscopy (SEM).

The metakaolins differed in the amounts of amorphous content. Mineralogical characteristics of the raw clay materials and of the metakaolins are shown in Tables 1 and 2.

Table 1. Mineralogical composition of the clay raw materials used to produce the different metakaolin qualities in wt%

Mineralogical characteristics	Low quality MK	Medium quality MK	High quality MK
Kaolinite	19	95	95
Quartz	33	4	4
Anatase	–	1	1
Feldspar	39	–	–
Mica	9	–	–

Table 2. Mineralogical composition of the metakaolins produced using an electric furnace in the laboratory, VSK and produced from the red brick clay

Mineralogical composition	Low quality MK	Medium quality MK	High quality MK
Amorphous phase	23	69	90.64
Quartz	31	4	5.65
Anatase	–	1.5	1.57
Mullite	–	4.9	0.73
Cristobalite	–	9.2	–
Mica	9	–	1.41
Feldspar	37	–	–

CEM I OPC 52.5 N was used in this study. The cement paste/MK blends prepared had MK contents of 10% and 50% for each quality of metakaolin. CEM I OPC 52.6 N was used for the control paste. The water/binder ratio used for all pastes was 0.35. No superplasticizer was used in this study. Cube specimens of 100 × 100 × 100 mm were cast in steel molds. They were de-molded after 1 day and cured in water at 22 °C. The compressive strengths of the pastes were determined at the ages 2 and 7 days.

3 Results

3.1 Compressive Strength

The results of the compressive strength are shown in Figs. 1 and 2, where each value is the average of three measurements.

The cement pastes containing 10% MK showed marginal decrease in strength when compared to the control at 2 days. The cement paste containing 10% high-quality MK had higher compressive strength than the control at 7 days. However, the cement pastes containing low and medium quality MK had compressive strength similar to the control. It was concluded that at early ages, the low and medium quality MK used in this study hardly contributed to the compressive strength of cement pastes. Addition of a superplasticizer could have improved consistency; workability and avoidance of

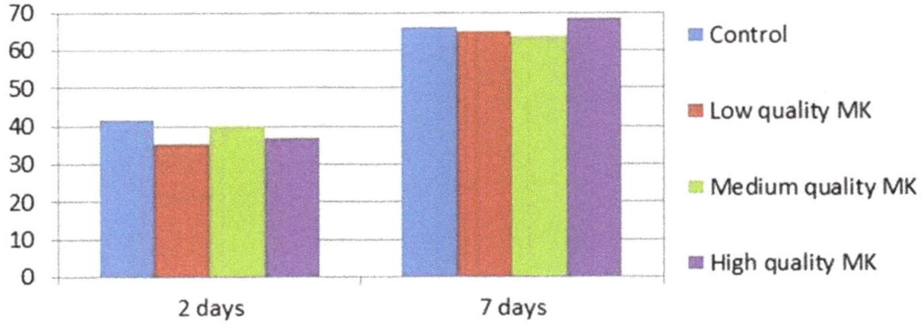

Fig. 1. Compressive strength of cement pastes containing 10% MK

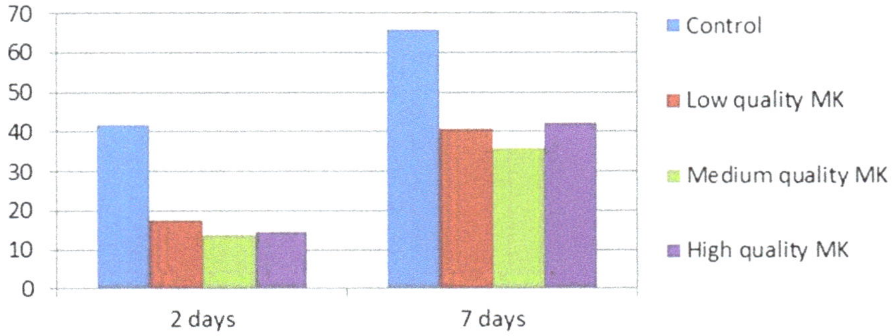

Fig. 2. Compressive strength of cement pastes containing 50% MK

agglomeration during mixing which resulted in lower strength [10]. Amazingly, cement paste with low quality MK performed very well in comparison to medium and high qualities MK as shown in Figs. 1 and 2. Juenger et al. [14] obtained similar results when clay shale as raw material was used to produce MK.

Literature shows that OPC replacement levels of 10% or 15% are optimal for maximum compressive strength [12, 13]. However, in this study cement some pastes contained 50% MK of different qualities. The cement pastes containing 50% MK had lower compressive strength than the control at both 2 and 7 days. However, the results indicate the compressive strength of cement pastes containing 50% MK increased rapidly from 2 days to 7 days.

3.2 Microstructure

Figures 3, 4, 5, 6, 7, 8 and 9 show the images of the control, cement pastes with MK contents of 10% and 50% for each metakaolin of different qualities at 2 days.

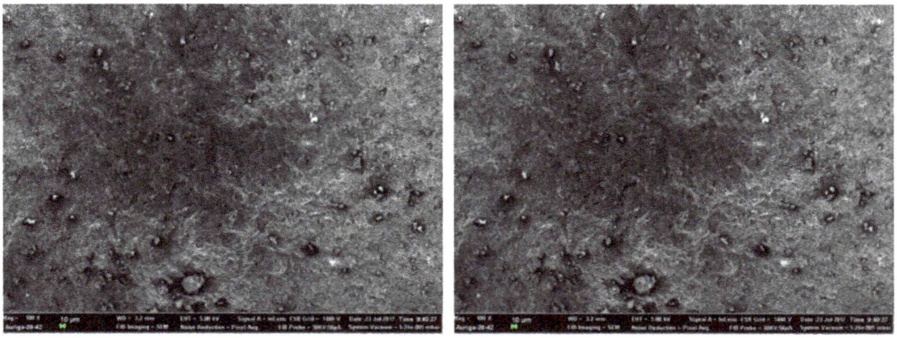

Fig. 3. 100% OPC magnification of 100 and 500X

Fig. 4. 10% low quality MK magnification of 100 and 500X

Fig. 5. 10% medium quality MK magnification of 100 and 500X

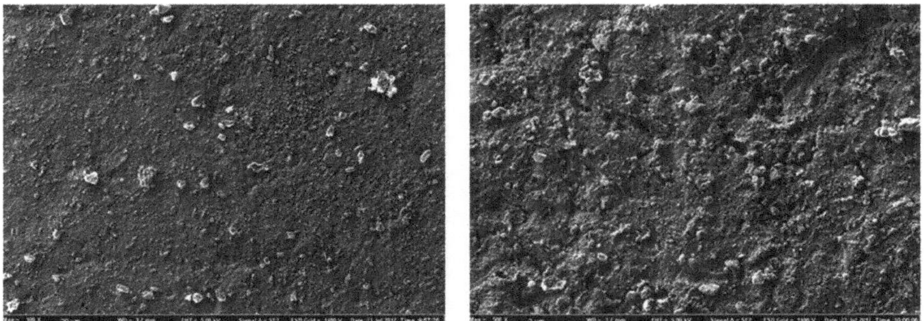

Fig. 6. 10% high quality MK magnification of 100 and 500X

Fig. 7. 50% low quality MK magnification of 100 and 500X

Fig. 8. 50% medium quality MK magnification of 100 and 500X

Fig. 9. 50% high quality MK magnification of 100 and 500X

The presence of lumps suggests agglomeration during mixing. As seen in Fig. 6, the paste with 50% medium quality shows some agglomeration. Similar results were observed for the paste incorporated with 50% high quality MK as shown in Fig. 9.

The results obtained from SEM indicate that the pastes with high content of MK had higher degree of agglomeration in comparison with lower content of MK.

4 Conclusion

The addition of metakaolin influences the compressive strength of cement pastes. The quality of MK affected the compressive strength obtained in MK-blended cement pastes. The cement pastes with high quality MK had high compressive strength compared to cement pastes with low and medium quality due to its high initial reactivity. The absence of superplasticizer in MK-blended cement pastes resulted to lower compressive strength and high degree of agglomeration. The cement pastes with low quality MK performed well in comparison with cement pastes containing high and medium qualities MK. Further work will be performed to obtain the compressive strengths of the MK-blended cements pastes at the age 14 and 28 days.

References

1. Ramezanianpour, A.A., Jovein, H.B.: Influence of metakaolin as supplementary cementing material on strength and durability of concretes. Constr. Build. Mat. **30**, 470–479 (2012)
2. Suryawanshi, Y.R., Kadam, A.G., Ghogare, S.S., Ingale, R.G., Patil, P.L.: Experimental study on compressive strength of concrete by using metakaolin. Int. Res. J. Eng. Tech. **2**, 235–239 (2015)
3. Juenger, M.C.G., Winnefeld, F., Provis, J.L., Ideker, J.H.: Advances in alternative cementitous binders. Cem. Concr. Res. **41**, 1232–1243 (2011)
4. Torgal, F.P., Shasavandi, A., Jalali, S.: Using metakaolin to improve the compressive strength and the durability of fly ash based concrete. In: INVACO2: Int. seminar, Innov. Valor. Civ. Eng. Constr. Mat. (2011)
5. Justice, J.M.: Evaluation of metakaolins for use as supplementary cementitious materials. MSc thesis. Georgia Institute of Technology (2005)
6. Wild, S., Khatib, J.M., Jones, A.: Relative strength, pozzolanic activity and cement hydration in superplasticised metakaolin concrete. Cem. Concr. Res. **26**, 1537–1544 (1996)
7. Si-Ahmed, M., Belakrouf, A., Kenai, S.: Influence of metakaolin on the performance of mortars and concretes. Int. J. Civ. Arch. Struct. Constr. Eng. **6**, 1010–1013 (2012)
8. Sabir, B.B., Wild, S., Bai, J.: Metakaolin and calcined clays as pozzolans for concrete: a review. Cem. Concr. Comp. **23**, 441–454 (2001)
9. Aiswarya, S., Arulraj, P.G., Dilip, C.: A review on use of metakaolin in concrete. Eng. Sci. Tech. Int. J. **3**, 592–597 (2013)
10. Dinakar, P., Sahoo, P.K., Sriram, G.: Effect of metakaolin content on the properties of high strength concrete. Int. J. Concr. Struct. Mat. **7**, 215–223 (2013)
11. Caldarone, M.A., Gruber, K.A., Burg, R.G.: High reactivity metakaolin (HRM): a new generation mineral admixture for high performance concrete. Concr. Int. **16**, 37–40 (1994)
12. Poon, C.S., Lam, L., Kou, S.C., Wong, Y.L., Wong, R.: Rate of pozzolanic reaction of metakaolin in high-performance cement pastes. Cem. Concr. Res. **31**, 1301–1306 (2001)

13. Li, Z.J., Ding, Z.: Property improvement of Portland cement by incorporating with metakaolin and slag. Cem. Concr. Res. **33**, 579–584 (2003)
14. Juenger, M., Seraj, S., Cano, R., Ferron, R.: Calcined shale as a low cost supplementary cementitous material. The University of Texas, USA paper presented at 1st Calcined Clays for Sustainable Concrete Switzerland (2015)

Influence of Initial Water Curing on Strength and Microstructure Development of Blended Cements

A.C. Emmanuel(✉), H. Talluru, S. Krishnan, and S. Bishnoi

Indian Institute of Technology Delhi, New Delhi, India

Abstract. Hydration of cement is a long term chemical process which requires water. Construction industry often prefers cement with high early strength and short curing period. Limestone calcined clay cement (LC3), which can replace clinker up to 50%, has potential benefits in this regard due to its low clinker factor and rapid reaction kinetics. In this study, ordinary portland cement, fly ash based portland pozzolanic cement, limestone calcined clay cement and composite cement (OPC with slag and fly ash) were prepared in laboratory. Mortar cubes and paste specimens were cast at constant water to binder ratio. The samples then exposed to different curing regimes by changing it from water curing to air curing (R.H ≈ 10–20%) at selected intervals. The curing temperature was kept constant as 27 °C. Compressive strength of mortar cubes were taken at standard ages. Effect of curing conditions on phase assemblage were quantified using, quantitative x-ray diffraction.

1 Introduction

Hydration of cement requires water. The process of curing ensures the presence of water in the system during hydration process, by providing moisture externally or stopping evaporation of water from specimen. Generally, the rate of hydration is high during the initial days and then decreases, making the initial days critical from curing point of view. But, this depends on type of cement. The industry is keen to reduce the use of clinker in cement, by replacing it with alternative materials called supplementary cementitious materials (SCMs). The cement blends with lower clinker factor have many advantages over conventional ordinary Portland cement (OPC). These blends can drastically reduce environmental pollution that comes from production of clinker through lowering the clinker factor by utilizing industrial by-products such as fly ash and slag. As per the Indian standards, based on quality, up to 30% of clinker can be replaced by fly ash and, up to 70% of clinker can be replaced using slag. Generally, these binary blends have low early strength and longer curing period [1]. The construction industry is interested in high early strength as well as lower curing period due to economic factors. In order to reduce these disadvantages, ternary cements were proposed using more than one SCMs [2].

The curing requirement may be different for different type of cements. It is expected that cements with high early age strength demands high amount of water. Ternary cements using alternative raw materials may influence the curing requirements. Many studies have

F. Martirena et al. (eds.), *Calcined Clays for Sustainable Concrete*, RILEM Bookseries 16,
https://doi.org/10.1007/978-94-024-1207-9_24

shown that the effect of curing conditions is more influential on blended cements. A study on fly ash based cement indicated that effect of curing largely depends on physical and chemical properties of clinker and w/b ratio rather than that of fly ash properties and replacement factor [3].

This study aims to identify the effect of curing conditions: relative humidity and curing duration on strength development of mortar from different blends prepared by replacing clinker with various SCMS.

2 Materials and Methods

2.1 Raw Materials and Blends

The raw materials were ground individually in laboratory ball mill. Blends were prepared by interblending of raw materials. The chemical composition of raw materials from X-ray fluorescence and composition of raw materials in blends are shown below (Tables 1 and 2).

Table 1. Chemical composition of raw materials

Chemical composition (%)	Clinker	Gypsum	Fly ash	Slag	Cal. Clay	Limestone
SiO_2	21.07	2.77	67.66	32.26	54.67	11.02
Al_2O_3	4.65	0.62	22.18	23.16	27.69	1.55
Fe_2O_3	4.32	0.36	5.32	1.93	4.93	2.53
CaO	65.16	32.62	0.32	33.88	0.06	44.24
MgO	2.13	1.2	0.18	7.01	0.13	1.96
SO_3	0.77	38.75	0.02	0.00	0.10	0.00
Na_2O	0.38	0.06	0.43	0.01	0.12	0.00
K_2O	0.20	0.037	1.6	0.37	0.25	0.00
LOI	0.96	23.02	1.74	1.08	10.28	36.96

Table 2. Raw material composition of blends (by % of weight)

Name of the blend	Abbr.	Clinker	Gypsum	Fly ash	Slag	Calcined clay	Lime-stone
Ordinary portland cement	OPC	95	5	–	–	–	–
Portland pozzolanic cement	FA30	65	5	30	–	–	–
Composite cement	CCS	50	5	15	30	–	–
Limestone calcined clay cement	LC3	50	5	–	–	30	15

2.2 Specimen Preparation

For compressive strength testing, mortar cubes of 70.6 mm size were cast as per IS 4031 part-6. The blends were dry mixed with standard sand [4] with a ratio of 1:3. A fixed water to binder ratio-0.45 was used for all blends. Paste samples were prepared by mixing cement blend with water at w/b ratio of 0.45. Proper mixing was ensured using high speed paste mixer.

2.2.1 Curing Conditions

The temperature during the whole process: material conditioning, casting of specimen and curing was maintained at 27 ± 2 °C. The prepared specimens (mortar cubes and paste samples) kept in an environmental controlled room (temperature: 27 ± 2 °C, RH: 65 ± 5%) for 24 h. After first day, specimens were demoulded and shifted to different curing regimes. Two different relative humidity (RH) was considered for this study: A higher RH of 100% (HH) and a lower RH of 10–20% (LH). HH was obtained by placing samples in lime saturated water and LH was obtained by placing samples in a closed box with silica gel inside. Fresh silica gel, which is blue in colour absorbs moisture from surroundings and turns colourless. Old silica gel was continuously replaced by fresh one at regular intervals to keep the humidity low. The curing duration was planned by shifting specimens from HH to LH at specific time durations. The details of curing durations and corresponding notation used in this paper are shown in Table 3.

Table 3. Curing regimes and notations

Curing regime	Notations
Specimen cured in HH	HH
Specimen cured in HH till 3 day and shifted to LH after 3 days	HH(3)_LH
Specimen cured in HH till 7 day and shifted to LH after 7 days	HH(7)_LH
Specimen cured in LH	LH

The cubes were tested in compression testing machine of capacity 500 kN with a loading rate of 2.40 kN/s. Average compressive strength of 3 cubes was considered for representation. The test is performed as per the Indian standard [5]. The compressive strength tests were carried out on the specimens in mortar cubes at 28 days.

For x-ray diffraction analysis, hardened Cement paste slices of 3 mm were cut and one surface of the slice was polished using 1200 silicon carbide paper. The XRD scanning was carried out in Bruker D8 Advance Eco X-ray diffractometer at 40 mV and 25 mA with Cu target (K_α = 1.54 A°). The range of scanning is 5° to 65° and rate of scanning is 0.019°/0.3 s. The whole running time for a single sample is around 17 min. In order to avoid carbonation of samples, scanning was carried out immediately after slicing and polishing. Quantitative analysis of phases was carried out using the software TOPAS v5 from Bruker. Major clinker phases and hydration products were quantified by external standard method using rutile.

3 Results

The 28-day compressive strength of mortar cubes cured under various curing regimes are shown in Fig. 1 (Fig. 2).

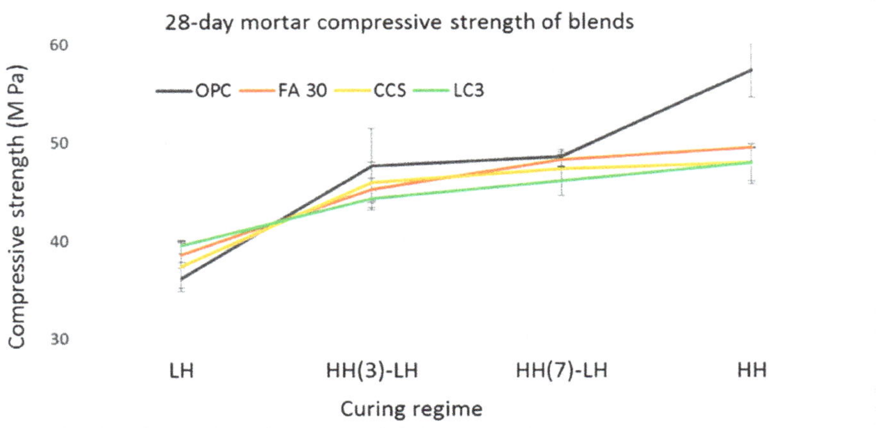

Fig. 1. 28-day mortar compressive strength of blends at different curing regimes

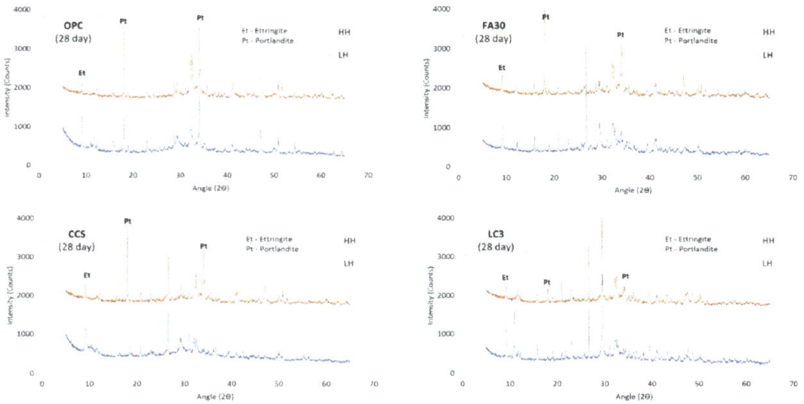

Fig. 2. XRD pattern of hydrated cement pastes at 28 days for both LH and HH

4 Conclusion

- It is evident the loss of strength occurs due to improper curing in all blends irrespective of clinker replacement factor.
- Both RH and curing period affects the strength development. This is more predominant in OPC where the clinker factor is high. The hydration of clinker phases especially belite, is highly influenced by presence of water in the system.

- It is to be noted that blended cements including FA30 has shown almost similar trend. This can be due to the particle size distribution fly ash and slag. The fine nature of these materials increases the pore refinement. The refinement of pores at specimen surface at early ages helps to maintain internal RH, by stopping evaporation during later age hydration.

The internal RH of specimen can impact the effect of curing, leads to possible influence of specimen size. Further work is required to check the influence of w/b ratio. Once the surface pores of the specimen stop percolation, the influence of external RH on properties of specimen as a whole stops. The time of occurrence of this transformation, after which curing has no influence, is depends on physical and chemical properties of raw materials, temperature history, w/b ratio etc.

References

1. Ozer, B., Ozkul, M.H.: The influence of initial water curing on the strength development of ordinary portland and pozzolanic cement concretes. Cem. Concr. Res. **34**(1), 13–18 (2004)
2. Menendez, G., Bonavetti, V., Irassar, E.F.: Strength development of ternary blended cement with limestone filler and blast-furnace slag. Cement Concr. Compos. **25**, 61–67 (2003)
3. Termkhajornkit, P., Nawa, T., Kurumisawa, K.: Effect of water curing conditions on the hydration degree and compressive strengths of fly ash-cement paste. Cement Concr. Compos. **28**(9), 781–789 (2006)
4. IS 650-1: Standard Sand for Testing Cement. Bureau of Indian Standards, Part 1 (1991)
5. IS 4031-6: Methods of Physical Tests for Hydraulic Cement. Bureau of Indian Standards, pp. 1–3 (1988)

Initial Performance Evaluation of Calcined Clay Based Ternary Blended Cement Under Various Climatic Conditions in India

A.C. Emmanuel(✉), G. Mishra, and S. Bishnoi

Indian Institute of Technology Delhi, New Delhi, India

Abstract. Portand pozzolanic cement using fly ash (FA30), which replaces 30% of clinker, is the most produced cement in India. But, rules due to the fast depletion of natural resources and stricter on emissions, the industry is keen to reduce the clinker factor further. New ternary blended cement, limestone calcined clay cement (LC^3) can be a good option due to its high clinker replacement and use of low grade raw materials. The cement is aimed for general purpose. India, having wide range of climatic conditions, producing general purpose cement for the entire country is always a hard task for cement companies. For a new cement, it is worthy to check the performance under different climatic conditions. In this regard, concrete columns were cast in both FA30 and LC^3, and placed at 8 locations. NDT test was performed on columns to understand the performance concrete.

Keywords: Field performance · Limestone calcined clay cement · Climatic conditions

1 Introduction

As one of the fastest growing economy in the world area with massive population, India has tremendous potential for growth in infrastructure sector including construction in which cement is an essential commodity. Forecasting the growth of the industry versus its impact on environment and sustainability; industry is in search for new technologies. Production of cement through clinker manufacturing emits significant amount of CO_2 due to chemical process and energy requirement involved in it. In the new scenario of climate change, the world is looking for emission reduction from industries who are more liable; and cement industry is one of them. The Indian cement industry which is second behind China in terms of production, practices best technologies available in market, and uses supplementary cementitious materials (SCMs) and mineral additives as clinker replacement in cement. The presence of large number of coal based thermal power plants in the country leads to use of fly ash, a by-product from coal combustion as the major SCM in the country. Currently more than 65% of market share goes to portland pozzolanic cement with fly ash which can replace 30% of clinker as per Indian standards. Due to the same emission issues, the Indian government is moving to alternate

F. Martirena et al. (eds.), *Calcined Clays for Sustainable Concrete*, RILEM Bookseries 16,
https://doi.org/10.1007/978-94-024-1207-9_25

sources for power generation, which can lead to closure of coal based thermal power plants, thus can drastically reduce the availability of fly ash in future.

At this stage, the major way for emission reduction is the use of alternate raw materials which can reduce the clinker factor further. The recently developed ternary blended cement known as limestone calcined clay cement (LC^3), which can replace clinker up to 50% is a good option [1]. LC^3, which uses the synergy between clinker, calcined clay and limestone has benefits such as high early strength, better durability, and can be produced from lower grade raw materials [2]. The cement is aimed to be used as a general-purpose cement. The details of raw material processing and production can be found in literature [3, 4].

India is a one of the largest countries in the world, with a wide range of climatic conditions. Cement which is proposed to be used for structural applications need to satisfy the necessary durability criteria in all the climatic conditions within the country. To evaluate this, a real field, long term experimental programme was proposed. The entire process of the experimental programme - material preparation, selection of locations, and initial tests are explained in the next section.

2 Methodology

Concrete columns of size 230 × 230 × 900 mm were cast using two blends; FA30 and LC^3. The size of column was designed in such a way that it can be considered as reinforced concrete structure and can be transported easily. The specimens were kept for 3 months under same conditions and placed in different locations across the country. It is planned to retrieve the concrete columns after 5 years for necessary durability testing in the laboratory.

2.1 Specimen Preparation

Both LC^3 and OPC produced by industrial trial production [3] were used for this study. FA30 blend was prepared in laboratory ball mill by intergrinding commercial fly ash and OPC. Water to binder ratio was kept at 0.45. A PCE based super plasticizer- Glenium MasterSky B8233 was used with a dosage of 0.7% of total binder content. Concrete mix was prepared using a drum mixer of 400 L capacity, and mix was poured into the prepared moulds. Concrete cubes of size 150 mm were also prepared to obtain compressive strengths. The specimens were wet cured for 14 days under ambient conditions, and kept for further 90 days at same conditions. Each set of columns were transported to the selected locations for installation.

2.2 Selection of Locations and Installation of Columns

Two climatic classifications were considered for determining the locations. The national building code of India (NCBI) [5] and Koppen classification [6]. The selected locations and respective climatic zones are explained in Table 1. The columns were installed in open areas. Efforts were taken to avoid any kind of shades including proximity of trees.

The half (lengthwise) of the column was placed below the ground level. In this way, it is possible to understand the effect of soil conditions on properties on concrete.

Table 1. Selected locations of installing concrete specimens

Nos	Place	State	Climatic condition	
			NBCI	Koppen
1	Delhi	Delhi	Composite	Semi-arid
2	Chennai	Tamil Nadu	Warm humid	Tropical wet and dry
3	Srinagar	Jammu & Kashmir	Cold	Mountain
4	Jodhpur	Rajasthan	Hot and dry	Arid
5	Bangalore	Karnataka	Temperate	Semi-arid
6	Jhansi	M. P	Composite	Humid Subtropical
7	Thangskai	Meghalaya	Warm Humid	Humid Subtropical
8	Surathkal	Karnataka	[a]Marine	[a]Marine

[a]Not defined by the classification systems. Selected for studying the effect of marine exposure

2.3 Tests on Columns

After 15 months, field visits were conducted to four locations. Non-destructive testing was carried out at this stage including visual observation. Studies have shown the relationship between electrical resistivity and pore refinement in concrete specimens, and can be used as a measure of quality of concrete against its resistance to ionic ingress. It is also an indication of rate of corrosion and chloride resistance of concrete [7, 8]. Concrete resistivity can be measured using a Wenner four-electrode technique commonly used for measuring soil resistivity [9]. The ACI committee report establishes the relationship between concrete resistivity and corrosion rate [10].

In order to observe the quality of concrete below the ground level, soil strata was removed from two side of the column till the bottom end. The selected surface portion was cleaned properly. Surface resistivity measurement were taken using a 4 probe 39 mm fixed probe Proceq surface resistivity meter. Moisture presence in the surface of specimen can drastically affect the readings. To eliminate this variation, the surface moisture was normalized by pouring water in concrete surface and monitored using a moisture meter. Measurements were taken from two adjacent vertical surfaces, with an interval of 10 cm between two readings.

3 Results

The compressive strength of concrete cubes were obtained as per Indian standard [11]. The test results are shown in Fig. 1. The surface resistivity of measured concrete columns is described in Fig. 2.

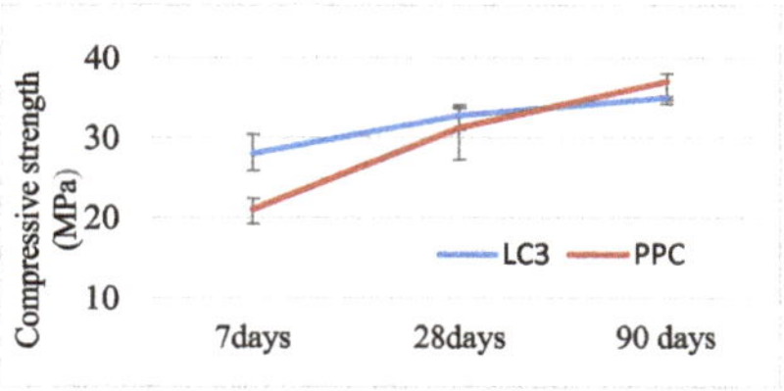

Fig. 1. Compressive strength of concrete

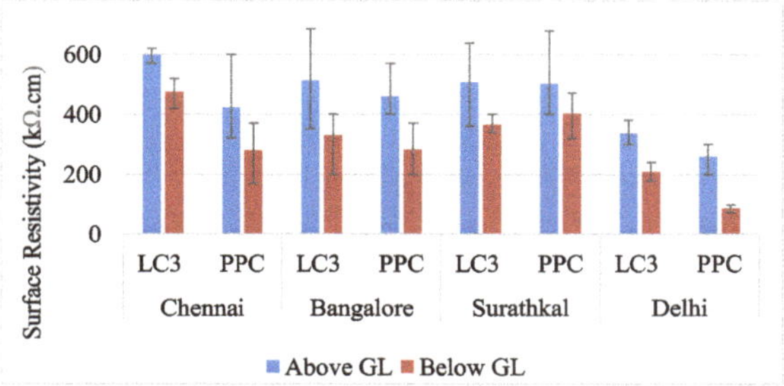

Fig. 2. Surface resistivity of concrete columns

4 Discussions

- No visible crack was observed in both PPC and LC^3 specimens.
- LC^3 and PPC specimens have shown good surface resistivity. At this point of time, LC^3 have shown a better surface resistivity. This can be due to better pore refinement in LC^3 system. All the tested samples have very low corrosion rate as per the ACI guidelines.
- There is a significant variation in the observed values. This can be attributed to presence of reinforcements as well as the variation of moisture in specimen. Specimen moisture depends on local conditions like rain, sunlight, humidity, temperature especially during the recent days of testing period. The conditions during testing: temperature and humidity can also affect the readings.
- It has to be noted that one year of duration not sufficient to understand the effect of climatic conditions on durability performance of concrete. Moreover, further testing such as chloride ingress, carbonation, corrosion and sulphate attack in which many of them are destructive in nature to be carried out for a proper conclusion. It is also advised to carry out these tests t under same conditions.

Acknowledgements. The Swiss Agency for Development and Cooperation is gratefully acknowledged for funding and supporting the study on Limestone Calcined Clay Cement in India. Faculties of research institutions: IIT Madras, IISc. Bangalore, IIT Jodhpur, NIT Surathkal; TARA, Col. Chandra Prakash, Dalmia cement Ltd are also acknowledged for their support.

References

1. Fernandez, R.: Calcined clayey soils as a potential replacement for cement in developing countries, vol. 4302, p. 178 (2009)
2. Sanchez Berriel, S., Favier, A., Rosa Domínguez, E., Sanchez Machado, I.R., Heierli, U., Scrivener, K., Martirena Hernandez, F., Habert, G.: Assessing the environmental and economic potential of limestone calcined clay cement in Cuba. J. Clean. Prod. **124**, 361–369 (2016)
3. Emmanuel, A.C., Haldar, P., Maity, S., Bishnoi, S.: Second pilot production of limestone calcined clay cement in India: the experience. Indian Concr. J., no. Clay 2 (2016)
4. Bishnoi, S., Maity, S., Mallik, A., Joseph, S., Krishnan, S.: Pilot scale manufacture of limestone calcined clay cement: the Indian experience. Indian Concr. J., 1–12 (2014)
5. Bureau of Indian Standards: National Building Code of India 2005, p. 883 (2005)
6. Kottek, M., Grieser, J., Beck, C., Rudolf, B., Rubel, F.: World map of the Köppen-Geiger climate classification updated. Meteorol. Zeitschrift **15**(3), 259–263 (2006)
7. Feliú, S., González, J.A., Andrade, C.: Electrochemical methods for on-site determinations of corrosion rates of rebars. In: ASTM International, vol. STP 1276, pp. 107–118 (1996)
8. Ghosh, P., Tran, Q.: Correlation between bulk and surface resistivity of concrete. Int. J. Concr. Struct. Mater. **9**(1), 119–132 (2015)
9. ASTM: Standard test method for field measurement of soil resistivity using the Wenner four-electrode method. In: ASTM, vol. 6, no. reapproved, pp. 1–6 (2012)
10. ACI Committee 222: Protection of Metals in Concrete Against Corrosion, vol. 1 (2001)
11. Bureau of Indian Standards: 191IS 516 -1959: Method of Tests for Strength of Concrete (2004)

Alkali Silica Reaction and Sulfate Attack: Expansion of Limestone Calcined Clay Cement

A. Favier(✉) and K. Scrivener

LMC, Ecole polytechnique fédérale de Lausanne, Lausanne, Switzerland

Abstract. In this paper, two durability issues: Alkali Silica Reaction (ASR) and external sulfate attack were studied. The results of expansion were presented for a Normal Portland cement and a blended cement LC3 containing calcined clay and limestone as supplementary cementitious materials (SCMs). The systems containing SCMs did not expand after several years of exposure showing a high resistance to ASR and sulfate attack.

1 Introduction

Concrete structures are often subject to degradation due to the alkali silica reaction (ASR). ASR is a reaction between amorphous silica from aggregates and the alkalis from the pore solution of the cement paste. An expansive gel is formed, which induces cracks until a complete deterioration of the concrete. For different reasons, the ingress of sulfate ions in concrete structures called sulfate attacks can also lead to major deterioration of civil engineering structures such as piers, bridges, concrete pipes. Damage due to sulfate interaction with concrete can result in the expansion then cracking and softening with loss of strength. The sulfate ions, which come from soil, seawater, react with the cement hydrates (calcium hydroxide, C-S-H. AFm phases) to form ettringite, gypsum and thaumasite in presence of carbonates [1–3].

The use of supplementary cementitious materials to control the expansion is well established and a number of reviews have been published [4–7]. However, the role of SCMs is strongly dependent on their chemical composition and replacement level. Their mechanisms in mitigating ASR and reducing sulfate ingress remain unclear. Therefore, in ASR condition, it has been also suggested that the presence of alumina in SCMs contributed some way to prevent the release of alkali back to the pore solution. However, Chappex and Scrivener [8] showed that this effect is extremely small. Alumina is an inhibitor of silica dissolution [9]. Based on these observations, the use of alumina-rich SCMs, such as calcined clays containing metakaolin should be favoured for mitigating ASR effects. Blended cement with fly ash or/and silica fume shows a better resistance to sulfate attack due to the additional formation of C-S-H and the reduced amounts of portlandite and monosulfate. This induces a refined porosity thus retarding the penetration of sulfate ions [6]. It was report that limestone addition up to 5%wt reduces porosity but this effect is reversed for larger amount (>10%wt) [6, 7]. On the other hand, metakaolin or natural pozzolans gives a significant reduction in expansion and an increasing resistance to sulfate attack with the increasing metakaolin replacement level [10–12].

F. Martirena et al. (eds.), *Calcined Clays for Sustainable Concrete*, RILEM Bookseries 16,
https://doi.org/10.1007/978-94-024-1207-9_26

In this paper, we studied the effect of the combination of two SCMs: limestone and calcined clay at high replacement level to improve the resistance to sulfate attack and Alkali Silica Reaction of concrete structures by measuring the length change.

2 Materials and Testing Procedures

2.1 Materials for Alkali Silica Reaction

Portland cement CEM I 42.5 N, calcined clay containing approximatively 50% of calcined kaolinite and commercial limestone were used to prepare blended cement LC3 containing 30% calcined clay and 15% limestone. Two types of aggregates with the same particle size distribution were tested, a highly reactive North American sand (NA sand) and a slowly reactive Swiss aggregate (CH sand).

2.2 Materials for Sulfate Attack

Portland cements CEM I 42.5 N with C3A content (5–6%), two different calcined clays with approximatively the same content in calcined kaolinite (50%) and limestone powder were used to prepare ternary blended cement with three replacement levels: LC3-50 with 45% replacement – LC3-65 with 30% replacement – LC3-80 with 15% replacement.

2.3 Procedure of Testing Alkali Silica Reaction

Mortars bars were cast with the same Water/Binder ratio of 0.46, the amount of plasticizer was adjusted to obtain a workable mixture. Mortar bars were cast with inset metal measurement studs to monitor expansion. The aggregate to cement ratio was 3. The mortars bars were first cured for 28 days at 20 °C and 95%RH. They were then put in alkaline solution containing 0.32 mol/l NaOH as it is close to the pore solution and at 38 °C to accelerate the expansion. The length was measured at regular intervals at 38 °C.

2.4 Procedure of Testing Sulfate Attack

Mortars were prepared according to EN196-1 with water to binder ratio of 0.5 and a sand/binder ratio of 3 to 1. Mortar bars were cast with inset metal measurement studs to monitor expansion. All mortars were cured for 90 days in saturated limewater at 20 °C before exposure to 30 g/l Na_2SO_4 solution. Prisms for expansion after curing were cut to 2 × 2 × 16 mm size. Then, they were immersed in 30 g/l sodium sulfate solution with dynamic mode i.e. the solution was renewed regularly.

3 Expansion Results

3.1 Expansion Due to Alkali Silica Reaction

Figure 1 shows the expansion as function of time for the Portland cement in NaOH solution with two concentration 0.32 M and 1.6 M, for ternary blend with calcined clay and limestone and 2 types of aggregates. We can observe that the expansion was more important in Portland cement than in LC3 systems. In the 1.6 M, the expansion seemed to be faster but stabilized after a while. However, in the lower concentration solution, the expansion increased slower without stabilisation. The limit defined by standard ASTM C1293 on concrete and C1260/C1367 on mortars (different exposure and curing conditions) are respectively 0.04% at 1 year and 0.1% at 14 days. All LC3-50 did not expand significantly and passed the requirements.

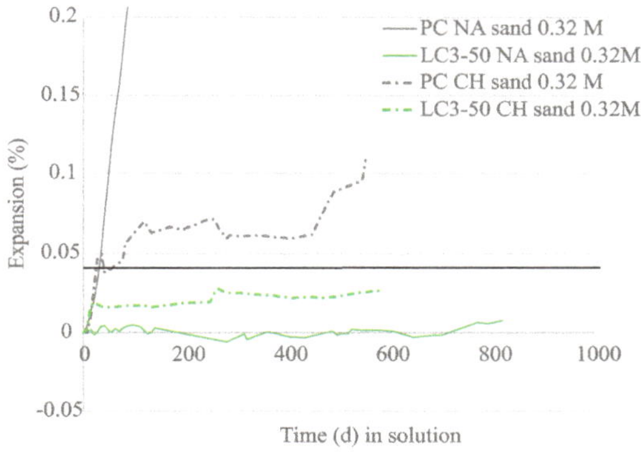

Fig. 1. Expansion as a function of time after immersion in NaOH solution for two different cements with NA sand and Ch sand. PC 0.32 M is immersed in 0.32 NaOH solution whereas PC 1.6 M is immersed in 1.6 M NaOH solution. The dotted line indicates the limit of innocuous behavior on concrete at one year according to ASTM C1293-08b

3.2 Expansion Due to Sulfate Attack

Figure 2 shows the linear expansion data of mortars in sulfate solution for both cements. These curves indicate that Portland cement is not sulfate resistant. Its expansion value at 200 days was close to 1% whereas the expansion of ternary blended cements was 0%. For information purposes, the high limit value described in standard ASTM C1012 (different exposure conditions) is 0.1%. However, after 200 days, the system containing less calcined clays and limestone LC3-80 started to expand. To mitigate completely the sulfate attack, a minimum content of calcined clay and limestone must be considered.

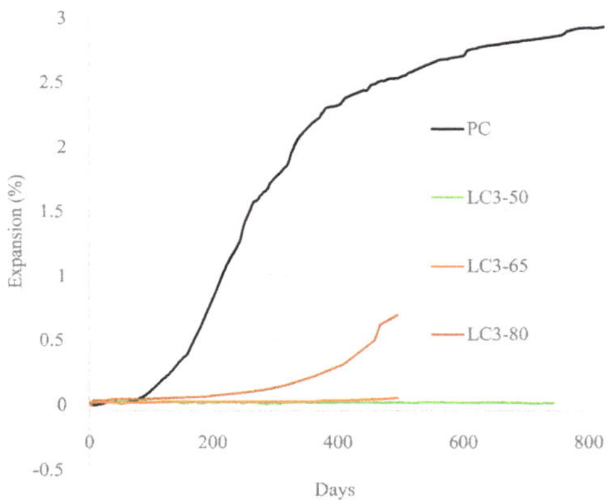

Fig. 2. Expansion curve overtime for mortars exposed in 30 g/l Na_2SO_4 solution.

4 Conclusion and Perspectives

LC3 cements could mitigate both ASR and sulfate attack. However, a minimum amount of calcined clay and limestone was required (around 30%) to mitigate the expansion caused by sulfate attack.

Extended researches were carried out to monitor sulfate ingress and damages of aggregates caused by ASR. Mechanisms of Sulfate and ASR mitigation are strongly investigated.

Acknowledgements. Dr Pawel Durdzinski and Dr Cyrille Dunant are thanked for the scientific support. The Swiss agency for Development and Cooperation (Project n°: 7F08527) is gratefully thanked for its financial support of LC3 project (www.lc3.ch).

References

1. Gollop, R.S., Taylor, H.F.W.: Microstructural and microanalytical studies of sulfate attack. I. Ordinary portland cement paste. Cem. Concr. Res. **22**, 1027–1038 (1992). doi:10.1016/0008-8846(92)90033-R
2. Marchand, J., Odler, I., Skalny, J.P.: Sulfate Attack on Concrete. CRC Press, New York (2003)
3. Marchand, J., Skalny, J.P.: Materials Science of Concrete: Sulfate Attack Mechanisms. Wiley-American Ceramic Society, New York (1999)
4. Thomas, M.: The effect of supplementary cementing materials on alkali-silica reaction: a review. Cem. Concr. Res. **41**, 1224–1231 (2011). doi:10.1016/j.cemconres.2010.11.003
5. Shafaatian, S.M.H., Akhavan, A., Maraghechi, H., Rajabipour, F.: How does fly ash mitigate alkali–silica reaction (ASR) in accelerated mortar bar test (ASTM C1567)? Cem. Concr. Compos. **37**, 143–153 (2013). doi:10.1016/j.cemconcomp.2012.11.004
6. Schmidt, T., Lothenbach, B., Romer, M., Neuenschwander, J., Scrivener, K.: Physical and microstructural aspects of sulfate attack on ordinary and limestone blended portland cements. Cem. Concr. Res. **39**, 1111–1121 (2009). doi:10.1016/j.cemconres.2009.08.005

7. Lee, S.T., Hooton, R.D., Jung, H.-S., Park, D.-H., Choi, C.S.: Effect of limestone filler on the deterioration of mortars and pastes exposed to sulfate solutions at ambient temperature. Cem. Concr. Res. **38**, 68–76 (2008). doi:10.1016/j.cemconres.2007.08.003
8. Chappex, T., Scrivener, K.: Alkali fixation of C–S–H in blended cement pastes and its relation to alkali silica reaction. Cem. Concr. Res. **42**, 1049–1054 (2012). doi:10.1016/j.cemconres.2012.03.010
9. Chappex, T., Scrivener, K.L.: The influence of aluminium on the dissolution of amorphous silica and its relation to alkali silica reaction. Cem. Concr. Res. **42**, 1645–1649 (2012). doi:10.1016/j.cemconres.2012.09.009
10. Khatib, J.M., Wild, S.: Sulphate resistance of metakaolin mortar. Cem. Concr. Res. **28**, 83–92 (1998). doi:10.1016/S0008-8846(97)00210-X
11. Al-Akhras, N.M.: Durability of metakaolin concrete to sulfate attack. Cem. Concr. Res. **36**, 1727–1734 (2006). doi:10.1016/j.cemconres.2006.03.026
12. Irassar, E.F., González, M., Rahhal, V.: Sulphate resistance of type V cements with limestone filler and natural pozzolana. Cem. Concr. Compos. **22**, 361–368 (2000). doi:10.1016/S0958-9465(00)00019-6

The Effect of Limestone on the Performance of Ternary Blended Cement LC3: Limestone, Calcined Clays and Cement

Aurélie Favier(✉), Franco Zunino, Ioannis Katrantzis, and Karen Scrivener

Laboratory of Construction Materials, EPFL, Lausanne, Switzerland

Abstract. This project explores the influence of different particle size distributions of limestone on the workability and early-age strength development of LC^3 materials. To quantify, and subsequently optimize, the packing, the modified Andreasen and Andersen model was used. Different systems were selected and studied using minicone tests and compressive strength (at 2, 7 and 28 days) to see which parameters govern workability and strength.

1 Introduction

It is very common today to find cements where part of the clinker is substituted with Supplementary Cementitious Materials (SCM) in order to reduce the ecological impact of the concrete [1]. More recently, research has demonstrated the feasibility of using limestone and calcined clays as SCM [2–4]. The combined use of both allows greater clinker substitution and a marked improvement in properties [5–9]. These materials are known as LC^3, for Limestone Calcined Clay Cement. On concern in these systems is the impact of the fine clay on the workability.

There are several methods to tune the workability of a concrete, the simplest is to add water, but this will negatively affect the strength. It is also possible to work with superplasticizers. The third method is to optimize the granular skeleton of concrete aggregates, which has been done for decades [11–14]. The idea of this work is to optimize the particle packing at the cement scale directly using the Andreasen and Andersen model [13, 14]. The scope of this study is to define the packing parameters and to understand their links with the components and properties of LC^3.

2 Materials and Characterisation Methods

2.1 Materials

In this study, LC^3-50 blends (Limestone Calcined Clay Cement) with ~45% replacement of cement were prepared. Six commercial limestones, a CEM I 42.5 N and a calcined clay from India with 50% calcined kaolinite were used. The limestones were combined together as explained in Table 1. The limestone LS1-o is a very fine limestone containing organics compounds to improve workability.

F. Martirena et al. (eds.), *Calcined Clays for Sustainable Concrete*, RILEM Bookseries 16,
https://doi.org/10.1007/978-94-024-1207-9_27

Table 1. Nomenclature of raw materials

Names	Series
LS 1-o	Limestone B
LS 4	
LS 2	Limestone D
LS 3	
LS 4	
LS 5	Limestone J
LS 6	
CC	Calcined clay
PC	CEM I 42.5 N

Three different ratios between calcined clay and limestone, respectively 2:1, 1:1 and 1:2, were tested as replacement materials.

The particle size distribution of the raw materials was measured by laser granulometry following procedure given in [15] and presented on Fig. 1.

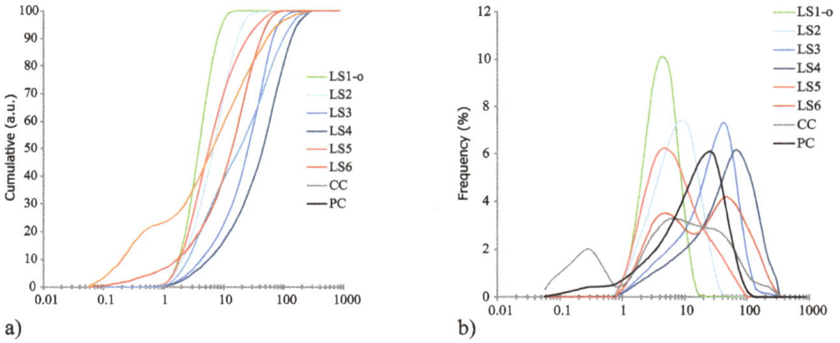

Fig. 1. (a) Cumulative Particle size distribution (PSD) of raw materials (b) Frequency plot of PSD for the raw materials

2.2 Methods

The workability was assessed using a mini-slump cone on paste with a water to binder ratio of 0.5 by mass. The mini-slump cone has a bottom diameter of 38 mm, a top diameter of 19 mm, and a height of 57 mm. The compressive strength was measured on 20 × 20 × 20 mm cubes of paste with a water to binder ratio of 0.5 by mass.

3 Andreasen and Andersen Model

The model of Andreasen and Andersen modified by Brouwers [13, 14] was used to design our experimental plan. Some compositions boundaries were defined. In this study, we focused on LC^3-50 blends: the amount of cement was fixed and the ratios between calcined clay to limestone were 2:1, 1:1 and 1:2. Equation 1 of the theoretical particle size distribution $F_{th}(d)$ is:

$$F_{th}(d) = \frac{d^\alpha - d^\alpha_{min}}{d^\alpha_{max} - d^\alpha_{min}} \, for \, \alpha \neq 0 \tag{1}$$

Brouwers et al. reported in [14] that the highest packing density can be obtained when $\alpha = 0$. Equation 1 becomes Eq. 2:

$$F_{th}(d) = \frac{\ln d - \ln d_{min}}{\ln d_{max} - \ln d_{min}} \, for \, \alpha = 0 \tag{2}$$

This theoretical distribution was compared to the distribution $F_{ex}(d)$, defined from the particle size distributions of raw materials $F_{i,ex}$ and described by Eq. 3.

$$F_{ex}(d) = \sum_i c_i.F_{i,ex} \, with \, c_i \, the \, volume \, fraction \, of \, each \, LC3 \, components \tag{3}$$

The c_i of the limestones were then optimized to get the higher correlation coefficient R^2, which translates how well the F_{ex} fits the F_{th}. The higher the R^2, the more packed the system is. It was a strong assumption accepted in this work to take $\alpha = 0$, describing a linear behaviour in log scale far from the one obtained in normal Portland cement. It was why we used another packing parameter ϕ_c from experiments, which accounts for the packing but also the water trapped in the calcined clay (see Fig. 2). A linear relationship between ϕ_c and the amount of calcined clay is observed. This parameter was calculated from Eq. 4, measuring V_{liq} by centrifugation at high speed of a binder with a (water + dispersant)/solids ratio of 3.

$$\varphi_c = \frac{V_{solids}}{V_{solids} + V_{liquid}} \tag{4}$$

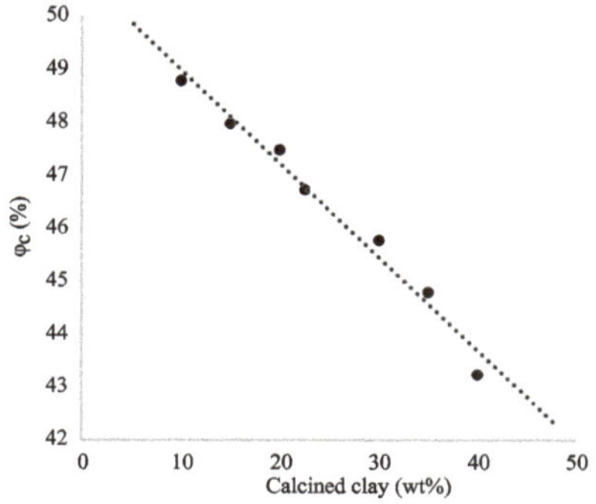

Fig. 2. φ_c as a function of calcined clay content, all series considered.

4 Main Results

4.1 Workability

From Fig. 3, a strong correlation between workability, calcined clay content and φ_c was revealed. Two sets of data can be identified, one corresponding to systems containing organic substances which enhance flowability (B) and another without such organic molecules (i.e. D, J). Workability increased with increasing φ_c. Therefore, φ_c is actually a measure of the water that is available as lubricant in the system, taking into account both:

- the water that is trapped by the calcined clays,
- the packing of the system.

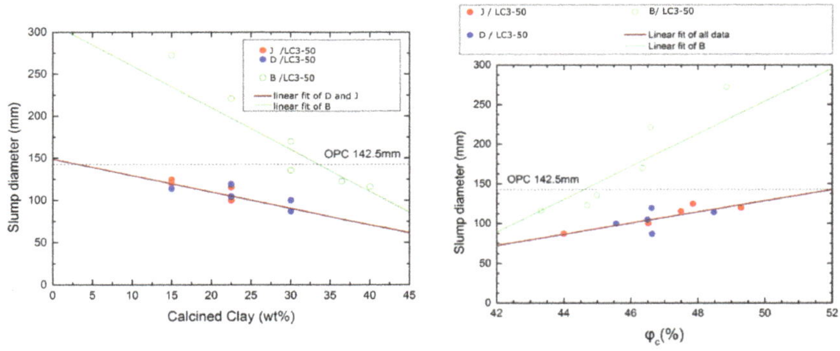

Fig. 3. (a) Slump diameter versus calcined clay content for all the systems. (b) Slump diameter versus φ_c for all the systems.

However, as it has been shown, the main factor is the amount of clay in the system, which overwhelmed the effect of packing.

4.2 Compressive Strength

Figure 4 shows the compressive strength of LC3-50 blends over time. An increase in strength was observed as a function of calcined content as suggested by Avet et al. [9], more significantly at 7 days and 28 days. At early age, the strength is attributed to the hydration of Portland cement. The same trend was shown as a function of R^2 and ϕ_c (Fig. 5) suggesting that an increase of the quality of packing affected positively the compressive strength at all ages.

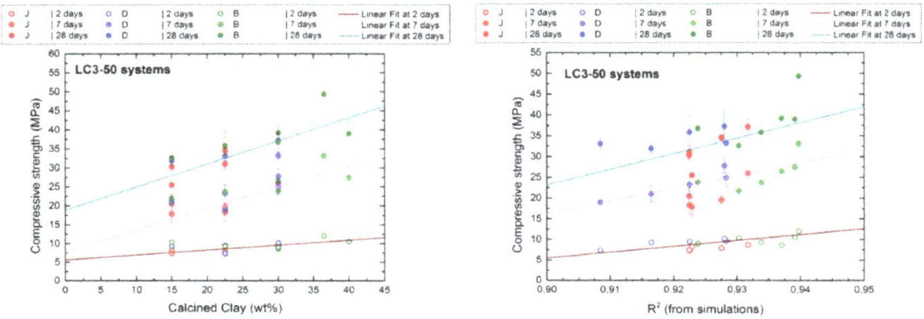

a) b)

Fig. 4. (a) Compressive strength versus Calcined Clay content (wt%) in LC3-50 systems at 2, 7 and 28 days. (b) Compressive strength versus R^2 (packing parameter) in LC3-50 systems at 2, 7 and 28 days. Linear fitting has been applied on the data corresponding to the same age.

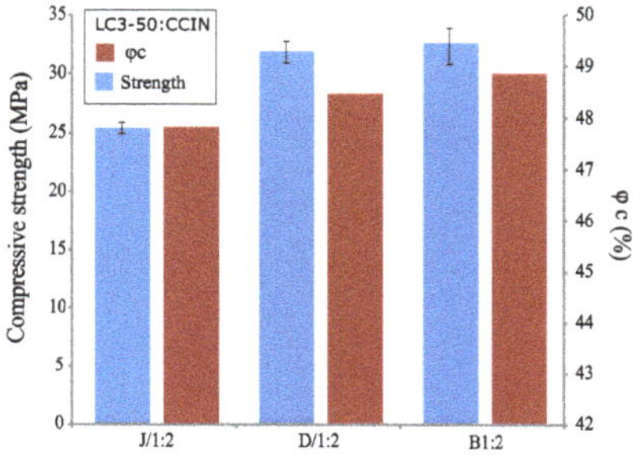

Fig. 5. Effect of the LS group on the φc and the strength for LC3-50 systems with 1:2 calcined clay/limestone ratio

5 Conclusion

This work contributed to the definition of two parameters R^2 and ϕ_c, representing respectively the quality of packing and the volume occupied by solids in the system.

Dependence of the compressive strength and the workability on these two parameters was studied. This demonstrated that workability is controlled by the calcined clay, which traps water reducing the effective water to binder ratio. However, the use of limestone containing organic components significantly increases the workability of the system. The other important observation is that it is possible to improve the compressive strength by improving the packing.

Acknowledgements. The authors gratefully acknowledge financial support from the Swiss Agency for Development and Cooperation (Project number 7F08527).

References

1. Lothenbach, B., Scrivener, K., Hooton, R.D.: Supplementary cementitious materials. Cem. Concr. Res. **41**, 1244–1256 (2011). doi:10.1016/j.cemconres.2010.12.001
2. De Weerdt, K., Sellevold, E.J., Kjellsen, K., Justnes, H.: Fly ash–limestone ternary cements: effect of component fineness. Adv. Cem. Res. **23**, 203–211 (2011)
3. Lothenbach, B., Le Saout, G., Gallucci, E., Scrivener, K.: Influence of limestone on the hydration of Portland cements. Cem. Concr. Res. **38**, 848–860 (2008). doi:10.1016/j.cemconres.2008.01.002
4. Sabir, B.B., Wild, S., Bai, J.: Metakaolin and calcined clays as pozzolans for concrete: a review. Cem. Concr. Compos. **23**, 441–454 (2001). doi:10.1016/S0958-9465(00)00092-5
5. Antoni, M., Rossen, J., Martirena, F., Scrivener, K.: Cement substitution by a combination of metakaolin and limestone. Cem. Concr. Res. **42**, 1579–1589 (2012). doi:10.1016/j.cemconres.2012.09.006
6. Favier, A., Avet, F., Antoni, M., Scrivener, K.: Limestone Calcined Clays Cement for a sustainable development. In: Guillaume Habert, Arno Schlueter, Zurich, Switzerland (2016). doi:10.3218/3774-6
7. Scrivener, K.L., Laffely, J.D., Favier, A.: Limestone calcined clay cement. Cement Plant Environmental Handbook, 2nd edn., 159 (2014)
8. Scrivener, K.: Options for the future of cement. Indian Concr. J. **88**(7), 11–21 (2014)
9. Avet, F., Snellings, R., Alujas Diaz, A., Ben Haha, M., Scrivener, K.: Development of a new rapid, relevant and reliable (R3) test method to evaluate the pozzolanic reactivity of calcined kaolinitic clays. Cem. Concr. Res. **85**, 1–11 (2016). doi:10.1016/j.cemconres.2016.02.015
10. Akhlaghi, O., Aytas, T., Tatli, B., Sezer, D., Hodaei, A., Favier, A., Scrivener, K., Menceloglu, Y., Akbulut, O.: Modified poly(carboxylate ether)-based superplasticizer for enhanced flowability of calcined clay-limestone-gypsum blended Portland cement. Cem. Concr. Res. **101**, 114–122 (2017)
11. de Larrard, F., Sedran, T.: Mixture-proportioning of high-performance concrete. Cem. Concr. Res. **32**, 1699–1704 (2002). doi:10.1016/S0008-8846(02)00861-X
12. Roussel, N.: Understanding the Rheology of Concrete. Woodhead Pub. Ltd. (2012)
13. Andreasen, A.H.M.: Ueber die Beziehung zwischen Kornabstufung und Zwischenraum in Produkten aus losen Körnern (mit einigen Experimenten). Kolloid-Z. **50**, 217–228 (1930). doi:10.1007/BF01422986
14. Brouwers, H.J.H.: Particle-size distribution and packing fraction of geometric random packings. Phys. Rev. E **74**, 031309 (2006). doi:10.1103/PhysRevE.74.031309
15. Scrivener, K., Snellings, R., Lothenbach, B.: A Practical Guide to Microstructural Analysis of Cementitious Materials. CRC Press, New York (2016)

Influence of Clay Type on Performance of Calcined Clay – Limestone Portland Cements

S. Ferreiro(✉), D. Herfort, and J.S. Damtoft

Cementir Holding S.p.A, Sølystvej 18, 9220 Aalborg, Denmark

Abstract. The present work focuses on key parameters which determine the best strength performance of two calcined clay mineral additions (1:1 and 2:1 type) and mixtures with limestone. Additionally, alternative options rather than increased admixture dosage have been investigated to mitigate the rheological problems typically found when calcined clays are incorporated to cement.

Isothermal heat development measurements have shown that SO_3 content of blended cements can be effectively optimized based on early available alumina from clinker, but also from calcined clay. Flow measurements before mortar casting demonstrated that an adequate mixing procedure with a delayed addition of chemical admixtures can maximize their efficiency for particle dispersion. In order to reveal the influence of each main cement constituent on both workability and strength, multiple blended cements with different replacement levels, ratios of calcined clay to limestone, fineness of calcined clay and w/c ratios were tested. Results showed that the best strength development is achieved by calcined 1:1 clay for a lower ratio of calcined clay to limestone rather than calcined 2:1 clay for any given replacement level. On the other hand, workability is more strongly affected by the calcined 1:1 clay reducing the efficiency of superplasticizer (SP) required for achieving acceptable flow even at higher w/c ratio.

The addition of fly ash on top of blended cements with 35% replacement of limestone and either calcined 1:1 or 2:1 clay certainly improves flow at high clinker replacements. However, strength performance of these new binders is only similar to corresponding calcined 2:1 blended cements with same replacement level, whist is significantly lowered in those containing calcined 1:1 clay. Therefore, w/c ratio may be potentially reduced to get even better strength performance with combinations of calcined 2:1 clay, limestone and fly ash.

1 Introduction

Properties and performance of calcined clay-limestone blended cements are affected by key parameters such as, substitution degree, SCM/s used and reactivity of those, which is closely related to composition and fineness. It is well known that metakaolin is a highly reactive SCM capable to provide high strength performance in low clinker blended cements, though inconvenient rheological problems are found with this mineral addition [1]. On the other hand, less attention has been paid to smectite clays, which may show inferior strength development than metakaolinite but better workability features. The aim of this work is to compare technically these two calcined clay

F. Martirena et al. (eds.), *Calcined Clays for Sustainable Concrete*, RILEM Bookseries 16,
https://doi.org/10.1007/978-94-024-1207-9_28

mineral additions, evaluate how they affect to performance of blended cements with limestone and investigate the potential of fly ash to improve workability in these low clinker binder systems.

2 Materials and Methods

The binder constituents used in this study were two Ordinary Portland Cements (OPC_1 and OPC_2, CEM I), two calcined clay mineral additions (1:1 and 2:1 type), two limestone fillers (L_1 and L_2), hemihydrate (H) and a typical siliceous fly ash used in concrete (FA). Both OPCs were produced with same grey clinker type in an industrial scale mill and they mainly differ on Blaine fineness (OPC_1: 590 ± 10 and OPC_2: 420 ± 10 m^2/kg). It should be also noted that OPC_2 contains 4.6% of L_2 on its own that was accounted in blended cement compositions and clinker replacements as estimated accordingly to EN 197-1. Two raw clays, one containing smectite and illite 2:1 clay minerals and the other, consisted mainly of kaolinite were flash calcined at different temperatures ($C_{1:1}$ and $C_{2:1}$) and ground to several fineness measured as the residue on 45 μm mesh sieve determined by jet air sieving in accordance with EN 196-6. $C_{1:1}$ was only ground up to 45 μm residue of 30.4%, whilst $C_{2:1}$ was ground up to four distinct 45 microns residues (89, 45.1, 18.9 and 0.8%). Unless specified in the legend of the graphs, the finest $C_{2:1}$ sample was used. Regarding limestone fillers, L_1 is a high purity Maastrichtian chalk and L_2 is a raw meal dust produced at Aalborg Portland cement plant. Both limestone fillers show Blaine fineness of 1250 ± 50 m^2/kg and meet the requirements of EN 197-1 to be considered as LL. The SO_3 content and density of hemihydrate is 55.35% and 2610 kg/m^3, respectively. The chemical composition determined by XRF and density of the main constituents are resumed in Table 1.

Table 1. Chemical composition and density of main blended cement constituents

Constituent	SiO_2	Al_2O_3	Fe_2O_3	CaO	MgO	K_2O	Na_2O	SO_3	TiO_2	Cl	P_2O_5	Cr_2O_3	L.O.I.	Density (kg/m^3)
OPC_1	19.01	5.51	3.81	64.49	0.99	0.43	0.28	3.69	0.27	0.01	0.35	0.01	1.10	3180
OPC_2	19.06	5.30	3.88	62.96	0.85	0.45	0.27	3.36	0.28	0.03	0.28	0.01	2.34	3120
L_1	3.92	0.33	0.14	53.73	0.35	0.05	0.08	0.05	0.02	0.01	0.10	0.00	41.80	2700
L_2	10.20	2.70	1.44	40.35	0.41	0.37	0.12	0.30	0.15	0.12	0.17	0.01	34.84	2710
$C_{1:1}$	54.53	41.11	0.85	0.05	0.24	0.12	0.00	0.23	1.47	0.00	0,13	0,01	1.45	2570
$C_{2:1}$	61.99	17.02	9.55	1.55	2.97	2.91	1.28	0.15	0.89	0.02	0.22	0.02	1.67	2430
FA	55.04	19.92	5.53	4.48	1.81	2.16	1.12	0.38	0.90	0.004	0.53	0.02	8.40	2210

Upon mixing the mortar constituents, CEN standard sand as described in EN 196-1 and polycarboxylate ether (PCE) based superplasticizer (SP) manufactured by BASF were employed to achieve measurable mortar flow of all the binder compositions. The dosage of SP was adjusted accordingly to calcined clay content. The w/c and sand-to-cementitious ratios were kept constant to 0.5 and 3.0, respectively.

All mortars were prepared in a Hobart mixer always at low speed for 4 min and 30 s, but mostly following different mortar mixing procedures rather than specified by EN 196-1 to consider the delay addition of SP over mixing water. Prior to casting, mortar flow was determined in accordance with EN 1015-3, but 20 jolts were applied and the diameter was measured in four directions, instead of two. Compressive strengths were determined as specified in EN 196-1, but all reported results have been normalized to 2% air content [2].

3 Results

3.1 Optimum SO_3 Content of Blended Cements

The isothermal heat development of mortars shows that early hydration of blended cements may be affected by calcined clay type and content at different clinker replacements. The increase of the $C_{1:1}$ content causes a greater formation of monosulfate and faster sulfate depletion (Fig. 1a), as shown by the time difference between initial sulfate depletion and maximum peak corresponding to alite hydration (Fig. 1b), whilst L_2 or $C_{2:1}$ blended cements show analogous heat development profiles to the control cement.

It was found that increased amounts of hemihydrate (H) added accordingly to $C_{1:1}$ content in cement can delay sulfate depletion to shown by control cement (Fig. 1c) and therefore, boost early strengths of the blended cements containing this calcined clay.

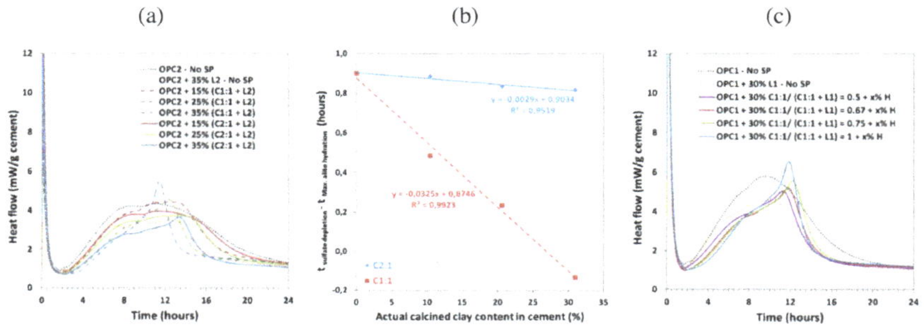

Fig. 1. (a) Isothermal heat development of calcined clay blended cements, (b) Time difference between sulfate depletion and maximum peak of C_3S hydration and (c) Heat development of calcined 1:1 clay-limestone blended cements with extra dosage of hemihydrate

3.2 Workability of Calcined Clay-Limestone Blended Cements

Workability of blended cements and the effect of delayed addition of superplasticizer (SP) was screened by the determination of mortar flow. Some preliminary experimental trials demonstrated that in practice there is generally a good correlation between calcined clay content in cement and SP dosage required to achieve a certain consistency for a given w/c. Thus, it was found out that $C_{1:1}$ requires around 3 times more SP than finely ground $C_{2:1}$ to show comparable mortar flows for any addition time at w/c = 0.5 (Fig. 2a).

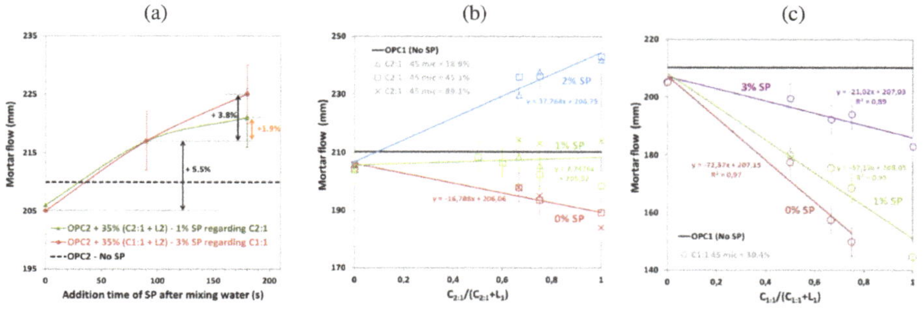

Fig. 2. (a) Effect of time addition of SP on mortar flow of blended cements and mortar flow of: (b) calcined 2:1 clay – limestone blended cements and (c) calcined 1:1 clay – limestone blended cements

It was observed that mortar flows of blended cement with $C_{2:1}$ and $C_{1:1}$ additions for 35% clinker replacement are comparable to control cement without any SP when addition of admixture is delayed 30 s after mixing water (Fig. 2a), whilst longer times for delayed addition clearly improve flow above control cement. The optimum time for delay for the $C_{2:1}$ was found around 90 s, whilst workability of $C_{1:1}$ blended cements seems to be enhanced after 3 min (Fig. 2a).

Additional flow tests were carried out considering 30 s of delayed addition of SP to evaluate the influence of $C_{2:1}$ fineness (Fig. 2b) and the effectiveness of different dosages of SP in blended cements with mixes of limestone and calcined clay mineral additions (Fig. 2b and c). OPC_1 was selected in purpose as control cement to investigate the workability and strength performance for the whole range of $C/(C + L_1)$ ratios from zero to one for 30% of cement substitution.

Despite OPC_1 shows similar mortar flow compared to OPC_2 (Fig. 2a), the higher Blaine fineness of OPC_1 affected to workability of their corresponding blended cements, in particular in higher degree for $C_{1:1}$ (Fig. 2c). Mortar flow is noticeably lowered with increasing $C_{2:1}$ content without any SP, but drastically drops for $C_{1:1}$, even with up to 3% SP. Both calcined clay types (Figs. 2b and c) give mortar flows linearly related to the $C/(C + L_1)$ ratios for any dosage of SP. However, the dispersion effectiveness is noticeably different between calcined clay types at this range of SP dosages and w/c. Furthermore, for any given $C/(C + L_1)$ ratio, mortar flow of $C_{1:1}$ increases linearly with SP dosage, whilst the $C_{2:1}$ shows a non-linear trend.

3.3 Compressive Strength Development

Relative compressive strengths of blended cements determined at 1 and 2 days compared to the control OPC_1 shows that $C_{1:1}$ reacts faster and results in greater early strengths (Fig. 3 left) than L_1 or $C_{2:1}$ (Fig. 3 right), enhanced by addition of hemihydrate. Furthermore, higher 28 days strengths are observed for $C_{1:1}$ at any C/(C + L_1), specially for the optimum ratio around 0.67 [3].

In contrast, plain $C_{2:1}$ blended cements show lower performance than L_1 at early term and they appear to be largely independent of $C_{2:1}$ fineness (Fig. 3 right). These facts indicate the slower pozzolanic reactivity of $C_{2:1}$ and lower influence on cement hydration by nucleation effect than L_1. However, reactivity and long-term performance can be strongly improved by increasing the $C_{2:1}$ fineness, in particular for the optimum $C_{2:1}/(C_{2:1} + L_1)$ ratio around 0.75.

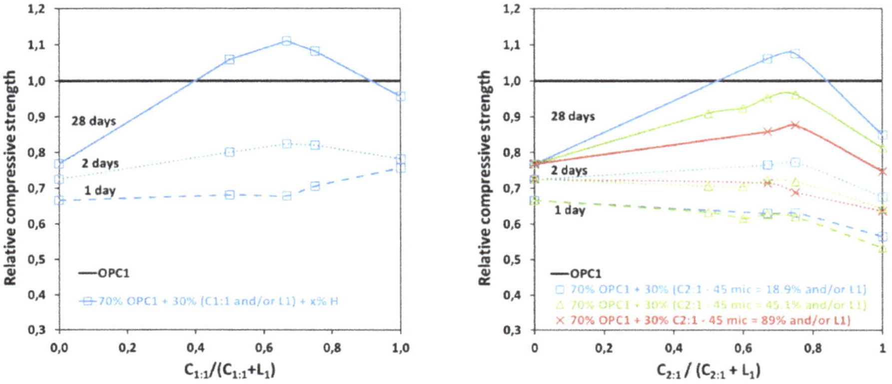

Fig. 3. Relative compressive strength at 1, 2 and 28 days of blended cements containing 30% of several combinations of L1 and C1:1 (left) or C2:1 (right)

3.4 Performance of Low Clinker Blended Cements and Influence of Fly Ash

It is well known that the incorporation of fly ash particles to OPC improves the workability of concrete with low w/c ratio [4] and therefore a similar trend may be expected when added to calcined clay-limestone blended cement. Two C/(C + L_2) ratios slightly lower than optimum found for maximum strength performance (Fig. 4) and extra series of samples with addition of fly ash on top of blended cements with 35% clinker replacement were tested. To exclude any other external influencing factors, mortars were prepared following the mixing procedure of EN 196-1, e.g. without SP.

It is observed that $C_{2:1}$ gives comparable workability to the control cement for replacements up to 30% (Fig. 4 left), whilst mortar flow is reduced for higher or any replacement with $C_{1:1}$. In addition, it can be seen the addition of fly ash on top of cements has a positive effect, with same trend independently of calcined clay type or C/(C + L_2) ratio, mainly due to dilution of calcined clay content in binder by the fly ash particles. It is remarkable that addition of fly ash on top $C_{2:1}$ blended cements for a total 50% of clinker replacement achieves the same flow as control cement without any admixture. Regarding 28 days strengths, further measurements (Fig. 4 right) reaffirmed that $C_{1:1}$ provides much higher strength than $C_{2:1}$, showing the same performance with the coarser control OPC_2 for 55% of clinker replacement. Higher strengths are found at $C/(C + L_2)$ ratio close to the optima determined in the previous section, but performance drops more sharply at replacements levels higher than 30% due to lack of portlandite.

The addition of fly ash on top of $C_{2:1}/(C_{2:1} + L_2) = 0.5$ provides comparable performance to the corresponding plain calcined clay-limestone blended cement up to 50% clinker replacement, but is lowered for the higher $C_{2:1}/(C_{2:1} + L_2)$ ratio or the other $C_{1:1}$ blended cements. So, the addition of fly ash on top of blended cements with low $C_{2:1}/(C_{2:1} + L_2)$ ratio maximizes flow at high clinker replacements, so the w/c ratio may be potentially reduced to get even better strength performance with SP.

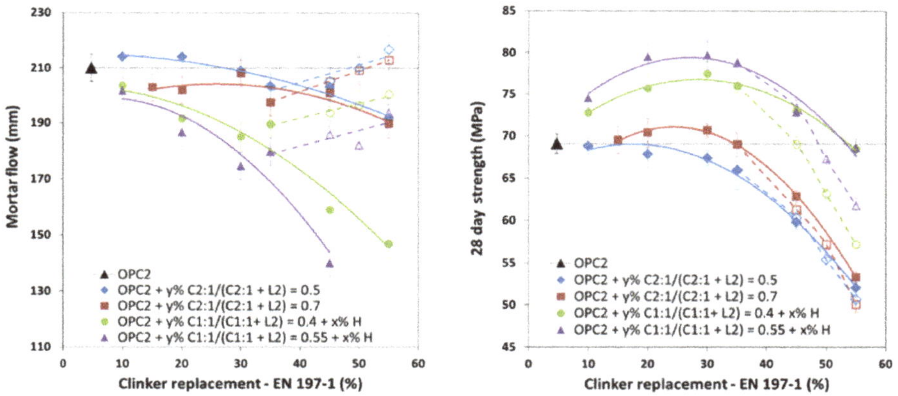

Fig. 4. Mortar flow (left) and 28 days strength (right) of calcined clay – limestone blended cements (solid lines and filled markers) and with an addition of fly ash (dashed lines and not filled markers)

4 Conclusions

There is no doubt that $C_{1:1}$ enables higher limestone contents and achieves best strength development for the same water to cement ratio at any replacement than $C_{2:1}$, but they may demand more water compared to $C_{2:1}$ in order to achieve an acceptable workability for same SP dosage. Both delayed addition of SP and fly ash can significantly improve the rheology of any binder containing calcined clay. Blended cements with slightly lower $C_{2:1}/(C_{2:1} + L_2)$ ratio than optimum for strength with fly ash addition seems to be the best overall solution for sustainable concrete production, since lowers CO_2 emissions due to clinker reduction and calcination of the clay, whilst providing acceptable workability for low w/c ratios and typical dosages of SP used in concrete.

Acknowledgments. The Danish National Advanced Technology Foundation is acknowledged for the financial support to the SCM project and FLSmidth R&D Centre Dania, Mariager for the flash calcination of clay samples.

References

1. Antoni, M., Baquerizo, L., Matschei, T.: Investigation of ternary mixes made of clinker limestone and slag or metakaolin: importance of reactive alumina and silica content. In: Scrivener, K., Favier, A. (eds.) Calcined Clays for Sustainable Concrete, pp. 545–553. Springer, Dordrecht (2015)
2. Osbaeck, B.: The influence of air content by assessing the pozzolanic activity of fly ash by strength testing. Cem. Concr. Res. **15**, 53–64 (1985). doi:10.1016/0008-8846(85)90008-0
3. Antoni, M., Rossen, J., Martirena, F., Scrivener, K.: Cement substitution by a combination of metakaolin and limestone. Cem. Concr. Res. **42**, 1579–1589 (2012). doi:10.1016/j.cemconres.2012.09.006
4. Thomas, M.D.A.: Optimizing the use of fly ash in concrete, IS548. Skokie, IL. Portland Cement Association, 24p (2007)

Metakaolin-Based Geopolymers for Nuclear Waste Encapsulation

D.A. Geddes[1(✉)], X. Ke[1], S.A. Bernal[1], M. Hayes[2], and J.L. Provis[1]

[1] Immobilisation Science Laboratory, Department of Materials Science and Engineering, University of Sheffield, Sir Robert Hadfield Building, Sheffield S1 3JD, UK
[2] National Nuclear Laboratory, Chadwick House, Warrington Road, Warrington WA3 6AE, UK

Abstract. The UK nuclear industry has a significant and challenging stockpile of nuclear wastes, and geopolymers produced from activation of the calcined clay metakaolin offer a valuable alternative to Portland cement-based systems. The characteristics of different formulations of metakaolin-based geopolymers, reacted with sodium and potassium silicate, are therefore of interest. As preliminary steps, the compressive strength and rheology of some metakaolin geopolymer grouts have been studied. This work showed that a potassium silicate-based geopolymer binder, with a sufficiently high water content can be produced to be highly workable. These grouts have a shear stress of approximately 80 Pa at a shear rate of 110 s^{-1} and can achieve compressive strengths of up to 40 MPa after 7 days of curing. This study is to be expanded in future and compared to the results produced from further analysis performed on the chemical structure of the geopolymer, as well as the overall physical characteristics achieved, to support the immobilisation, incorporation and retention of metal and oil based nuclear wastes.

1 Introduction

Since the introduction of nuclear power to the UK in the 1950s, the question of how to safely handle the potentially harmful radioactive waste produced has always been posed. The UK nuclear industry continues to develop and produce new ideas and concepts for dealing with the diverse waste streams that are produced in operational and decommissioning phases of the lifecycle of a nuclear power plant, as well as the complex legacy wastes resulting from experimental and historical operations.

The UK has three basic classifications of nuclear waste: High-level waste (HLW), Intermediate-level waste (ILW) and Low-level Waste (LLW). ILW is by far the most diverse and problematic of these waste types; it is defined as waste that does not generate significant heat but emits radioactivity greater than 4 GBq per tonne of alpha activity or 12 GBq per tonne of beta/gamma activity. This waste can be generated from many sources across the operational lifecycle of a nuclear reactor. This includes, but is not limited to, plant maintenance and decommissioning, fuel cladding, effluents from plant operation, and sludges formed from liquid effluents and corrosive metallic elements [1]. Geopolymers based on calcined clays, produced through the reaction of the calcined clay powder with an alkali source, could be ideal for the encapsulation and

F. Martirena et al. (eds.), *Calcined Clays for Sustainable Concrete*, RILEM Bookseries 16,
https://doi.org/10.1007/978-94-024-1207-9_29

immobilisation of ILW due to the possibility of tailorable chemistry, such as alkalinity and corrosion resistance, and physical properties, such as the refined pore structure and adaptable viscosity in the fresh state [2].

The gel chemistry of metakaolin (MK) is based on a highly disordered, highly cross-linked structure containing high levels of Si and Al, and low levels of Ca. This leads to the formation of a $(Na,K)_2O$-Al_2O_3-SiO_2-H_2O (or in abbreviated notation (Na,K)-A-S-H) gel, where Na and/or K are alkali cations supplied by the activating solution. The chemical characteristics of the gel can be closely related to the Si/Al ratio and mineralogy of the aluminosilicate precursor [2].

An (Na,K)-A-S-H gel is formed of Al- and Si-centred tetrahedra, bridged by oxygen [3], with the alkali cations present in a charge balancing role associated with the Al-centred tetrahedra. The initial setting of a geopolymer takes a few hours, and a relatively strong microstructure takes around 1–7 days to form depending on the precursor characteristics and curing temperature [4]. As the reaction advances over longer time periods, the structure begins to rearrange, and allows the formation of zeolites [5]. The formation of zeolites generates host sites in the crystal structure that are ideal hosts for some of the metal cations present within nuclear waste.

The International Atomic Energy Agency (IAEA) has examined the use of clays for the encapsulation of ILW: natural clays, including kaolinite, were used to encapsulate Cs contaminated liquid waste streams [6], with moderate success. However, after many years of intermittent research in the use of calcined clay-based geopolymers in the nuclear industry, Kuenzel et al. [7–10] and Frizon et al. [11–13] have produced useful data that has generated much interest within the nuclear sector.

Kuenzel et al. demonstrated some initial success for the encapsulation of Cs/Sr [7] and Al [9], and studied how composition can affect the shrinkage and cracking characteristics of geopolymers [10]. Their work has shown that the metallic Al present in some ILW streams has the potential to be encapsulated in MK-based geopolymers [9]. Cs can also be bound tightly to the aluminosilicate phase replacing Na, while Sr has a limited degree of uptake; beyond 0.4 mol of Sr per mol of Al in the geopolymer, the Sr was observed to precipitate out [7].

Frizon et al. have also moved towards the solidification/stabilisation of oil-based waste streams in geopolymers. This work was carried out by producing an emulsion of oil and sodium silicate, mixing this with the MK and allowing to set. This produced very limited oil leaching and a retention of the mechanical properties at up to 20 vol.% oil fraction [12]. Oil is incompatible with Portland cement, and so if a valid alternative arose this could provide the answer to many problems, not just within the nuclear sector. Work from the same group also focused on Mg-Zr waste streams, showing that a sodium silicate activated MK-geopolymer with appropriate fluoride corrosion inhibitors can reduce the release of hydrogen as a product of Mg metal corrosion [13].

The brief summary, presented above shows that the use of MK-based geopolymers to encapsulate materials of specific interest to the nuclear sector is viable. This report demonstrates that the work that has been undertaken can be used and advanced to promote the use of geopolymers within the nuclear industry. A discussion will be provided showing how the results from the literature can be used in conjunction with new results to provide key formulations for immobilisation of problematic nuclear waste streams in clay-based geopolymers.

2 Experimental Programme

Flash-calcined MK (Argeco, France), with a particle size (d_{50}) of 25 μm and a SiO_2/Al_2O_3 of 5.1 (elevated due to the presence of quartz in the pre-calcined kaolin), was used to produce geopolymers using sodium or potassium silicate solutions (provided by PQ Silicates) blended with reagent grade MOH (M = Na or K; Sigma Aldrich UK). The activating solutions were designed with SiO_2/M_2O molar ratios of 0.5, 1 and 1.5, and H_2O/M_2O molar ratios of 11 and 13. The geopolymers were produced with Al_2O_3/M_2O = 1, by high shear mixing of the MK and activating solution, at 20 °C.

The geopolymer grouts were cast in 50 mm cubes for compressive strength analysis. Compressive strength tests were performed using a Control 5000 kN Compression Tester, at 3, 7 and 28 days after casting. The rheological experiments were undertaken using a Haake Viscotester with a 6-bladed vane impeller and a sample volume sufficient to approximate an infinite medium.

3 Results and Discussion

This section will describe some characteristics that must be understood, to determine the suitability of geopolymers for the solidification and stabilisation of waste from the nuclear industry.

The compressive strengths of sodium and potassium silicate activated MK geopolymers are shown in Fig. 1a and b, respectively. These strength curves show that binders where the activator contains more silica, up to an $SiO_2/(Na,K)_2O$ molar ratio of 1.5, tend to have greater compressive strengths. The low compressive strength determined for the sodium silicate formulation, $H_2O/Na_2O = 11$ and $SiO_2/Na_2O = 0.5$, is believed to be due to the formation of sodium metasilicate crystals within the activating solution before or during shear mixing. This has been described by Provis et al. [14], where low modulus sodium silicate solutions, $SiO_2/Na_2O \leq 1$, crystallised and solidified during storage. This formulation has therefore been removed from further studies. A compressive strength of 75 MPa was reached after 28 days for the lowest alkalinity potassium-based system, $H_2O/K_2O = 11$ and $SiO_2/K_2O = 1.5$. When the $H_2O/(Na,K)_2O$ ratio is increased to 13, the compressive strength is approximately 40 MPa for both formulations. This is well within the acceptable range for nuclear waste immobilisation.

The next step in explaining this geopolymer system is to perform rheological analysis on the different grouts produced. These grouts are described by the Herschel-Bulkley (H-B) Model, which states that shear stress (τ) of these materials are related to the shear rate ($\dot{\gamma}$) by a power law, Eq. 1. This is different from Portland cement based systems that often follow the Bingham model (constant plastic viscosity once the yield stress is overcome) [15].

$$\tau = \tau_0 + K\dot{\gamma}^n \quad (1)$$

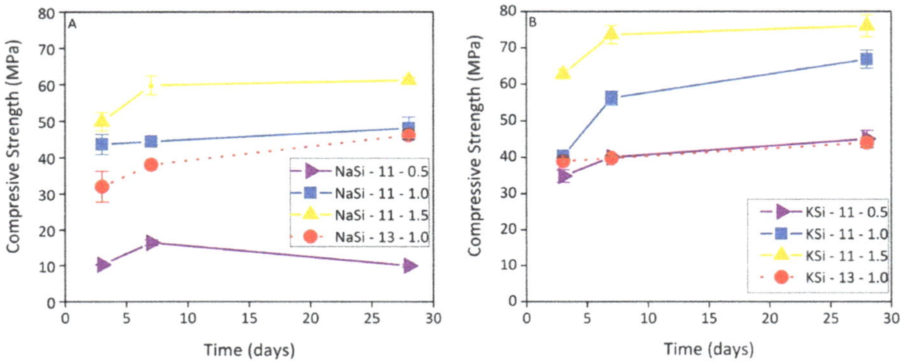

Fig. 1. Compression strength curves showing a comparison between sodium silicate (a) and potassium silicate (b) activated geopolymers. The legends show the activator, H2O/(Na,K)2O molar ratio and activator SiO2/(Na,K)2O ratio for each formulation. The curves for $H_2O/(Na,K)_2O$ ratios of 11 and 13 are shown with solid and dotted lines, respectively. Error bars show the standard deviation among triplicate samples.

The viscosity data and the calculated theoretical H-B curve are given in Fig. 2a and b for sodium and potassium silicate activated solutions, respectively. The rheology curve for the grout with $SiO_2/Na_2O = 1.0$ and $H_2O/Na_2O = 11$ was unable to be obtained as the paste formed was too stiff to be measured, and formed a solid skin-like layer on the top of the grout mix upon exposure to air, possibly through a combination of carbonation and drying effects. This is therefore not a valid formulation for the applications outlined in this study.

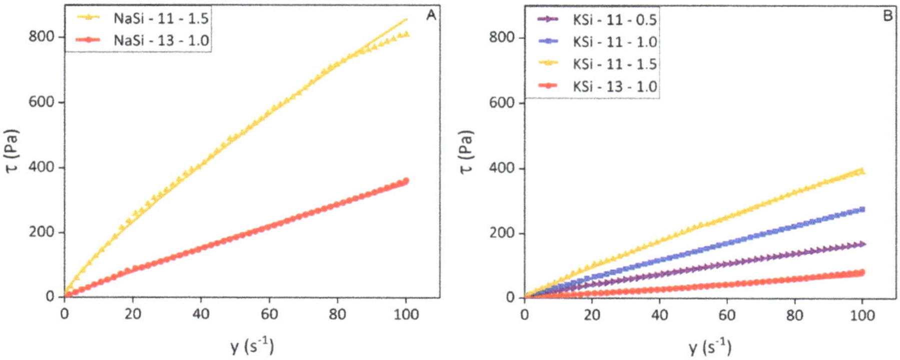

Fig. 2. Flow curves for the alkali-activated MK grouts. The sodium and potassium activated systems are given in A and B, respectively, and the solid lines represent the H-B model fits to each data set.

All the grouts produced appear to show some degree of shear thinning in this shear rate range (H-B parameter $n < 1$), which can be of use in and beyond the nuclear industry.

The potassium-activated systems show a lower viscosity than the sodium-activated systems, and particularly when the water content of the grout is increased to $H_2O/K_2O = 13$, low to negligible yield stress and highly fluid behaviour is observed. This is therefore the ideal formulation to examine further, both chemically and structurally, to determine its ability to encapsulate nuclear waste streams whilst complying with the relevant UK guidelines for waste management and disposal.

4 Conclusions and Future Work

It has been demonstrated that the production of MK-based geopolymers using sodium and potassium silicate activation can yield desirable material characteristics for the encapsulation of nuclear waste streams: sufficient strength and high flowability. A moderate activator $SiO_2/(Na,K)_2O$ ratio of 1.5 gives the highest overall compressive strength at low water content, but the viscosity of the sodium silicate-activated composition at the studied water content is too high to give the high fluidity needed for nuclear wasteform grouts. Two effective ways of reducing the viscosity were to use an increased water ratio, $H_2O/M_2O = 13$, or to replace sodium silicate with a potassium silicate activator. A combination of these parameters yielded high fluidity whilst keeping a compressive strength of 40 MPa. This is a good formulation on which to base further studies of the key physical and chemical characteristics of geopolymers for nuclear waste immobilisation, to allow ideal formulations for different problematic waste streams to be validated.

Acknowledgements. The PhD project of D. Geddes is funded by the UK Nuclear Decommissioning Authority with input from the National Nuclear Laboratory. This research was performed in part at the MIDAS Facility, at the University of Sheffield, which was established with the support of the Department of Energy and Climate Change. The authors gratefully acknowledge the donation of alkali silicates by PQ Silicates.

References

1. Radioactive Wastes in the UK: A Summary of the 2016 Inventory. Department for Business, Energy & Industrial Strategy, 20 pp (2016)
2. Provis, J.L., van Deventer, J.S.: Alkali Activated Materials. Springer/RILEM, Dordrecht (2014)
3. Li, C., Sun, H., Li, L.: A review: the comparison between alkali-activated slag (Si+Ca) and metakaolin (Si+Al) cements. Cem. Concr. Res. **40**, 1341–1349 (2010). doi:10.1016/j.cemconres.2010.03.020
4. Walkley, B., et al.: Phase evolution of Na2O-Al2O3-SiO2-H2O gels in synthetic aluminosilicate binders. DTr **45**, 5521–5535 (2016). doi:10.1039/C5DT04878H
5. Duxson, P., et al.: Geopolymer technology: the current state of the art. J. Mater. Sci. **42**, 2917–2933 (2007). doi:10.1007/s10853-006-0637-z
6. Barney, G.S.: Immobilization of aqueous radioactive cesium wastes by conversion to aluminosilicate minerals. Atlantic Richfield Hanford Co. (1975)

7. Kuenzel, C., et al.: Encapsulation of Cs/Sr contaminated clinoptilolite in geopolymers produced from metakaolin. J. Nucl. Mater. **466**, 94–99 (2015). doi:10.1016/j.jnucmat.2015.07.034
8. Kuenzel, C., et al.: Influence of metakaolin characteristics on the mechanical properties of geopolymers. Appl. Clay Sci. **83–84**, 308–314 (2013). doi:10.1016/j.clay.2013.08.023
9. Kuenzel, C., et al.: Encapsulation of aluminium in geopolymers produced from metakaolin. J. Nucl. Mater. **447**, 208–214 (2014). doi:10.1016/j.jnucmat.2014.01.015
10. Kuenzel, C., et al.: Ambient temperature drying shrinkage and cracking in metakaolin-based geopolymers. J. Am. Ceram. Soc. **95**, 3270–3277 (2012). doi:10.1111/j.1551-2916.2012.05380.x
11. Berger, S., Frizon, F., Joussot-Dubien, C.: Formulation of caesium based and caesium containing geopolymers. Adv. Appl. Ceram. **108**, 412–417 (2009). doi:10.1179/174367609X422072
12. Cantarel, V., et al.: Solidification/stabilisation of liquid oil waste in metakaolin-based geopolymer. J. Nucl. Mater. **464**, 16–19 (2015). doi:10.1016/j.jnucmat.2015.04.036
13. Rooses, A., et al.: Encapsulation of Mg–Zr alloy in metakaolin-based geopolymer. Appl. Clay Sci. **73**, 86–92 (2013)
14. Provis, J.L., et al.: Stabilization of low-modulus sodium silicate solutions by alkali substitution. Ind. Eng. Chem. Res. **51**, 2483–2486 (2012). doi:10.1021/ie202143j
15. Banfill, P.: Rheology of fresh cement and concrete. Rheol. Rev. 61–130 (2006)

The Influence of Recycled Concrete and Clay Brick Particles on the Strength and Porosity of Cement-Based Pastes

T.M. Grabois[1(✉)], G.C. Cordeiro[2], and R.D. Toledo Filho[1]

[1] Programa de Engenharia Civil, COPPE, Universidade Federal do Rio de Janeiro, Rio de Janeiro, Brazil

[2] Laboratório de Engenharia Civil, Universidade Estadual do Norte Fluminense Darcy Ribeiro, Campos dos Goytacazes, Brazil

Abstract. Experimental results of pore size distribution and compressive strength on hardened cement-based pastes containing fine ground particles from the recycling of construction and demolition waste (CDW) are reported. Supplementary cementitious material (SCM) coming from the crushing of laboratory concrete beams and clay bricks were used at a replacement level of 10% by weight of Portland cement. A reference paste without any addition was designed with water-cement ratio of 0.4 and the pastes containing SCM kept the water-cementitious material ratio of 0.4 as well. Compressive strength tests were performed on water-cured samples for various ages from 24 h to 90 days and mercury intrusion porosimetry was performed at 28 days. The general results suggest pozzolanic activity by the clay material while the concrete waste appears to behave similar to an inert quartz inclusion.

1 Introduction

Besides the fact that construction and demolition waste (CDW) might combine different material types (*e.g.* steel, wood, ceramics, etc.), it is well known that an amount greater than 70% of the CDW total mass is composed by concrete and ceramic materials [1]. Therefore, reusing them always yields to questions about the effects of their heterogeneous composition. In the present work, the recycled materials originate from two construction components of known material sources: the demolition of laboratory concrete beams and bricks from the red ceramic industry.

Construction and demolition waste can be potentially used as fine and coarse aggregates in concrete, that is to say, its production as well as its acceptance have been increasing as demonstrated by state-of-the-art reports [1–3] that date back the '80 s and '90 s. Nevertheless, to find alternatives to recycle the remaining dust, i.e. the micro particles, generated during the crushing procedure of the debris – when producing recycled aggregates – deserves reflection too. For instance, a procedure of recycling CDW with only one crushing action (e.g. using jaw or impact crusher), generates coarse and fine aggregates in similar proportions of approximately 50 wt% exhibiting high levels of these remaining dust [4]. A few relevant studies have been carried out to characterize

F. Martirena et al. (eds.), *Calcined Clays for Sustainable Concrete*, RILEM Bookseries 16,
https://doi.org/10.1007/978-94-024-1207-9_30

the effects of recycled fine ground particles coming from ceramic wastes in cement-based materials [5–9]. Of particular interest, some of them reported about the pozzolanic activity of these particles [7, 9, 10]. On the other hand, if we consider a more heterogeneous construction and demolition waste, *e.g.* debris from a building demolition or a singular concrete structure, not but a few researches have addressed to the re-use of these micro particles as a valuable alternative [4, 11].

In this study, we restrict the objective on the behavior of these fine particles as supplementary cementitious material (SCM). First, their physical and chemical properties are assessed experimentally. In order to understand the robustness of our propositions, we confront Portland cement pastes containing recycled concrete or ceramic particles against a mixture with inert quartz and a reference paste without SCM. In particular, this paper discusses about the effects of the recycled and inert particles on the mechanical development and porosity of the cement-based pastes.

2 Materials and Methods

A Brazilian class G cement with no mineral addition [12] was adopted to better qualify the effects of the recycled additions as supplementary cementitious materials. The recycled raw materials come from two sources with known information: (i) the rubble pile of some laboratory concrete beams, and (ii) two types of clay bricks commercially for masonry construction in Brazil, namely, face and hollow. Quartz fragments, concrete rubble, and ceramic bricks were crushed to produce coarse and fine aggregates using an automatic jaw crusher for small grinding plant. In this study, we focus only on the remaining dust generated during this procedure (i.e. the particles $\leq$75 μm) as supplementary cementitious material. Furthermore, quartz was used to produce one mixture with a recognized inert material and confront its influence against the ceramic and concrete wastes. Deionized water was used in the production of cement pastes.

The raw materials were experimentally characterized using the following techniques: oxide composition measured semi-quantitatively using X-ray fluorescence; specific gravity from helium gas pycnometer; BET specific surface area using a nitrogen adsorption apparatus; X ray diffraction analysis to characterize the crystalline phases; and particle size distribution using a laser diffraction analyzer.

Of particular interest is the high levels of silicon dioxide, calcium oxide and aluminium oxide of the recycled concrete particles (44.6, 29.6, and 12 wt%, respectively). The particles from ceramic bricks reveal high levels of silicon dioxide, aluminium oxide, and ferric oxide besides some other metal oxide impurities. Regarding the chemical composition of quartz, it is a material of high level of purity (94.4 wt% of SiO_2). While the SCMs present density values raging between 2.5 to 2.8 g/cm^3, the cement presents 3.3 g/cm^3. The BET results indicate that the recycled particles have a higher specific surface area than the quartz - of same size range - and cement particles, especially the ones coming from clay bricks (>20 m^2/g). The X ray diffraction data of the concrete waste show that quartz is the major crystalline phase and that portlandite and calcite phases confirm the existence of hydrated cement particles. The X ray diffraction patterns of the ceramics present quartz as their major crystalline phase and illite as a common phase between both clay materials.

Concluding the raw materials characterization, Fig. 1 displays their particle size distribution. Overall, the same crushing and sieving procedures generate similar fraction sizes of SCMs, which contributes to evaluate their particular effects when including each of them into cement mixtures. In particular, their characteristic size D_{50} varies between 30 to 40 μm, which is approximately 55% larger than the cement particles.

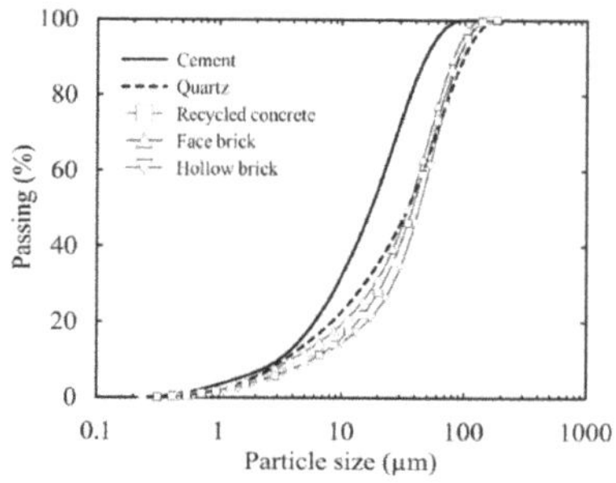

Fig. 1. Particle size distribution of the supplementary cementitious materials and cement type G

A reference paste without any addition was prepared with a water-cement (w/c) ratio of 0.4. For the pastes with SCMs, a 10% substitution rate (in mass) of Portland cement was determined maintaining the water-cementitious materials (w/cm) ratio to 0.4. The cement pastes were produced in a planetary mixer (2-liter capacity) under lab-conditions with a controlled temperature of 21 ± 1 °C. Finally, the specimens were sealed with a plastic wrap and maintained under saturated limewater for curing up to the test time.

Uniaxial compressive strength *fc* tests were conducted using cylindrical specimens of 25 mm in diameter and 50 mm in height [13] for each mixture after 1, 7, 14, 28 and 90 days of curing. The tests were carried out using a universal test machine with displacement rate of 0.1 mm/min. These results were validated by statistical testing using analysis of variance (ANOVA) and Duncan's Multiple Range Tests [14], if necessary. Significant differences were considered when the probability ≤0.05. All results are expressed as average of four specimens (± standard deviation).

Mercury intrusion porosimetry (MIP) measurements were performed according to ISO 15901-1 [15]. In particular, the mercury under applied pressure ranging from 0.1 to 400 MPa permits measurement of pore sizes down to 3.5 nm [16]. One sample of about 1 cm^3 was extracted from the central region of cylindrical specimens (25 × 50 mm) for each of the five cement-based pastes at 28 days of curing.

3 Results

Figure 2a shows the average compressive strength results with standard deviation as function of the curing time. As expected, replacing 10% of Portland cement by quartz reduces the compressive strength independently of the curing time. At 1 and 7 days of curing, the effects of the quartz inclusion are high, when *fc* is about 35% lower than the reference paste. Once we have now less cement, the addition is considered inert and the mixture is then diluted, this behavior is expected due to the intensive reactions at the early hydration period [16] of Portland cement blends.

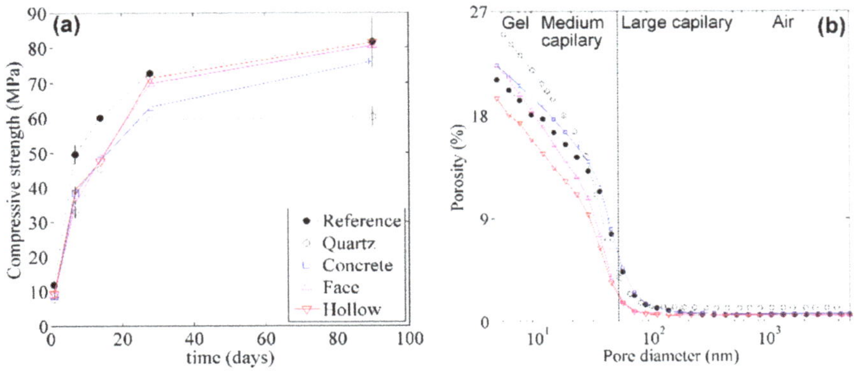

Fig. 2. (a) Compressive strength evolution with time of all cement pastes. The fc results are the average value (with the error bars representing the standard deviation) of four specimens tested from 24 h to 90 days of wet curing. (b) Mercury intrusion porosimetry measurement of all cement pastes with identification marks of the pore size classification according to [16, 17]

According to the analysis of variance, the *fc* results of the paste containing recycled concrete were lower than the reference for all ages evaluated. When substituting cement by recycled concrete, the compressive strength decreases due to none (or low) reactivity of this component. Let us now evaluate the 90 days *fc* of the paste with concrete; when its value increases from around 63 MPa at 28 days to 76 MPa at 90. Thus, considering that the pastes containing recycled concrete and quartz behaved similarly up to 28 days, when comparing them at these ages, we observe that while quartz paste presents a strength evolution rate in time from around 1.4%, the concrete paste strength increases by about 17%, which is even higher than the trend presented by the reference paste (around 11%). It leads us to guess about possible reactivity of non-hydrated cement particles present in the recycled concrete waste (see *e.g.* Refs. [4, 11]).

Up to 14 days, the *fc* of both mixtures with brick particles are about 25% lower than the of the reference paste. This can be explained by the low reactivity of the material. However, after that a significant enhancement of the compressive strength is observed. They reach the same range of compressive strength of the reference paste at 28 and 90 days. This suggests that the fine particles from red ceramic can works as a pozzolan [6]. From this first analysis on the mechanical behavior of the recycled cement pastes, the SCMs from demolished concrete and ceramic bricks appear as a feasible option to replace partially Portland cement.

Figure 2b displays the porosity as function of the pore sizes for all pastes at 28 days identifying the pore size classification (see Refs. [16, 17]). We observe that the paste with quartz provides a pore refinement in the capillary scale, i.e. it increases the amount of medium capillary pores (10 nm > diameter >50 nm), while decreases the amount of the larger ones in comparison to the reference paste. However, let us consider now that the amount of large pores and air content (diameter > 1 μm) increases by about 80% and its total porosity is 25% (*i.e.* 4% higher than the reference). Regarding to the pore size distributions of both the reference and the paste with concrete waste, their results remain nearly the same. The curves representing the pastes with both ceramic wastes display a discrepancy between their total porosity (about 2%). However, the pore size distribution reveals a similar refinement of their porous structure in comparison to all the other pastes. Thus, the data present an amount of gel and medium capillary pores around 25 and 30% higher than the paste without SCMs, respectively. Let us insist here that this fact indicates, as expected and harmonizing with the previous discussion, the existence of pozzolanic activity when introducing these clays particles. Nevertheless, to confirm this hypothesis, complementary experiments are necessary (e.g. pozzolanic activity index or thermogravimetric analysis).

4 Conclusion

An experimental investigation has been conducted on the strength development and porosity of cement-based pastes containing recycled concrete and ceramic bricks as SCM. Overall, it seems viable to apply both of these wastes as SCM. Some of the main findings are summarized below.

The cement substitution by SCMs affected the compressive strength of the cement-based pastes at the early ages (from 1 to 14 days). This was expected since the most intense hydration reactions occurs during this period [16]. At 28 and 90 days the effects of the (expected) pozzolanic activity of the ceramic [6, 9, 10] wastes enhanced the compressive strength of the pastes reaching the levels of the plain paste. Although the compressive strength of the paste containing concrete waste evolved similarly to the paste containing quartz, it increased 17% from 28 to 90 days. This trend was comparable to the one showed by the pastes with ceramics but it needs further investigation. The total porosity of the pastes at 28 days were all around 20%. An exception was the paste with quartz that showed 25%. The reference and concrete pastes presented almost the same pore size distribution. Despite the fact that the total porosity has been similar to the reference, the refinement of the pore size distribution in the pastes with ceramic material confirmed the first evidences of pozzolanic activity of these clay-based SCMs. The amount of capillary pores reduced about 20% and, consequently, the medium capillary and gel pores augmented about 25 and 4%, respectively.

References

1. Hansen, T.C.: Recycling of Demolished Concrete and Masonry, 1st edn. Taylor & Francis Group, London (1992)
2. Nixon, P.J.: Recycled concrete as an aggregate for concrete- a review. Matériaux Constr. **11**, 371–378 (1978). doi:10.1007/BF02473878
3. Hansen, T.C.: Recycled aggregates and recycled aggregate concrete second state-of-the-art report developments 1945–1985. Mater. Struct. **19**, 201–246 (1986). doi:10.1007/BF02472036
4. Schoon, J., De Buysser, K., Van Driessche, I., De Belie, N.: Fines extracted from recycled concrete as alternative raw material for Portland cement clinker production. Cem. Concr. Compos. **58**, 70–80 (2015). doi:10.1016/j.cemconcomp.2015.01.003
5. O'Farrell, M., Wild, S., Sabir, B.B.: Pore size distribution and compressive strength of waste clay brick mortar. Cem. Concr. Compos. **23**, 81–91 (2001). doi:10.1016/S0958-9465(00)00070-6
6. Amorim, L.V., Lira, H.L., Ferreira, H.C.: Use of residential construction waste and residues from red ceramic industry in alternative mortars. J. Environ. Eng. **129**, 916–920 (2003). doi:10.1061/(ASCE)0733-9372(2003)129:10(916)
7. Gonçalves, J.P., Filho, R.D.T., de Fairbairn, E.M.R.: Estudo da hidratação de pastas de cimento Portland contendo resíduo cerâmico por meio de análise térmica, Ambient. Construído. **6**, 83–94 (2006). http://www.seer.ufrgs.br/index.php/ambienteconstruido/article/view/3721
8. Puertas, F., García-Díaz, I., Barba, A., Gazulla, M.F., Palacios, M., Gómez, M.P., Martínez-Ramírez, S.: Ceramic wastes as alternative raw materials for Portland cement clinker production. Cem. Concr. Compos. **30**, 798–805 (2008). doi:10.1016/j.cemconcomp.2008.06.003
9. Pacheco-Torgal, F., Jalali, S.: Reusing ceramic wastes in concrete. Constr. Build. Mater. **24**, 832–838 (2010). doi:10.1016/j.conbuildmat.2009.10.023
10. Silva, J., de Brito, J., Veiga, R.: Incorporation of fine ceramics in mortars. Constr. Build. Mater. **23**, 556–564 (2009). doi:10.1016/j.conbuildmat.2007.10.014
11. Gastaldi, D., Canonico, F., Capelli, L., Buzzi, L., Boccaleri, E., Irico, S.: An investigation on the recycling of hydrated cement from concrete demolition waste. Cem. Concr. Compos. **61**, 29–35 (2015). doi:10.1016/j.cemconcomp.2015.04.010
12. ABNT NBR 9831, Oil well cements - Specification and test methods, Brazilian Association of Technical Standards, Rio de Janeiro (2008). (in Portuguese)
13. Ambroise, J., Murat, M., Pera, J.: Hydration reaction and hardening clays and related minerals. IV. Experimental conditions for strength improvement on metakaolinite minicylinders. Cem. Concr. Res. **15**, 83–88 (1985)
14. Montgomery, D.C.: Design and Analysis of Experiments. Wiley, New York (1997)
15. ISO 15901-1, Evaluation of pore size distribution and porosity of solid materials by mercury porosimetry and gas adsorption - Part 1, Mercury porosimetry, Geneva (2016)
16. Taylor, H.F.W.: Cement Chemistry. Academic Press Limited, London (1990)
17. Mehta, P.K., Monteiro, P.J.M.: Concrete: microstructure, properties, and materials (2006). doi:10.1036/0071462899

The Effect of Kaolinite Content of China Clay on the Reactivity of Limestone Calcined Clay Cement

P.K. Haldar(✉), S. Mithia, K. Mukherjee, N.R. Dhabarde, E. Bansal, P. Phulwari, A. Kumar, S. Kesh, and S. Maity

Technology and Action for Rural Advancement, Delhi, India

Abstract. Limestone calcined clay cement or LC^3 is a ternary blend of clinker, calcined clay and limestone, usually with the clinker factor as low as between 0.4 to 0.5. Calcined clay acts as a source of reactive pozzolanic component of the ternary cement blend. The extent of reactivity of calcined clay depends on the kaolinite content of raw china clay. The present study explores the effect of kaolinite content of clay on the properties of the cement blend. The clays with different kaolinite content were collected from Bhuj area of Gujarat. The ranges of kaolinite content were broadly classified as 20–29%, 30–39%, 40–49%, 50–59%, 60–69% and 70–85%. LC^3 blends were prepared using respective clays with addition of similar quality of clinker, limestone and gypsum. Identical conditions and process parameters for the production of the blend were maintained. Isothermal reactivity of calcined clays and LC^3 blends were measured. Finally, the compressive strength of the LC^3 mortar was tested. It was evident that the reactivity of calcined clay, LC^3 blend and compressive strength varied almost linearly. However no definitive correlation of these properties could be established with kaolinite content of the respective clay. It was concluded that the quality and the performance of calcined clay in LC^3 not only depends on kaolinite content of the clay but also on other properties like chemical composition, presence of various mineralogical phases and geology of the china clays.

Keywords: Ternary cement blend · Kaolinite · Calcined clay · Limestone · Clinker

1 Introduction

China clay or kaolin is formed by weathering of feldspars which mainly consists of Kaolinite ($Al_2O_3.2SiO_2.2H_2O$), associated with other clay minerals. Kaolin is commercially valued for its whiteness and fine particle size which distinguishes it from other clays. Other physical characteristics that influence commercial utility include brightness, glossiness, abrasiveness and viscosity. Crude china clay is usually used in cement industry and processed china clay in ceramic, paint, paper, pigment industries etc.

According to UNFC system as on 1.4.2010, China clay resources in the country is 2,705.21 million tonnes. The reserves constitute only about 7% of the resources at 177.16 million tonnes. Out of the total reserves, 70% (about 124 million tonnes) reserves are under proved category whereas 30% (about 53 million tons) reserves fall under probable

F. Martirena et al. (eds.), *Calcined Clays for Sustainable Concrete*, RILEM Bookseries 16,
https://doi.org/10.1007/978-94-024-1207-9_31

category. The resources are spread mainly in the states of Kerala (about 25%), followed by West Bengal and Rajasthan (16% each) and Odisha and Karnataka (10% each). Jharkhand also has quite a substantial deposit of china clay. Out of the 21 states only 9 states have invested in an in-depth study on the clay deposits where Rajasthan and Gujarat are the leading states in china clay exploration and utilizations.

The quality of china clay varies in a wide range depending on % Kaolinite content, geology and mineralogy. However, out of total resources, about 22% of resources is ceramic/pottery grade, 4% are for chemical, paper filler and cement grades and about 73% or 1,980 million tonnes resources fall under mixed grade and unclassified categories [1].

It has been widely known that the basic components of ordinary Portland cement are clinker and gypsum. Additionally, clinker is produced through a combination of lime-stone and minor amounts of clay. In India to reduce clinker factor, supplementary cementitious materials (SCM's) e.g. fly ash and blast furnace slag are commonly used. In fact, a majority of the cement produced in India is PPC or Portland Pozzolana cement made up of at least 25% substitution of clinker by fly ash. However, above the threshold substitution of about 30%, these materials reduce the mechanical strength of mortar particular at early age [2]. On the other hand, fly ash and blast furnace slag are limited to local availability, retarding their widespread uses. Consequently, alternative sources of SCM's such as calcined clays are of interest [3, 4]. These are widely available in the earth's crust and can easily be dehydroxylated at temperatures ranging between 700–800 °C to produce metakaolin [5, 6]. Metakaolin or calcined clay exhibits excellent pozzolanic properties [7, 8] and is one of the major raw materials for producing the ternary cement blend. The cement is also known as limestone calcined clay cement (LC^3) consisting clinker, calcined clay and raw limestone with clinker factor as low as 0.4 to 0.5.

It has been shown that 45% of substitution of clinker by 30% metakaolin and 15% of limestone give comparable mechanical properties at 7 and 28 days to that of the 100% Portland cement reference. Results show that calcium carbonate reacts with alumina from the metakaolin, forming supplementary AFm phases and stabilizing ettringite [2]. It has also been shown that gypsum addition should be carefully balanced when using calcined clays because it considerably influences the early strength by controlling the rapid reaction of aluminates. In this study, the clays with different kaolinite contents were collected from Gujarat and studied their performance in limestone calcined clay cement.

2 Objective

The main objective of the study is to find out the correlation between kaolinite content of china clay and performance of limestone calcined clay cement in terms of compressive strength of cement mortar.

3 Experimental Method

Six samples of china clays with different kaolinite contents were collected from Bhuj area of Gujarat. Chemical composition of the china clay was determined by X-ray fluorescence (XRF) method. Mineralogical phases were identified by X-ray diffraction (XRD) with X'Pert Pro diffractogram system. Scanning was done at a speed of 2°/min for a 2θ ranging from 0° to 70°. The phases were identified by JCPDS, ICDD-PDF2 data base. % wt. loss of clay sample was measured by thermo gravimetric analysis (TGA). From the weight loss measurement, % kaolinite content of the clay sample was calculated. Raw clays were calcined by static method using a temperature controlled muffle furnace at 850 °C with a soaking period of 1 h. Pozzolanic behaviour of calcined clay was measured by lime reactivity test. Amount of released heat during hydration reaction was determined by isothermal calorimetric study. LC^3 blend was formulated by using a standard composition of 50% clinker, 30% calcined clay, 15% limestone and 5% gypsum. Total six LC^3 blends were produced. The nomenclature of blends followed similarly to that of clay series. Compressive strength of the mortar cubes was measured by standard compression testing device.

4 Results

Figure 1 shows the TGA plots of the raw clays. % Kaolinite contents of raw clays have been derived from the weight loss behaviour of respective clay (Table 1). From TGA analysis of calcined clays, it has been quite clear that the calcination was complete at 850 °C. Mineralogical phases were obtained from XRD analysis. XRD patterns reveal that kaolinite and quartz are present as major phases in all raw clays with minor amount of hematite, muscovite and anatase. However, in some cases, significant amount of

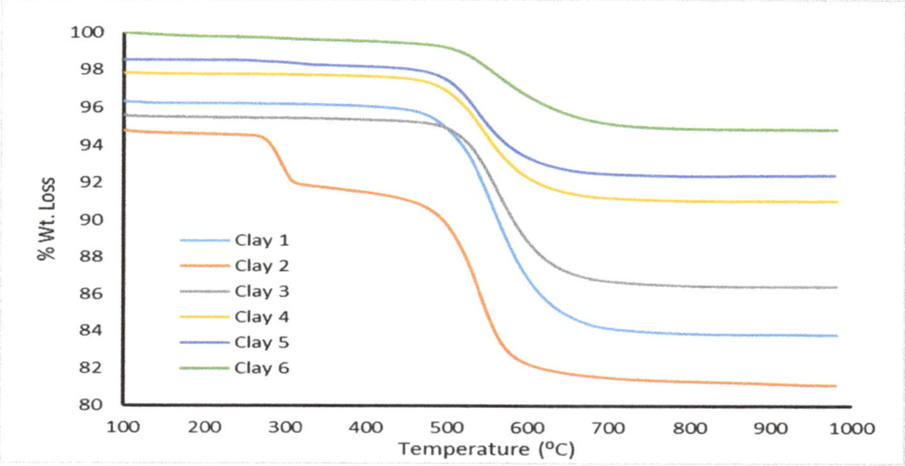

Fig. 1. TGA plots of raw clays

hematite and anatase (>4%) are also present in the clay minerals as evident from XRF analysis. Endothermic peak at 250–300 °C (TGA) and characteristic XRD peak at 18.4 °(2θ) indicate the presence of gibbsite in clay 2. Again, the clay 1 contains a significant amount of hematite (6%) as impurity. Clay 5 (kaolinite content ~ 40%) shows the highest pozzolanicity as evident from lime reactivity test. Blaine value of LC3 is limited within a range of 540 m2/kg to 580 m2/kg. Figure 2 illustrates the development of compressive strength of mortars from respective cement blends compared to Portland cement prepared from the same clinker. Blend 5 containing clay 5 with kaolinite content 39% shows highest compressive strength among all the LC3 blends. On the other hand, Blend1 with clay1 having highest amount of kaolinite content (81%) is showing relatively lower compressive strength which is usually not in agreement to kaolinite content. Even the Blend 6 containing clay 6 (kaolinite content 29%) shows higher strength compared to Blend 1. This non-linearity between kaolinite content and quality of LC3 blend may be attributed to complex behaviour of impurities present in the clay minerals which create a negative effect on kaolinite dehydroxylation and ultimately leading to reduced reactivity.

Table 1. Thermal behaviour of raw and calcined clay

Parameters	Different clays					
	Clay 1	Clay 2	Clay 3	Clay 4	Clay 5	Clay 6
Kaolinite content (%)	81	70	59	45	39	29
Degree of calcination at 850 °C (%)	>99	>99	>99	>99	>99	>99

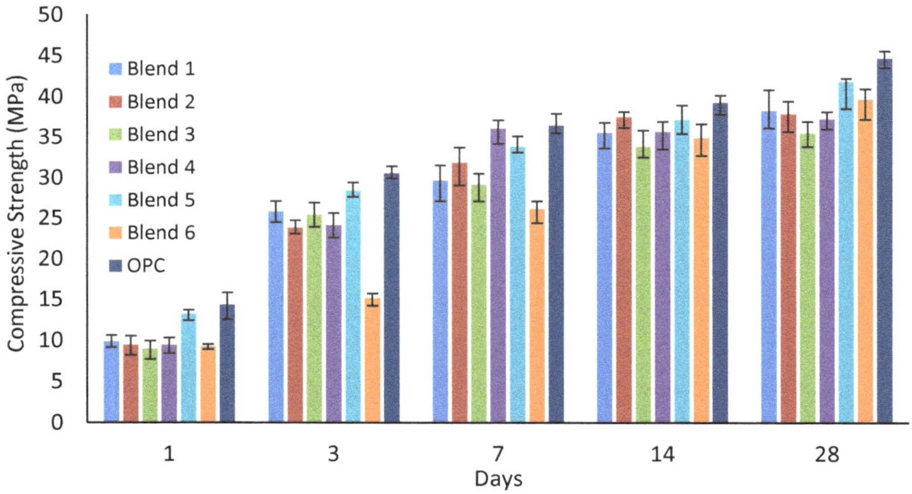

Fig. 2. Compressive strength of LC3 blends

5 Conclusion

It is commonly known that kaolinite content of china clay is a key parameter which controls the reactivity or quality of limestone calcined clay cement to a greater extent. But there are other factors like geology, mineralogy, microstructure, presence of impurities etc. which can significantly influence the quality of LC^3 blend irrespective of kaolinite content of china clay. Presence of impurities in significant amount might affect the stacking pattern of hexagonal layers and disrupt complex amorphous structure of metakaolin leading to reduced reactivity.

References

1. Indian Minerals Yearbook 2013, part III: Mineral Reviews, 52nd edn. IBM (2015)
2. Antoni, M., Rossen, J., Martirena, F., Scrivener, K.: Cement substitution by a combination of metakaolin and limestone. Cem. Concr. Res. **42**, 1579–1589 (2012)
3. Badogiannis, E., Kakali, G., Tsivilis, S.: Metakaolin as supplementary cementitious material: optimization of kaolin to metakaolin conversion. J. Therm. Anal. Calorim. **81**, 457–462 (2005)
4. Scrivener, K., Fernando, J., Martirena, F., Antoni, M.: Tackling social housing through the commercial use of low clinker cementitous systems. In: Innovations on the Use of Calcined Clay as Supplementary Cementitious Material (2012)
5. He, C., Osbaeck, B., Makovicky, E.: Pozzolanic reactions of six principal clay minerals: activation, reactivity assessments and technological effects. Cem. Concr. Res. **25**(8), 1691–1702 (1995)
6. Kakali, G., Perraki, T., Tsivilis, S., Badogiannis, E.: Thermal treatment of kaolin: the effect of mineralogy on the pozzolanic activity. Appl. Clay Sci. **20**, 73–80 (2001)
7. Murat, M., Comel, C.: Hydration reaction and hardening of calcined clays and related minerals III, influence of calcination process of kaolinite on mechanical strengths of hardened metakaolinite. Cem. Concr. Res. **13**(5), 631–637 (1983)
8. Souza, P.S.L., Molin, D.C.C.D.: Viability of using calcined clays, from industrial by-products, as pozzolans of high reactivity. Cem. Concr. Res. **35**(10), 1993–1998 (2005)

The Effect of Alkali on the Properties of Limestone Calcined Clay Cement (LC^3)

W. Hanpongpun(✉) and K. Scrivener

Ecole Polytechnique Fédérale de Lausanne, Lausanne, Switzerland

Abstract. The impact of alkali content on the properties of Limestone Calcined Clay Cement (LC^3) was investigated in this study. KOH was added in order to increase the alkali content of the blended cements from 0.44% Na_2O_{eq} to 1.21% Na_2O_{eq}. The increase of alkali accelerates the degree of clinker hydration at 1 day but it slows down for later ages. Moreover, increasing alkalinity enhances the formation of carboaluminate phases at the early ages but the highest alkali content shows the smallest amount these phases at the later ages. The porosity decreases with increasing the percentage of alkali at 3 days but it is opposite at 28 days. The compressive strength can be boosted at early ages but there is loss gain strength at the later age for LC^3-65 (2:1)-1.21% Na_2O_{eq}.

1 Introduction

The use of limestone and calcined clay to replace clinker in Limestone Calcined Clay Cement (LC^3) is becoming widespread in recent years because of environmental concerns. However, the replacement of clinker with Supplementary Cementitious Materials (SCMs) may result in slow early-age hydration. Alkali can boost the hydration of cement particularly in the early ages but it may have an adverse effect at the later ages [1]. Although there are many publications on the investigation of the alkali level on the reactivity of clinker phases [2, 3] and a few work has been carried out on the blended cements such as fly ash [4] and metakaolin [5]. This paper contributes to the understanding of the effect of alkali on the hydration and the properties of LC^3 blended cement.

2 Materials and Methods

In this work, the mix used is a LC^3-65 (2:1) which is a blended cement containing 65% of clinker, 30% of calcined clay and limestone with a ratio of calcined clay to limestone to 2:1 and 5% of gypsum. Portland cement (PC) with 0.5% Na_2O_{eq} ($Na_2O_{eq} = Na_2O + 0.658 \cdot K_2O$) was used as a plain cement and KOH was added to adjust the equivalent alkali content of the LC^3-65 (2:1) blended cement from 0.44% Na_2O_{eq} to 0.83% Na_2O_{eq} and 1.21% Na_2O_{eq}. Calcined clay with approximately 48.5% of kaolinite calcined at 750 °C for 5 s in fluidized bed combustor was used. The chemical composition of PC and calcined clay are shown in Table 1. The hydration reactions of the blended cement pastes were investigated using water to solid ratio of 0.4. The heat of

F. Martirena et al. (eds.), *Calcined Clays for Sustainable Concrete*, RILEM Bookseries 16,
https://doi.org/10.1007/978-94-024-1207-9_32

hydration was measured by using isothermal calorimeter for 24 h. Phase assemblage was investigated using X-Ray Diffraction (XRD) and the degree of clinker hydration at 1, 3, 7 and 28 days was calculated using Rietveld analysis. The porosity and the pore size distribution of cement paste samples were obtained by using Mercury Intrusion Porosimetry (MIP). The compressive strengths of mortar were tested according to EN 196-1 at 1, 3, 7, 28 days.

Table 1. Chemical composition of PC and calcined clay

Oxides	SiO_2	Al_2O_3	Fe_2O_3	CaO	MgO	SO_3	Na_2O	K_2O	TiO_2	LOI
PC	19.21	4.92	3.26	65.24	1.24	2.37	0.22	0.47	0.29	2.56
Calcined clay	40.70	30.36	20.13	2.07	0.55	0.01	0.31	0.12	2.67	2.49

3 Results and Discussion

3.1 Heat of Hydration

The heat flow of PC and the LC^3-65 (2:1) systems with varying alkali content is shown in Fig. 1(a). All data were normalized per gram of PC. It can be observed that the increase of alkali content accelerates the reaction of silicate phase and aluminate phase when increasing the alkali content from 0.44%Na_2O_{eq}–1.21%Na_2O_{eq}. The acceleration of clinker phases reaction shows the increase of the degree of clinker hydration at 1 day but the excess of alkali content which is LC^3-65 (2:1)-1.21%Na_2O_{eq} provides the lower degree of clinker hydration at the later ages as seen in Fig. 1(b).

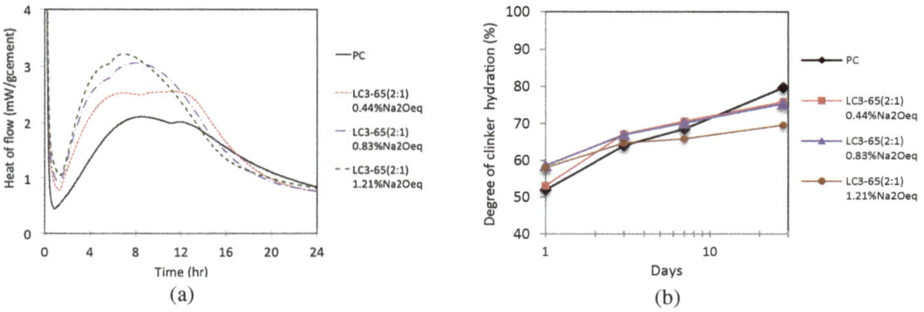

Fig. 1. Heat flow (a) and degree of clinker hydration (b) of cement pastes

3.2 Phase Assemblage

The XRD patterns of LC^3 blended cement pastes at 1, 3, 7 and 28 days are shown in Fig. 2. Generally, the presence of limestone and calcined clay enhances the formation of hemicarboaluminate (Hc) and monocarboaluminate (Mc) for the blended cement [6]. In this study, it can be observed that increasing alkalinity stimulates the formation of Hc at 1 day and Mc at 3 days due to the boost of aluminate phase hydration by alkali.

Moreover, it seems likely that the formation of carboaluminate phases (Hc and Mc) in LC^3-65 (2:1)-1.21%Na_2O_{eq} is less than in the other systems at the later ages.

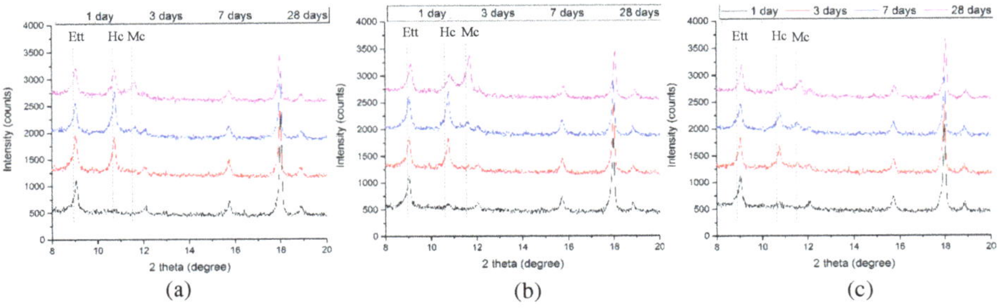

Fig. 2. XRD patterns of LC^3-65 (2:1)-0.44%Na_2O_{eq} (a), LC^3-65 (2:1)-0.83%Na_2O_{eq} (b)

3.3 Porosity

Pore size distribution of cement pastes measured by MIP are shown in Fig. 3. It is found that the increase of alkali content from 0.44%Na_2O_{eq} to 1.21%Na_2O_{eq} shows the smaller pore entry size diameter at 3 days. On the contrary, the alkali content 1.21%Na_2O_{eq} shows the larger pore entry size diameter at 28 days. This reduction in porosity refinement is indeed expected due to a slowing down of the degree of clinker hydration and the less formation of Hc/Mc leading to the generation of the higher porosity.

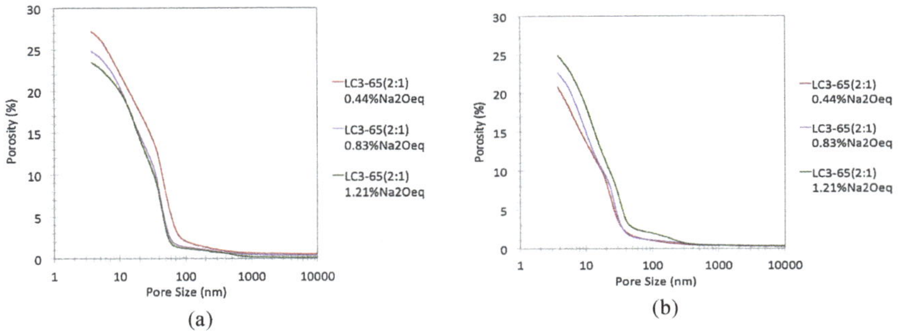

Fig. 3. Pore size distribution of cement pastes at 3 days (a) and 28 days (b)

3.4 Compressive Strength

The degree of clinker hydration and the porosity affect the compressive strength development of blended cement. As seen in Fig. 4, owing to the higher reaction degree of clinker phases, the compressive strength of LC^3-65 (2:1) is increased from 13.0 MPa to 16–16.6 MPa at 1 day when the percentage of alkali equivalent is increased from 0.44% to 1.21%. Moreover, the strength of LC^3-65 (2:1)-0.83%Na_2O_{eq} is comparable to the plain PC from 3 days onwards. However, LC^3-65 (2:1)-1.21%Na_2O_{eq} shows the lowest compressive strength development at the later ages due to the lower reaction degree of clinker phases and the presence of a high volume of porosity at 28 days.

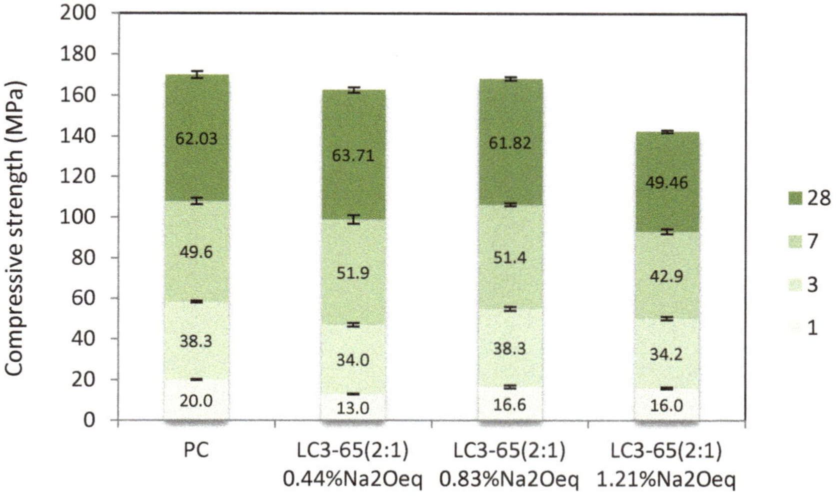

Fig. 4. Compressive strength of mortar

4 Conclusions

The proper alkali content in PC can improve the early ages properties of LC^3 blended cement. In this study, KOH was used to adjust the alkali equivalent from 0.44% Na_2O_{eq} to 0.83% Na_2O_{eq} and 1.21% Na_2O_{eq}. The increase of alkali content increases the degree of clinker hydration and compressive strength at 1 day. The optimum alkali content is 0.83% Na_2O_{eq} which shows the strength value of LC^3-65 (2:1) similar to the PC at 3 days. However, the excess alkali content, 1.21% Na_2O_{eq}, adversely affects at 28 days due to a lower degree of clinker reaction and a higher porosity.

Acknowledgements. I would like to thank SCG Cement-Building Materials Co., Ltd. for the raw materials and the financial support.

References

1. Jawed, I., Skalny, J.: Alkalies in cement: a review. Cem. Concr. Res. **8**, 37–51 (1978)
2. Kumar, A., Sant, G., Patapy, C., Gianocca, C., Scrivener, K.L.: The influence of sodium and potassium hydroxide on alite hydration: experiments and simulations. Cem. Concr. Res. **42**, 1513–1523 (2012)
3. Mota, B., Matschei, T., Scrivener, K.: The influence of sodium salts and gypsum on alite hydration. Cem. Concr. Res. **75**, 53–65 (2015)

4. Diamond, S.: Effects of two Danish flyashes on alkali contents of pore solutions of cement-flyash pastes. Cem. Concr. Res. **11**, 383–394 (1981)
5. Lagier, F., Kurtis, K.E.: Influence of Portland cement composition on early age reactions with metakaolin. Cem. Concr. Res. **37**, 1411–1417 (2007)
6. Antoni, M., Rossen, J., Martirena, F., Scrivener, K.: Cement substitution by a combination of metakaolin and limestone. Cem. Concr. Res. **42**, 1579–1589 (2012)

On the Reactivity of Calcined Clays from Lower Lusatia for the Production of Durable Concrete Structures

Klaus-Juergen Huenger(✉), Robert Gerasch, Ingolf Sander, and Maria Brigzinsky

Brandenburg University of Technology Cottbus-Senftenberg, Cottbus, Germany

Abstract. Metakaolin used as a SCM in concrete is obviously a very good tool to improve the resistance against acid and alkali attack too. Even if, finally, the mechanisms are not fully understood until now, lower mass loss (acid attack) or lower expansions (ASR) are showing their efficiency.

Metakaolin, burnt by using relatively pure natural kaolin clays, contains Al_2O_3 and SiO_2 only with a ratio of approx. one. The question is: Are there any other clays maybe also in mixtures which are suitable for use as an admixture for concrete or even as a binder?

This research work has a strong regional reference. Three clays from Lower Lusatia were selected. The clays and a wide range of mixtures too were burnt at different temperatures (between 600 to 700 °C) to find out the "best" results for such materials. The mixing process has also the background to eliminate fluctuations in the compositions of the clays. A continuous working rotary kiln with a continuous supply of clay materials was used for the production of calcined clay samples. Using this equipment, the rate of heating and the duration of stay of the material under almost practical conditions can be varied. Mineralogical compositions, measured before and after heating, confirm the formation of amorphous phases already under relatively low temperature conditions in dependence on the clay mineral species and the mixing relations. Reactivity (activity index and solubility in alkaline solutions) of each sample were determined and mortar bars were produced. In dependence on their reactivity parameters the so produced calcined clay samples influence mechanical and durability properties of concrete structures. The aim of this research project is to produce concrete bars with such calcined clay as a SCM and store them under extreme conditions in some of the Lusatian lakes.

Keywords: Clay mixtures · Meta clays · Activity index · Solubility · Pozzolanic behavior · Dynamic modulus of elasticity

1 Introduction

The formation of metakaolin, the common material for use as a pozzolanic reactant in cement or concrete structures, is connected to relatively pure raw clays consisted of kaolinite with a content of more than 90%. Because of the calcination process at approx. 750 to 800 °C, metakaolin is formed by destruction of the crystal structure of

F. Martirena et al. (eds.), *Calcined Clays for Sustainable Concrete*, RILEM Bookseries 16,
https://doi.org/10.1007/978-94-024-1207-9_33

the kaolinite mineral. Destruction means that the crystalline water bound in the octahedral sheet is expelled, the structure collapses and a metastable but energy rich condition arises. All other constituents, from which a natural clay can still exist, interfere with the structure formation process to metakaolin [1–3] and can lead to a decrease of the reactivity. Other researchers deal with natural clays with a small kaolinite content to investigate their pozzolanic activities in combination with different cements [4, 5]. The focus of this work is to select regional clays with different compositions but in defined mixtures to investigate the influence of the annealing temperature of the pure clays and of course of the mixtures of clays. The aim is to find optimum burning and mixing conditions of clay mixtures to improve special mortar and concrete properties. The calcined clays were produced in a rotary kiln in a pilot plant to have enough pozzolans for further experiments. At this stage of investigation, some clay mixtures could be selected based on reactivity parameters.

2 Materials and Methods

Three different clays were used for the investigations. The chemical and mineralogical composition are summarized in Table 1. Especially the SiO_2 and the Al_2O_3 contents are different, which is reflected in the mineralogical composition. While clay PL1 is rich on quartz (with approx. 60wt% the main part of this clay), is clay PL7 absolute rich on kaolinite (with 60wt% is it here the main part). PL2 lays between. Characteristically are also the illite contents of PL1 and PL2, which between 15 wt% (PL2) and more than 20 wt% (PL1) are not negligible.

Table 1. Chemical and mineralogical composition of starting materials

Chemical composition					Mineralogical composition				
	PL1 wt%	PL2 wt%	PL7 wt%	Remarks	PL1 [%]	PL2 [%]	PL7 [%]	Error	Remarks
Na_2O	0.27	0.15	0.11		Amorphous	0.1	0.6	0.1	No amor-phous phases can be detected
MgO	0.97	0.49	0.06		Hematite	1.7	1.8	1.7	
Al_2O_3	17.54	20.83	27.19	Al_2O_3 is a measure of the clay mineral content, highest for PL7	Illite	22.7	14.9	5.5	
SiO_2	75.93	74.17	70.95	In addition to the clay mineral content, SiO_2 is a measure of the quartz content	Kaolinite	15.5	25.5	60.1	
K_2O	2.60	1.46	0.38	K_2O is a measure of the feldspare and illite content, highest for PL1	Microcline		4.6	2.5	
CaO	0.19	0.61	0.19		Quartz	60.1	52.6	30.1	
Fe_2O_3	2.47	2.70	1.12	coloured part					

In a first step a completely annealing program was performed, based on the pure clays but also on mixtures between PL1 and PL2 on the one hand and PL7 on the other hand. Three different temperature (600 °C, 650 °C and 700 °C) were chosen to produce calcined clays. A typical temperature profile is demonstrated in Fig. 1. The rotary kiln of the pilot plant can be seen in Fig. 2. It is a tube with a length of approx. 3 m and a diameter of 10 cm. The mass transport is 1 kg per hour, so that a continuous operation is possible.

After grinding and sieving the so produced calcined clay materials were used for further investigations.

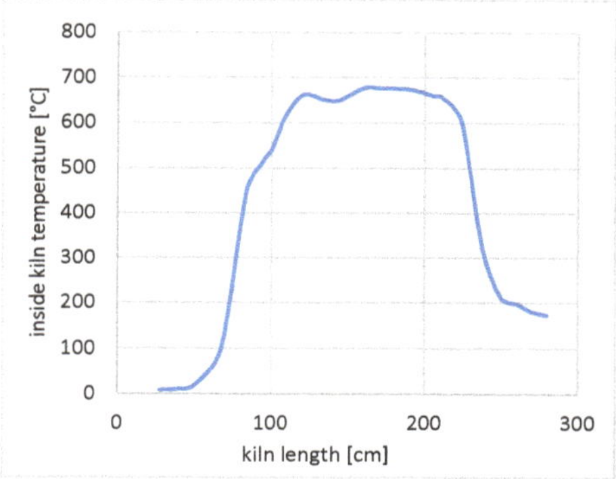

Fig. 1. Temperature profile for 650 °C

Fig. 2. Rotary kiln

2.1 Characterization of Annealed Product by Solubility Experiments

A main part of the investigations was to characterize the burned reaction products. Especially the focus here lays on the question: What happens with the materials while mixing? The background is to substitute the relatively high expensive clay PL7 (but which has the best properties) by the cheaper clays PL1 and PL2. To answer this question the solubility in alkaline solution (1 Mol/L sodium hydroxide solution at 40 °C) was determined.

2.2 Determination of the Activity Index

Additionally, mortar prisms were produced to measure the activity indices of the pozzolanic calcined clay materials in comparison to a reference mortar produced with an OPC CEM I 52.5.

2.3 Evaluation of the Dynamic Modulus of Elasticity

To describe the hardening process of so produced mortar bars the evaluation of the dynamic modulus of elasticity was measured by using the so-called GRINDOSONIC method.

3 Results

The mineralogical phase content after burning is included in Table 2. As it can be seen in the table, the amorphous content increases significantly, especially PL7 reaches with more than 50 wt% a high value, followed by PL2 and then PL1. Kaolinite can be determined only in traces. In opposite to this fact the illite content decreases slightly or the value is in the detection limit.

What is the amorphous phase? Due to the fact that not only pure kaolinite is annealed, maybe other clay minerals can play an important role to influence the composition of the amorphous phase. Since the chemical composition of each clay has not changed, the composition of the amorphous phase can be calculated from the knowledge of mineralogy after burning based on a method by NIGGLI [6]. Obviously only PL7 has an acceptable composition with a ratio between Si and Al of approx. 1. The amorphous phase of the other samples can be characterized as too Al rich. Maybe this is a criterion to differentiate the pozzolanic activities of calcined clays.

Selected results of the dissolution experiments are summarized in Fig. 3. 10 g clay material (raw and calcined) was added to 1000 ml sodium hydroxide solution (1 Mol/L) and stored at 40 °C up to 14 days. The figure shows the dissolution behavior of SiO_2 and Al_2O_3 of the pure clays PL1 and PL7 at 600 °C, 650 °C and 700 °C, and the mixtures between PL1 and PL7 in 20 wt% steps too. It was expected that the samples, in gradation to the mixture compositions, exhibit a graduated solubility behavior. But this is obviously not the case. It can be found that an addition of up to 60 wt% PL1 to the clay PL7 provide approx. the same dissolution curves as the pure PL7. For SiO_2 this effect is valid more or less for all temperatures investigated. The aluminum solubility obviously requires a temperature of 650 °C or more to initiate this effect. In comparison to the solubility of the raw materials the rates of the calcined clays are three to four times higher. In comparison to a common metakaolin sample the solubilities of the produced meta clays are approx. in the same range.

Table 2. Mineralogical composition after annealing (650 °C)

	PL1 wt%	PL2 wt%	PL7 wt%	Error	Remarks
Amorphous	**9.2**	**19.2**	**50.7**	8	amorphous content increases significantly
Hematite	2.5	1.5	1.5	1	
Illite	18.9	13.0	6.1	4	Illite content decreases slightly
Kaolinite	3.0	5.3	3.9	3	Kaolinite content decreases significantly
Microcline		6.0	1.7	2	
Quartz	63.6	52.4	36.1	4	
Composition of the amorphous phase after NIGGLI calculation method					
SiO_2	42	47	56		
Al_2O_3	58	53	44		
Si/Al ratio	0.6	0.8	1.1		

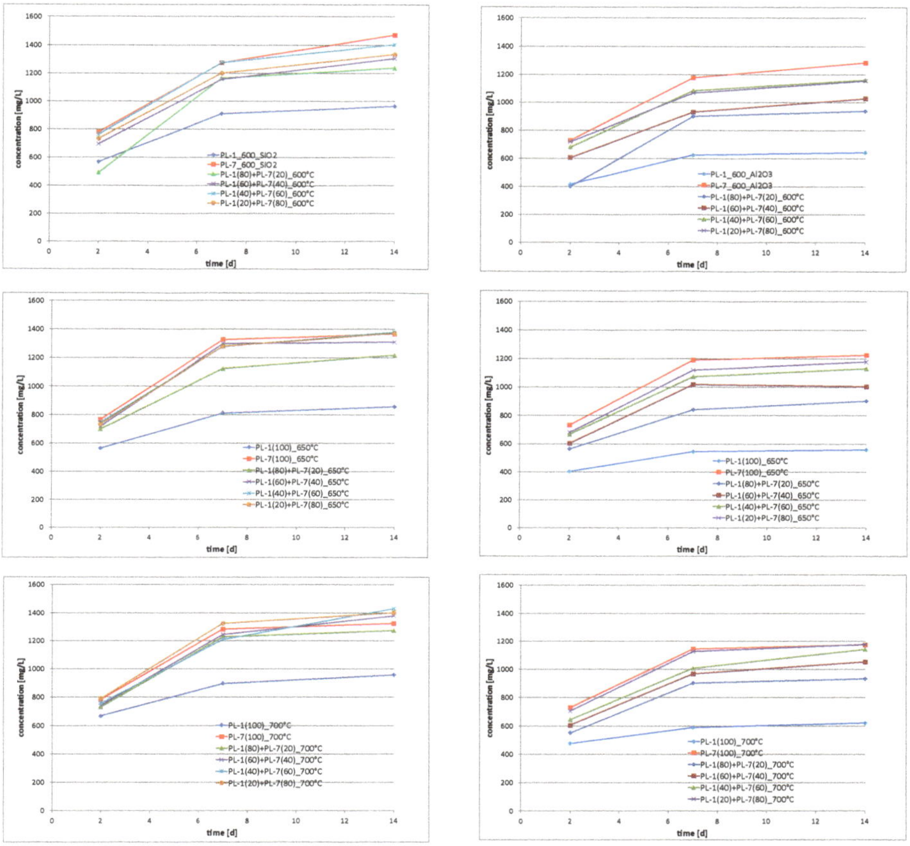

Fig. 3. Dissolution behavior of selected samples at 600 °C (above), 650 °C (middle) und 700 °C (below)

To confirm these results mortar prisms were produced and certain properties were determined. An important value, according to the European standard EN 450 [7], is the activity index (ai). In a reference mixture 25 wt% of the cement has to substitute by other materials to determine their activity index in comparison to 100% of cement. This procedure was transferred to the calcined clay samples. The results are demonstrated in Fig. 4. It can be seen that very often the temperature of 650 °C delivers the best results. However, differences exist. Some samples are below the required value of 75% index after 28 days storage, but many samples lay above this value. PL7 with approx. 110% ai shows the highest, mixtures between PL1 and PL7 in a range of 80 to 20% provide the lowest values.

Values of dynamic modulus of elasticity confirm in principle the found activity indices of the samples annealed. Selected curves are shown in Fig. 5. Using certain clay samples, the values of the reference material cement were reached.

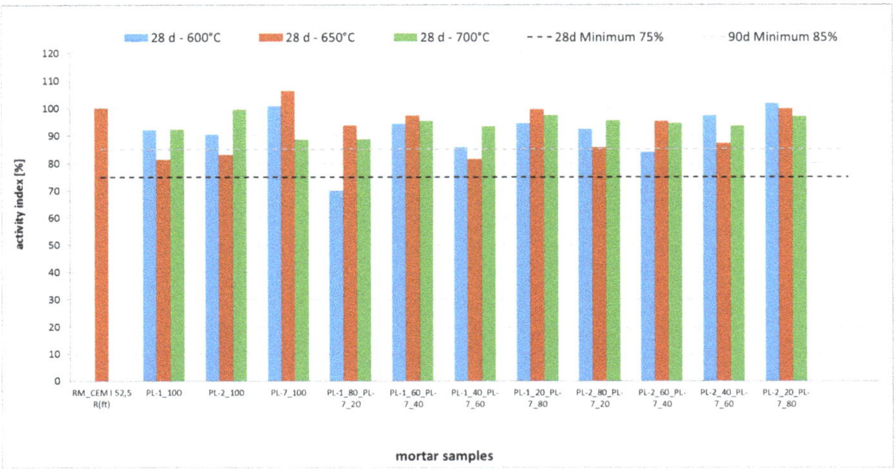

Fig. 4. Activity indices of calcined clay samples

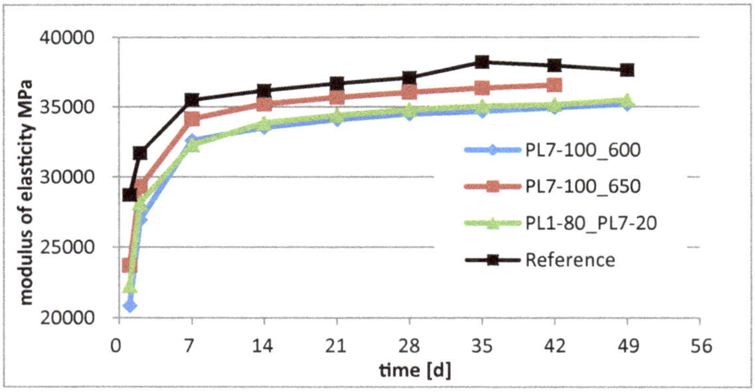

Fig. 5. Selected curves of elasticity modulus

Some "good" mixtures were selected to produce mortar and concrete prisms for the determination of special durable properties (behavior against acid and alkaline attack). The investigations here stand at the beginning. However so far only very less mass loss can be measured.

4 Conclusions

The investigations show that it is possible to produce so-called meta clays by mixing of pure clays with different compositions. Basic characterization studies give a good overview on the properties of materials annealed at different temperatures and different mixture proportions. At this stage of studies, it can be summarized that a mixing ratio between 60% quartz rich clays with 40% kaolinite rich clays burnt at a temperature of approx. 650–700 °C allowed the production of qualitatively good calcined clays, which can be used as a pozzolanic material in cementitious systems. First results to determine durability properties (storage of samples in acid solutions) allow the cautious conclusion that such materials may well be suitable for this application. This is a very important result. The mixing of a "good" and effectiveness clay with a cheaper and less good clay leads to satisfactory results regarding the pozzolanic behavior.

References

1. He, C., et al.: Pozzolanic reactions of six principal clay minerals: activation, reactivity assessments and technological effects. Cem. Concr. Res. **25**, 1691–1702 (1995)
2. Sabir, B.B., et al.: Metakaolin and calcined clay as pozzolans for concrete: a review. Cem. Concr. Res. **23**, 441–454 (2001)
3. Fernandez, R., et al.: The origin of pozzolanic activity of calcined clay minerals: a comparison between kaolinite, illite and montmorillonite. Cem. Concr. Res. **41**, 113–122 (2011)
4. Thienel, K.-C., Beuntner, N.: Effects of calcined clay as low carbon cementing materials on the properties of concrete. In: Proceedings of the 8th International Conference: Concrete in the Low Carbon Era, University of Dundee – Concrete Technology Unit, pp. 504–517 (2012)
5. Beuntner, N., et al.: Efficiency of calcined clay in cementitious systems. In: Proceedings of the 12th International Conference on Recent Advances in Concrete Technology and Sustainability Issues, Prag, pp. 413–424 (2012)
6. Burri, C.: Petrochemical calculations on equivalent basis (Method by Paul Niggli), Birkhäuser Verlag, Basel, Stuttgart, pp. 47–103 (1959). (in german)
7. EN 450-1:2012, Fly ash for concrete - Part 1: Definition, specifications and conformity criteria, German version, Beuth Verlag, 34 pp (2012)

Compressive Strength Improvements of Cement-Based Composites Achieved with Additional Milling of Metakaolin

Biljana Ilić[1(✉)], Aleksandra Mitrović[1], Vlastimir Radonjanin[2], Mirjana Malešev[2], and Miodrag Zdujić[3]

[1] Institute for Testing of Materials, Bulevar vojvode Mišića 43, 11000 Belgrade, Serbia
biljana.ilic@institutims.rs

[2] Department of Civil Engineering, Faculty of Technical Sciences, University of Novi Sad, Trg Dositeja Obradovića 6, 21000 Novi Sad, Serbia

[3] Institute of Technical Sciences of the Serbian Academy of Sciences and Arts, Knez Mihailova 35, 11000 Belgrade, Serbia

Abstract. Kaolin from a Serbian deposit, having a high content of mica and quartz, disordered kaolinite and a high specific surface area, was used to prepare metakaolin (MK). The calcination at temperatures of 700 °C or 750 °C for 30 to 180 min resulted in MK having high pozzolanic activity, but also significant aglomeration of particles. In order to disperse aglomerates, MK was milled, which resulted in increased pozzolanic activity and reduced particle size. The effects of MK and additionally milled metakaolin (MKmill) on the composite strengths and microstructure of the pastes were compared. We prepared and investigated composites in which ordinary Portland cement (OPC) was replaced with 10% to 50% of MK or MKmill, as well as the representative samples of paste for determination of microstructure.

Compressive strengths higher than the control were obtained for composites having up to 30% of MK and up to 40% of MKmill, respectively. Increase of composite strengths with MKmill was more pronounced at lower cement replacement levels (10% and 20%).

Compressive strength of composites containing agglomerated MK were satisfactory, which suggests that milling of MK, as well as purification of kaolin, may not be necessary.

Keywords: Calcination · Metakaolin · Milling · Cement-based composites

1 Introduction

Metakaolin is a recently man-made pozzolanic material formed by controlled thermal activation of kaolin. Its usage in concrete has many technical and environmental advantages, such as improvement of mechanical properties and durability, as well as contribution to the reduction of greenhouse gas emissions. Despite its sustainability and environmental significance, the use of MK as pozzolanic material for modern cement-based composites has not been widely accepted yet, probably due to relatively high cost of MK, compared to other materials, such as fly ash or blast furnace slag.

F. Martirena et al. (eds.), *Calcined Clays for Sustainable Concrete*, RILEM Bookseries 16,
https://doi.org/10.1007/978-94-024-1207-9_34

Nevertheless, the utilization of MK is well documented in literature [1]. However, a clear conclusion regarding the optimum MK percentage for obtaining maximum strength of concrete/mortar is still not available. After a comprehensive literature review, Rashad [2] concluded that the highest 28 days compressive strength of mortars could be attained with 20% MK (w/b of 0.5–0.53), followed by 15% and 10% respectively.

The aim of this study was to investigate compressive strength of cement-based composites with MK, obtained in a single-stage process (only by calcination) and improvements of composites achieved with additional milling of metakaolin (MKmill), from the aspect of sustainable development. In this work, compressive strength of cement composites, containing up to 50% of MK or MKmill, and relationship between the strength and microstructure of the pastes was analyzed.

2 Materials and Methods

Composite mixtures were prepared with ordinary Portland cement (CEM I 42.5R), metakaolins MK or MKmill, CEN standard sand, distilled water and superplasticizer Sika ViscoCrete TECHNO 20S.

MK was produced in IMS laboratory, by thermal activation of Serbian kaolin at 700 °C for 60 min [3]. MK was additionally milled to break particles, agglomerated during thermal activation. Milling 10 min, performed in the low-energy ball mill, resulted in 38% increased pozzolanic activity, 60% reduced particle size (d_{50}) and slight changes in specific surface area.

The chemical composition and main physical properties of MK and MKmill are given in Table 1.

Table 1. The main properties of metakaolin

Chemical composition, (mass%)				Physical properties	MK	MKmill
SiO_2	55.22	Na_2O	0.33	Particle size (d_{50}), μm	15.924	6.432
Al_2O_3	33.07	K_2O	1.06	Specific surface area (BET), m^2/g	26.6	25.4
Fe_2O_3	5.23	SO_3	0.01	Pozzolanic activity, MPa	15.8	21.8
CaO	0.85	LOI	1.47			
MgO	0.61	Reactive silica	38.40			

For compressive strength measurements, composite mixtures, where 0%, 10%, 20%, 30%, 40% and 50% of cement was replaced by MK or MKmill, were prepared with the water-to binder ratio 0.50, and the sand to binder ratio 3.0. In the composite mixture with over 30% of MK or MKmill, hydrated lime was added. For the estimation of the hydrated lime quantity, it was assummed that 20% of calcium hydroxide (CH) was released during the cement hydration, and that the best mechanical properties could be achieved when MK was reacted with CH in the ratio MK/CH=2 [4, 5]. To maintain a constant workability superplasticizer was added to mixtures. The samples were cured in water under standard curing condition for 28 days. Compressive strength measurements were carried out according to EN 196-1.

For microstructure analysis representative pastes, with 20%, 30% and 50% of MK or MKmill and reference paste (CTRL P) were prepared at a water-to binder ratio (w/b) of 0.4. Suitable amount of hydrated lime was added to the pastes having 30% and 50% of MK or MKmill. Hydration products were determined by thermal analysis (DTA/TG) and porosity and pore size distribution by mercury intrusion porosimetry (MIP). Detailed description of the paste preparation, as well as applied methods, are given previously [6].

The CH content, as an indicator of pozzolanic reaction, was estimated by comparing CH dehydroxylation peak areas in all pastes, with the DTA peak area obtained for the decomposition of pure laboratory grade $Ca(OH)_2$, p.a., as shown previously [6].

3 Results and Discussion

3.1 Compressive Strength

Compressive strengths of MK and MKmill composites after 28 days are given in Fig. 1.

Compressive strengths higher than the control were achieved using 10% and 20% of MK. Composite strength with 30% of MK was equal to the control, while MK content over 30% reduced the strength.

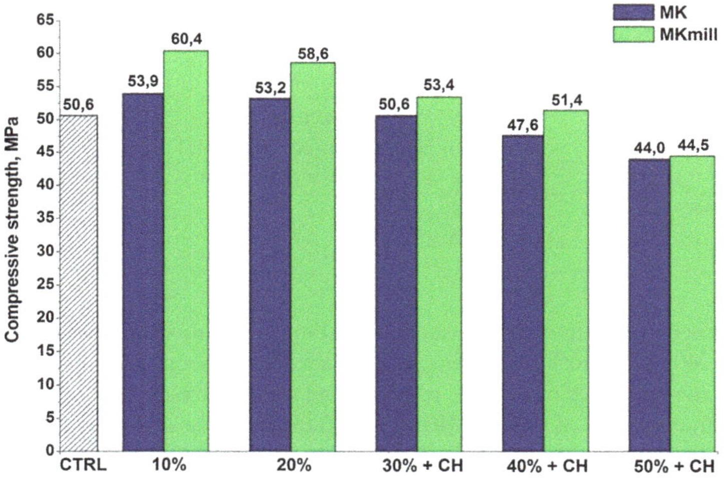

Fig. 1. Compressive strength of composites with MK or MKmill

An increase in compressive strenth was observed in all composites with MKmill, compared to MK composites. The strength increase was more pronounced at lower cement replacement levels, i.e. 19.4% for 10% of MKmill and 15.8% for 20% of MKmill. The additional milling has the smallest influence on the strenth of composite with 50% of MK.

The results are in accordance with findings that MK with a finer particle size shows a greater improvement in the mechanical properties, due to a better packing, easier access of smaller particles for the reaction with the CH, and thus greater acceleration of cement hydration and a faster pozzolanic reaction rate [7].

These cement replacement levels (up to 30% of MK and up to 40% of MKmill), both from an engineering and an environmental perspective, are excellent and contribute to the sustainable production of cement composites. The results also suggests that milling of MK, as well as the purification of kaolin, may not be necessary.

3.2 Microstructure Analysis

3.2.1 Thermal Analysis

Figure 2 shows the thermograms obtained using DTA/TG analysis from hydrated pastes at 28 days.

The first endothermic peak up to 250 °C, corresponds to the dehydration of hydrated phases, such as C-S-H and C-A-S-H. The endothermic peak assigned to the dehydroxylation of the CH is between 400 and 500 °C. Exothermic peaks at around 890 °C (respectively ~915 °C) and 975 °C, appeared as a consequence of the formation of high temperature phase precursors, from unreacted MK. From the thermograms it is evident that the area and intensity of the peaks, characteristic for C-S-H/C-A-S-H and CH, which represent the main hydration products, are similar for MK and MKmill pastes. It means that the difference between strengths did not originate from the different hydration product, but as the result of the filler effect.

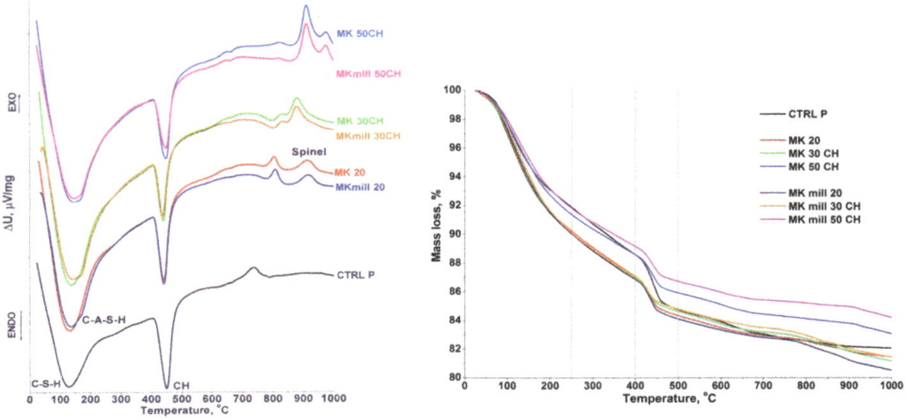

Fig. 2. DTA/TG analysis of pastes

TG analysis (Fig. 2) showed that the mass loss of the pastes occured through a number of stages and that almost all curves coincided, except for the pastes with 50% of MK or MKmill.

The amount of the hydration products C-S-H/C-A-S-H and CH may be approximately determined by the mass loss up to 250 °C and from 400 °C to 500 °C, respectively (Table 2). The C-S-H/C-A-S-H content in MK pastes increases, while CH content decreases compared to the control paste. The mass loss of MK and MKmill pastes are similar, except for the pastes with 50% of MKmill which has lower content of C-S-H/C-A-S-H phases and higher consumption of CH.

The CH content, estimated using the method described previously [6], is shown in Table 2. Consumption of CH is almost identical in MK and MK mill pastes. There is a difference only in the paste having 50% of MKmill, which indicates that the reduction of particle size leads to greater CH consumption and increase of the pozzolanic reaction. The utmost CH consumption is in the paste with 30% of MKmill (57% of CH is consumed).

Table 2. Analysis of MK and MKmill pastes and composites

	Mass loss, %		CH dehydroxylation peak area, a.u.	CH content, %	Total porosity, %	Compressive strength, MPa
	up to 250 °C	400–500 °C				
$Ca(OH)_2$, p.a.	–	–	13.87	–	–	–
CTRL P	8.11	3.80	1.36	9.8	24.58	50.6
MK 20	9.77	2.73	0.80	5.8	22.61	53.2
MK 30 CH	9.94	2.31	0.59	4.3	21.10	50.6
MK 50 CH	8.62	2.66	0.83	6.0	14.17	44.0
MKmill 20	9.84	2.76	0.83	6.0	18.86	58.6
MKmill 30 CH	9.88	2.26	0.58	4.2	19.94	53.4
MKmill 50 CH	8.17	2.42	0.67	4.8	13.13	44.5

3.2.2 Pore Size Distribution

Pore size distribution of pastes is shown in Fig. 3., while the total porosity is given in Table 2.

It is evident that MKmill pastes have smaller total porosity compared to the both control and MK pastes. However, incorporation of MK and MKmill led to the pore structure refinement, compared to the control sample, as a consequence of pozzolanic reaction. MKmill has higher volume fraction of smaller pores and is more effective in the pore structure refinement compared to the MK. As a consequence, the compressive strength of MKmill composites is higher than that of MK composites.

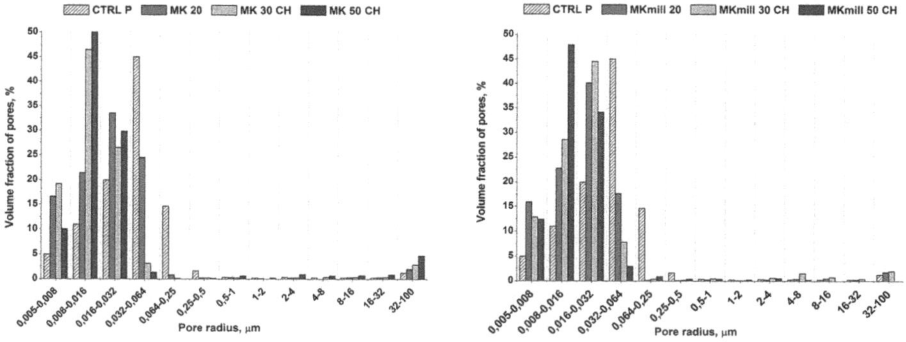

Fig. 3. Pore size distribution of pastes

4 Conclusions

Results showed that cement substitution with up to 30% of MK, obtained by thermal activation, increased compressive strengths of composites, compared to the control.

Additional milling of MK ensures higher cement replacement level, up to 40%, while maintaining a compressive strength higher than that of the control.

Replacement with 50% of both MK and MKmill, gives composites with comparable strength, which are still acceptable for ordinary concrete.

Microstructure analysis showed that there were no significant differences between MK and MKmill pastes relating to the hydration products and CH consumption. Smaller total porosity and higher effectiveness in the pore structure refinement of MKmill, resulted in the higher compressive strength, through the filler effect.

Acknowledgments. The work reported in this paper is a part of the investigation within the research projects TR 36017 and 45001 supported by the Ministry of Education, Science and Technological Development of the Republic of Serbia. This support is gratefully acknowledged.

References

1. Sabir, B.B., Wild, S., Bai, J.: Metakaolin and calcined clays as pozzolans for concrete: a review. Cem. Concr. Compos. **23**, 441–454 (2001). doi:10.1016/S0958-9465(00)00092-5
2. Rashad, A.M.: Metakaolin: fresh properties and optimum content for mechanical strength in traditional cementitious materials - a comprehensive overview. Rev. Adv. Mater. Sci. **40**, 15–44 (2015)
3. Ilić, B., Radonjanin, V., Malešev, M., Zdujić, M., Mitrović, A.: Effects of mechanical and thermal activation on pozzolanic activity of kaolin containing mica. Appl. Clay Sci. **123**, 173–181 (2016). doi:10.1016/j.clay.2016.01.029
4. Murat, M.: Hydration reaction and hardening of calcined clays and related minerals. I. Preliminary investigation on metakaolinite. Cem. Concr. Res. **13**, 259–266 (1983). doi:10.1016/0008-8846(83)90109-6

5. Moropoulou, A., Bakolas, A., Aggelakopoulou, E.: Evaluation of pozzolanic activity of natural and artificial pozzolans by thermal analysis. Thermochim. Acta **420**, 135–140 (2004). doi:10.1016/j.tca.2003.11.059
6. Ilić, B., Radonjanin, V., Malešev, M., Zdujić, M., Mitrović, A.: Study on the addition effect of metakaolin and mechanically activated kaolin on cement strength and microstructure under different curing conditions. Constr. Build. Mater. **133**, 243–252 (2017). doi:10.1016/j.conbuildmat.2016.12.068
7. Justice, J.M., Kurtis, K.E.: Influence of metakaolin surface area on properties of cement-based materials. J. Mater. Civ. Eng. **19**, 762–771 (2007). doi:10.1061/(ASCE)0899-1561(2007)19:9(762)

Properties of the Cement-Based Composites with High Content of Metakaolin

Biljana Ilić[1(✉)], Vlastimir Radonjanin[2], Mirjana Malešev[3], Miodrag Zdujić[3], and Aleksandra Mitrović[1]

[1] Institute for Testing of Materials, Bulevar vojvode Mišića 43, 11000 Belgrade, Serbia
biljana.ilic@institutims.rs

[2] Department of Civil Engineering, Faculty of Technical Sciences, University of Novi Sad, Trg Dositeja Obradovića 6, 21000 Novi Sad, Serbia

[3] Institute of Technical Sciences of the Serbian Academy of Sciences and Arts, Knez Mihailova 35, 11000 Belgrade, Serbia

Abstract. Environmental concerns and sustainable development require increased replacement of cement. Most of previous studies have shown that the compressive strength of cement-based composites is maximized with a 20% content of metakaolin. We investigated composites prepared by replacing ordinary Portland Cement (OPC) with 30 to 50% of metakaolin (MK) and addition of appropriate amount of hydrated lime, which were ordinary cured for 2, 28 or 90 days. Hydration products and microstructure of the pastes were determined by X-ray diffraction (XRD), differential thermal analysis/thermal gravimetry (DTA/TG) and mercury intrusion porosimetry (MIP). MK was produced by calcination of kaolin from a Serbian deposit, which contained a high level of impurities.

Replacement of OPC with 30% of MK achieved 28 days compressive strength equivalent to that of the control mix. Higher replacement levels, 40% and 50%, combined with the addition of hydrated lime, achieved satisfactory relative strengths of 94% and 87%, respectively. The positive contribution was particularly pronounced after 90 days for a composite containing 50% of MK. The results clearly showed a possibility of obtaining composites having acceptable compressive strength with reduced cement content in accordance with environmental and sustained development requirements.

Keywords: Metakaolin · Hydrated lime · Cement-based composites · Compressive strength

1 Introduction

The advantages of use of supplementary cementitious materials (SCMs), such as fly ash, granulated blastfurnace slag, silica fume and natural pozzolana are well known, but their supplies are quite limited compared to the worldwide production of cement. Thus, the main goal of recent studies is to develop new SCMs, such as metakaolin (MK).

The effects of MK addition on the strength of mortar and concrete has been reported by many investigators. It was shown that better performance of cement-based composites is achieved with addition of 10% to 25% of MK, compared to the reference.

F. Martirena et al. (eds.), *Calcined Clays for Sustainable Concrete*, RILEM Bookseries 16,
https://doi.org/10.1007/978-94-024-1207-9_35

In most of these studies cement was replaced with up to 30% MK, due to the limiting quantity of available calcium hydroxide (CH) that could participate in the pozzolanic reaction and thus contribute to the strength. Properties of the composites with MK content over 30%, which are of particular interest with respect to environmental benefits, are rarely reported [1]. In this study, we investigated strength of composites prepared by replacing OPC with 30% to 50% of MK. Influence of hydrated lime addition was also determined. Factors affecting strength were studied by examining the microstructure of pastes by different analytical methods.

2 Materials and Methods

OPC (Lafarge CEM I 42.5R), MK (produced in the laboratory furnace by thermal treatment of Serbian kaolin) [2], sand, distilled water, hydrated lime and superplasticizer were used. The OPC and MK chemical composition and main physical properties are given previously [3].

Composite mixtures, with 0%, 30%, 40% and 50% of MK, were prepared. The mix proportions, with hydrated lime (designation CH) or without, are shown in Table 1. For estimation of the hydrated lime quantity it was assumed that 20% of CH was released during the cement hydration, and that the best mechanical properties could be achieved when MK was reacted with CH in the ratio MK/CH = 2 [4, 5].

Table 1. Composite mixture proportions (in g)

Designation	Cement	MK	Sand	Water	CH
CTRL	450	–	1350	225	
MK 30	315	135	1350	225	–
MK 30 CH	315	135	1350	225	5
MK 40	270	180	1350	225	–
MK 40 CH	270	180	1350	225	36
MK 50	225	225	1350	225	–
MK 50 CH	225	225	1350	225	68

The workability was adjusted using superplasticizer. The samples were cured in water under standard curing condition until testing periods (2, 28 and 90 days). Compressive strength measurements were carried out according to EN 196-1.

For microstructure analyses, selected pastes with 30 and 50% of MK, with the addition of appropriate hydrated lime quantity, were prepared. Paste without MK was prepared as the control (CTRL P). Blends of cement and MK were mixed with distilled water at a water-to-binder ratio (w/b) of 0.4. Pastes were cured under standard curing condition for 28 days. Microstructure was determined using XRD, DTA/TG and MIP analyses. The preparation of paste samples and analytical methods are thoroughly described in work [3].

3 Results and Discussion

3.1 Compressive Strength

The compressive strength of composites ordinary cured for 2, 28 and 90 days are presented in Fig. 1.

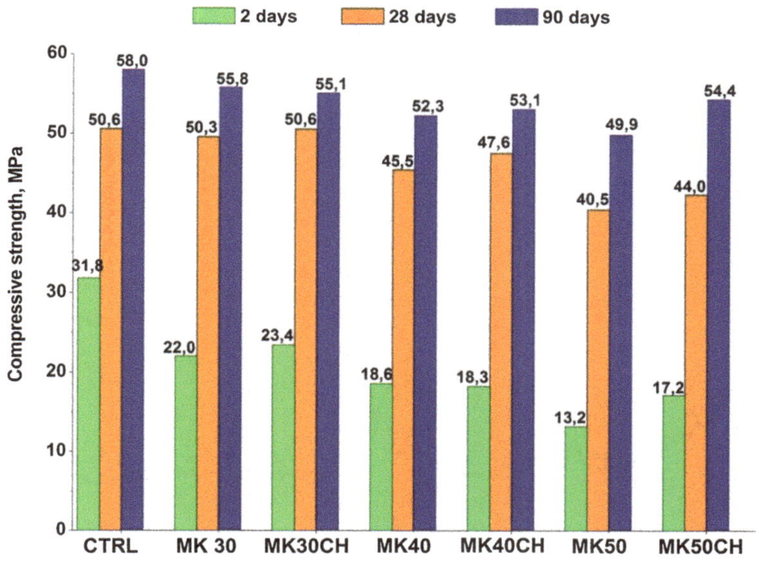

Fig. 1. Compressive strength of MK composites

At 2 days increased content of MK, either with or without hydrated lime, caused significant compressive strength decrease, compared to the control sample, which could be explained by a low degree of pozzolanic reaction and high dilution effect. The hydrated lime addition resulted in a small increase of compressive strength.

At 28 days composite MK 30 achieved the strength equivalent to that of the control. Higher MK content reduced the strength compared to the control. For composites containing hydrated lime this strength reduction was lower. Very good results, in terms of relative strengths of 94% and 87%, were obtained for MK 40 CH and MK 50 CH, respectively. At 28 days pozzolanic reaction significantly contributed to the strengths, but together with the MK filler effect, it was not sufficient to overcome the dilution effect [6].

At 90 days compressive strengths were lower than control. The positive contribution of hydrated lime addition was particularly pronounced for the composite MK 50 CH, where the relative strength was 94%.

It is evident that for all ages hydrated lime addition increases compressive strengths, and that further investigations are needed to optimize its quantity.

3.2 Microstructure

3.2.1 XRD Analysis

Figure 2 shows the XRD patterns of pastes.

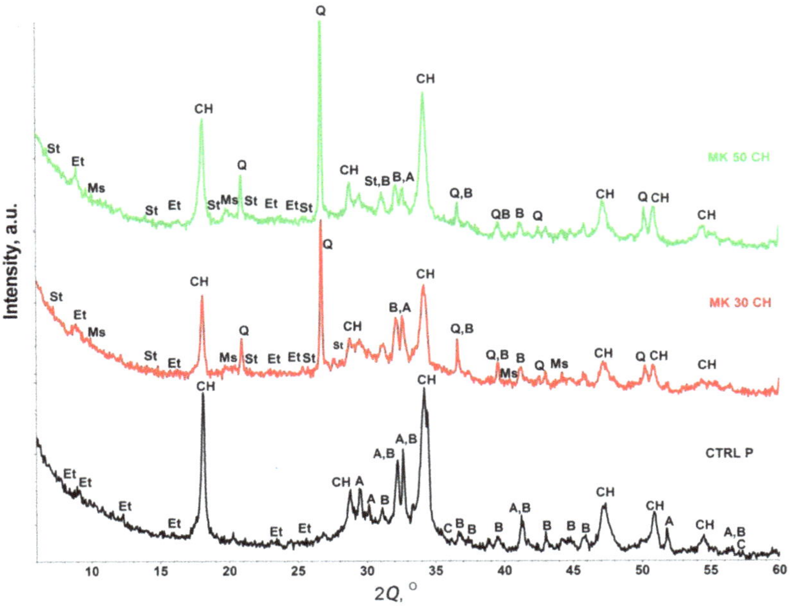

Fig. 2. XRD patterns of the pastes

The major crystalline minerals (hydrated phases) detected on the control paste were portlandite (CH), alite (A) and belite (B). As minor phases, ettringite (Et) and small content of calcite (C) were detected.

A significant reduction of the portlandite (CH) peaks for MK pastes can be observed compared to the control paste. It was less pronounced at composite MK 50 CH due to the higher hydrated lime addition.

New phases identified in MK pastes, which have positive effect on compressive strength, were strätlingite (St) and hardly noticeable monosulfoaluminate (Ms). Intensity of strätlingite and ettringite reflections increases with higher MK content and they are most pronounced in the paste MK 50 CH. Quartz (Q) reflections arise from impurities present in MK. Detected hydrated phases in MK pastes are similar to the findings reported previously [7, 8].

3.2.2 Thermal Analysis (DTA/TG)

DTA/TG curves corresponding to thermal decompositions of different phases in MK pastes are given in Fig. 3.

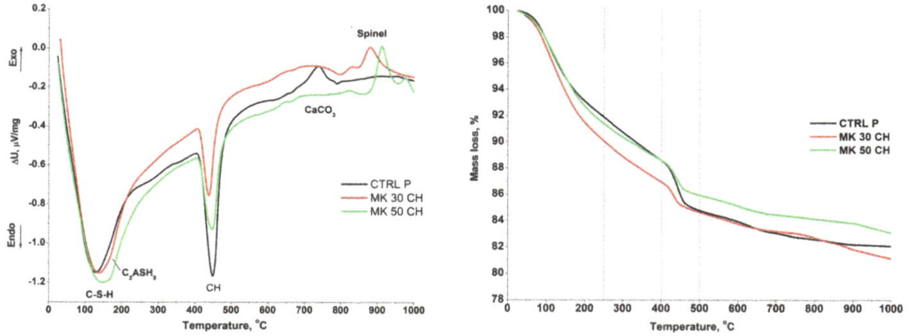

Fig. 3. DTA/TG curves of pastes

The endothermic peak at aproximatelly 120–150 °C is a result of decomposition of amorphous C-S-H phase, as well as the possible presence of calcium sulphoaluminate hydrates (ettringite and monosulfate hydrates) [9, 10]. The peak increases with higher quantity of MK, indicating the formation of additional amounts of C-S-H, which in turns leads to higher strength.

Endotherm between 400 and 500 °C is characteristic for the decomposition of CH. Significant decrease in the CH peak intensity is notable in MK pastes, compared to the control paste, due to the consumption of CH in the pozzolanic reaction. This thermal effect is most pronounced in MK 30 CH.

Exothermic peaks at around 890 °C (respectively ~915 °C) and 975 °C could be attributed to the formation of precursors (alumina or aluminium-silicon spinel) of high temperature phases (mullite and cristobalite) from unreacted MK [7]. This peak is most pronounced in MK 50 CH paste.

TG curves can be divided into four regions. A mass loss in the temperature range up to 250 °C originates from dehydration of the C-S-H phases and possible presence of ettringite and monosulfoaluminate. Between 250 °C and 400 °C a dehydration reaction takes place due to the loss of water from the C-A-S-H phases. Mass loss between 400 °C and 500 °C is a result of dehydroxylation of CH, while decarbonation of calcium carbonate takes place in the region between 600 and 800 °C. The results of TG analysis, showed that content of C-S-H phases, as well as the consumption of CH, is higher in MK pastes (up to 40% in MK CH 30), compared to the control, which positively affects compressive strength of MK composites.

3.2.3 Pore Size Distribution

Figure 4 shows the pore size distribution curve for the studied pastes.

In general, pore radius higher than 0.025 μm influences the compressive strength [11]. MK pastes have a higher volume fraction of pores with radius smaller than 0.032 μm compared to the control paste. This result indicates that the substitution by MK reduces the volume fraction of larger pores. However, almost no pores larger than 0.25 μm are detected in any of the pastes, with the exception of a small fraction of the

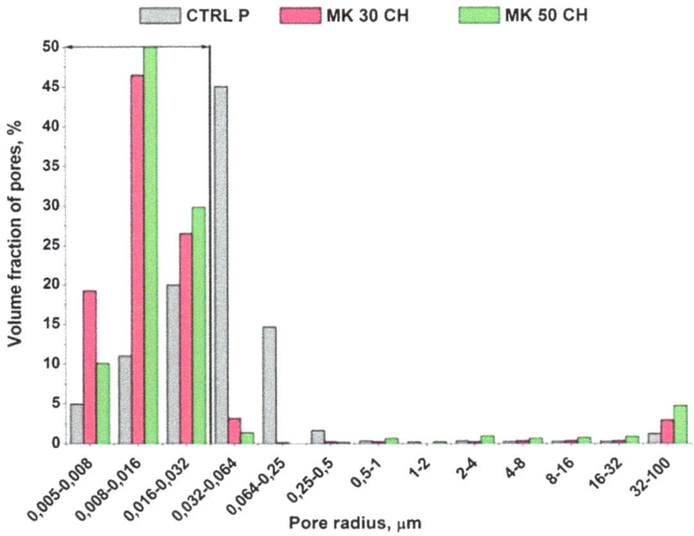

Fig. 4. Pore size distribution of pastes

largest pores (32–100 μm). The results are in accordance with the conclusions that MK is effective in the refinement of pore structure [6, 12].

4 Conclusions

Metakaolin could be used for higher cement substitution level, from 30% to 50%, in order to achieve satisfactory high compressive strengths. The positive contribution was particularly pronounced after 90 days for the composite MK 50 CH where the relative strengths was 94%, which is specially important with respect to environmental protection.

Microstructure analyses showed the pore structure refinement and the presence of C-S-H/C-A-S-H phases, which contributes to the strength, as well as the existence of remained CH, that might be the reason for the strength decrease.

Further improvements of compressive strength could be expected with optimization of quantity of the hydrated lime addition.

Acknowledgments. The work reported in this paper is a part of the investigation within the research projects TR 36017 and 45001, supported by the Ministry of Education, Science and Technological Development of the Republic of Serbia. This support is gratefully acknowledged.

References

1. Gonçalves, J.P., Tavares, L.M., Toledo Filho, R.D., Fairbairn, E.M.R.: Performance evaluation of cement mortars modified with metakaolin or ground brick. Constr. Build. Mater. **23**, 1971–1979 (2009). doi:10.1016/j.conbuildmat.2008.08.027
2. Ilić, B., Radonjanin, V., Malešev, M., Zdujić, M., Mitrović, A.: Effects of mechanical and thermal activation on pozzolanic activity of kaolin containing mica. Appl. Clay Sci. **123**, 173–181 (2016). doi:10.1016/j.clay.2016.01.029
3. Ilić, B., Radonjanin, V., Malešev, M., Zdujić, M., Mitrović, A.: Study on the addition effect of metakaolin and mechanically activated kaolin on cement strength and microstructure under different curing conditions. Constr. Build. Mater. **133**, 243–252 (2017). doi:10.1016/j.conbuildmat.2016.12.068
4. Murat, M.: Hydration reaction and hardening of calcined clays and related minerals. I. Preliminary investigation on metakaolinite. Cem. Concr. Res. **13**, 259–266 (1983). doi:10.1016/0008-8846(83)90109-6
5. Moropoulou, A., Bakolas, A., Aggelakopoulou, E.: Evaluation of pozzolanic activity of natural and artificial pozzolans by thermal analysis. Thermochim. Acta **420**, 135–140 (2004). doi:10.1016/j.tca.2003.11.059
6. Tironi, A., Castellano, C.C., Bonavetti, V.L., Trezza, M.A., Scian, A.N., Irassar, E.F.: Kaolinitic calcined clays – Portland cement system: hydration and properties. Constr. Build. Mater. **64**, 215–221 (2014). doi:10.1016/j.conbuildmat.2014.04.065
7. Cyr, M., Trinh, M., Husson, B., Casaux-Ginestet, G.: Effect of cement type on metakaolin efficiency. Cem. Concr. Res. **64**, 63–72 (2014). doi:10.1016/j.cemconres.2014.06.007
8. Antoni, M., Rossen, J., Martirena, F., Scrivener, K.: Cement substitution by a combination of metakaolin and limestone. Cem. Concr. Res. **42**, 1579–1589 (2012). doi:10.1016/j.cemconres.2012.09.006
9. Sha, W., O'Neill, E.A., Guo, Z.: Differential scanning calorimetry study of ordinary Portland cement. Cem. Concr. Res. **29**, 1487–1489 (1999). doi:10.1016/S0008-8846(99)00128-3
10. Amin, M.S., Abo-El-Enein, S.A., Abdel Rahman, A., Alfalous, K.A.: Artificial pozzolanic cement pastes containing burnt clay with and without silica fume. J. Therm. Anal. Calorim. **107**, 1105–1115 (2012). doi:10.1007/s10973-011-1676-5
11. Mehta, P.K., Monteiro, P.J.M.: Concrete. Microstructure, Properties, and Materials, 3rd edn. McGraw-Hill, New York (2006). doi:10.1036/0071462899
12. Khatib, J.M., Wild, S.: Pore size distribution of metakaolin paste. Cem. Concr. Res. **26**, 1545–1553 (1996). doi:10.1016/0008-8846(96)00147-0

Calcined Clays – Performance as Composite Material

Christian Kalb(✉)

Thyssenkrupp Industrial Solutions AG, Essen, Germany

Abstract. Cements with calcined clays were tested for their performance in mortar, and compressive strength data was collected to quantify the reaction. References were run in all experiments to quantify relative performance differences.

- Calcined fineness becomes relevant for strength development at high Blaine values of calcined clay.
- 20% calcined clay substitution shows the best results
- High substitution rates (35%) make the early strengths worse – compared to OPC
- Due to an increase of the clinker fineness the performance can be improved
- Water demand of the blends is not critical for the strength development
- BM and VRM both show good results – considered in absolute terms, the VRM shows the highest strengths

1 Introduction

Cement production is an energy- and carbon-intensive process and therefore a major contributor to global anthropogenic CO2 emissions. Clinker substitution is an effective way to reduce CO2 emissions from cement production. Due to the worldwide availability and the suitability for cement production, calcined clays are expected to be major constituents of future composite cements. They can be used as a pozzolan.

Colour and water demand of calcined clays are identified as limitations in terms of application requirements and market acceptance for cements with major additions of calcined clays.

From the economic as well as from the competitive points of view, the objective of cement production is to achieve the highest and most uniform product quality possible. One important quality criterion is the cement strength. Therefore, the most important parameters had been investigated with the object of providing a basis for improved control strategies in cement production.

F. Martirena et al. (eds.), *Calcined Clays for Sustainable Concrete*, RILEM Bookseries 16,
https://doi.org/10.1007/978-94-024-1207-9_36

This paper discusses the results of several investigations concerning the performance of calcined clays. This study compares in particular the results concerning production parameters, with special emphasis on separate grinding and intergrinding. The effects of the thermal treatment and the grinding process will be discussed.

2 Experiment

Against this background, pilot plant tests were carried out. The clay to be analysed was calcined in a "pilot scale flash calciner" and characterised by means of X-ray diffractometry (see Fig. 1). After the calcination, the respective clay was processed into three fineness levels by means of ball mill and vertical roller mill. During the process, the project samples were analysed regarding their cement performance (water demand and strength). The proportions of the blended cements were kept at a constant level with clinker contents of 95, 80 and 35 wt%, a sulphate carrier content of 4.5 wt% (50/50 hemihydrate and anhydrite) and calcined clay contents of 5, 20 and 35 wt%. The current project is also intended to allow a direct comparison of interground and separate ground

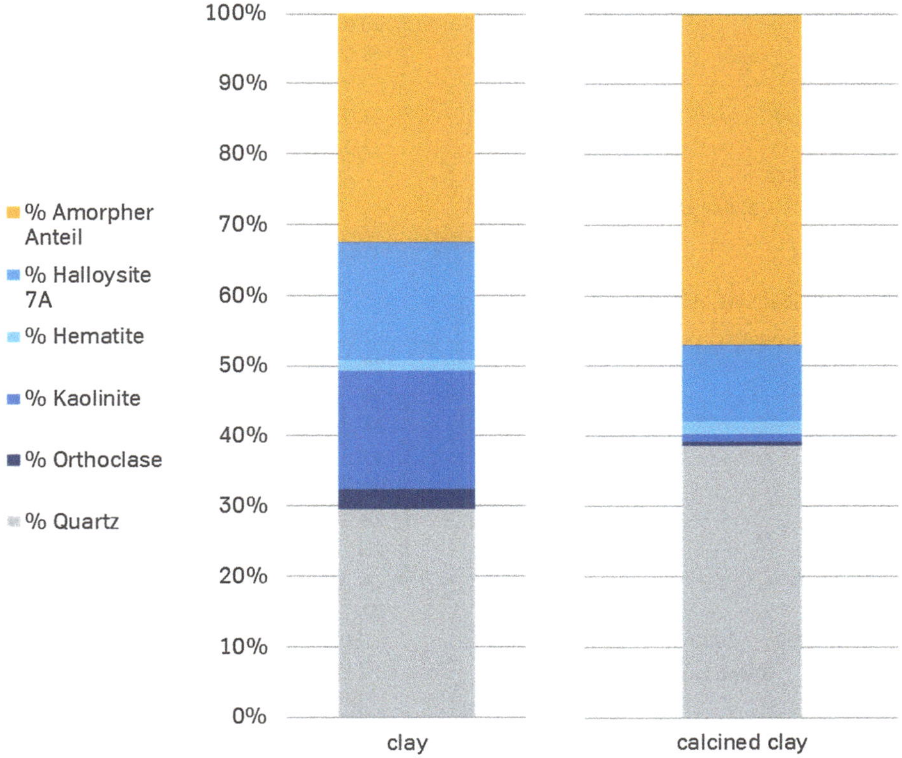

Fig. 1. XRD

cements with calcined clay. The samples can be used to assess the impact of clinker fineness, calcined clay fineness and grinding system on cements with calcined clay.

3 Water Demand

Analysis confirms that calcined clay in binders has a small effect on the water demand (see Fig. 2). Relative to the reference sample, the water demand increases by roughly 12% if 35% are substituted with calcined clay, depending on the calcined clay fineness. With increasing calcined clay fineness water demand rises. Fineness drives the water demand. The good results allow the conclusion that the packing density is favourable.

5% calcined clay substitution

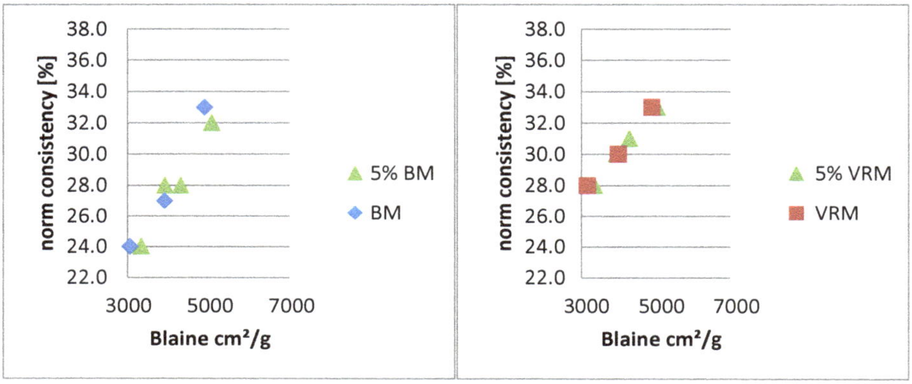

35 % calcined clay substitution

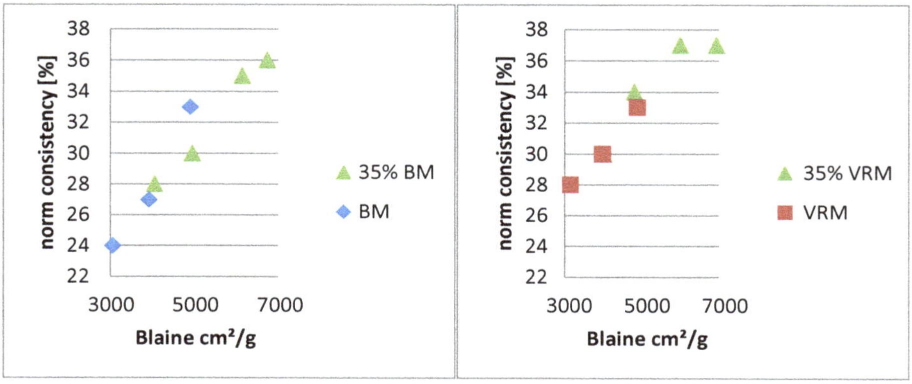

Fig. 2. Data on water demand

4 Compressive Strength

Figure 3 shows the relative compressive strength data. All samples with calcined clay substitution perform poorer than the reference cement (OPC). This was expected. A substitution of 5% calcined clay does not show any deterioration of the performance compared to the reference (OPC). In the case of 35% substitution, the early strengths in particular decrease strongly. In comparison with the OPC, max. 50% of its strength can be achieved. This corresponds to the expectations. A pozzolanic reaction is expected

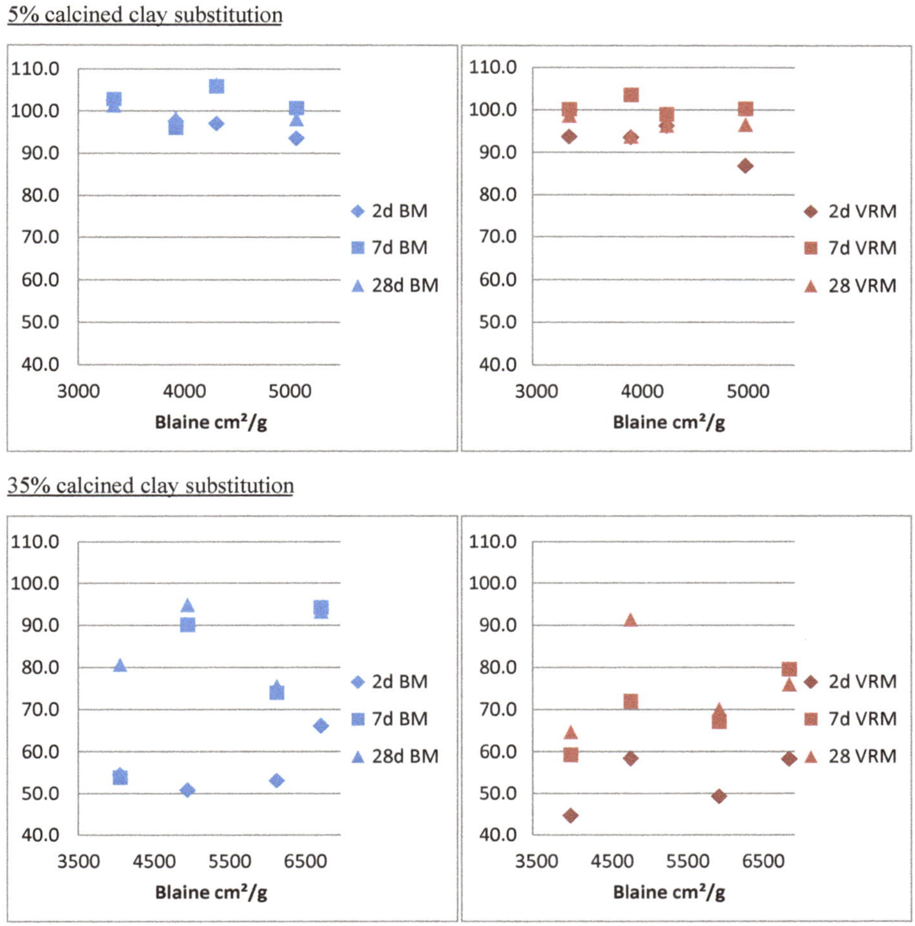

Fig. 3. Data on compressive strength

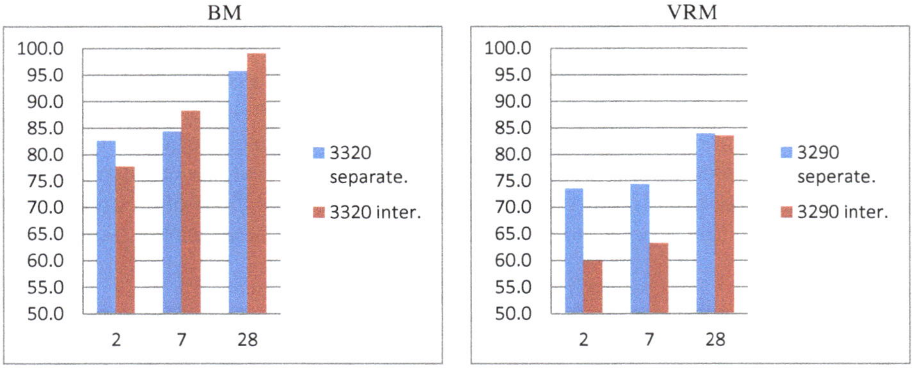

Fig. 4. Intergrinding vs. separate grinding

after 7 days at the earliest. The ultimate strengths of the samples with 35% calcined clay show positive results. 70% of the strength can be achieved in comparison with the reference. The analyses with 20% calcined clay content show the best results (see Fig. 4). In the case of increasing substitution, a deterioration of the performance is to be expected.

5 Separate Grinding vs. Intergrinding

The difference between joint and separate grinding in the ball mill is scarcely apparent. Both variants show good results and display a 28-day compressive strength above 80% – compared to the reference (OPC) (see Fig. 4).

6 Outlook

Subsequently, a grouping of all analysed cements took place in accordance with DIN EN 197 (see Fig. 5). This shows that a strength class of 42.5N is achievable with high substitution rates (20%). Substitution rates of 35% show losses in the early strength. By increasing the clinker grinding fineness this can be compensated.

As a result of the worldwide availability of calcined clays, this composite may play a significant role during the reduction of the clinker factor in cements. The thermal activation, which is necessary to gain pozzolanic properties, is a negative aspect. During this, some available minerals are transferred into a largely amorphous state. tkIS offers a solution for calcining clays.

For the future application, as main component in the cement, it is necessary to put the focus on the strongly contaminated clays. Such contaminated clays are very common and possibly contain montmorillonite and/or illite. These clay types have a different structure and can influence the product qualities.

Fineness of clinker		Cement	Mill	Fineness of calcined clay		
VRM	BM			3000 cm^2/g	7600 cm^2/g	12200 cm^2/g
3000 cm^2/g		CEM I, 5%	ATROL II	42.5 N	42.5 N	42.5 N
			LKM	42.5 N	42.5 N	42.5 N
		CEM II/A, 20%	ATROL II	42.5 N	42.5 N	42.5 N
42.5 N	42.5 N		LKM	42.5 N	42.5 N	42.5 N
		CEM II/B, 35%	ATROL II	32.5 N	32.5 N	42.5 N
			LKM	–	32.5 N	32.5 N
4000 cm^2/g		CEM I, 5%	ATROL II	42.5 N	42.5 N	42.5 N
			LKM	42.5 N	42.5 N	42.5 N
		CEM II/A, 20%	ATROL II	42.5 N	42.5 N	42.5 N
52.5 N	42.5 N		LKM	42.5 N	42.5 N	42.5 N
		CEM II/B, 35%	ATROL II	32.5 N	42.5 N	42.5 N
			LKM	32.5 N	32.5 N	42.5 N
5000 cm^2/g		CEM I, 5%	ATROL II	52.5 N	52.5 N	52.5 N
			LKM	52.5 N	52.5 N	52.5 N
		CEM II/A, 20%	ATROL II	42.5 N	42.5 N	42.5 N
52.5 N	52.5 N		LKM	42.5 N	42.5 N	42.5 N
		CEM II/B, 35%	ATROL II	32.5 N	42.5 N	42.5 N
			LKM	42.5 N	42.5 N	42.5 N
52.5 N also achieves 32.5 N and 42.5 R/ 42.5 N also achieves 32.5 N and 32.5 R						

Fig. 5. Performance of calcined clays

Structural Ordering of Aged and Hydrothermally Cured Metakaolin Based Potassium Geopolymers

Xinyuan Ke, John L. Provis, and Susan A. Bernal(✉)

Immobilisation Science Laboratory, Department of Materials Science and Engineering, The University of Sheffield, Sir Robert Hadfield Building, S1 3JD Sheffield, UK
s.bernal@sheffield.ac.uk

Abstract. This study evaluates the potential correlation between natural aging and hydrothermal curing (accelerated aging), related to the crystallisation of zeolites in potassium-based metakaolin geopolymer binders. 7-year old sealed-cured specimens, formulated with varying silicate contents, were evaluated. The effect of different accelerated aging durations on the mineralogy of these potassium-based geopolymers was also assessed. The results show that although zeolite formation is favoured under both natural and accelerated aging in potassium-based geopolymers, different types of zeolites are formed depending on the silicate content added to the mix, and the curing conditions of the specimens.

1 Introduction

Geopolymer materials, produced from the chemical reaction between a poorly crystalline aluminosilicate source and a highly alkaline solution, have been the object of study for over half a century [1]. Metakaolin-based geopolymers were developed as 'inorganic polymers' with high fire resistance properties, but they have been gaining acceptance for a wide range of applications including as innovative matrices for radioactive waste cementation [2]. Particularly, geopolymer materials based on calcined clays provide highly desirable performance in the immobilisation of heavy metals, and key radionuclides including ^{137}Cs and ^{90}Sr.

In the context of nuclear waste disposal, cementitious grouts need to withstand hundreds of years of in-service conditions, with high stability under moderate temperature and/or pressure changes, depending on the final method adopted for their disposal. In metakaolin-based geopolymers, it is well known that the main binding phase (an alkali aluminosilicate gel), develops short range ordering with a pseudo-zeolitic structure [3]. Formation of crystalline phases has been identified in these materials over time, and this is accelerated by changes in temperature and/or pressure [4]. The bulk chemical composition of the mix design, and the type and dose of activator, also appear to have a significant influence on the crystalline phase formed. The mechanism of zeolite growth in metakaolin-based geopolymers is likely to resemble a gel-solid transformation via rearrangement of the precursor structure. Therefore, the rates of dissolution and polycondensation within these materials are expected to have a significant impact on the zeolite species present and the degree of crystallisation [5].

F. Martirena et al. (eds.), *Calcined Clays for Sustainable Concrete*, RILEM Bookseries 16,
https://doi.org/10.1007/978-94-024-1207-9_37

In this study, the mineralogy of two K-based metakaolin geopolymers cured for 7 years, under controlled laboratory conditions, is examined, and compared with that of geopolymers produced with similar formulations but cured under hydrothermal conditions, with the aim to simulate accelerated aging of the specimens, and determine potential structural changes in the long-term.

2 Experimental Methods

A commercial grade metakaolin (Metastar 402, Imerys UK) was used in this study, which contains 44.87 wt% Al_2O_3 and 53.16 wt% SiO_2. Potassium hydroxide and silicate solutions (i.e. $SiO_2/K_2O = 0.0$ and 0.25 respectively), with $H_2O/K_2O = 11$ were prepared by dissolving potassium hydroxide flake and amorphous silica in Milli-Q water until clear. Solutions were stored for a minimum of 24 h before use. Geopolymer samples were prepared by mixing stoichiometric amounts of metakaolin and activator solution to give a product with bulk $Al_2O_3/K_2O = 1$. For all mix designs, the reagents were mechanically mixed for between 5 to 15 min until a homogeneous paste was achieved.

For geopolymers cured under ambient conditions, the freshly prepared geopolymer pastes were cured in a laboratory oven at 40 °C first for 20 h, and then transferred into sealed containers for storage at ambient temperature (20–25 °C) for 5 years until testing. For hydrothermal curing, a standard pressure cooker was used to achieve an internal temperature of 120 °C and an internal pressure of ~2 atm, in a saturated steam environment. The freshly prepared geopolymer samples were firstly sealed in plastic containers containing 100 mL of water (no direct contact with the samples), and then put into the pressure cooker for hydrothermal curing. After 24, 48, 72, 120 and 168 h of curing, the geopolymer samples were assessed.

Prior to testing, geopolymer samples were crushed and sieved (<67 μm). A Philips PW 1800 diffractometer with Cu Kα radiation generated at 20 mA and 40 kV was used to evaluate the hydrothermally cured specimens, while a Bruker D8 Advance diffractometer with Cu Kα radiation and a nickel filter was used was used to evaluate the naturally aged samples. All data were collected with a step size of 0.020°, over a 2θ range of 5° to 70°.

3 Results and Discussion

Figure 1 shows the diffractograms of two K-based metakaolin geopolymers cured under hydrothermal conditions for different durations. Between the potassium hydroxide ($SiO_2/K_2O = 0$) activated geopolymer and potassium silicate ($SiO_2/K_2O = 0.25$) activated geopolymer, the structural evolution over the curing time is comparable, where the K-feldspar was formed after 24 h of curing in both specimens, along with a transition zeolite phase, K-exchanged zeolite F (K-EDI), with a less ordered structure in the samples containing silicate after 48 h of curing. Traces of potassium chabazite (K-CHA) (powder diffraction file (PDF) #00-044-0250 and/or PDF#00-012-0194) are also observed in both specimens.

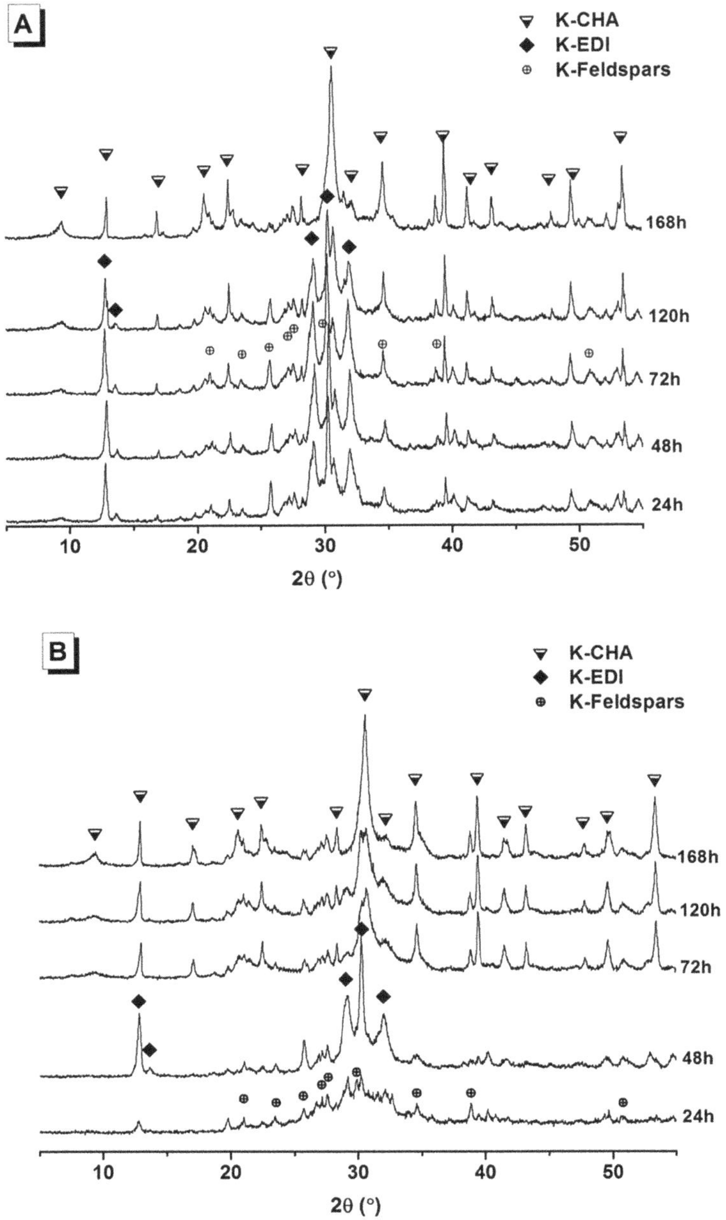

Fig. 1. X-ray diffraction patterns of hydrothermally cured potassium-based metakaolin geopolymer with a SiO_2/K_2O of (A) 0, and (B) 0.25; where K-CHA refers to potassium chabazite (similar to PDF#00-044-0250), K-EDI refers to K-exchanged zeolite F (PDF# 00-038-0216), and K-Feldspars refer to either sanidine (PDF# 00-025-0618) or orthoclase (PDF# 00-031-0966), both of which are monoclinic

These results elucidate the important role of silicate in crystallisation kinetics of the geopolymers assessed, where ordered K-EDI and K-Feldspar are observed in the potassium hydroxide activated geopolymer (Fig. 1A) after 24 h, while the potassium silicate activated geopolymer remained mostly amorphous at this curing duration, with a small amount of K-Feldspar formed. However, at advanced times of hydrothermal curing, both samples present similar mineralogy. In the silicate-containing samples, the additional amorphous Si dissolved in the activator increases the viscosity of the solution, slowing down the mobility of dissolved ionic species [1]. This might be responsible for the delayed formation of crystalline phases. However at later times in the gel nucleation/growth reaction under hydrothermal curing conditions, the additional monomeric silica together with continuously dissolved ionic species favours the further polycondensation process of the geopolymer gel [5]. The formation of K-EDI (zeolite K-F, Si/Al = 1) as a transient zeolite phase is likely related to the sufficient presence of alkali in the pore solution prior to the fully reaction of metakaolin [7]. Also, the potassium chabazite (zeolite K-G) formed in the two geopolymer specimens is likely to have different Si/Al ratios (Si/Al between 1 and 2) in the aluminosilicate framework of the chabazite phase formed, as a result of the differences in bulk chemical compositions [8].

Under ambient curing conditions for up to 7 years, as shown in Fig. 2, zeolite EDI was also identified; however neither K-feldspar nor chabazite was observed. Instead, other zeolite group minerals are formed. Notable differences in crystallinity are observed between the potassium hydroxide and the potassium silicate activated geopolymer; the

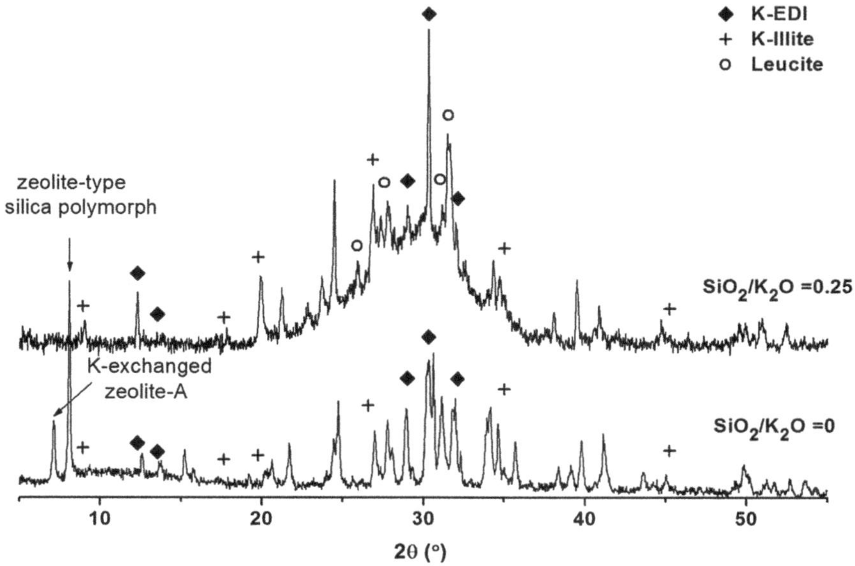

Fig. 2. X-ray patterns of potassium geopolymer cured under ambient condition for 7 years, where K-EDI refers to K-exchanged zeolite F (PDF# 00-038-0216), and other phases are K-illite (PDF# 00-026-0911), leucite (PDF# 00-049-0619), K-exchanged zeolite A (PDF# 01-082-2070), and zeolite-type silica polymorph (PDF# 01-085-0462)

specimens containing silicate are notably less crystalline. A zeolite-type silica polymorph [9] and K-exchanged zeolite A are identified in the potassium hydroxide activated geopolymer, while leucite is forming in the potassium silicate activated geopolymer. The potassium silicate activated geopolymer also maintained a significant amount of poorly crystalline alkali aluminosilicate type gel ('geopolymer gel') after 5 years, observed as a diffuse hump at ~30° (2θ).

4 Conclusion

Formation of zeolite type phases is favoured in potassium-based metakaolin geopolymers, particularly those produced using a hydroxide type activator. Using silicate-containing activators favours formation of a less ordered structure, which remains almost unchanged after 5-years of curing. Under hydrothermal curing conditions, using a silicate activator delays the formation of zeolite type phases. The fact that different types of zeolites were identified in naturally aged and hydrothermal cured samples, indicates the marked influence of differences in time, temperature, and potentially also pressure during curing influencing the nucleation stage of the crystallisation process of these materials, which is of great importance within the context of nuclear waste disposal.

Acknowledgments. The work of David G. Brice and Peter Duxson (both formerly at University of Melbourne) in producing specimens and conducting hydrothermal curing is acknowledged. The participation of X. Ke, J.L Provis and S.A. Bernal was sponsored by the UK Engineering and Physical Sciences Research Council (EPSRC) through grant EP/P013171/1. Initial work in sample formulation was also partially sponsored by the Australian Research Council (ARC) through Discovery Project and Linkage Project Grants, including co-funding of the Linkage project by Zeobond Pty Ltd.

References

1. Provis, J.L., Bernal, S.A.: Geopolymers and related alkali-activated materials. Annu. Rev. Mater. Res. **44**, 299–327 (2014). doi:10.1146/annurev-matsci-070813-113515
2. Khalil, M.Y., Merz, E.: Immobilization of intermediate-level wastes in geopolymers. J. Nuclear Mater. **211**, 141–148 (1994). doi:10.1016/0022-3115(94)90364-6
3. Provis, J.L., Lukey, G.C., van Deventer, J.S.J.: Do geopolymers actually contain nanocrystalline zeolites? A reexamination of existing results. Chem. Mater. **17**, 3075–3085 (2005). doi:10.1021/cm050230i
4. Bernal, S.A., Rodriguez, E.D., Mejia de Gutierrez, R., Gordillo, M., Provis, J.L.: Mechanical and thermal characterisation of geopolymers based on silicate-activated metakaolin/slag blends. J. Mater. Sci. **46**, 5477–5486 (2011). doi:10.1007/s10853-011-5490-z
5. Cundy, C.S., Cox, P.A.: The hydrothermal synthesis of zeolites: precursors, intermediates and reaction mechanism. Micro. Meso. Mater. **82**, 1–78 (2005). doi:10.1016/j.micromeso.2005.02.016
6. Provis, J.L., Kilcullen, A., Duxson, P., Brice, D.G., van Deventer, J.S.J.: Stabilization of low-modulus sodium silicate solutions by alkali substitution. Ind. Eng. Chem. Res. **51**, 2483–2486 (2012). doi:10.1021/ie202143j

7. Rios, A.C., Reyes, C.: Denver Williams, hydrothermal transformation of kaolinite in the system K_2O-SiO_2-Al_2O_3-H_2O. DYNA **77**, 55–63 (2010)
8. Barrer, R.M., Baynham, J.W.: The hydrothermal chemistry of the silicates. Part VII. Synthetic potassium aluminosilicates. J. Chem. Soc. 2882–2891 (1956). doi:10.1039/JR9560002882
9. Camblor, M.A., Corma, A., Villaescusa, L.A.: ITQ-4: a new large pore microporous polymorph of silica. Chem. Comm. **8**, 749–750 (1997). doi:10.1039/A700672A

Carbonation of Limestone Calcined Clay Cement Concrete

M.S.H. Khan(✉), Q.D. Nguyen, and A. Castel

School of Civil and Environmental Engineering, Centre for Infrastructure Engineering and Safety, The University of New South Wales, Sydney, NSW 2052, Australia

Abstract. This study aims to investigate the carbonation resistance of limestone calcined clay (LC^3) concrete. Ordinary Portland Cement (OPC) substitution rates considered were 15%, 30% and 45%. Low grade calcined clay was used with about 50% amorphous phase. Phenolphthalein indicator test was carried out in order to assess the carbonation depth of concrete exposed to natural and accelerated carbonation. Overall, results show that the carbonation front penetration is consistently increasing with the increase in OPC substitution rate Generally, LC^3 concrete under performs compared to reference OPC concrete having similar 28 days compressive strength. At 30% OPC substitution rate, performance of LC^3 concrete is close to that of OPC counterpart concrete, both having an average 28 days compressive strength of 36 MPa. Only a marginal increase in concrete cover would be enough to maintain the same service life. This is an important outcome showing that LC^3 concrete with OPC substitution up to 30% is suitable for a large range of applications including outdoor exposure (assuming no other aggressive ions are involved). However, LC^3 concrete with OPC substitution beyond 30% is not suitable in exposure condition where carbonation induced steel reinforcement corrosion can be an issue.

1 Introduction

The utilisation of supplementary cementitious materials (SCMs) to partially replace Ordinary Portland Cement (OPC) in concrete has been a rational approach over the last decades to limit CO_2 emission. Among the promising SCMs, calcined clays have received a considerable attention nowadays due to its global availability. Calcined clay is essentially composed of silica and alumina (more than 90%) and have a SiO_2/Al_2O_3 ratio between 1.5 and 2.5, depending on the purity of raw materials used in the production [1]. Calcined clay provides fine pore networks and is able to increase packing of cementitious particles due to its pozzolanic nature [2, 3]. OPC blend with calcined clay and limestone is referred to as Limestone Calcined Clay Cement (LC^3) [4]. It has been reported that the synergic between calcined clay and limestone can lead to an improvement of the mechanical performance of LC^3 based concrete [5]. LC^3 blends present an extremely promising option to the increasing demand for cement in coming decades without increasing CO_2 emissions and costs [4]. This study aims to investigate the carbonation resistance of LC^3 concrete. OPC substitution rates considered were 15%, 30% and 45%. Low grade calcined clay was used with about 50% amorphous phase. Accelerated carbonation tests were carried out using 1% CO_2.

F. Martirena et al. (eds.), *Calcined Clays for Sustainable Concrete*, RILEM Bookseries 16,
https://doi.org/10.1007/978-94-024-1207-9_38

2 Experimental Program

2.1 Materials

OPC, calcined clay and limestone were used as cementitious materials in this study. OPC was obtained from Cement Australia and limestone was supplied by Boral Construction Materials Limited, Australia. The calcined clay used was made with a ''flash'' calcination process and supplied by Argeco, France. The chemical composition of all cementitious materials determined by X-ray fluorescence (XRF) is summarized in Table 1. The particle size distribution curves for the cementitious materials are shown in Fig. 1. XRD analysis of raw calcined clay showed that the main crystalline phase is quartz, indicating that all the kaolinite in the clay is dehydroxylated during calcination. The quartz and the amorphous content in the calcined clay are 49.1 wt% and 50.9 wt% respectively, measured using Rietveld method [6]. Therefore, the calcined clay used in this study may be categorised as low-grade calcined clay [7–10]. Sydney sand was used as fine aggregate with specific gravity of 2.65 and water absorption of 3.5%. The coarse aggregate was 10 mm nominal size crushed basalt with a specific gravity of 2·8 and water absorption of 1·6%. To obtain saturated surface dry (SSD) condition, all aggregates were first oven dried at 105 °C for 24 h to eliminate any moisture content, then the exact amount of water required was added prior to concrete casting.

Table 1. Chemical composition of cementitious materials (wt%)

Chemical composition	OPC	Calcined lay	Limestone
SiO_2	18.83	70.42	0.36
Al_2O_3	4.97	22.34	0.11
Fe_2O_3	2.84	2.34	0.1
CaO	63.82	0.49	57.51
MgO	1.03	0.16	0.29
Na_2O	0.3	0.1	–
K_2O	0.66	0.19	–
TiO_2	0.27	1.1	–
SO_3	3	0.02	–
Loss on ignition (LOI)	4.45	1.76	42.61

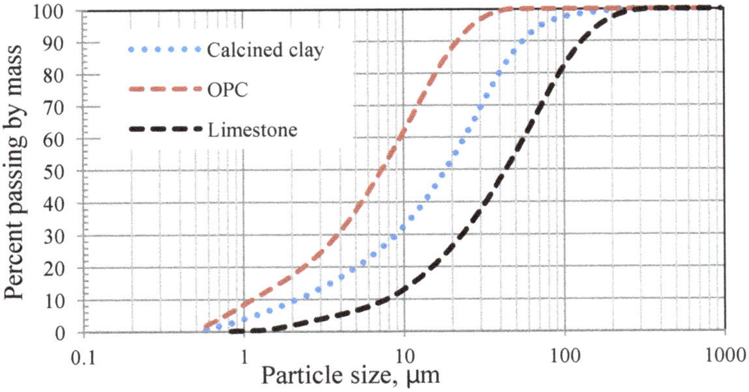

Fig. 1. Particle size distribution of OPC, calcined clay and limestone

2.2 Concrete Mix Design and Batching Procedure

Four concrete mixes have been fabricated to investigate the properties of LC^3 concrete in this study. All LC^3 mixes contain calcined clay and limestone with a ratio of 2:1 by mass. The details of the mixes are shown in Table 2. Cylindrical moulds were filled with fresh concrete in two layers and compacted using a vibrating table. After surface finishing, all moulds were covered by using lids to prevent moisture loss. All specimens were demoulded after 24 h and were placed into a lime saturated water bath for 7 days. After that, the specimens were stored in the controlled room at a fixed temperature of 23 ± 2°C and relative humidity of 55% until the testing dates.

Table 2. Mix details of LC^3 concrete (kg/m^3)

Materials	LC3-15	LC3-30	LC3-45	OPC_50	OPC_30
Coarse aggregate	1221	1221	1221	1221	1265
Fine aggregate	620.8	620.8	620.8	620.8	545
Total binder	388	388	388	388	380
OPC	329.8	271.6	213.4	388	380
Metakaolin	38.8	77.6	116.4	0	0
Limestone	19.4	38.8	58.2	0	0
Water/binder ratio	0.45	0.45	0.45	0.45	0.55
Water	174.5	174.5	174.5	174.5	210

2.3 Accelerated Carbonation Test

At 28 days, 25 mm section was removed from top and bottom of the cylinders and remaining segment was cut into 50 mm sections for subsequent carbonation testing. 50 mm discs were sealed using aluminium tapes along the perimeter leaving top and bottom sides exposed for CO_2 diffusion. The specimens were placed in a carbonation chamber with a CO_2 concentration of 1%. Temperature and relative humidity were 23°C and 55% respectively. To assess the carbonation depth, each specimen was split and 1% phenolphthalein was sprayed on the fractured surface.

3 Results and Discussions

3.1 Mechanical Properties of Concrete

Table 3 presents the compressive strength and modulus of elasticity of LC^3 concrete at 28 days. LC^3-15 had a compressive strength of 58 MPa which is 15.2% higher than the reference OPC_50 mix. The synergy between calcined clay and limestone led to an increase in compressive strength of LC3-15 mix. But the compressive strength of LC^3-30 and LC^3-45 mixes was 35.5 MPa and 36.3 MPa, respectively. These results suggest that the compressive strength is significantly reduced when the OPC replacement rate exceeds 30% by mass and this is not in agreement with previous study by Antoni et al. [5]. It is important to mention that high purity calcined clay was used in Antoni et al. experiments. The calcined clay used in this study is low grade containing 49.1 wt% of quartz. Therefore, at high replacement level, the synergy between calcined clay and limestone could not compensate the negative impact of the large quartz contain, leading to a significant reduction in the compressive strength of LC^3-30 and LC^3-45 mixes. Tironi et al. [10] reported a reduction in compressive strength of LC^3 mortar when utilising similar grade of calcined clay. Nicolas et al. [11] also observed lower strength at 28 days when replacing 25% OPC by low grade calcined clay.

Table 3. Compressive strength and elastic modulus of LC^3 concrete at 28 days

Concrete type	Compressive strength (MPa)	Modulus of elasticity (GPa)
LC3-15	58	32.6
LC3-30	35.5	29.3
LC3-45	36.3	30.7
OPC_50	52.3	31.1
OPC_30	36	28.6

3.2 Carbonation Depth of Concrete

Figure 2 shows the carbonation depths deduced from phenolphthalein indicator test together with the scatters up to 8 weeks exposure to 1% CO_2 and after 30 weeks of natural carbonation. OPC-50 specimens exposed to 1% CO_2 did not show any clear carbonation front. No carbonation front could be reported in Fig. 2 for OPC-50. Overall, results show that the resistance of LC^3 concrete against carbonation is reducing when increasing the OPC substitution rate. But, importantly, at 30% OPC substitution, performance of LC^3 concrete is close to that of OPC counterpart concrete, both having similar 28 days average compressive strength. Only a marginal increase in concrete cover would be enough to maintain the same service life. The 28 days compressive strength of OPC-30 is 36 MPa making it suitable, in most of international standards, for outdoor exposure where carbonation induced corrosion can be an issue (assuming that no other aggressive ions are involved). This is an important outcome showing that LC3 concrete with OPC substitution up to 30% is suitable for a large range of applications including outdoor exposure. However, LC^3 concrete significantly under performs if the OPC substitution rate is superior to 30%. Results obtained in natural condition confirm the vulnerability of LC^3-45 to carbonation. Indeed, for all other concretes, only few millimetres of slightly faded purple colour could be observed which could not be considered as fully carbonated regions whereas 7 mm of fully carbonated depth was observed in LC^3-45 concrete in natural condition.

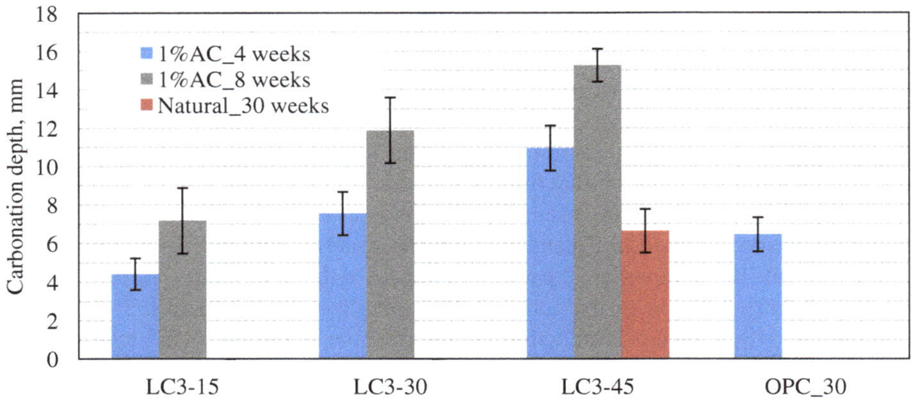

Fig. 2. Carbonation depth of concrete

4 Conclusions

Overall, results show that the carbonation front penetration is consistently increasing with the increase in OPC substitution rate. Generally, LC^3 concrete under performs compared to reference OPC concrete having similar 28 days compressive strength. At 30% OPC substitution rate, performance of LC^3 concrete is close to that of OPC counterpart concrete, both having an average 28 days compressive strength of 36 MPa. Only a marginal increase in concrete cover would be enough to maintain the same service life. This is an important outcome showing that LC^3 concrete with OPC substitution up to 30% is suitable for a large range of applications including outdoor exposure (assuming that no other aggressive ions are involved). However, LC^3 concrete with OPC substitution beyond 30% is not suitable in exposure condition where carbonation induced steel reinforcement corrosion can be an issue.

Acknowledgements. This research project was supported by the Australian Research Council through ARC Discovery Project DP160104731. The experiments were conducted in the materials and structures laboratory in the School of Civil and Environmental Engineering at the University of New South Wales. The assistance of the laboratory staff is acknowledged here.

References

1. Nicolas, R.S., Cyr, M., Escadeillas, G.: Characteristics and applications of flash metakaolins. Appl. Clay Sci. **83–84**, 253–262 (2013)
2. Siddique, R., Klaus, J.: Influence of metakaolin on the properties of mortar and concrete: a review. Appl. Clay Sci. **43**, 392–400 (2009)
3. Shi, Z., Lothenbach, B., Geiker, M.R., Kaufmann, J., Leemann, A., Ferreiro, S., Skibsted, J.: Experimental studies and thermodynamic modeling of the carbonation of Portland cement, metakaolin and limestone mortars. Cem. Concr. Res. **88**, 60–72 (2016)
4. Scrivener, K.L.: Options for the future of cement. Indian Concr. J. **88**(7), 11–21 (2014)
5. Antoni, M., Rossen, J., Martirena, F., Scrivener, K.: Cement substitution by a combination of metakaolin and limestone. Cem. Concr. Res. **42**, 1579–1589 (2012)
6. Rietveld, H.M.: A profile refinement method for nuclear and magnetic structures. J. Appl. Crystallogr. **2**, 65–71 (1969)
7. Badogiannis, E., Kakali, G., Dimopoulou, G., Chaniotakis, E., Tsivilis, S.: Metakaolin as a main cement constituent. Exploit. Poor Greek kaolins Cem. Concr. Compos. **27**, 197–203 (2005)
8. Badogiannis, E., Papadakis, V.G., Chaniotakis, E., Tsivilis, S.: Exploitation of poor Greek kaolins: strength development of metakaolin concrete and evaluation by means of k-value. Cem. Concr. Res. **34**, 1035–1041 (2004)
9. Vizcaíno, L., Antoni, M., Alujas, A., Martirena, F., Scrivener, K.: Industrial manufacture of a low-clinker blended cement using low-grade calcined clays and limestone as SCM: the cuban experience. In: Scrivener, K., Favier, A. (eds.) Calcined Clays for Sustainable Concrete. RB, vol. 10, pp. 347–358. Springer, Dordrecht (2015). doi:10.1007/978-94-017-9939-3_43
10. Tironi, A., Scian, A.N., Irassar, E.F.: Blended cements with limestone filler and kaolinitic calcined clay: filler and pozzolanic effects. J. Mater. Civ. Eng. (2017). doi:10.1061/(ASCE)MT.1943-5533.0001965
11. Nicolas, R.S., Cyr, M., Escadeillas, G.: Performance-based approach to durability of concrete containing flash-calcined metakaolin as cement replacement. Constr. Build. Mater. **55**, 313–322 (2014)

Grinding of Calcined Clays and Its Effects on Cement Properties

W. Kluge(✉) and B.O. Assmann

Thyssenkrupp Industrial Solutions AG, Essen, Germany

1 Introduction

The production of cement with more than one main component plays a key role in prospective development in the cement industry. Targets steered by political and economic interests give rise to the demand to substitute the clinker component and the need to reduce energy consumption and greenhouse gas emissions without any constraints in terms of cement quality. The properties of cement and concrete are determined by the chemical-mineralogical composition and by the specific surface area (acc. to Blaine), as well as by the location parameter x' and the slope n of the RRSB particle size distribution of every single main component. This reinforces the optimization of the cement grinding process. Questions concerning separated-grinding or inter-grinding and the influence of the type of grinding machine in connection with product quality and energy consumption appear increasingly frequently. The most common grinding machines for cement grinding are ball mills, HPGRs and vertical roller mills. Each system has different characteristics with regard to energy consumption and particle size distribution, and hence with regard to cement properties. The slope n for a product of a vertical roller mill is typically situated between the values of a ball mill and an HPGR.

2 Method

The following investigation considers the comminution of calcined clay and clinker to CEM II/A-Q and CEM II/B-Q in different ways, based on a statistical design of experiment. Cement mixtures produced by separated-grinding in a vertical roller mill and a classic ball mill in the tkIS technical center (Fig. 1) will be verified with regard to cement properties according to DIN EN 196 and with regard to specific energy consumption. Influencing variables are fineness of clinker, fineness of calcined clay, content of calcined clay up to 35 wt% and the type of mill (Table 1). On the basis of the results, it is possible to find an optimum for the composition of a mixture and for the configuration of grinding machines, taking into account the mentioned effect variables in comparison with inter-grinding.

F. Martirena et al. (eds.), *Calcined Clays for Sustainable Concrete*, RILEM Bookseries 16,
https://doi.org/10.1007/978-94-024-1207-9_39

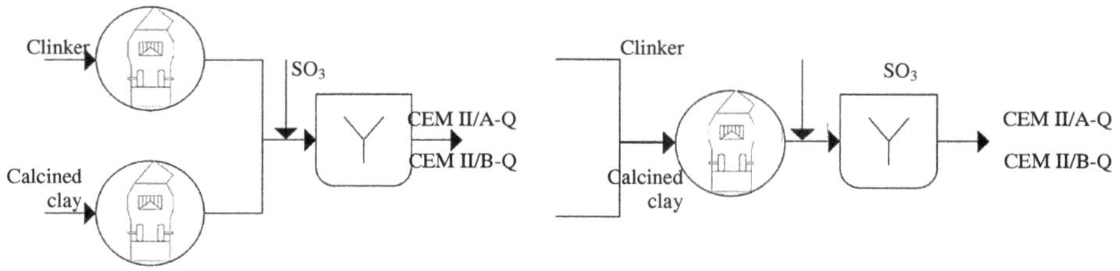

Fig. 1. Experimental setup - schematic sketch of separated-grinding (left) and inter-grinding (right) by a vertical roller mill with subsequent mixing process

Table 1. Influencing variables for design of experiment

Fineness clinker acc. Blaine	Fineness calcined clay acc. Blaine	Content of calcined clay
3,000 cm^2/g	3,000 cm^2/g	5.0 wt%
4,000 cm^2/g	7,600 cm^2/g	20.0 wt%
5,000 cm^2/g	12,000 cm^2/g	35.0 wt%

To avoid the influence of sulfate in relation to content and composition, it is necessary to add standard grounded sulfate subsequently, taking into account the sulfate content of the clinker.

3 Results

The theoretical context of slope *n* of RRSB particle size distribution for the ball mill and the vertical roller mill was confirmed by investigations in the tkIS technical center. The vertical roller mill produces steeper particle size distribution than a ball mill [1].

Earlier determination of grindability according to ZEISEL for calcined clay and clinker indicates a lower grindability for calcined clay in comparison with clinker (Table 2), but no evidence was found during grinding analysis.

Table 2. Grindability of calcined clay and clinker according ZEISEL-test

	Calcined clay	Clinker
w_m (3,000 cm^2/g) [kWh/t]	8.4	30.1

Calcined clay generates high Blaine values with little effort compared to clinker. This implies low grindability, but the Blaine value does not represent the granulometry property for calcined clay. The right-hand diagram in Fig. 2 shows the context in detail, based on sieve analysis. The progress of calcined clay fineness is limited irrespective of energy input in the ZEISEL device. Increased agglomeration can be seen as the main reason. From this point of view, it is much easier to grind clinker than calcined clay. Therefore, in the experiment, the fineness of calcined clay according to Blaine was selected at a much higher level than the fineness of clinker.

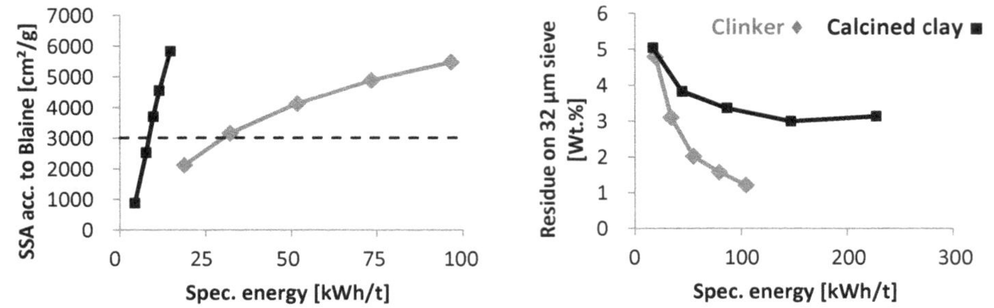

Fig. 2. Progress of clinker and calcined clay fineness in the ZEISEL device, based on Blaine value (left) and sieving residue (right)

3.1 Separated-Grinding

Separated-grinding gives the opportunity to find an optimum in terms of cement composition and the fineness of each composition in relation to cement quality and specific energy consumption.

Figure 3 shows the main parameter effects on seven-day compressive strength. The main effects are similarly represented by the fineness of clinker and the amount of calcined clay in the cement. The fineness of the calcined clay in the cement plays an insignificant role compared to the others. The slope of the curves underlines this statement. The results of two-day compressive strength show a similar tendency.

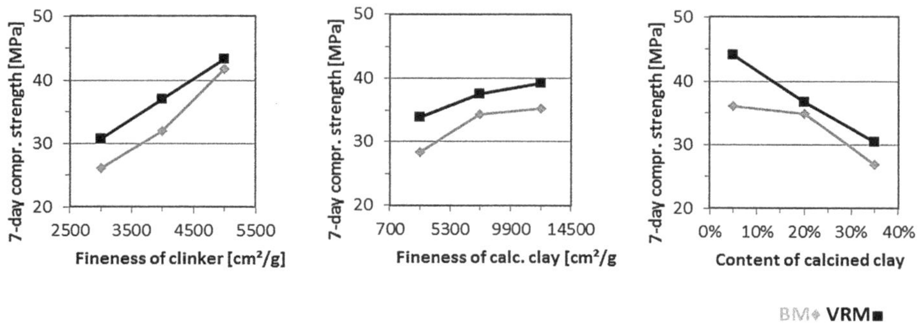

Fig. 3. Main effects on 7-day compressive strength

Initial examination of early compressive strength shows that there is no need to grind clinker and calcined clay separately. This was also confirmed by results regarding the main effects on specific energy consumption for grinding (Fig. 4). The influence of clinker fineness and calcined clay fineness on specific energy consumption for grinding is more or less the same. Therefore an improvement of specific energy consumption for grinding due to substitution of clinker with calcined clay is not directly recognizable (Fig. 4; right).

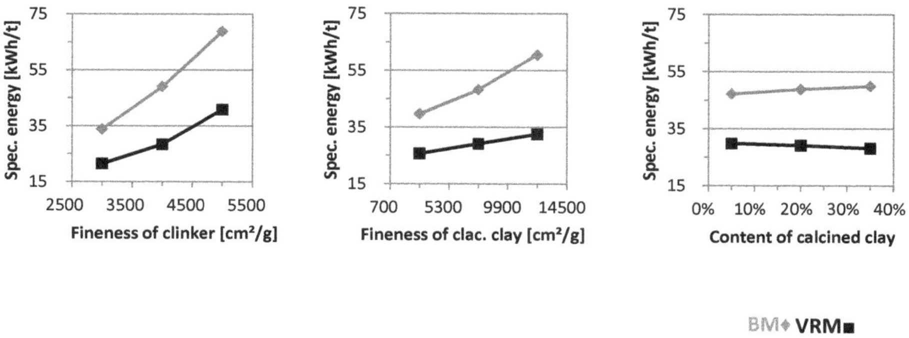

Fig. 4. Main effects on specific energy consumption; BM and VRM in comparison

Furthermore, a comparison between a ball mill and a vertical roller mill in terms of cement performance and energy consumption for grinding shows advantages when grinding cements containing calcined clay using a vertical roller mill. Based on the results of the investigation – for instance Figs. 3 and 4 – it was possible to make energy savings of 30.0% to 50.0%. An improvement of mortar compressive strength with a vertical roller mill was observed, compared to a ball mill producing similar products. Only in terms of water demand does the product of a vertical roller mill show a disadvantage in comparison with the product of a ball mill. On average, water demand was 2.0% higher than for the product of a ball mill.

3.2 Inter-grinding

Selected samples of separated-grinding investigations were also ground in inter-grinding mode by keeping the same fineness of mixture according to Blaine in the ball mill and the vertical roller mill.

Samples were analyzed by MLA (Mineral Liberation Analysis). MLA provides an opportunity to evaluate the particle size distribution of the main components in cement samples [2]. Those results provided evidence of similar grindability of clinker and calcined clay. The particle size distributions of clinker and calcined clay were almost identical after inter-grinding.

Significant savings in specific energy consumption of up to 50.0% were found in relation to grinding, with a negligible reduction of strength development of, on average, 7.5% compared to separated-grinding samples.

4 Conclusion

The specious low grindability of calcined clay according Zeisel was not found in the experiment. It was necessary to grind calcined clay using the same energetic effort as for clinker in separated-grinding. Furthermore, separated-grinding of calcined clay hides problems with regard to the agglomeration of particles, the progress of comminution and the handling of materials.

There are advantages with separated-grinding in terms of cement quality for cements containing calcined clay, but these advantages do not compensate for the higher operating and investment costs. In conclusion, inter-grinding of cements containing calcined clay is recommended.

References

1. Komorek, D.: Zementmahlung, Seminar: Mahlbetrieb und Produkteigenschaften. VDZ, Düsseldorf, 25 November 2015
2. Bornefeld, M., Kluge, W.: New way of evaluation of composite cements. Cem. Int., 63–65, February 2017

Hydration and Mechanical Properties of Limestone Calcined Clay Cement Produced with Marble Dust

Sreejith Krishnan(✉), Arun C. Emmanuel, Swadesh Kumar Kanaujia, and Shashank Bishnoi

Department of Civil Engineering, Indian Institute of Technology, Delhi, New Delhi, India

Abstract. Nearly 17 million tonnes of dimensional stones, such as marble, are produced in India annually. Around 7 million tonnes of stone dust waste is generated annually from the excavation and processing of these stones with an estimated wastage of 30–60% depending on the technology used. Stone dust waste generated, usually in the form of a slurry is dumped on open land, resulting in the wastage of productive land. The drying up of slurry leads to generation of fine air borne particles that can result in severe environmental pollution and health problems. The study shows that stone dust waste primarily consists of carbonate stones such as calcite, dolomite and magnetite. This study aims to understand the potential of utilizing stone dust waste in limestone calcined clay cement by replacing limestone with marble dust waste. Formation of carboaluminates was observed in the using marble dust as a replacement in LC^3. It was found that stone dust waste can be used as a limestone substitute for the production of limestone calcined clay cement.

1 Introduction

Ternary blended cements in which a high level of clinker is being replaced by a combination of multiple supplementary cementitious materials (SCMs) have emerged as a sustainable alternative to ordinary portland cement (OPC). Various potential combinations of SCMs are possible, which can be designed to exploit the synergetic interactions between OPC and SCMs. Limestone calcined clay cement (LC3) is one such ternary cement blend that combines OPC with calcined clay and limestone. In presence of carbonates, additional hydration products (calcium monocarboaluminate and calcium hemicarboaluminate are formed). Presence of these phases stabilizes ettringite and there by prevents it conversion to monosulphate. The improved volume efficiency caused due to the presence of ettringite can improve the mechanical properties of cement [1].

Marble and kota stone are two popular dimensional stones used in India. Every year, millions of tonnes of waste dust is produced during the processing of these stones, leading to health issues as well as major environmental hazards such as loss of productive land. Marble is a metamorphic carbonate rock whereas kota stone is a quartzitic limestone [2].

The aim of this study was to investigate whether it is possible to utilize the stone dust waste as a replacement for limestone to produce limestone calcined clay cement.

F. Martirena et al. (eds.), *Calcined Clays for Sustainable Concrete*, RILEM Bookseries 16,
https://doi.org/10.1007/978-94-024-1207-9_40

Till date, the majority of attempts to utilize the stone dust waste in cement was as an unreative filler [3], whereas in LC^3 systems reactivity of carbonates is important so as to produce hemicarboaluminate and monocarboaluminate phases.

2 Materials and Methods

Two marble dusts, MD1 and MD2, were procured from Kisangarh, Rajasthan and kota stone dust K was procured from mining unit in Kota, Rajasthan. The XRD analysis of the marble dust are shown in Fig. 1. It was seen that MD1 contains dolomite as the major phase whereas MD2 contains dolomite, quartz and calcite phases. Kota stone dust do not contain any dolomite. Commercially available metakaolin, MetaCem85 was used as calcined clay. LC^3 blends where prepared by grinding clinker (50% by wt.), calcined clay (31% by wt.), stone dust waste (15% by wt.) and gypsum (4%) in a laboratory ball mill (Table 1). The particle size distribution of the blends produced is shown in Fig. 2. The physical properties of the LC^3 blends where calculated as per IS 4031 and are shown in Table 2.

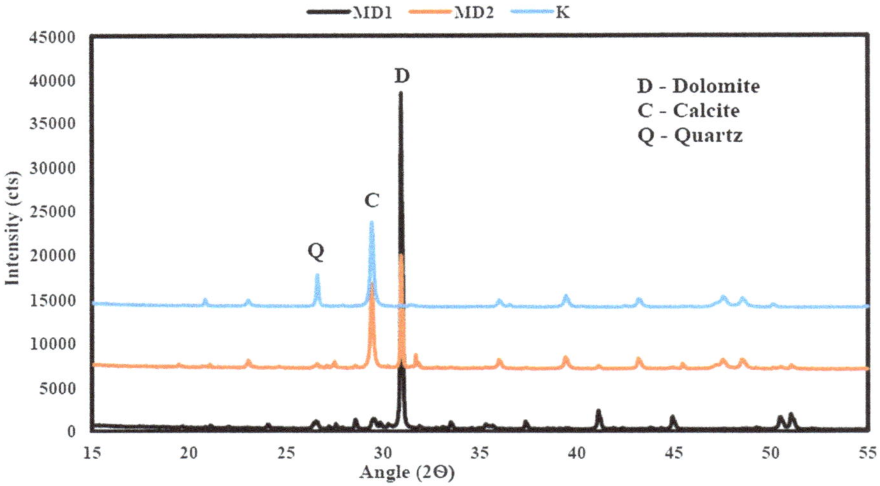

Fig. 1. XRD of the stone dust wastes – Only major peaks identified

Table 1. Details of the blends used in the study

Blend name	Stone dust
LC^3 – MD1	MD1
LC^3 – MD2	MD2
LC^3 – K	K

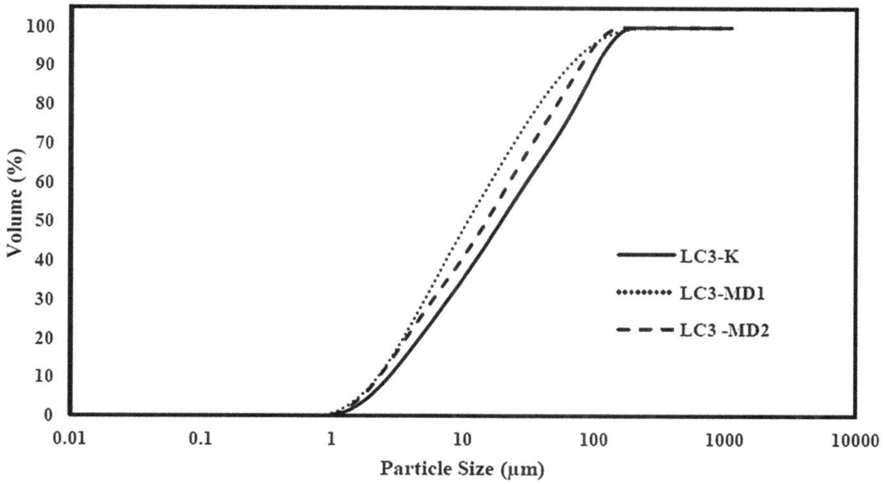

Fig. 2. Particle size distribution of LC^3 blends produced

Table 2. Physical properties of LC3 cement blends

	LC^3 - MD1	LC^3 - MD2	LC^3 - K
Standard consistency (%)	35	35.5	33.5
Initial setting time (min)	60	45	40
Final setting time (min)	450	420	330
Specific gravity (g/cc)	2.81	2.76	2.74

Compressive strength was determined by casting cubes of size 70.6 * 70.6 * 70.6 mm with a w/c ratio of 0.45. XRD of the hydrated paste specimen (prepared with a w/c 0.45) was carried out at 28 days for the qualitative analysis of the phases formed.

3 Results

The standard consistencies of the LC^3 was found to be 35%, 35.5% and 33.5% for blends MD1, MD2 and K respectively. This is higher than typical value of ordinary portland cement which is normally around 28% to 30%. Similarly, the initial setting times of the LC^3 blends where found to be between 40 and 60 min which is again lower than ordinary portland cement. The increases consistency and reduction in initial setting time can be attributed to the presence of calcined clay [4]. It is to be noted that the setting times satisfies the requirements prescribed by the Indian Standards (minimum 30 min of initial setting time, maximum 600 min for final setting time). The X-ray diffractogram of the hydrated LC^3 paste are shown in Figs. 3 and 4 respectively. Formation of carboaluminate phases are clearly seen from as early as 3 days. Strong peaks of ettringite peaks are visible even at 28 days. The compressive strength of the LC^3 blends are shown in the

Fig. 5. Similar strengths are observed for all the blends at 3, 7 and 28 days. The improvement in the strength can be attributed to the formation of carboaluminates and the stabilization of ettringite. The compressive strengths of the LC^3 blends also satisfies the Indian Standard for portland pozzolana cement i.e. 33 MPa at 28 days.

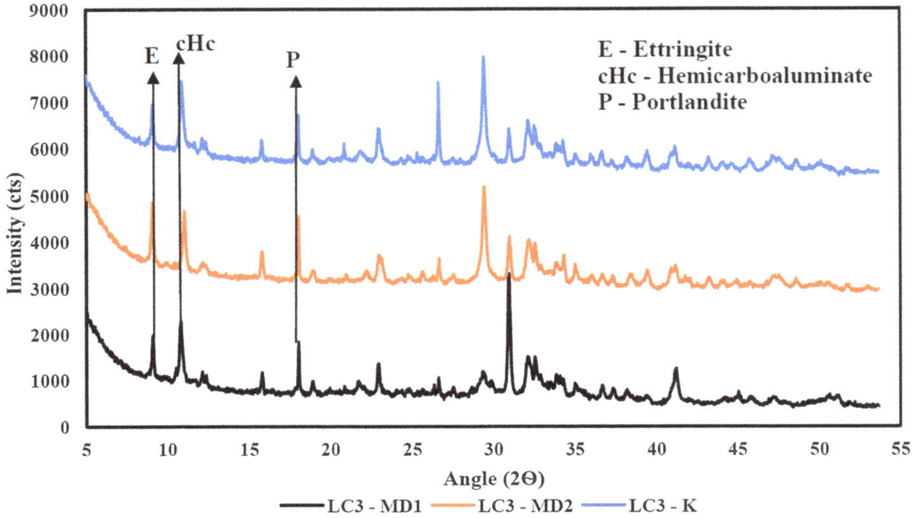

Fig. 3. Hydrated XRD of LC^3 blends at 3 days

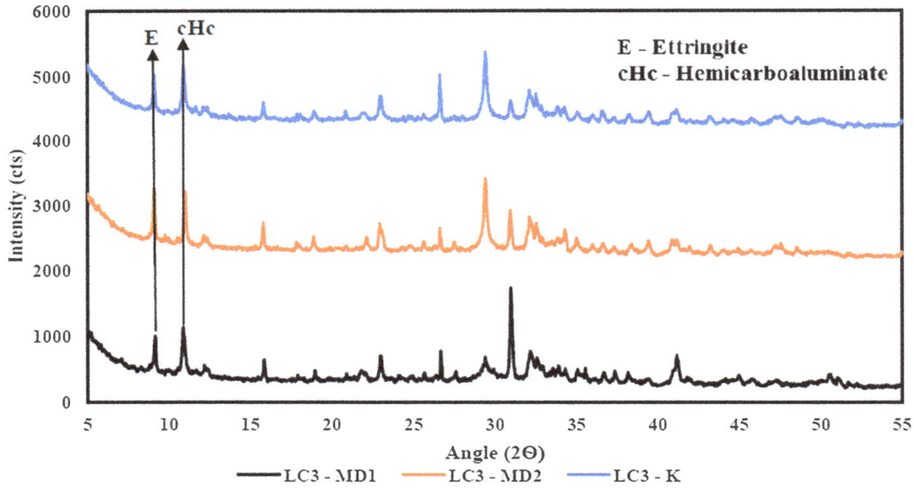

Fig. 4. XRD of 28 days hydrated LC^3 hydrated paste

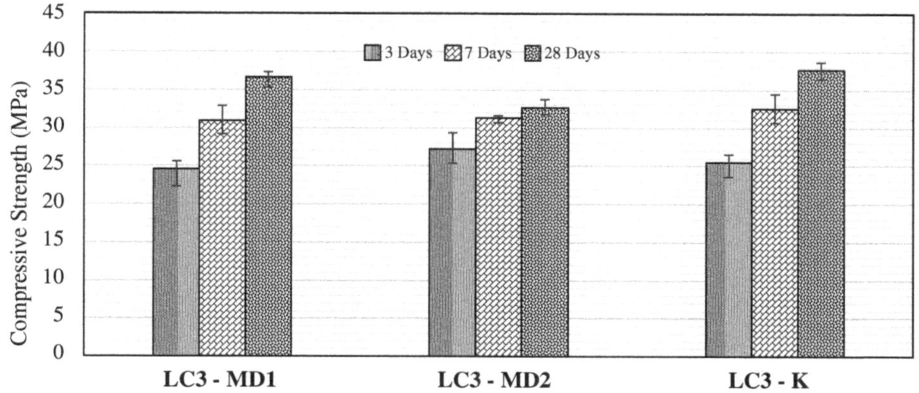

Fig. 5. Compressive strength of LC^3 blends

4 Conclusions

It is seen that marble dust and kota stone dust can be used as a carbonate substitute for limestone in Limestone calcined clay cement. The major phases present in the stone dust are calcite and dolomite. The formation of carboaluminate phases were confirmed in presence of marble dust and kota stone dust with stabilization of ettringite.

Acknowledgements. The authors would like to acknowledge the Swiss Agency for Development and Co-operation for the support of Limestone Calcined Clay cement project in India.

References

1. Antoni, M., Rossen, J., Martirena, F., Scrivener, K.: Cement substitution by a combination of metakaolin and limestone. Cem. Concr. Res. **42**, 1579–1589 (2012). doi:10.1016/j.cemconres.2012.09.006
2. Centre for Development of Stones, Stones of Rajasthan, Centre for Development of Stones, Jaipur (2011)
3. Aliabdo, A.A., Abd Elmoaty, A.E.M., Auda, E.M.: Re-use of waste marble dust in the production of cement and concrete. Constr. Build. Mater. **50**, 28–41 (2014). doi:10.1016/j.conbuildmat.2013.09.005
4. Snelson, D., Wild, S., O'Farrell, M.: Setting times of Portland cement–metakaolin–fly ash blends. J. Civ. Eng. Manag. **17**, 55–62 (2011). doi:10.3846/13923730.2011.554171

Performance-Based Design Procedure Applied to the Selection of Low-CO_2 Binder Systems Including Calcined Clay

Wilson R. Leal da Silva[1(✉)], Lars N. Thrane[1], Thomas L. Svensson[1], Sergio Ferreiro[2], Duncan Herfort[2], Claus Pade[1], and Jesper S. Damtoft[2]

[1] Danish Technological Institute, Gregersensvej 4, 2630 Taastrup, Denmark
[2] Cementir Holding S.p.A, Sølystvej 18, 9220 Aalborg, Denmark

Abstract. The reduction of the clinker content in concrete by replacing cement with supplementary cementitious materials (SCM) is widely used to lower the environmental impact of concrete. However, for some SCMs such as calcined clay this often leads to higher water requirements for the same concrete consistence. This severely limits the widespread implementation of green concrete technologies, both in practical terms, and in meeting the requirements of relevant concrete standards such as the national annexes to EN 206. In order to overcome this, a shift in mind-set from prescriptive to performance-based design criteria would be beneficial. Thus, we have proposed a performance-based design approach for the selection of binder systems, which we have successfully applied it to several combinations of SCMs. In this paper the procedure is described for a binder system consisting of calcined clay, limestone filler and fly ash. A parametric study was carried out on mortars, which included rheological properties, compressive strength and chloride migration. The requirements for each parameter was chosen based on the intended concrete application (highway bridge) and values obtained for a reference mix typically used for the application, i.e. a concrete that fulfils current Danish regulations. The proposed alternative binder system features a CO_2 footprint reduction of 30% compared to the reference concrete mix. However, the alternative concrete mix does not conform with current Danish regulation on the minimum w/c ratio.

1 Introduction

Market estimates suggest that by 2050 the amount of concrete produced globally will be about twice that of today [1]. Thus, the demand for environmental friendly concrete structures is increasing and even a small reduction in CO_2 emissions can make a significant difference. The reduction of clinker content in concrete by replacing cement with supplementary cementitious materials (SCM) is pointed out as one of the key areas that can deliver substantial reductions in global CO_2 emissions associated with to cement and concrete [2–4].

In that context, a vast amount of research work has been done on SCMs, e.g. fly ash (FA), to partially replace cement, reducing the environmental impact of concrete [4–6]. Though FA is widely used in concrete production, its availability cannot be guaranteed

F. Martirena et al. (eds.), *Calcined Clays for Sustainable Concrete*, RILEM Bookseries 16,
https://doi.org/10.1007/978-94-024-1207-9_41

in the years to come. In several countries, the long-term energy production plan is to transfer the energy production method from coal-fired power plants to more environmental-friendly technologies. This is the case of Denmark, where all major power plants will have shifted from coal to biomass as primary source of fuel by the end of 2023.

Thus, despite the potential of using high amounts of FA to substitute cement clinker and reduce the carbon footprint, this may not be a sustainable solution. Therefore, alternative SCMs to supplement or replace FA needs to be investigated. From this perspective, calcined clay and limestone filler are promising candidates given their availability worldwide [5].

The potential of using calcined clay and limestone filler is investigated in an ongoing Danish Innovation Consortium "Green Transition of Cement and Concrete Production", a project supported by the Danish Innovation Fund, which aims to create the foundation for a green transition of cement and concrete production in Denmark. As part of the project, a performance-based design procedure has been proposed for selection of low-CO_2 binder systems.

In the project, the procedure has been applied to propose binder systems in which calcined clay and limestone filler are used in combination with and without fly ash, respectively. In both cases, studies have been carried out to optimize binder combinations for building and civil engineering applications. To illustrate the design framework, this paper presents the procedure applied for optimisation of the concrete binder composition for a highway bridge in Denmark using calcined clay, limestone filler and fly ash.

2 Performance-Based Optimisation of Binder Compositions

The suggested framework of a performance-based design procedure for the selection of concrete binder systems with low-environmental impact consists of four steps:

Step 1 Define the target application and a reference concrete mix design. Relevant properties are selected and target criteria are defined based on the performance of the reference mixture.

Step 2 Select SCMs, and define and execute an experimental test programme at mortar level.

Step 3 Step 3 Propose the lowest environmental impact binder composition with equivalent or improved technical performance over the reference mix design.

Step 4 Step 4 Implement the low environmental impact concrete at industrial scale by fine-tuning the Step 3 proposal at the concrete batch plant.

Figure 1 illustrates the principles of a traditional prescriptive mix design procedure versus a performance-based design procedure. It shows a certain material property, e.g. the resistance to chloride ingress, as a function of the W/C ratio. In accordance with a prescriptive design, a maximum W/C ratio is usually specified for a given exposure class. In a performance-based approach, the W/C is variable, while the performance is fixed, i.e. a limit acceptance criterion set on the performance.

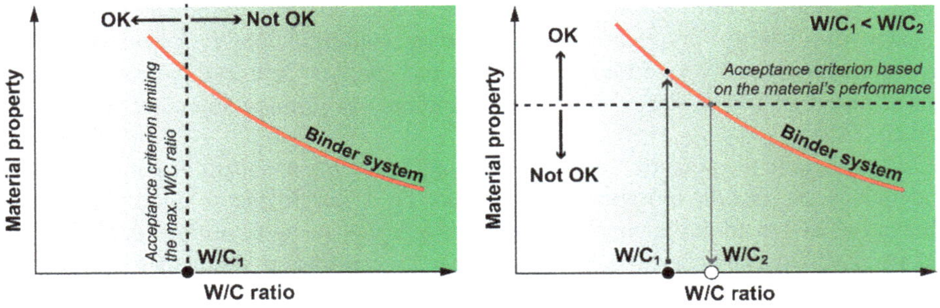

Fig. 1. Design method based on a prescriptive (left) and a performance-based design (right).

To determine the target criteria, we propose to measure the performance of the reference mixture. The example illustrated in Fig. 1 shows that the alternative binder system performs better than reference at the prescribed water to cement ratio (W/C_1). Ideally, the application of performance-based design will allow higher W/C for mix designs (W/C_2) with lower environmental impact without compromising performance.

3 Target Application and Reference Criteria

The highway bridge is constructed in 2017. Part of the structure will be cast with concrete consisting of an alternative low-CO_2 binder system. The reference mix design, provided by Danish concrete manufacturer Unicon A/S, is currently used for civil engineering applications and is in full compliance with the existing standards, DS2426 [7], and rules specified by the Danish Road Directorate, i.e. cement type, maximum w/c ratio, and minimum cement content.

For the target properties, we selected: μ (Pa.s) - plastic viscosity, $f_{c,28d}$ (MPa) - compressive strength at 28 days, and D ($\times$ 10^{-12} m^2/s) - chloride migration coefficient at 91 days.

4 Experimental Program

Table 1 summarises the binder compositions of the investigated mortars in % b.w. and kg per m^3 of concrete, respectively. To obtain as comparable and useful results as possible, the volumetric proportions of paste and fine aggregate (S) were kept constant at 56% and 44%. The superplasticiser content was adjusted for each mix to obtain mortars with yield stress corresponding to similar slumps at concrete level. This provided a sound basis for comparing the plastic viscosity measured for the tested paste compositions. The calcined clay-to-limestone ratio and SCM substitution were kept

Table 1. Mix composition (FA is accounted with an activity factor of 0.5 for the W/C ratio. According to the Danish rules, limestone filler and calcined clay cannot be included in the equivalent cement content for civil engineering applications).

ID binder	C	FA	Q	L	SCM b.w. of binder	FA/C	Q/C	L/C	ID mortar	C	FA	Q	L	Water	S	W/C	$CO_{2\text{-equiv}}$
	Binder composition % b.w.				%	b.w	b.w.	b.w.		Kg per m^3 of concrete						b.w.	Kg/m^3
Ref.	86.9	13.1	–	–	13.1	0.15	–	–	Ref.	359	54	–	–	143	580	0.37	333
A	67.6	19.5	10.8	2.2	32.4	0.29	0.16	0.03	A1	235	68	38	8	161	580	0.60	208
									A2	290	83	46	9	133	580	0.40	255
B	67.6	12.2	16.9	3.4	32.4	0.18	0.25	0.05	B1	242	44	61	12	159	580	0.60	220
									B2	298	54	75	15	130	580	0.40	270

a) b) c)

Fig. 2. (a) Mortar samples for compressive strength - DS/EN 196-1, (b) Chloride penetration testing - NT Build 492 and (c) Plastic viscosity - 4C-mini rheometer device.

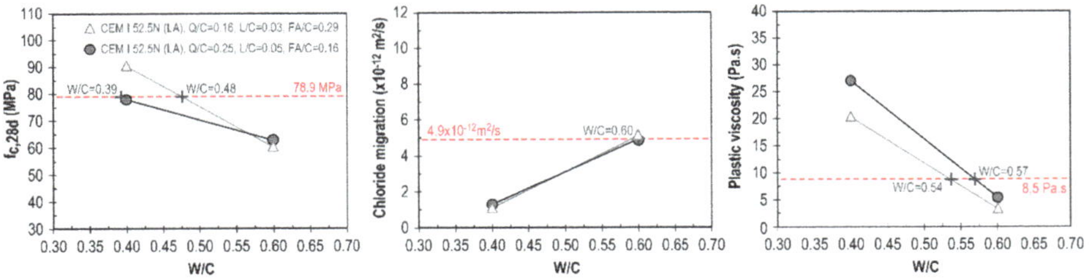

Fig. 3. Experimental and target results for (a) fc, 28d (MPa), (b) J (× 10–12 m2/s) and (c) μ (Pa.s).

constant at 5.0 and 32.4% by weight of the binder, respectively, and then two variations of binder composition was tested at two different w/c ratios, 0.40 and 0.60 respectively.

The compressive strength of the mortar samples is tested in accordance with the technical standard DS/EN 196-1. Note, that the strength is not normalised to the same air content. The chloride migration coefficient was measured according to test method NT Build 492 on samples taken from Ø 100 mm x 200 mm cylinders after 91 days of curing in water. Figure 2b shows the test samples after 492 days of exposure to chloride. The rheological properties were measured using the 4C-mini-Rheometer® developed at the Danish Technological Institute (Fig. 2c).

The target values for μ, $f_{c,28d}$ and J are based on the mortar results of the reference mixture, which was produced with a CEM I 42,5 N – SR 5 (EA). The mixtures A1, A2, B1 and B2 (Table 1) were produced with CEM I 52,5 N (LA). The experimental results for the reference mixture are $\mu = 8.5\ Pa.s$ and $J = 4.9 \times 10^{-12}\ m^2/s$, and $fc,_{28d} = 78.9\ MPa$. The experimental results for the mixtures with a binder system comprising calcined smectitic clay (Q), limestone filler (L) and fly ash (FA) are displayed in Fig. 3.

5 Results and Discussion

The results show that increasing the amount of calcined clay and limestone filler over fly ash results in a compressive strength that is lower at w/c ratio of 0.40 but the same or slightly higher at a w/c ratio of 0.60. In contrast, the two binder compositions perform similarly at both w/c ratios of 0.40 and 0.60 in terms of chloride migration. For the plastic viscosity, the increase of calcined clay and limestone over fly ash content results in higher viscosities of approximately 30%. Furthermore, the mixes with higher calcined clay/limestone filler content seemed stickier.

To find the optimal solution compared to the acceptance criteria, Table 2 shows the mix designs as a result of interpolation when the compressive strength is fixed at the reference strength of 78.9 MPa.

When comparing the obtained W/C intervals for each criterion, there exist no solution that fulfils all criteria at once. It is possible to fulfil compressive strength and chloride migration up to a w/c ratio of 0.48, however, at this w/c ratio, the plastic viscosity is 13.3 Pa.s, which is larger than the reference value of 8.5 Pas. To overcome this, the next step could be to lower the substitution ratio of the cement and to change the relative ratios of the supplementary cementitious materials. For instance, the effect of changing the calcined clay-to-limestone filler ratio from 5.0 to 1.5 was tested for both mixes at w/c ratio of 0.60. The results showed similar or slightly improved performance on strength, chloride migration and plastic viscosity. The plastic viscosity dropped by approximately 10%, but also the stickiness was significantly improved. Despite not able to reach a plastic viscosity equal or lower than the reference, practical

Table 2. Mix design in the range from W/C = 0.39 to 0.48 for fc, 28d = 78.9 MPa (Criterion 1).

Mix ID	Ref.	B1&B2	← Interpolation for $f_{c,28d}$ = 78.9 MPa →					A1&A2	Proposal
CEM I 42.5 N	359.0	–	–	–	–	–	–	–	
CEM I 52.5 N	–	301.8	293.0	288.7	280.4	276.4	272.4	264.8	**264.8**
FA	54.0	55.4	61.0	63.7	68.8	71.2	73.6	78.0	**78.0**
Q	–	74.6	66.5	62.6	55.1	51.5	48.0	41.3	**30.5**
L	–	14.9	13.1	12.3	10.7	9.9	9.2	7.7	**20.4**
Water	143.0	128.5	132.6	134.6	138.5	140.4	142.2	145.8	**145.8**
Sand (0–4 mm)	580.0	580.0	580.0	580.0	580.0	580.0	580.0	580.0	**580.0**
FA/C	0.15	0.18	0.21	0.22	0.25	0.26	0.27	0.29	**0.29**
Q/C	–	0.25	0.23	0.22	0.20	0.19	0.18	0.16	**0.12**
L/C	–	0.05	0.04	0.04	0.04	0.04	0.03	0.03	**0.08**
W/C	0.37	0.39	0.41	0.42	0.44	0.45	0.46	0.48	**0.48**
$f_{c,28d}$ (MPa)	78.9	78.9	78.9	78.9	78.9	78.9	78.9	78.9	*~78.9*
J ($\times 10^{-12} m^2/s$)	4.9	1.1	1.5	1.6	2.0	2.2	2.3	2.7	*~2.7*
μ (Pa.s)	8.5	28.0	24.3	22.6	19.3	17.7	16.2	13.3	*~10–13*
kg-CO_2/m^3	333	273	264	259	250	245	241	233	**231**
CO_2 reduction	–	18%	21%	22%	25%	26%	27%	30%	**30%**

observations indicate that the production and execution properties can be satisfactory even in the range from 10–15 Pa.s. Thus, allowing for a slightly higher plastic viscosity, the first proposal for a binder composition could be the one indicated to the right in Table 2, which shows CO_2 reduction of approximately 30% compared to the reference.

The resulting w/c ratio is 0.48 when calculated according to DS2426, which prescribes a limit value of 0.40. Notice that if the proposed binder system were to fulfil the requirement w/c $\leq$ 0.40, the concrete would be less environmental-friendly and the plastic viscosity would be much greater than that of the reference mixture. This may result in poor production and execution properties, most like making it impossible to use in practice.

The final concrete mix design of the highway bridge would be based on large-scale trials at concrete plant.

6 Conclusions

A performance-based design procedure for selecting low-CO2 binder systems have been proposed. The procedure has been applied to propose a binder system for a bridge structure containing calcined clay, limestone filler and fly ash as supplementary cementitious materials. The procedure results in proposed mix design with a CO2 reduction of 30% compared to a reference mix. The alternative concrete mix shows equivalent or better performance on compressive strength and chloride migration. However, the plastic viscosity is slightly higher, but acceptable. The results show that despite of current standard constrains it is technically possible to manufacture greener concrete at equivalent performance than produced up to date. In the future standards, other specific requirements may be considered for the use of calcined clay – limestone blended cements in concrete.

Acknowledgements. This work is part of the project "Green transition of cement and concrete production". The financial support from the Danish Innovation Fond (InnovationsFonden) and the contribution from project partners is acknowledged. Also, the supply of calcined clay from Saint Gobain – Leca International is acknowledged.

References

1. Wray, P., Scrivener, K.: Straight talk with Karen Scrivener on cements, CO_2 and sustainable development. Am. Ceram. Soc. Bull. **91**, 47–50 (2012)
2. Glavind, M., Munch-Petersen, C.: Green concrete: a life cycle approach. In: Proceedings of International Congress on Challenges of Concrete Construction, pp. 642–50, Scotland (2002)
3. Schneider, M., Romer, M., Tschudin, M., Bolio, H.: Sustainable cement production at present and future. Cem. Concr. Res. **41**(7), 642–650 (2011). doi:10.1016/j.cemconres.2011.03.019
4. Damtoft, J.S., Lukasik, J., Herfort, D., Sorrentino, D., Gartner, E.M.: Sustainable development and climate change initiatives. Cem. Concr. Res. **38**, 115–127 (2008). doi:10.1016/j.cemconres.2007.09.008

5. Scrivener, K.L., John, V.M., Gartner, E.: Eco-efficient cements: potential, economically viable solutions for a low-CO_2, cement based materials industry, United Nations Environment Program (2016)
6. Lothenbach, B., Scrivener, K., Hooton, R.D.: Supplementary cementitious materials. Cem. Concr. Res. **41**(12), 1244–1256 (2010). doi:10.1016/j.cemconres.2010.12.001
7. DS2426 - EN 206-1:2011: Concrete – Materials – Rules for application of EN206-1 in Denmark, The Danish Standards Association (2011)

Thermal Processing of Calcined Clay

J. Lemke(✉) and C. Berger

Thyssenkrupp Industrial Solutions AG, Essen, Germany

Abstract. Various clayey materials containing different clay minerals were calcined using muffle furnace and pilot scale flash calciner. While pure clay minerals can be easily calcined using soak calcination, mixed systems, especially Ca/Mg-rich materials, benefit from flash calcination. High process temperatures enable a complete calcination while the short retention time suppresses the formation of inert high-temperature phases.

1 Introduction

Since 2010, various types of clayey material have been investigated at the thyssenkrupp cement technology research center. Based on tests with rather pure clay substances like kaolinite, illite or montmorillonite, a focus was set on mixed clay materials. These materials contain more than one clay mineral species and also other components like lime or dolomite. Further work was done with respect to the clay color. While the color does not affect the strength development, the brownish or reddish tint is considered as an aesthetic nuisance. The application of alternative fuels is being discussed in an ongoing concept study.

2 Calcination Trials

For the various calcination trials, a muffle furnace (600…1600 °C) and the thyssenkrupp pilot flash calciner POLCAL2 (600…1200 °C) were employed [1].

For soak calcination with a muffle furnace, corundum crucibles were used for bulk calcination. For tests with short retention times less than 10 min, a high temperature steel crucible was used. For such trials, a thin bulk layer of less than 5 mm in height ensured sufficient heat transfer.

For flash calcination pilot trials, the POLCAL2 flash calciner including a four-stage preheater was used (see Fig. 1). With a nominal throughput of 50 kg per hour, also larger amounts of material for subsequent grinding and mortar tests were produced. The material for flash calcination trials is usually dried and desagglomerated to particle size less than 2 mm.

F. Martirena et al. (eds.), *Calcined Clays for Sustainable Concrete*, RILEM Bookseries 16,
https://doi.org/10.1007/978-94-024-1207-9_42

Fig. 1. Natural gas fired pilot flash calciner POLCAL2.

3 Main Findings

Mixed clays can be well activated by flash calcination. The main advantage of the short heat treatment is the prevention of high temperature phases like mullite, gehlenite and pyroxenes, even if the material is treated above the optimal calcination temperature.

However, carbonates like lime, dolomite and magnesia narrow the possible temperature window for calcination as their presence significantly lowers the sintering temperature and melting point of the material [2]. Despite a temperature gap between clay calcination and sintering temperature, uneven heat distribution can lead to difficulties in case of local overheating.

Screening tests with various clayey materials revealed that the reactions of Ca and Mg highly depend on the chemical and mineral composition of the material. Materials that initially contain phases like pyroxene tend to form increased amounts of such phases if sufficient calcium is available. Contrarily, clays which are free of feldspar show only little to no tendency in forming pyroxenes [3].

Free CaO and MgO in flash calcined clays show usually a very high reactivity. Consequently, there is little to no risk for CaO/MgO induced expansion reaction in the hardened concrete.

During flash calcination, the product color can be effectively controlled via the combustion atmosphere inside the calciner. Using a near-stoichiometric combustion process which is principally based on the so-called multi-stage combustion in cement production, the formation of colorful iron (III) oxide and iron hydroxide can be prevented.

The reaction of sulfur components during the thermal treatment depends on the sulfur species and CaO/MgO content. Sulphidic materials cause SO_2 emission unless sufficient CaO/MgO is available to form sulphates. SO_2 emission from common sulphidic materials like iron sulphide can be hardly absorbed as the expulsion already occurs at low temperatures. Consequently, secondary abatement measures might be necessary.

For clay calcination, common fuels like coal, pet-coke, and natural gas can be used, usually burnt in dedicated hot gas generators in order to avoid direct flame contact to the clay. Natural gas can also be burnt inside the calciner employing swirl burners with high recirculation rates and low flame temperatures. The use of alternative fuels is possible but requires suitable combustion equipment. Well proven grate combustors allow to burn coarse solid fuels under controlled atmospheric conditions [4].

4 Conclusion

Mixed clay systems with different clay minerals and increased concentration of alkali and earth alkali elements can be converted to pozzolanic materials. However, such mixed systems require a dedicated heating regime in order to achieve sufficient pozzolanic activity on one hand and prevent the formation of high temperature phases on the other hand.

Consequently, flash calcination is considered as the most suitable process compared to soak calcination for maximising the pozzolanic reactivity of calcined clay regardless of its mineral composition.

References

1. Truemer, A., Ludwig, H.-M., Rohloff, K.: Investigations into the application of calcined clays as composite material in cement. ZKG Int. **9**, 52–57 (2014)
2. Snellings, R., Cizer, Ö., Horckmans, L., Durdziński, P., Dierckx, P., Nielsen, P., Van Balen, K., Vandewalle, L.: Properties and pozzolanic reactivity of flash calcined dredging sediments. Appl. Clay Sci. **129**, 35–39 (2016)
3. Huang, T.: Bewertung von tonhaltingen Sekundärrohstoffen zur Puzzolanherstellung (DE), seminar paper, Institute of Non-Metallic Materials, Clausthal University of Technology (2017)
4. Yousif, A.: Generating a sub-Stoichiometric Producer Gas for an Entrained Flow Calciner based on Solid Fuels, unpublished master thesis, Offenburg University (2016)

Thermal Transformation of Illitic-Chlorite Clay and Its Pozzolanic Activity

Roxana Lemma(✉), Cristina C. Castellano, Viviana L. Bonavetti, Monica A. Trezza, Viviana F. Rahhal, and Edgardo F. Irassar

Facultad de Ingeniería, CIFICEN (UNCPBA-CICPBA-CONICET), B7400JWI Olavarría, Argentina

Abstract. Illite-chlorite clay from quarry located at Buenos Aires Province (Argentine) was characterized by XRD, FTIR and TG-DTA. Mineralogical transformations during clay firing under oxidizing conditions were studied from 100 to 1100 °C by XRD and FTIR. From select temperatures, calcined clay was ground (85% passing to 45 μm sieve) and the pozzolanic activity of blended cements (25% w/w) was evaluated by the Frattini test and the strength activity index (SAI). Finally, the hydration phase assemblage of blended cements was studied using XRD analysis.

The solid-state phase transformations of clay during thermal treatment involves: water loss at low temperature; the partial dehydroxylation of chlorite resulting in the "modified chlorite structure" between 500 and 600 °C and its collapse at 800 °C; the dehydroxilation of illite is completed and its structure collapsed up to at 900 °C. Up to 1000 °C, the clay minerals are collapsed thoroughly, with formation of amorphous compounds.

Results of Frattini tests indicate that materials present pozzolanic activity after 7 days when they are fired up to 900 °C, however the best SAI ($\sim$1.00) at 28 days occurs for clay calcined at 1100 °C. For blended cements, the hydration products assemblage is similar to plain Portland cement used at all ages.

1 Introduction

For the reduction in global CO_2 emissions related to cement and concrete with minimum investment over the next 20–30 years, two keys areas are identified by UNEP report [1]:

1. Increased use of low-CO_2 supplementary cementitious materials (SCMs) as partial replacements for Portland cement clinker.
2. More efficient use of Portland cement clinker in mortars and concretes

The construction building materials are classified as non-renewable industry and the large volume of materials used requires determining the best option for the use the local available resources to select the SCM. Among the SCMs, calcined clays appear as a good opportunity to produce reactive aluminosilicate that contribute to the first key objective of report. Thermal activation of natural clays is an old process [2] that is well established for main clayed minerals such as kaolinite; illite, or montmorillionite [3, 4]. However, shales or common clays are a complex mixture of clayed minerals and

F. Martirena et al. (eds.), *Calcined Clays for Sustainable Concrete*, RILEM Bookseries 16,
https://doi.org/10.1007/978-94-024-1207-9_43

associated mineral (quartz, feldspars, anatase, etc.) that determine the possible thermal treatment. In recent years to produce SCMs, the focus is on low cost and abundant raw materials, such as common clays, with an appropriated thermal treatment. The thermal treatment of common clays generally is described by three stages: the dehydration, the clay dehydroxylation transforming the crystalline phases into metastable or amorphous phases and the neoformation of new crystalline phases or the glassy phase.

This study deals with the potential of illite-chlorite clays to produce SCM after appropriated thermal treatment to be used in blended cements. The aim of this study is to understand the thermal transformation of illlite-chlorite natural clay, determined the firing window, the grinding behavior and the potential pozzolanic activity for selected firing temperatures.

2 Experimental Procedure

A clay stone obtained from quarry La Cabañita, San Jacinto, near to Olavarría city, Buenos Aires Province, Argentine was studied. It corresponds to *Cerro Negro Fm* characterized by the presence of marls in its basal section and a continue set of rocks with reddish brown with greenish tones and frequent lamination of fine sandstone and illitic-chlorite argillite [5]. The chemical composition obtained by XRF of whole rock shows that SiO_2 (61.07%) and Al_2O_3 (17.80%) were the main elements. It has high content of Fe_2O_3 (7.50%), K_2O (4.05%), MgO (2.61%) and a medium Na_2O (1.45%). Minor elements are Ti, Ca, Mn, Zn, Ba, Cr, S, P and Sr. LOI at 950 °C was 3.68%

The mineralogical composition of natural and calcined (from 100 to 1100 °C) clays was determined by XRD (Philips X'Pert PW 3710) using CuKα radiation and Rietveld analysis (PANalytical HighScore Plus®). TG-DTA (Netzsch STA 409) was run from 20 to 1200 °C (10 °C/min) in dynamic air atmosphere. The FTIR (Nicolet Magna 500) was obtained in the range of 250–4000 cm^{-1}. Sample were calcined at different temperatures at the same heating and cooling rate, and ground in the laboratory ball mill until the retained on 45 μm sieve was lower than 15% (ASTM C 430). The PSD using laser granulometer (Mastersize 2000, Malvern), the relative density (ASTM C 188) and Blaine specific surface (ASTM C 204) were determined.

A Portland cement (PC) with 5.0% of limestone as minor constituent was used. The mineralogical composition of clinker was 63.8% C_3S, 13.7% C_2S, 3.0% C_3A and 12.8% C_4AF and its specific surface Blaine was 337 m^2/kg. The pozzolanic activity of calcined clays (25% w/w) was evaluated using the Frattini test (EN 196-5) and the strength activity index (SAI). Standard mortar (sand:cm = 3:1; w/cm = 0.50) was used to study the flow (ASTM C 230) and the compressive strength (CS) at 2, 7, 28 and 90 days curing in water at 20 °C. SAI was calculated as the ratio of CS of blended cement and plain PC at the same test age. The hydration products in cement pastes (w/cm = 0.5) cured in plastic bags at 20 °C during 2, 7, 28 and 90 days were identified by XRD.

3 Results and Discussion

3.1 Characterization of Clay Mineral

The XRD patterns for clay stone is shown in the bottom of Fig. 1. The main clay mineral is illite (I) accompanied by chlorite (Ch). The associated minerals are quartz (Q) and plagioclase in the albite (A) form ($NaAlSi_3O_8$). The mineralogical composition obtained by Rietveld method is approximately 45% of illite, 14% chlorite, 30% quartz, 10% feldspars and very low calcite content (<0.6%). In FTIR spectra (Fig. 2), illite is identified by the bands at 3624 and 3423 cm^{-1} assigned to the stretching and absorption vibration in the O–H bonds, respectively. The Si-O stretching at 1017 cm^{-1} and the doublet at 825–750 cm^{-1} are assigned to the vibration of Si-O-Al bonds [6]. Finally, the band at 467 cm^{-1} is characteristic of Si–O–Si bending vibrations in quartz accompanied with the stretching bands of Si-O located at 693, doublet at 780-800, 1090 and 1170 cm^{-1} [7].

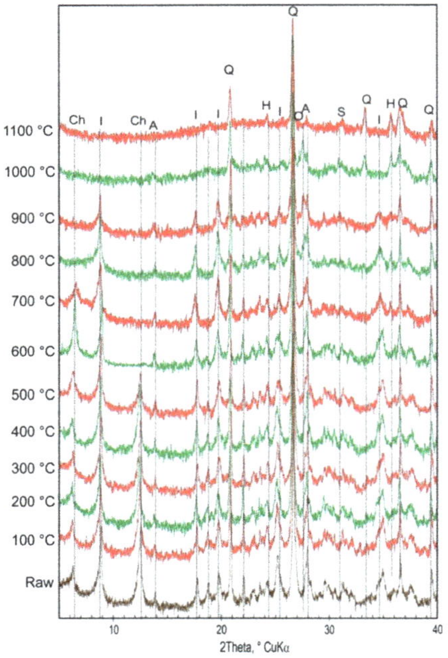

Fig. 1. XRD of clay treated at different temperatures

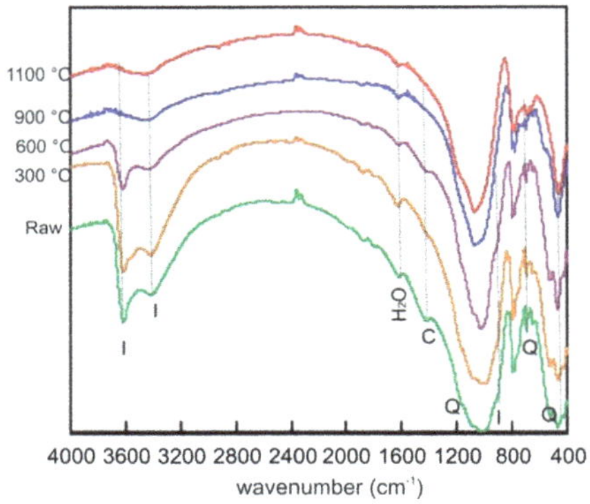

Fig. 2. FTIR of clay treated at different temperatures

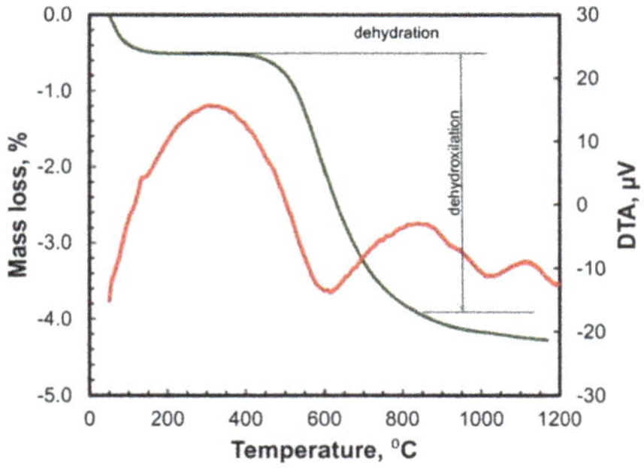

Fig. 3. TG-DTA of clay

3.2 Thermal Transformation of Clays

TG-DTA analysis is presented in Fig. 3. The mass loss between 30 and 200 °C is attributed to the evaporation of absorbed water and interlayer water called as clay dehydration. The mass loss between 200-400 °C was practically null. The greatest mass loss occurs between 400 and 800 °C with DTA endo-peak at ~600 °C that is attributed to the dehydroxylation of illite and chlorite. The mass loss between 350 and 1000 °C was 3.70%. From 900 to 1200 °C, the thermal transformation occurs without significant mass loss. Dissociation of carbonates was around 800 °C. Finally, the mass loss up to 1000 °C was 3.75%.

XRD patterns from 100 to 1100 °C are also shown in Fig. 1 and selected FTIR spectra are shown in Fig. 2. Between 500 and 600 °C, the peak at 12.4° 2θ corresponding to the (002) plane of the chlorite disappears and the peak at 6.4° 2θ of the (001) plane grows indicating the shift in the structure of chlorite [8]. Finally, the (001) plane peak is wide at 700 °C and all chlorite peaks disappear at 800 °C indicating that structure collapsed. Up to 700 °C, the spacing and intensity of illite peaks have little changes. At 800–900 °C, the intensity of peaks at 8.8° and 17.6° 2θ decreases while the intensity of peak at 19.7° 2θ increases. Finally, all peaks collapsed when the calcination temperature was higher than 900 °C. FTIR spectra show the gradual decrease of the intensity of O–H stretching vibration band assigned to octahedral hydroxyl groups illite (3630 and 3469 cm^{-1}) at 300 and 600 °C, and they disappear at 900 °C confirming the complete dehydroxylation before the collapse of structure [9]. The shoulder at 934 cm^{-1} assigned to the stretching deformation of O–H also disappears. The Si–O stretching band at 1027 cm^{-1} remains without change up to 900 °C and thereafter it decreases and shift to high wavenumber indicating that the frame structure remains [9]. At high temperature, the band at 463 cm^{-1}, characteristic of Si–O–Si bending is shaped indicating the increase of amorphous aluminum silicate phases. The transformation temperatures of clay minerals are similar to previously reported by Fernández et al. [3]. The XRD and FTIR analyses confirm that mass loss from 400 to 800 °C is attributed to illite dehydroxylation, but XRD patterns indicate that the d-spacing and intensity of its peaks remains until the collapsed up to 900 °C.

Regarding to neo-formations, hematite appears at 900 °C as the transformation of the structural iron release from decomposed chlorite and illite in oxidant atmosphere causing the reddish color. The formation of hercynite ($Fe^{2+}Al_2O_4$), also called iron spinel, appears up to 800 °C and the intensity of their peaks (30.8 and 35.7° 2θ) is more significant at 1000 and 1100 °C. Up to 700 °C, XRD reveals the transformation of albite causing that the intensity of peak at 27.9° 2θ decreases and the new peak at 27.6° 2θ attributed to oligoclase increases. Finally, all feldspars disappear at 1100 °C causing a glassy material that increases the amorphous halo between 18 and 32°. Quartz is the crystalline phase detected at this temperature [10].

3.3 Grinding

Figure 4 shows the **PSD curves** for calcined clays at 900, 1000 and 1100 °C. At 900 °C, the PSD curve has a trimodal distribution with particle size at: 100–350 μm attributed to aggregation and agglomeration; 20–30 μm attributable to quartz grains and 4–5 μm

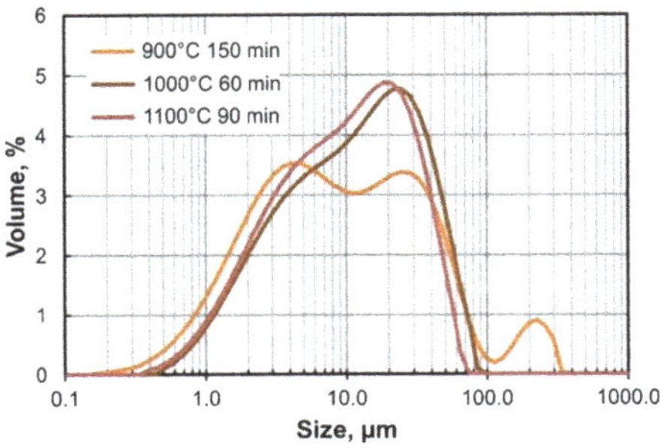

Fig. 4. PSD of calcined clays

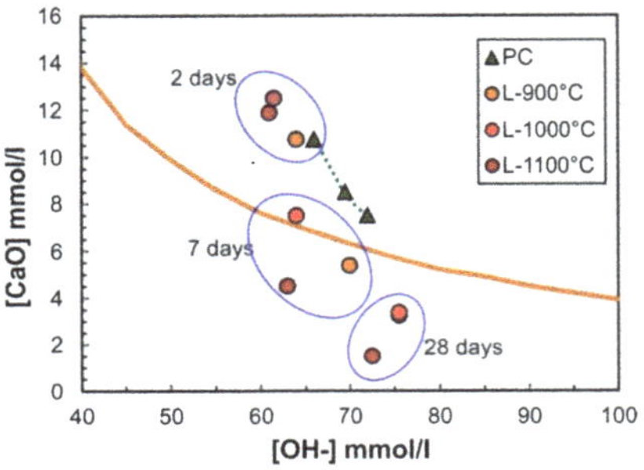

Fig. 5. Frattini test of calcined clays

attributable to clay fraction. For calcined clays at high temperature, the PSD curves show a pseudo unimodal distribution with a maximum at 20–30 μm and a shoulder at 6–8 μm. This change is attributable to the glass formation that produces more fragile and friable particles. The specific surface of calcined clay at 900 °C (678 m^2/kg) is larger than the corresponding to calcined clays at higher temperatures (525 and 529 m^2/kg) instead of the similar particle size distribution. It could be attributed to the partial restructuration of clayed minerals causing porous particles.

3.4 Pozzolanic Performance

The results of **Frattini test** (Fig. 5) show that all samples do not present pozzolanic activity at 2 days. At 7 days, calcined clays are on or below the solubility curve with great [CaO] reduction while [OH^-] remains approximately constant. At 28 days, the [CaO] vs. [OH^-] points are below the isothermal curve indicating the pozzolanic activity for calcined clays at all firing temperature.

The **mortar flow** made with blended cements at constant w/cm was between 124 and 141%. It was slightly lower than that to PC mortar (142%) indicating no significant water demand of calcined clays. Less extended flow (124%) was for calcined clays at 900 °C that has large specific surface (678 m^2/kg).

Compressive strength of blended cements is shown in Fig. 6. At 2 and 7 days, CS of calcined clay at different temperatures (18.9 and 27.1 MPa) has lower value than that to PC mortar (27.3 and 37.5 MPa). At 28 days, the CS was comparable for all blended cements (39.7 to 42.4 MPa), but it is lower than that of PC mortar (47.2 MPa), with exception of calcined clays at 1100 °C (44.7 MPa) with smaller particle sizes. Finally, all blended cements reach to comparable CS (52.0 to 57.4 MPa) with the PC mortar (55.1 MPa). Regarding the SAI (Fig. 6), it increases with the test age: 0.69 to 0.81 at 2 days up to values from 0.95 to 1.00 at 90 days. At 28 days, the SAI is greater than 0.85 for all calcination temperatures.

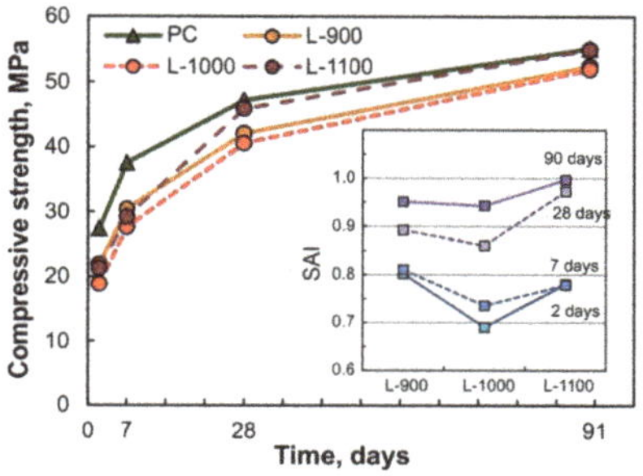

Fig. 6. Compressive strength and SAI

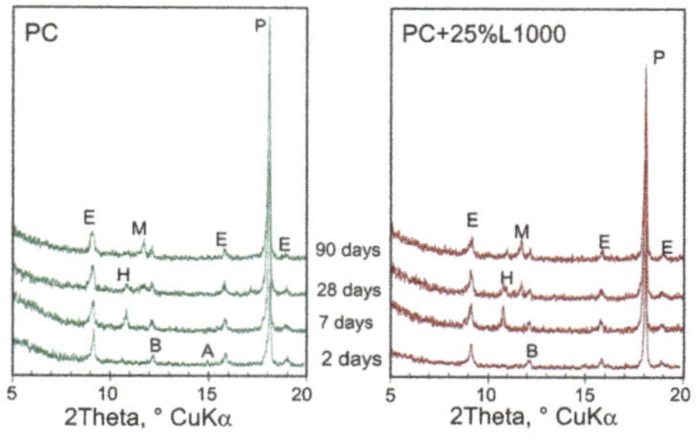

Fig. 7. XRD of hydrated pastes at different ages

3.5 XRD Analyses on Hydrated Paste

Figure 7 illustrates the XRD patterns for PC and blended cements with 25% by weight of calcined clays at 1000 °C. At 2 days, the PC paste shows the presence of ettringite (E), calcium hydroxide (P) and the peaks corresponding to un-hydrated cement phases (B = C_4AF and A = C_3S). For blended cement, the crystalline hydrated products detected are E and P as occurred for PC. At 7 days, the E and P are accompanied by the formation of AFm phase assigned as hemicarboaluminate (H) in PC. For blended cement, H is also revealed by the XRD analysis. At 28 days, the intensity of P peak increase for PC and decreases for blended cements. Monocarboaluminate (M) formation is insipient and it coexists with the presence of H. Finally, the M is a well-defined hydrated product at 90 days. The pozzolanic reaction of calcined clay is limited and the P was clearly detected in blended cement at 90 days.

4 Conclusions

The chlorite-illite clay is appropriate shale to produce SCM and the firing window is from 900 to 1100 °C. The solid-state phase transformations of clay during thermal treatment involves: water loss at low temperature; the partial dehydroxylation of chlorite resulting in the "modified chlorite structure" between 500 and 600 °C and its collapse at 800 °C when dehydroxilation of illite is completed and its structure collapsed up to 900 °C. At 1000 °C, the clay minerals are collapsed thoroughly, with formation of amorphous compounds and the formation of hematite and hercynite. Also, the feldspars collapsed and form a glass. The grindabillity increases at samples fired at high temperature. Results of Frattini tests indicates that materials present pozzolanic activity at 7 days when they are fired up to 900 °C, however the best SAI at 28 days occurs for clay calcined at 1000 °C. For blended cements, the hydration products assemblage is similar to the plain Portland cements used at all ages.

References

1. Scrivener, K.L., John, V.M., Gartner, E.M.: Eco-efficient cements: potential, economically viable solutions for a low-CO2, cement based materials industry. United Nations Environment Program Sustainable Building and Climate Initiative (UNEP-SBCI), 64 pp. (2016)
2. Mielenz, R.C., Witte, L.P., Glanz, O.J.: Effect of calcination on natural pozzolans. In: Symposium on Pozzolanic Materials in Mortars and Concrete, ASTM STP-99, pp. 43–92 (1950)
3. Fernandez, R., Martirena, F., Scrivener, K.L.: The origin of the pozzolanic activity of calcined clay minerals: a comparison between kaolinite, illite and montmorillonite. Cem. Concr. Res. **41**, 113–122 (2011). doi:10.1016/j.cemconres.2010.09.013
4. He, C., Osbaeck, B., Makovicky, E.: Pozzolanic reactions of six principal clay minerals: activation, reactivity assessments and technological. Cem. Concr. Res. **25**(8), 1691–1702 (1995). doi:10.1016/0008-8846(95)00165-4
5. Arrouy, M.J., Poiré, D.G., Gómez Peral, L.E., Canalicchio, J.M.: Sedimentología y estratigrafía del grupo La Providencia (NOM. NOV.): cubierta superior neoproterozoica, sistema de Tandilia, Argentina. Latin Am. J. Sedimentol. Basin Anal. **22**(2), 171–189 (2015)
6. Madejová, J., Komadel, P.: Baseline studies of the clay minerals society source clays: infrared methods. Clays Clay Miner. **49**, 410–432 (2001). doi:10.1346/CCMN.2001.0490508
7. Bensted, J., Varma, S.P.: Infrared and raman spectroscopy in cement chemistry-miscellaneous applications. World Cem. Technol. **8**, 18–20 (1977)
8. Zhan, W., Guggenheim, S.: The dehydroxylation of chlorite and the formation of topotactic product phases. Clays Clay Miner. **43**, 622–629 (1995). doi:10.1346/CCMN.1995.0430512
9. Jiang, T., Li, G., Qiu, G., Fan, X., Huang, Z.: Thermal activation and alkali dissolution of silicon from illite. Appl. Clay Sci. **40**, 81–89 (2008). doi:10.1016/j.clay.2007.08.002
10. Khalfaoui, A., Hajjaji, M.: A chloritic-illitic clay from Morocco: temperature–time–transformation and neoformation. Appl. Clay Sci. **45**, 83–89 (2009). doi:10.1016/j.clay.2009.03.006

In-situ Observation of Dissolution Behavior of Carbonatite in Water Glass Solution

Jing Li, Jianqin Lin, Qijun Yu, Jie Hu, and Suhong Yin(✉)

School of Materials Science and Engineering, South China University of Technology, Guangzhou 510640, People's Republic of China
imshyin@scut.edu.cn

Abstract. To investigate the reaction process of alkali-activated carbonatite, the dissolution behaviour of carbonatite in water glass solution was monitored by in-situ Polarizing Microscope (PM) and Digital Holographic Microscopy (DHM). Six types of carbonatites were mixed with water glass solution (M = 1.6), then observed constantly by PM and DHM. The polarizing microscope results showed that the size of all types of carbonate minerals was gradually reduced and the particles disappeared as the time prolonged, indicating that carbonatites reacted with water glass. The reaction rate of carbonatites with water glass was qualitatively characterized via the comparison of the dissolution time of carbonatite particles with the same size. The reaction rates between carbonatites and water glass solution followed the following order: calcium carbonate (chemical reagent) > carbonatite with 1.35% MgO > calcite > carbonatite with 11.37% MgO > carbonatite with 14.39% MgO > dolomite. DHM provides the quantitative results about the dissolution rate of carbonatite. However, the algorithm of three-dimensional reconstruction is still being improved, and the final results will be included in the full paper.

1 Introduction

The production of ordinary Portland cement(OPC) accounts for 5–8% of global CO_2 emission and 12–15% of energy cost in the world, which hinders the implementation of the sustainable development strategy [1, 2]. From the perspective of environmental protection, a new type of cementitious material should be devoloped as sustainable alternative material, while alkali-activated materials (AAM) is considered to be a promising one [3]. Alkali-activated carbonatite is one type of AAM and was proposed by Wen since 1990s [4]. Carbonatite occupies almost 20% of the total amount of sedimentary rocks on average, and mainly exists in forms of limestone and dolomite. In China, this value is increased to 55% [5]. Carbonatite with high MgO content (low graded limestone or dolomite) was not able to be applied as calcareous materials for cement production. The utilization of such high abundance of resource as alkali-activated materials was widely investigated in various aspects due to its low CO_2 emission and energy cost [4, 6–9]. Yin integrated thermo-dynamics with experimental investigations to study the reaction products and mechanism of alkali-activated carbonatite [6, 9]. However, the poor mechanical performance (with a compressive strength of 4 MPa at 28 days)

F. Martirena et al. (eds.), *Calcined Clays for Sustainable Concrete*, RILEM Bookseries 16,
https://doi.org/10.1007/978-94-024-1207-9_44

constrained its application as cementitious-grouting materials only. Later on, multiple attempts were made to improve its mechanical property: the compressive strength (28d) could be improved from 3.7 MPa to about 40 MPa with the addition of up to 30wt% of other admixtures such as slag, fly ash and metakaolin, concentration of water-glass of 41.5%, M = 1.8, and liquid to solid ratio of 0.8 by mass [8]. Other authors also summarized similar conclusions when they studied the effect of the slag content on the compressive strength of alkali-activated carbonatite material [10]. The huge gap in mechanical performance between alkali-activated carbonatite and other alkali-activated admixtures (slag, fly ash or metakaolin) systems and the difficulties in characterizing the reaction products due to their amorphous nature result in a raising doubt on whether the carbonatites really react with water glass solution or the whole reaction is a dedolomitization process of carbonatites [10–12]. To solve this problem, in this study the dissolution behavior of carbonatite in water glass solution was investigated by in-situ observation method to determine the character of the reaction. Polarizing microscope (PM) and digital holographic microscopy (DHM) were applied as in-situ observation methods.

2 Experiment

2.1 Raw Materials

Six types of carbonatites with different content of MgO (from 0 wt% to 21.6 wt%) were used as raw materials. Carbonatites were first ground into powders and sieved with 200 meshes sieve before use. Water glass with a modulus of 1.6 (two concentrations: 15% and 35%) was used. The detailed compositions and mixture design were shown in Table 1.

Table 1. Compositions and mix design for the experiments

No.	Carbonatites	CaO/wt%	MgO/wt%	Loss/wt%	c(water glass)/%	Method	Observing Times/d
S1	Calcium carbonate	60	/	40	15	PM	9
S2	Calcite	60.05	/	43.93	15	PM	12
S3	Dolomite Limestone*	53.45	1.35	43.43	15	PM	14
S4	Calcite Dolomite-1*	41.14	11.37	43.71	15	PM	22
S5	Calcite Dolomite-2*	37.15	14.39	43.45	15	PM	35
S6	Dolomite	30.5	21.6	47.9	15	PM	35
S7	Calcite Dolomite-1*	41.14	11.37	43.71	35	DHM	1

*Classified according to the "carbonate classification figure" [13].

2.2 Samples Preparation and In-situ Tests

Carbonatites and water glass were weighed and mixed in a tube with a solid/liquid ratio of 1/1500 by mass, and then shaken evenly by lab dancer MS-3 for 2 min. After that, few drops of mixed liquid (approximately 0.5 ml) were dripped onto the glass slide and sealed with coverslip. The wave plate was placed on the objective table and the change of carbonatite particles was recorded by taking pictures continuously. The whole device and sample were kept motionless throughout the whole observation process. The images of PM and DHM were recorded on CMM-55E PM and sCMOS camera (Zyla-5.5-CL3, Andor Tech), respectively. The details of experiments were listed in Table 1.

3 Results and Discussion

Due to the high similarity in dissolution behaviors of Specimen S1 to S6, only the image process of specimen S1 is shown in Fig. 1 as example. The polarizing microscope images showed that the size of all types of carbonate minerals was gradually reduced and the particles disappeared as the time prolonged, indicating that carbonatites reacted with water glass not just a dedolomitization process.

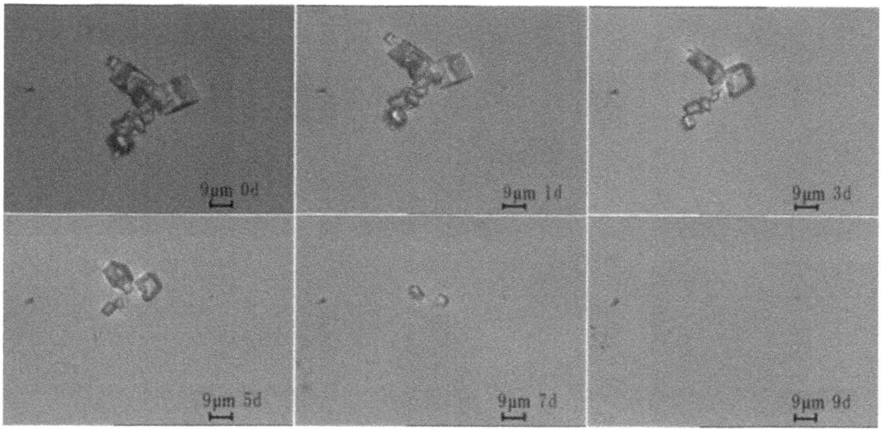

Fig. 1. Dissolution behavior of sample S1 by PM (MgO = 0%, c = 15%)

For specimen S1, carbonate particles dissolved in 9 days, presenting the fastest reaction rate among all specimens. Specimen S6 presented the slowest reaction rate, in which carbonate particles didn't completely dissolved after 35 days. The reaction rates between carbonatites and water glass solution followed the following order: calcium carbonate chemical reagent > carbonatite with 1.35% MgO > calcite > carbonatite with 11.37% MgO > carbonatite with 14.39% MgO > dolomite.

Polarizing microscope (PM) observation is a useful method to qualitatively characterize the reaction rate by comparing the dissolution time of carbonatite particles with the same size. However, polarizing microscope can't provide the information about the specific reaction rate and the detail morphology alterations of reaction products. Therefore, digital holographic microscopy (DHM) was also applied for quantitative analysis because it can record three-dimensional image and reconstruct the subject. DHM observation was only kept for 12 h (the used instrument was unable to be used continuously for days). In order to accelerate the reaction rate, water glass with a concentration of 35% was used and the dissolution process and reconstruct results are shown in Fig. 2. The thickness values of carbonatite particles were obtained by reconstructed images. According to the data determined by DHM in Table 2, the average dissolution rate of the tested particles is 65.33 $\mu m^3 \cdot h^{-1}$ within the

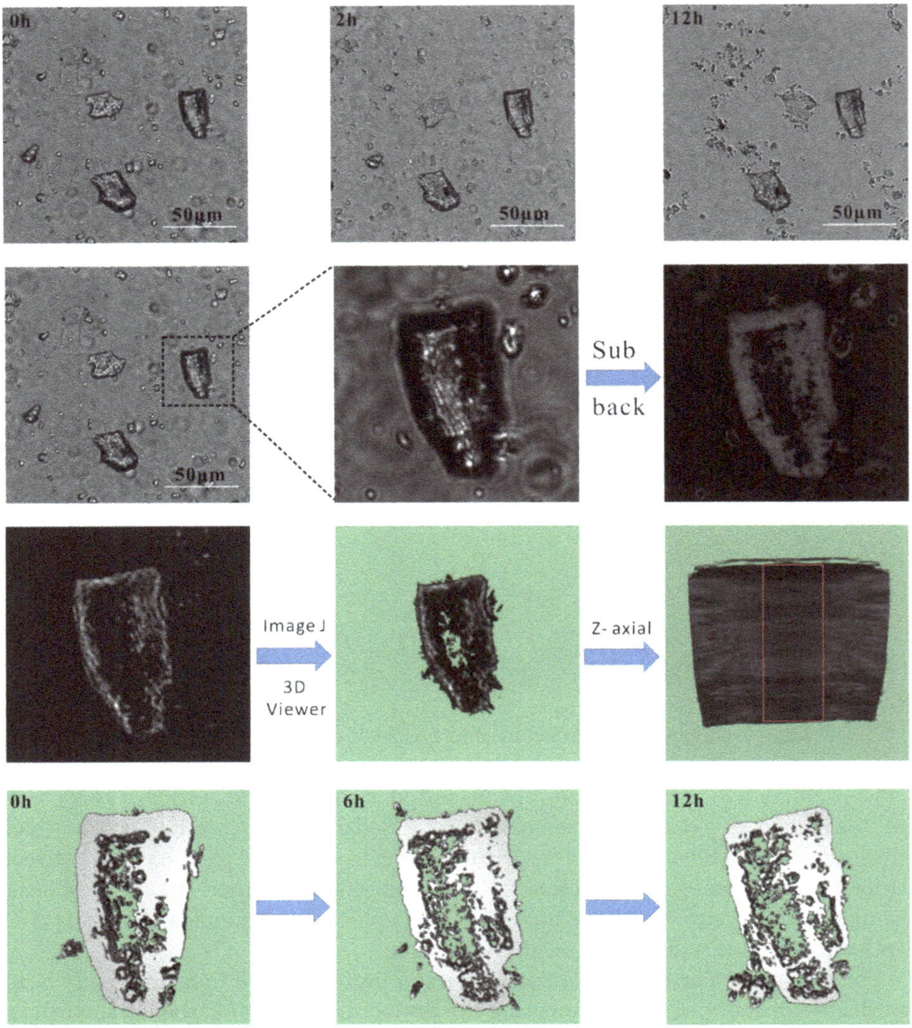

Fig. 2. 3D-reconstructed graphs of S7

initial 6 h, and then raised to 99.83 $\mu m^3 \cdot h^{-1}$ in the later 6 h, indicating that the dissolution of carbonatites in water glass is an accelerative process.

Table 2. Quantitative analysis results of sample S7

Sampling time/h	Thickness/μm	Volume /μm^3	Dissolution rate/$\mu m^3.h^{-1}$
0	6.0	2637	/
6	5.2	2245	65.33
12	4.3	1646	99.83

4 Conclusion

Base on the results and discussion above, a few conclusions can be achieved. The carbonatites do react with water glass. The reaction rates between carbonatites and water glass solution followed the following order: calcium carbonate (chemical reagent) > carbonatite with 1.35% MgO > calcite > carbonatite with 11.37% MgO > carbonatite with 14.39% MgO > dolomite. The preliminary result of DHM shows the dissolution rate of carbonatite in water glass solution is increased from 65.33 $\mu m^3/h$ to 99.83 $\mu m^3/h$ as sampling time prolonged from 6 h to 12 h. The algorithm of three-dimensional reconstruction is still being improved, and the final results will be included in the full paper.

Acknowledgements. The authors gratefully acknowledge the financial support of National Science Foundation of China 51561135012 and instrument support from X. Gong and G. Huang.

References

1. Olivier, J.G.J., Janssens-Maenhout, G., Peters, J.A.H.W.: Trends in Global CO2 Emissions, p. 40. The Hague, Netherlands, PBL Netherlands Environmental Assessment Agency (2012)
2. Unluer, C., Al-Tabbaa, A.: The role of brucite, ground granulated blastfurnace slag, and magnesium silicates in the carbonation and performance of MgO cements. Constr. Build. Mater. **94**, 629–643 (2015)
3. Schneider, M., Romer, M., Tschudin, M., Bolio, H.: Sustainable cement production- present and future. Cem. Concr. Res. **41**, 642–650 (2011)
4. Zhao, S.: Research on high performance alkali-activated carbonatite-slag composite cementitious materials of low environment load. South China University of Technology, Guangzhou (2005)
5. Wu, K., Zhang, X.: Building Materials. Tongji University Press, Shanghai (1996)
6. Yin, S., Wen, Z., Yu, Q., et al.: Investigation on the thermo-dynamics of alkali-activated carbonatite. J. Wuhan Univ. Technol. **30**(4), 774–780 (2015)
7. Yin, S., Wen, Z., Wang, Y., et al.: The study on alkali-activated-carbonatites cementitious-grouting materials. In: Proceedings of the 5th International Symposium on Cement and Concrete, pp. 917–924. Press of Tongji University, Shanghai (2002)
8. Zhao, S., Yu, Q., Qiao, F., et al.: Setting and strength characteristics of alkali-activated carbonatite cementitious materials with ground slag replacement. J. Wuhan Univ. Technol. **21**(1), 125–128 (2006)

9. Yin, S.: Study on alkali-activated-carbonatites cementitious-grouting material. South China University of Technology, Guangzhou (20040
10. Lin, J.: Basic research on structure formation mechanism and performance modification of alkali-activated carbonatite cementitious material. South China University of Technology, Guangzhou (2008)
11. Mingshu, T., Min, D.: Progress on the studies of alkali–carbonate reaction. In: Proceedings of the 12th International Conference on Alkali–Aggregate Reaction in Concrete, pp. 51–59 (2004)
12. Grattan-Bellew, P.E., Mitchell, L.D., Margeson, J., et al.: Is alkali–carbonate reaction just a variant of alkali–silica reaction ACR = ASR? Cem. Concr. Res. **40**(4), 556–562 (2010)
13. Liu, C., Wen, Z.: Dam Concrete Monograph (I): Alkali Aggregate Reaction in Concrete. South China University of Technology Press, Guangzhou (1995)

Analysis of the Mixing Performance Containing the LC3 as Agglomerant with Different Types of Calcined Clay

D. Lins(✉), J. Rêgo, and E. Silva

PECC (UnB), Universidade de Brasília, Brasília, Brazil

Abstract. Given the growth of the construction industry in recent years, Portland cement production has been increasing, but the cement industry is adversely affected by sustainability, accounting for about 5% of the world's CO_2 emissions. With the view to reduce these emissions, mitigating measures have been analyzed, and one of the most promising ones is the reduction of clinker/cement ratio, using as substitution mineral additions. This research aims to analyze the mechanical performance of mortars from the use of cement with addition of two types of metakaolin, one of these being more reactive, with the addition of limestone. From the characterization of these materials and the different substitution contents used from binary and ternary mixtures, it was verified that some of them have a good performance.

1 Introduction

The construction industry has been a growing and important department in the economy of many countries around the world, and with it, Portland cement, one of the most consumed products on the planet, has increasing amounts of production. Research shows that the construction industry produces about 3.6 billion tons of cement annually in the world, and predicts that this volume will increase to more than 5 billion tons by 2030 [1]. However, a cement industry is responsible for about 5% of the planet's CO_2 emissions, contributing significantly to greenhouse problems [2]. 60% of total emissions come from clinker formation processes, due to the decarbonation of calcium carbonate [3]. Given this large share of carbon emission contribution, a reduction in the ratio of clinker/cement becomes essential for the production process of its attenuated environmental impact.

In order to reduce the amount of clinker used in the manufacture of cement without losing the characteristics of strength and durability, it is necessary to use alternative materials, such as blast furnace slag and fly ash. However, these additions are not available in abundance, and will not be able to meet the demand for world cement production in the coming years. According to [3], the addition that simultaneously meets the required conditions of availability and performance is the calcined clay.

The calcined clay alone is already proven capable of replacing part of the clinker, however, recent researches such as [3–5] have shown that when limestone is added to the mixture, a synergistic reaction occurs between the mineral additions, providing an improvement in properties. This is due to a reaction of the calcium carbonate from the

F. Martirena et al. (eds.), *Calcined Clays for Sustainable Concrete*, RILEM Bookseries 16,
https://doi.org/10.1007/978-94-024-1207-9_45

limestone with the alumina present in the calcined clay, and the calcium hydroxide, forming a compound called hydrated carboaluminate, according to Eq. 1, a product capable of filling the pores and reduce existing gaps.

$$A + Cc + 3CH \rightarrow C_3A.Cc.H_{12} \tag{1}$$

This research has as general objective to evaluate the technical feasibility of LC^3 cement from the mineral additions available in Brazil. In addition, it will also analyze the influence of the type of calcined clay used as a function of its amorphousness and its alumina content, considering that this compound is one of the main ones for the synergic reaction that originates the hydrated carboaluminate. In order to verify the influence of the limestone on the performance of LC3, we chose to use a limestone that is poor in calcareous, in order to verify if this factor could result in the loss of efficiency of the new cement.

2 Experimental Procedure

For the manufacture of mortars with LC3, a common Portland cement (free of additions), two types of calcined clay, and one kind of limestone were used. In the case of calcined clay, calcined clay "a" (Ca) is commonly used in Brazil to produce Portland cements, and calcined clay "b" (Cb) is also marketed as a more amorphous mineral addition, in which a higher Al2O3 content was identified by the X-ray fluorescence assay. From this test, which can be analysed in Table 1, the limestone filer was also characterized, to prove its reduced limestone content, and the cement used in the standard mixture and in the replacement mixtures.

Table 1. XRF Composition of materials

Chemical composition (%)				
Oxides	Ca	Cb	Limestone	PC
SiO2	55,27	50,55	55,85	16,82
Al2O3	35,89	42,28	16,57	3,94
CaO	2,56	–	12,35	66,04
Fe2O3	2,08	3,24	7,27	3,74
TiO2	2,01	1,64	0,99	0,24
MgO	0,73	1,02	2,61	4,25
K2O	0,49	1,08	3,52	0,91
SO3	0,48	0,1	0,46	3,76
BaO	0,36	–	–	–
MnO	–	–	0,14	–
P2O5	–	–	0,11	–
SrO	–	–	–	0,21
Other oxides	0,14	0,1	0,12	0,1

In order to complement the characterization of the mineral additions used, the X-ray diffraction (XRD) test was also performed, as shown in Fig. 1 below.

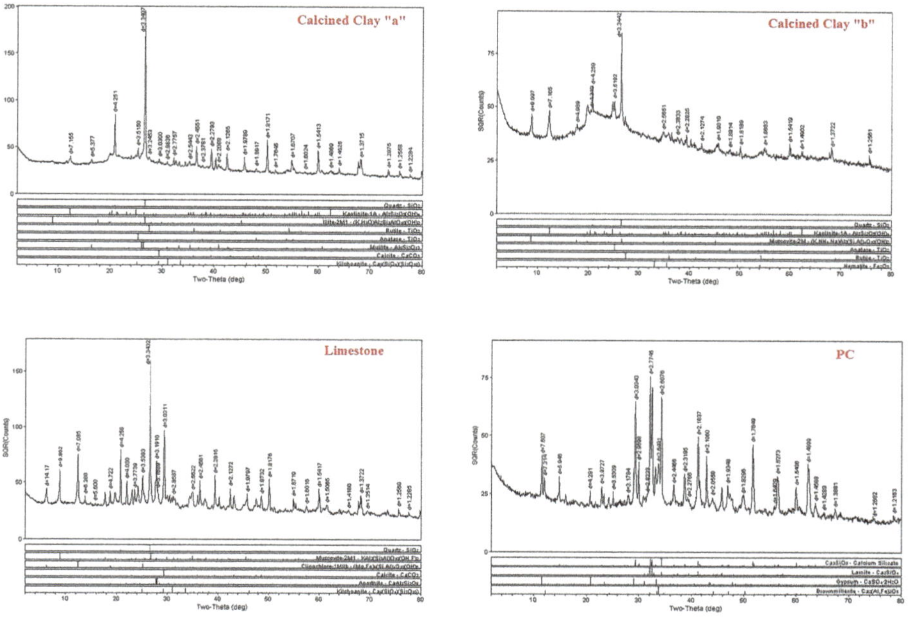

Fig. 1. X-ray diffraction analysis of materials

Twenty-one mortars of different mixtures were then produced, the first one as a reference, and the others obtained from binary and ternary mixtures, with total substitution levels of 15%, 30%, 45% and 60%. These substitution levels were determined in order to reproduce those used by [5], making it possible to compare the performance of LC3 cement in different regions of the planet. In the case of ternary mixtures, it was always chosen to use the ratio 2:1 (MK: LS), according to the optimum ratio indicated by [4]. In Table 2 it is possible to analyze the applied proportions.

The mortars were made with water/binder ratio of 0.48, and with the aid of superplasticizer additive, to maintain the same consistency and workability of all mixtures for the molding procedure, as established by [6]. The compressive strength of the mixtures was analyzed from the rupture of 16 cylindrical specimens measuring 5 × 10 mm, at the ages of 1, 3, 7 and 28 days, with 4 samples for each age.

Table 2. Composition of cements

Mixtures	PC	Total substitution (%)	Limestone	Calcined clay "a" (%)	Calcined clay "b" (%)
OPC	100	0	0	0	0
LS 1	85	15	15	0	0
LS 2	70	30	30	0	0
LS 3	55	45	45	0	0
LS 4	40	60	60	0	0
Ca 1	85	15	0	15	0
Ca 2	70	30	0	30	0
Ca 3	55	45	0	45	0
Ca 4	40	60	0	60	0
Cb 1	85	15	0	0	15
Cb 2	70	30	0	0	30
Cb 3	55	45	0	0	45
Cb 4	40	60	0	0	60
LC3a 1	85	15	5	10	0
LC3a 2	70	30	10	20	0
LC3a 3	55	45	15	30	0
LC3a 4	40	60	20	40	0
LC3b 1	85	15	5	0	10
LC3b 2	70	30	10	0	20
LC3b 3	55	45	15	0	30
LC3b 4	40	60	20	0	40

LS: PC + Limestone
Ca: PC + Ca
Cb: PC + Cb
LC3a: PC + Limestone + Ca
LC3b: PC + Limestone + Cb

3 Results

Depending on the fineness and the substitution content in each mixture, different amounts of superplasticizer additive were used. The amount of additive and the substitution content showed a directly proportional growth, as the quantity of the mineral additions increased, the dosage of superplasticizer was also increasing. When the filler was used in the substitutions there was not a high demand in the dosage of superplasticizer, however, when using the calcined clays, as a function of the fineness of these additions, the dosage of the polycarboxylate additive was necessary in greater quantities, so that the workability was maintained of mortars.

Regarding the mechanical performance, the averages obtained from the rupture of 4 cylindrical test pieces (5 × 10 mm) are presented in Fig. 2. Figure 3 shows the percentages of compressive strength at the ages of 1, 3, 7 and 28 days, when compared to the reference dosage.

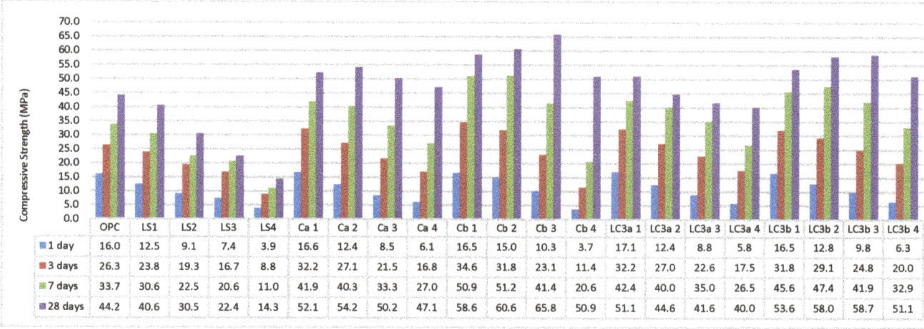

	OPC	LS1	LS2	LS3	LS4	Ca 1	Ca 2	Ca 3	Ca 4	Cb 1	Cb 2	Cb 3	Cb 4	LC3a 1	LC3a 2	LC3a 3	LC3a 4	LC3b 1	LC3b 2	LC3b 3	LC3b 4
1 day	16.0	12.5	9.1	7.4	3.9	16.6	12.4	8.5	6.1	16.5	15.0	10.3	3.7	17.1	12.4	8.8	5.8	16.5	12.8	9.8	6.3
3 days	26.3	23.8	19.3	16.7	8.8	32.2	27.1	21.5	16.8	34.6	31.8	23.1	11.4	32.2	27.0	22.6	17.5	31.8	29.1	24.8	20.0
7 days	33.7	30.6	22.5	20.6	11.0	41.9	40.3	33.3	27.0	50.9	51.2	41.4	20.6	42.4	40.0	35.0	26.5	45.6	47.4	41.9	32.9
28 days	44.2	40.6	30.5	22.4	14.3	52.1	54.2	50.2	47.1	58.6	60.6	65.8	50.9	51.1	44.6	41.6	40.0	53.6	58.0	58.7	51.1

Fig. 2. Compressive strength of blends mortars at 1, 3, 7 and 28 days.

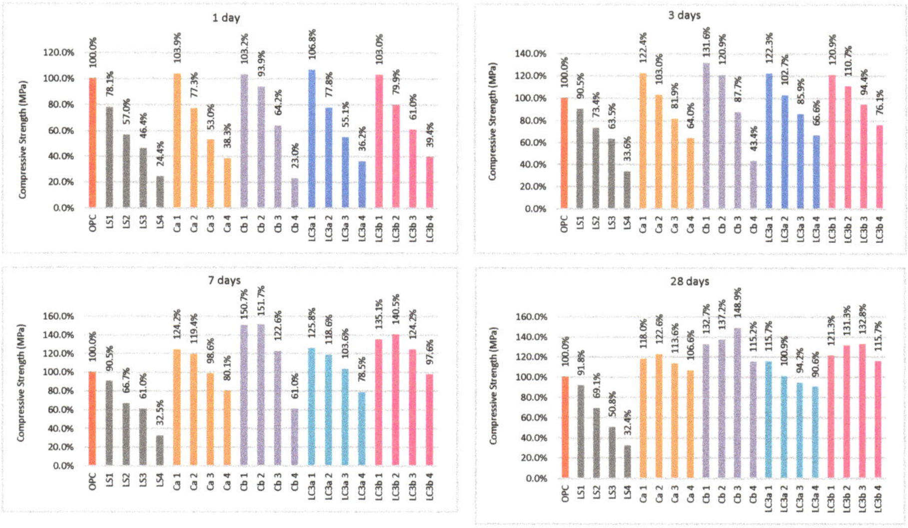

Fig. 3. Compressive strength of blends normalized to the strength of pure PC

Considering the results presented from the binary mixtures, it is possible to infer that when limestone is added to mixtures in different percentages, these presents lower results in the tests of compressive strength in all of the ages, when compared to the results of the reference cement.

Mortars that used cement with addition of both types of calcined clay at 45% and 60% substitution levels did not exceed the reference at the initial ages of 1 and 3 days, a result expected as a function of pozzolanic reactions occurring at a higher age or excessive use of superplasticizer in some of these mortars, a factor that can cause a delay in the initial resistance gain. However, the binary mixtures using these mineral additions presented higher results than the refining mixture at all levels at the age of 28 days.

With respect to the mixtures composed of the LC3 cement tested, it is possible to observe that at the age of 1 day, only the substitution of 15% exceeded the reference, however, at 3 days the mortars with 30% also exceeded the cement results common.

Even more positive results occurred at more advanced ages, in which mortars produced from LC3a, with substitution levels of 15% and 30%, exceeded the reference values, and when used 45% of mineral additions (30% calcined clay + 15% filer) and 60% (40% calcined clay + 20% filer) the mortars obtained 94.2% and 90.6% respectively in relation to the standard mix. In the case of mortars that were produced from LC3b, at 28 days all the results of compressive strength obtained were higher than the reference values.

As for the performance of LC3 using two different types of clay, it was verified that the mixtures made with calcined clay "b" obtained a more satisfactory performance, a result similar to that obtained in the study [7], and its premises are supported by researches as [3] which cite the importance of kaolinite content and amorphity for the best performance of the new cement.

4 Conclusions

Analyzing the results obtained, it can be concluded that the addition of calcined clay and limestone is perfectly feasible with respect to the compressive strength characteristic for percentages of addition of up to 30% in case of use of clay "a" (20% Calcined clay "a" + 10% filer, and up to 60% addition in the case of clay "b" (40% calcined clay "b" and 20% filer). Therefore, the importance cited by [3] regarding the quality of the calcined clay is remarkable, since from this it is possible to establish higher substitution levels, collaborating more efficiently in reducing the clinker/cement ratio, and consequently emissions. It was also found that the use of a low limestone filer did not impair the performance of the LC^3 mixtures.

In view of the research it is possible to say that the LC3 is a promising cement, that from the use of calcined clay, a material available in abundance, and limestone, also accessible in great quantity and that does not require calcination, becomes competitive technically, financially and environmentally, and should be studied in the coming years, and can replace in an equal way the conventional mineral additions currently.

References

1. Müller, N., Harnisch, J.: A blueprint for a climate friendly cement industry: how to turn around the trend of cement related emissions in the developing world. Gland, Switzerland: World Wide Fund for Nature (2008)
2. Gartner, E., Hirao, H.: A review of alternative approaches to the reduction of CO_2 emissions associated with the manufacture of the binder phase in concrete. Cem. Concr. Res. **78**, 126–142 (2015)
3. Scrivener, K.L.: Options for the future of cement. Indian Concr. J. **88**(7), 11–21 (2014)
4. Damidot, D., et al.: Thermodynamic sand cement science. Cem. Concr. Res. **41**(7), 679–695 (2011)
5. Antoni, M.: Investigation of cement substitution by blends of calcined clays and limestone. Tese de Doutorado. École Polytechnique Fédérale de Lausanne (2013)

6. Associação Brasileira de Normas Técnicas: ABNT NBR 5752: Materiais pozolânicos — Determinação do índice de desempenho com cimento Portland aos 28 dias, Rio de Janeiro (2014)
7. Bishnoi, S., et al.: Pilot scale manufacture of limestone calcined clay cement the Indian experience. Indian Concr. J. **88**(6), 22–28 (2014)

Evaluation of Calcined Clays from Boyaca-Colombia Containing Alunite as Supplementary Cementitious Materials

Ariam Lozano Perez(✉) and Mathieu Antoni

Holcim Colombia, Corporative Cement Industrial Performance,
Lafarge Holcim, Jona, Switzerland

Abstract. The goal of this study was to understand the behavior of clays from Iza, Boyaca, Colombia and their potential to be used in blended cement after calcination as pozzolan. The materials were characterized before and after thermal treatment by XRF, XRD, TGA and SEM. The performance in mortar was evaluated by compressive strength of mortar.

The XRD analysis showed that the main minerals present in the raw materials were Kaolinite, Alunite, Quartz, Opal and Feldspars, each one with different behavior after activation. The performance of mortar was investigated by measuring the compressive strength according to ASTM C109. The produced blended cements were done with 30% of SCM and 70% of Ordinary Portland Cement (OPC).

The optimization of the calcination temperature has been carried out using Thermal Gravimetric Analysis (TGA) as well as with X Ray Diffraction (XRD). After the calcination and depending on the temperature, both the Kaolinite as well as Alunite show structural changes with different effects on the performance. While kaolinite dehydroxylation significantly improves the performance, alunite decomposition into the new reactive phases $KAl(SO_4)_2$ and amorphous Al_2O_3 have strong effect on the early age of hydration, decreasing the workability, increasing the initial temperature of the mortar. This can nevertheless be controlled by optimizing the gypsum content.

1 Introduction

Cement producer are facing a challenge: meet the growing cement demand without increasing their environmental footprint. Production of one ton of clinker as main components of cement produced between 0.5–0.8 tons of Carbon Dioxide (CO_2) [1]. LafargeHolcim has set ambitious CO_2 reduction target in its "2030 Plan for Sustainability" aiming at reducing in 2030 by 40% the net CO_2 emissions compared to 1990 situation and further reducing clinker factor below 65. In order to achieve these objectives and due to the abundance in the world, calcined clays are promising options as supplementary cementitious materials(SCMs). Several results have showed that calcined clay can be used as substitute of clinker and that combined addition of limestone can allow minimizing even more the clinker factor.

F. Martirena et al. (eds.), *Calcined Clays for Sustainable Concrete*, RILEM Bookseries 16,
https://doi.org/10.1007/978-94-024-1207-9_46

As a result of thermal treatment of clays in a temperature range in which the layer structure of clay changes due to dehydroxylation yields amorphous alumina-silicate [2], phase with good pozzolanic properties. Improving the negative impact of clays, decreasing water demand and acting as high reactive pozzolan materials. However, as any natural material, clay is subject to local variation originating from its geological formation process and can vary greatly in term of type of clays, type and amount of others minerals, etc. that can affect calcined clays reactivity. This article intends to investigate a particular case of kaolinitic clays from a hydrothermal deposit in Colombia.

2 Materials and Methods

The materials under study were taken from a quarry of Pozzolan located in the city of Iza, State of Boyaca, Colombia. Currently the quarry is under exploitation and has a sub-product called Pozzolan type "clays" considered as a waste. An estimation of amount of this material is around 4 Million of Tons. The chemical composition and mineralogical investigation was performed with a X-ray Fluorescence(XRF) and X-Ray Diffraction spectrometer ARL TM 9900 X-Ray WorkStation. The morphology-microstructure was estimate by SEM [3], temperature of activation and thermal behaviour by Thermogravimetric Analysis(TGA-DTG). Sulphur content was measured by LECO equipment. The performance of mortar was investigated by measuring the compressive strength according to ASTM C109 at 1, 3, 7, 28 days.

2.1 Raw Materials

Four samples were investigated; the selection of the samples was according to the chemical composition and representability of materials in the field (Table 1).

Table 1. Chemical composition by XRF of samples

	Chemical composition %											
Sample	SiO_2	Al_2O_3	Fe_2O_3	CaO	MgO	Na_2O	K_2O	TiO_2	P_2O_5	Mn_2O_3	LOI	SO_3
A	61.50	20.64	2.11	0.00	0.35	0.03	1.42	0.26	0.07	0.02	12.02	2.60
B	54.22	14.78	5.30	2.59	0.12	0.05	3.39	0.32	0.13	0.08	16.34	11.31
C	66.22	17.53	2.75	0.21	0.43	0.22	2.70	0.27	0.05	0.04	8.35	0.58
D	61.78	19.60	2.66	0.00	0.35	0.04	2.26	0.23	0.16	0.03	10.59	2.24

In Fig. 1, the SEM image shows the characteristic layered structure of clays platelets stacked together. This structure is responsible to increase the water retention due to the hydroxyl group in the surface of the particles, which have electrostatic attraction with water molecules forming a hydrogen bond [4].

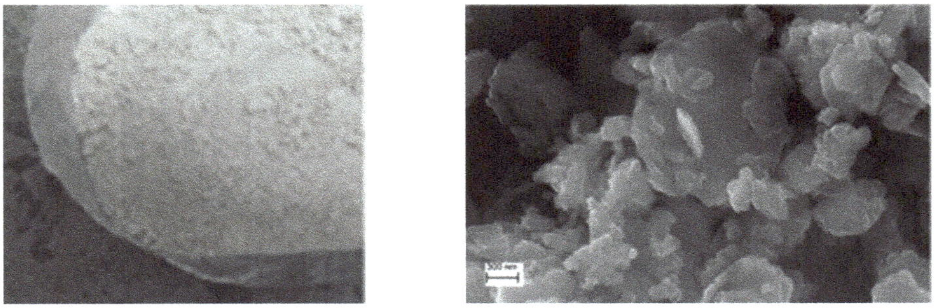

Fig. 1. Appearance and SEM image of sample A (Back-Scattered Electrons)

3 Results and Discussion

The chemical composition of the materials was not uniform, with the sulphur content expressed as SO_3 had values from 0.58% in sample C to 11.31% in sample B as well as variations for SiO_2 between 54% (sample B) and 66% for sample C. It demonstrates the broad variety of compositions present in the quarry.

3.1 Mineralogy

The main minerals phases identified were Quartz 2ϕ:31 and Opal (Amorphous phase), Kaolinite 2ϕ:14 $Al_2Si_2O_5(OH)_4$ and Alunite 2ϕ:34 $KAl_3(SO_4)_2(OH)_5$. Feldspar and Micas are also present 2ϕ≈ 26–30. The chemical composition can be correlated with mineral phases. SiO_2 is due to Quartz an Opal. SO_3 amount indicate the relative amount of Alunite and the Al_2O_3 content in clays can therefore be estimated, by deducting the amount in the alunite from the total Al_2O_3.

Therefore, having the XRD pattern (intensities) and the chemical composition, it is possible to assume that sample B has the highest amount of Alunite and lowest clay

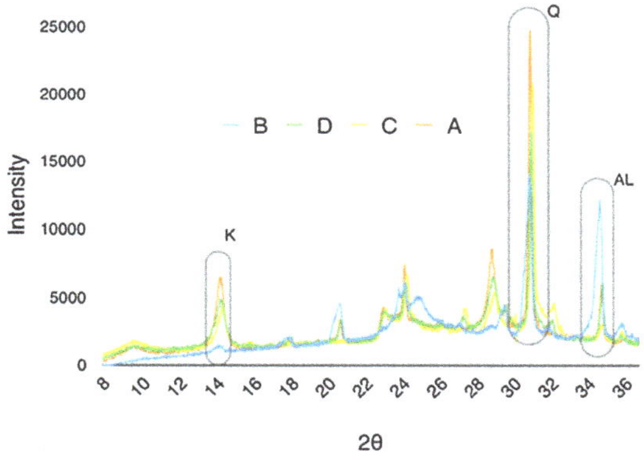

Fig. 2. XRD pattern of samples, before thermal treatment

content, sample C lowest alunite content, sample A highest Clay content and sample D intermediate composition (Fig. 2).

3.2 Thermal Treatment and Structural Changes

The samples were thermally treated by batch of 2.5 kg in a muffle kiln for 2 h at the temperature in which the dehydroxylation process takes place (800 °C) and later on cooled down to room temperature.

After calcination at 750 °C, the microstructure changed as can be show in Fig. 3. From an ordered layer structure to an amorphous phase, during the process the size of particles slightly increases (residue sieve increase). Above 783 °C the sintering process occurs, reducing the Blaine Value-Fig. 4 and this will affect the reactivity. The optimal temperature of activation should be between 600 °C to 783 °C.

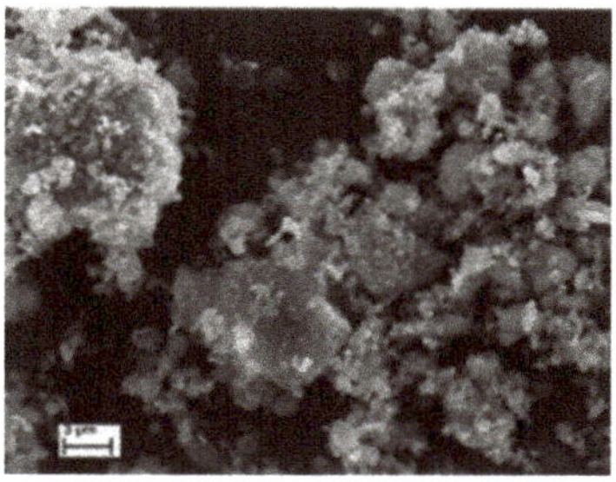

Fig. 3. SEM image of sample A after calcination

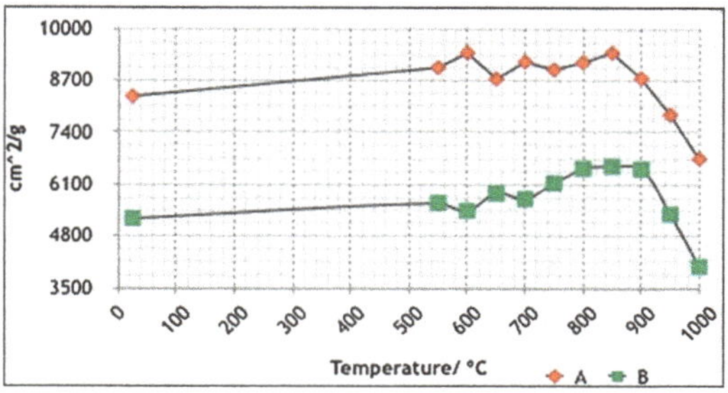

Fig. 4. Blaine value increasing the activation temperature

Figure 5 showed that loss of mass took place in two ranges at high temperature [5]:

1. 500 °C–600 °C the dehydroxylation of Kaolinite and Alunite.
2. 700 °C–800 °C, Alunite decomposition and release of SO_3.

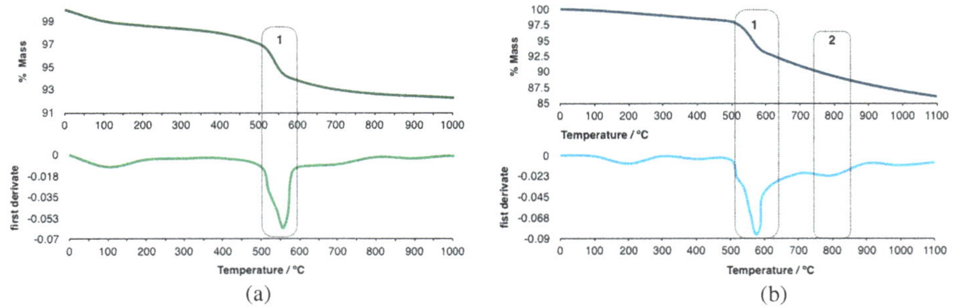

Fig. 5. TGA and first derivate analysis of sample C (a), sample B (b)

3.3 Performance Evaluation

Blended Cements were prepared with 70% Portland Cement and 30% of the Iza material before and after activation. The samples were prepared and tested according to ASTM C109. All the pozzolan were ground at constant residue sieve (R90 μm: 9–11%). Reference samples with only Portland cement and Portland cement with 30% Ottawa Sand (inert quartz) were also tested as reference to calculate hydraulic index.

Hydraulic index was calculated according to Eq. (1).

$$IH = \frac{R^D_{sample} - R^D_{sand}}{R^D_{opc} - R^D_{sand}} \tag{1}$$

Where R is the compressive strength of the sample at age D. The hydraulic index value is relative to the contribution of the material that was tested respective to inert material (Ottawa sand) and Ordinary Cement Portland and can be analyzed as:

- IH < 0, indicates no pozzolanic activity.
- 0 < IH < 1, indicate s moderate contribution to strength development better than Quartz(inert) but less than that OPC. The higher HI, better reactivity.
- IH > 1, indicate higher strength contribution than OPC indicating excellent pozzolanic activity.

The hydraulic index of samples at different ages can be seen in Fig. 6. The samples A, C, D before activation were considered as non-reactive (IH < 0), while sample B showed good HI of 60, in agreement with its high opal content and low clay content. However after calcination the sample (A,C,D ACT) showed highly improved in reactivity specially at later ages due to the new amorphous reactive amorphous metakaolin phase formed during thermal treatment.

On the contrary, sample B(act) showed decreasing the hydraulic index and bad performance at early ages after activation. 1d strength was also reduced after calcination for samples A & D for the same reason, they also contained alunite.

Thermal activation showed an antagonistic effect. Activated kaolinite(metakaolin) improved the compressive strength at 3, 7 and 28 days (Samples A, C, D). In contrast, decomposed alunite doesn't improve compressive strength and increases strongly the

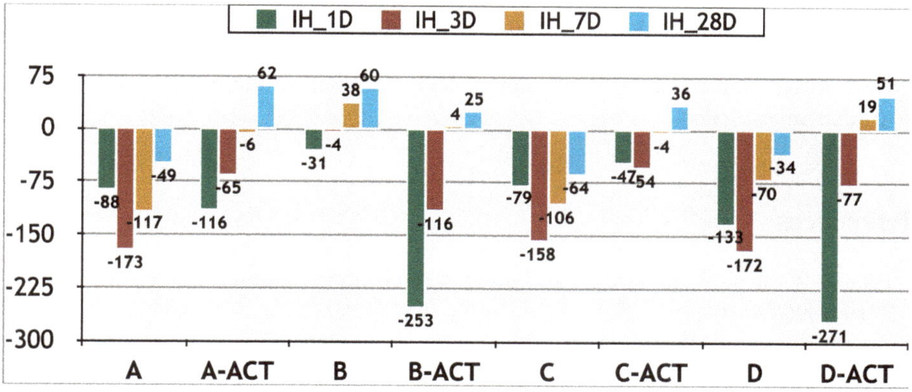

Fig. 6. Hydraulic Index of samples before and after activation

Table 2. Change in water demand after activation.

Sample	Δ A/C %
A	−7.0
B	4.2
C	−8.1
D	−6.8

water demand Table 2 (Sample B). It is attributed to the formation of new aluminates reactive phases $KAl(SO4)_2$ and amorphousAl_2O_3. An option to control it could be increase the gypsum amount.

The water demand after activation was decreased for samples A, C, D in which the negative effect of the clays disappears, however for sample B it increased due to its high alunite content, showing increased water demand and decreased mortar workability. An increase of mortar temperature after the mix was also observed, presumably because of the fast dissolution of the reactive potassium sulfate aluminate and/or amorphous Al_2O_3 phases.

4 Conclusions

- The performance of calcined clay from the quarry under investigation is promising in terms of hydraulic performance, the clinker replacement could be at least 20%.
- Thermal decomposition of Alunite produced new phases that can have negative impact on setting and early strength. When looking at new clay deposit, a particular attention should therefore be taken to alunite content, but also calcination temperature should therefore be carefully assessed and controlled. Alunite decomposition sequence and its effect on strength development, but also on emissions level, should be further studied.

- The calcined material with its high alumina reactive content, could have synergic effect with limestone [6] in ternary blended cement in order to produce a cement with lower clinker factor. First unpublished data confirm these data and it will be further investigated.

References

1. Gartner, E.: Industrially interesting approaches to low-CO2 cements. Cem. Croncr. Res **34**, 1489–1498 (2004)
2. Andrini, L., Gauna, M.R., Conconi, M.S., Suarez, G., Requejo, F.G., Aglietti, E.F., Rendtorff, N.M.: Extended and local structural description of a kaolinitic clay, its fired ceramics and intermediates: An XRD and XANES analysis. Appl. Clay Sci. **124–125**, 39–45 (2016). doi:10.1016/j.clay.2016.01.049
3. SEM Analysis were done by Victor Galvan. Universidad Nacional de Cordoba- Argentina, X ray and Electron Microscopy Laboratory (LAMARX), July 2015
4. Farrokhpay, S., Ndlovu, B., Bradshaw, D.: Behaviour of swelling clays versus non-swelling clays in flotation. Miner. Eng. **96–97**, 59–66 (2016). doi:10.1016/j.mineng.2016.04.011
5. Piga, L.: Thermogravimetry of a kaolinite-alunite ore. Thermochim. Acta **265**, 177–187 (1995)
6. Antoni, M., Rossen, J., Martirena, F., Scrivener, K.: Cement substitution by a combination of metakaolin and limestone. Cem. Concr. Res. **42**, 1579–1589 (2012). doi:10.1016/j.cemconres.2012.09.006

Improvement of the Environmental Energy Sustainability in the Production of Cement Portland with Addition of Thermally Activated Clays

I.L. Machado[1(✉)], H.I. Moya[2], S.B. Sánchez[3], and F. Martirena[1]

[1] Center for Research and Development of Structures and Materials (CIDem), Central University "Marta Abreu" of Las Villas (UCLV), Carretera a Camajuanì km 5 1/2, Santa Clara, Villa Clara, Cuba
ivanm@uclv.edu.cu

[2] Faculty of Mechanical Engineering, Central University "Marta Abreu" of Las Villas (UCLV), Santa Clara, Cuba

[3] Faculty of Economy, Central University "Marta Abreu" of Las Villas (UCLV), Santa Clara, Cuba

Abstract. The sustainability of the production of the Portland cement, among another factors, is directly related to the decrease of specific consumption of fuel and the decrease of the emissions of gases and contaminating particles. To decrease the clinker's factor, an important alternative is to improve the environmental efficiency. It is demonstrated in this work that with the addition of thermally activated clays, a substantial decrease of the consumption of energy and CO_2 emissions occurs.

In this sense, the Life Cycle Analysis (LCA) tool is used, and the methodology of baseline approved ACM0005: "Baseline consolidated methodology for increasing addition in cement production" (taken from IPCC, 2006) is used.

The work is carried out in the Siguaney factory (Republic of Cuba), and proves to be innovative for the ternary system (clinker – activated clay – limestone), establishing the rate of emissions in 1191.15 kg CO_2/t.ck. Corresponding results are obtained for other important pollutants. It is demonstrated that in direct relationship with the decrease of the clinker consumption for ton of cement, the energy consumption decreases by 18%, which leads to obtaining ternary blended cement.

Keywords: Life cycle · Emissions · Cement Portland · Activated clays

1 Introduction

The Industry of the Cement constitutes at the present time one from the most polluting industries to the environment, for its continuous emission of gases in the atmosphere associated to a high energy consumption, mainly in the form of fossil fuels, as well as the chemical reactions of dis - carbonization that comes from the process.

Although cement is recognized as one of the elements that has contributed the most to humanity's development (and today it is even associated with the level of

F. Martirena et al. (eds.), *Calcined Clays for Sustainable Concrete*, RILEM Bookseries 16,
https://doi.org/10.1007/978-94-024-1207-9_47

development of a country [1]), it is also one of the main factors responsible for the environmental degradation of the planet. This material is among those most employed and with the highest production level on a global scale (more than 3 500 million tons in 2015). Its high production volumes make it responsible of about 7% of the world emissions of CO_2 of origin anthropogenic and of 5% of the energy consumption in the industrial sector [2, 3].

Emissions of carbon dioxide can be reduced by [4]:

- Improvement of the energy efficiency of the process
- Replacing high carbon fuels by low carbon fuels
- Applying lower clinker/cement ratio (increasing the ratio additives/cement): blended cements.
- Application of alternative cements (mineral polymers)
- Removal of CO_2 from the flue gases

Amongst these alternatives, the employment of the supplementary cementitious materials (SCMs) can be one of the short-term and more promising roads for many production methods of cement. With this solution, the mechanical resistance of the cement and their durability in the presence of a certain number of environmental agents [5] are generally increased.

Preliminary studies show that the formulation of a cement based on the clinker substitution by thermally activated clays (AAT) in proportions up to 30% (and even up to 50%), called cement of low emission of carbon (LCC or LC^3), can represent an important reduction of the energy consumption and emission of CO_2 during its production [6, 7].

Determining the environmental feasibility through studies that allow to verify the current situation and to propose valid recommendations to diminish the pollution, is of utmost importance. The present investigation is carried out with the objective of evaluating the energy and environmental feasibility of the production of CPO with addition of Clays Activated Thermally, on the base of the production pilot in the factory Siguaney in the central region of Cuba.

2 Analysis and Discussion of the Results

It is well-known that the system of cement production consumes significant quantities of fossil fuel as main source of thermal energy, which has motivated the cement sector to aim for the reduction in the consumption of these resources. The environmental damage resulting from these energy consumptions are presented mainly in form of gasses emitted in the atmosphere, such as carbon dioxide (CO_2), nitrogen oxides(NO_X), sulfur oxides (SO_2) and others.

In this work, the relative cause of carbon dioxide emissions is exposed. The chemical reactions of calcinations of the limestone in the rotatory kiln of clinker production are an equally important part of the process in which CO_2 is generated.

The tool of Life Cycle Assessment (ACV) facilitates the evaluation of the environmental affectations that is caused the industry. It evaluates in this case the cycle that extends from the entrance from the materials to the clinker kilns, until the obtaining of the cement processed in the balls mills.

In this work, the following evaluation of the polluting emissions the focus starting from the calculation of the emission factors, matter balances and production indexes is used:

- Calculation of the Emission Factors (FE given in kg of CO_2/t.ck). It focuses starting from the mensuration carried out *in situ* in the factory.

$$FE = \frac{\Phi * \%CO_2 * P * 1000}{B} = 966.17\,kg.CO_2/t._{ck} \quad (1)$$

Where Φ is the volumetric flow of the gasses to the exit of the kiln (200000 m^3/h), P is the density of the CO_2 (1.87 kg/$m^{3)}$, B the clinker production (12000 kg/h) and % CO_2 the volumetric fraction (0.026)
- Calculation of the emissions. Focus starting from the pattern proposed by the IPCC.

2.1 Calculations of the Emissions Coming from the Clinker Production Process

(a) CO_2 emissions coming from calcination
The data of laboratory analysis are used:
C_aO = 65.33%, M_gO = 1.05%. Being obtained the following equation

$$BE_{calcin} = 0.785 * (outC_aO - inC_aO) + 1.092 * \left(outM_gO - inM_gO\right) \quad (2)$$

Where BE_{calcin} is the emission factor in the calcination of clinker, InC_aO is the percentage of CaO in the matter multiplied by the quantity of matter in t, $Out.C\,_aO$ is the percentage of C_aO in the clinker multiplied by the quantity of clinker in t, InM_gO is the percentage of M_gO in the matter multiplied by the quantity of matter in t and $OutM_gO$ is the percentage of M_gO in the clinker multiplied by the quantity of clinker in t.
0.785 accounts for the stoichiometric emission factor for CaO (t CO_2/t C_aO) and 1.092 for the stoichiometric emission factor for M_gO (t CO_2/t M_gO).
By substituting, we obtain the following:

$$\begin{aligned} BE_{calcin} &= 0.785 * (65.33 - 0) + 1.092 * (1.05 - 0) = 52.426 * \left(\frac{1000}{100}\right) \\ &= 524.6\,kg\ CO_2/t.ck \end{aligned}$$

(b) Emissions coming from fuel burning
It is necessary to keep in mind that the used fuel is cuban raw petroleum with a factor of emission of 3,28 kg CO_2/t.fuel. The consumption index in the Siguaney factory is 195.96 kg of fuel per ton of clinker, where 642.75 kg CO_2/t.ck is obtained.

(c) Emissions coming from electricity consumption
The established emission factor for the electric power generation in Cuba is of 744 kg of CO_2 for MWh, according to the official information of the National Energy System in the year 2015. For the clinker production, in the factory of Siguaney, the consumption index is of 0.032 MWh per ton of produced clinker.

$$\text{Emissions}\,CO_2 = \text{Emission factor} * \text{consumption index} = 23.8\,\text{kg}\ CO_2/\text{t.ck} \quad (3)$$

In the Tables 1 and 2, a summary is presented with the determination of the Factor of Emission according to the focus of the realization of calculations. These values are adopted for most of the considerations carried out in this work. Periodic data was obtained from the factory to carry out a statistical analysis.

Table 1. Factors of emission of CO_2 to produce clinker in Siguaney.

	Emission source	FE kg CO_2/t.ck
a	Calcination	524.6
b	Fuel	642.75
c	Electricity	23.80
Total		1191.15

Table 2. Factors of emission of CO_2 to produce AAT in Siguaney.

	Emission source	FE kg CO_2/t.AAT
d	Calcination	0
e	Fuel	257.09
f	Electricity	20.24
Total		277.33

2.2 Calculations of the Emissions Coming from the Obtaining of Thermally Activated Clays

(d) Emissions caused by calcination in the production of A.A.T.
In the clay, the content of $CaCO_3$ is present in very low quantities. Therefore, CO_2 is not emitted in the thermally activated clay process.

(e) The emissions caused from fuel burning
To obtain A.A.T. in a rotatory kiln of clinkerization without big modifications, we consider that the consumption of fuel decreases b < 60% with regard to the clinker production [8]. The following calculation can be made:

$$\frac{195.96\,\text{kg}_{\cdot\text{crudo}}}{\text{t}_{\text{A.A.T.}}} * 0.4 \quad (4)$$

By substituting we obtain:

$$\frac{78.38\,\text{kg}_{\cdot\text{crudo}}}{\text{t}_{\text{A.A.T.}}} * \frac{3.28\,\text{kg}.CO_2}{\text{kg}_{\cdot\text{crudo}}} = 257.09\,\text{kg}\ CO_2/\text{t}_{\text{A.A.T.}}$$

(f) Emissions coming from electricity consumption
In the clinker production, the consumption index is of 0.032 MWh/t.ck, we consider a reduction of 15% when obtaining AAT in the kiln modified with a calcinator, then 20.24 kg $CO_2/t_{A.A.T}$ calcinator are obtained.

2.3 Comparison Among Different Cement Types as for the Emissions of CO2 and Energy Consumption

For the calculation and comparison of the emissions of CO_2, each one of the sources that impact the life cycle assessment was considered. The results are shown in Table 3.

Compared to the PP25, a reduction of 24% in the emissions of CO_2 is obtained when considering a decrease of 228 kg of CO_2 for each ton of cement taken place with the addition of A.A.T., justifying even more the production of LCC in substitution to other blended cements. The environmental affectations that can produce the pozzolans of the cements type PP25 have not been considered, which makes the production of blended cements with AAT environmentally more attractive.

Table 3. Emissions of CO_2 for different cement types

Cementos	Kg CO_2/t_{ck} o A.A.T.			kg.CO_2/t.cement
	Calcination	Fuel	Electricity	Total[c]
P-35[a]	461.65	565.62	20.94	1130.92
PP-25[b]	393.45	482.06	17.85	976.09
LCC	251.81	385.65	17.49	737.69

[a] Half-high content of Clínker (NC-95 – 2011).
[b] Medium content of Clínker and pozzolans additions. (NC-96 – 2011).
[c] It is added in all the cases 82.73 kg.CO_2/t.cement, coming from the electricity consumption in the mill.

2.4 Calculation of the Energy Consumption for Each Cement Type

For the calculation of the energy required by each cement type, one keeps in mind the characteristics of the fuel according to historical data of the factory (caloric value 9457 kcal/kg) and the consumption index (195.96 kg/t.ck).

Obtaining AAT requires a much smaller grade of temperature that the clinkerization process, and thereby a lower energy consumption. Consequently, when substituting clinker levels by AAT in the cement, the required energy and the emissions are decreased. In Fig. 1, the energy level for each process is detailed for the obtaining of the different cement types.

It is proved that the LCC with regard to the PP25 offers a 18% reduction in the energy consumption. Increasing the substitution of the clinker (justified on the base of studies of mechanical resistance, durability and other) diminishes the energy consumption and the improvement of the environmental profile of the material.

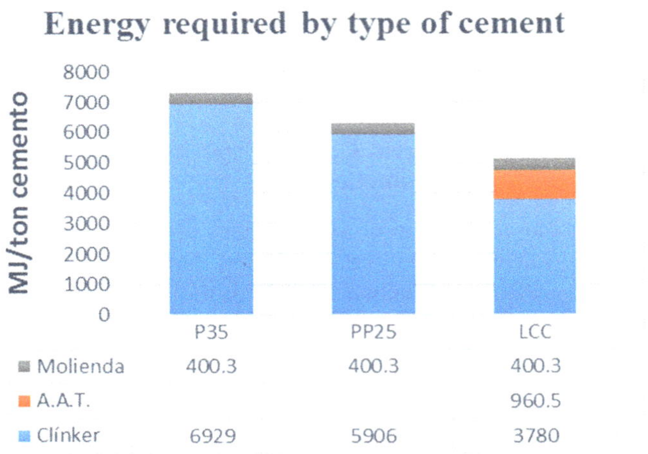

Fig. 1. Energy required by types of manufactured cements.

3 Conclusions

The theoretical calculations reflect an emission factor in correspondence with that approved by the EPA, the decrease is verified of up to 18% in the energy consumption when obtaining LCC with the addition of AAT, with an important decrease of the emissions of CO_2 for each ton of produced cement, demonstrating the environmental significance of this solution.

References

1. Pierre-Claude, A.: Cements of yesterday and today. Concrete of tomorrow. Cem. Concr. Res. **30**(9), 1349–1359 (2000)
2. Unidas, N.: Mecanismo de desarrollo limpio (2006)
3. Scrivener, K.L.: Options for the future of cement. Ind. Conc. J. **88**(7), 11–21 (2014)
4. Hendriks, C.A., Worrell, E., de Jager, D., Blok, K., Riemer, P.: Emission Reduction of Greenhouse Gases from the Cement Industry (2003)
5. Vizcaíno, L., Sánchez, S., Damas, S, Pérez, A., Scrivener, K., Martirena, F.: Industrial trial to produce a low clinker, low carbon cemen. Materiales de Construccion **65**(317) (2015)
6. Antoni, M., Rossen, J., Martirena, F., Scrivener, K.: Cement substitution by a combination of metakaolin and limestone. Cem. Concr. Res. **42**(12), 1579–1589 (2012)
7. Delgado, E.L.: Evaluación energético ambiental de la producción de cemento portland ordinario (CPO) con la adición de arcillas activadas térmicamente. In: Ingeniería Civil. Universidad Central Marta Abreu de Las Villas, Santa Clara (2013
8. FLSmidth: A one source supplier to the global minerals and cement industries (2013)

Resource Mapping of China Clay for LC^3 Application in India

A. Soumen Maity[1(✉)] and B. Santanu Mithia[2]

[1] Technology and Action for Rural Advancement, New Delhi, India
[2] Development Alternatives, New Delhi, India

Abstract. Low carbon cement is a ternary blended Portland cement with clinker factor as low as 0.50. It uses the synergetic hydration of clinker, calcined clay produced from mine waste and non-cement grade limestone to achieve properties comparable to commercial cements. In this limestone calcined clay cement (LC^3) one of the most important raw material is the china clay. The present paper discusses the availability and occurrence spread of china clay in India. It has been shown through a GIS map that china clay is available almost throughout the entire country. The resource map is based on a GIS platform to have real time data on the raw material availability. To enable an informed decision making by cement companies, the occurrences and availability of fly ash used as pozzolana has also been shown in addition to location of integrated cement plants and grinding units. The map shows that there are quite some regions in India where the production of LC^3 is feasible. This is contradictory to the view that almost the whole of India is feasible for fly ash pozzolana cement.

1 Introduction

The Indian cement industry is the 2nd largest after China accounting for about 8% of the total global production. It had a total capacity of over 360 million tonnes (MT) as of the financial year 2013–2014. The Indian cement industry comprises of 140 large units and more than 365 mini cement plants.

With the new government announcing plans for dedicated freight corridors and 100 "smart cities", cement makers are anticipating robust demand and have planned huge capacity expansion in North India. These companies include Jaypee Cement, JK Cement Ltd, Mangalam Cement Ltd, Birla Corporation Ltd, Ambuja Cements Ltd, Shree Cement Ltd, UltraTech Cement Ltd and India Cements Ltd. However, any further expansion in the region will have to be done by way of consolidation, experts say, as limestone reserves are limited in north and cost of transporting limestone from south are high. The north has enough reserves of low-grade limestone but supply of high-grade limestone is limited.

The Indian cement industry is one of the most efficient in the world [1], yet, because the manufacturing process relies on the burning of limestone (calcium carbonate), it still

F. Martirena et al. (eds.), *Calcined Clays for Sustainable Concrete*, RILEM Bookseries 16,
https://doi.org/10.1007/978-94-024-1207-9_48

produces 137 million tonnes of CO_2 in 2010 which is approximately 7% of India's total man-made CO_2 emissions. The Indian cement industry has made strong efforts in reducing its carbon footprint and has reduced it from a substantial average of 1.12 tCO_2/tonne of cement produced in 1996 to 0.719 tCO_2/tonne of cement produced in 2010. Even with this emission efficiency CO_2 emissions form the Indian Cement industry are projected to reach between 488 $MtCO_2$ and 835 $MtCO_2$ by 2050. This represents a 255% to 510% increase compared to current emissions.

Major cement manufacturers in India are also increasingly using alternate fuels, especially bioenergy, to fire their kilns. This is not only helping to bring down production costs of cement companies, but is also proving effective in reducing emissions. With increased efforts by cement companies on reducing their energy costs for being cost competitive, the long-term business sustainability in terms of resource availability is and has been overlooked till date. This is quite important since the earth's resources are finite and takes millions of years to regenerate.

2 Objectives of the Study

Current trends in energy supply and use are unsustainable – economically, environmentally and socially. It has been referred by WBCSD CSI report [1] that without decisive action, energy related emissions will more than double by 2050 and increased fuel demand will heighten concerns over the security of supplies. Till date the cement industry has been focussing on reducing its energy and emission due to increased global pressure. However, the issue of resource availability has been overlooked and not gained importance over the issue of emissions.

The present study looks at the issue of availability of selected raw materials to be used in Limestone Calcined Clay Cement (LC^3) production. This has been done through a GIS based decision support tool. The major objective is to enable the clay producers, cement manufacturers to look at capacity augmentation using same infrastructure or explore new production units based on raw material availability.

The present Resource Mapping study does not intend to be a comprehensive databank for providing resource information. It is expected to provide basic information for decision making only. It is also expected that interested cement companies and other stakeholders will further invest for more detailed information.

3 Approach and Methodology

The resource mapping study has been carried out in two parts:

- Availability of clay resources and its distribution in India based on available secondary data.
- Suitability map of kaolinitic clays across India for LC^3 production.

Whereas the china clay availability map based on GIS tools and techniques is a desk research with super-imposition of cement industry related information, a specific resource mapping in the two states of Rajasthan and Gujarat and its application in LC^3

technology was also carried out focussing mainly on use of china clay as the major raw material. For making the mapping exercise interactive and useful for future work, GIS system has been used for plotting of all data for accuracy of mapping and upgradation in the future as per needs and requirements.

4 Limestone Calcined Clay Cements and Raw Material Requirement

Low carbon cement is a ternary blend of supplementary cementitious material allowing a clinker substitution of around 50–60%, by a synergetic combination of calcined clay and limestone. The usefulness of this cement has been demonstrated in the collaborative research between the Laboratory of Construction Materials (LMC) at EPFL, Switzerland, CIDEM in Cuba, IIT Delhi, Madras, Bombay and TARA, New Delhi, India.

Usually the main ingredients of cements are limestone, clay, gypsum and coal. Clinker is produced by chemical reaction between limestone (around 90%) and clay at high temperature (>1400 °C). Coal is used as fuel during clinkerization. The raw materials are abundant in earth crust but the reserves of these materials are limited and going to be depleted with unlimited human demand which will affect the production of cement in near future. Hence the need to find alternative resources to produce cement in the form of supplementary cementitious materials (SCM'S) [2].

5 Geographic Information System (GIS)

A geographic information system or (GIS) is a system designed to capture, store, manipulate, analyze, manage, and present spatial or geographical data.

The Resources Mapping Decision Support System (RM-DSS) is composed of two main elements: the graphical user interface (GUI) and the database. It has five functional components: Query, Direction, Analysis, Measurement and Base Map Gallery for the visual result. The system design is illustrated with particular emphasis on the development of the GUI, database queries and map management design with ready-to-use widgets. The application of the RM-DSS is useful for both commercial and research purposes to be applied by commercial companies or institutions.

The main objective of this GIS based RM-DSS is to provide a solution through the given set of indicators and offer the most appropriate solution which helps the user (companies and institutions) to do the strategic planning and cost benefit analysis. This system, integrate all indicators of china clay (quality, quantity, total deposit, location), thermal power plant (capacity, fly ash quality and location) and cement manufacturing units (type of cement unit, distance from the china clay deposit, distance from thermal power plant and production capacity) and also the surrounding market place. This GIS based system provides the transport network and resources availability through spatial interactive and scale dependent maps. This is a tool for companies and institutions to quantitatively assess the feasibility of the resources through visual interpretation and identify the challenges for a suitable solution.

6 The Decision-Making Map

The decision-making map for selection of potential areas for LC^3 have been based on secondary literature and data collected from various agencies.

Step 1

As a first step, location of all the major integrated cement plants and grinding units were plotted in GIS map. The data of cement plants and grinding units were collected from Cement Manufacturing Association. In addition, all the locations of the cement units were individually verified from the respective locations given in the cement plant website. Where confusion occurred, it was cross-checked with the respective cement companies.

Figure 1 shows the location of the cement plants. The green dots are the integrated cement plants, purple the large grinding units and blue are blending units. All the data plotted can be blown up and actual unit can be seen. A drop-down list also gives details of the cement plant including type of cement produced and capacity.

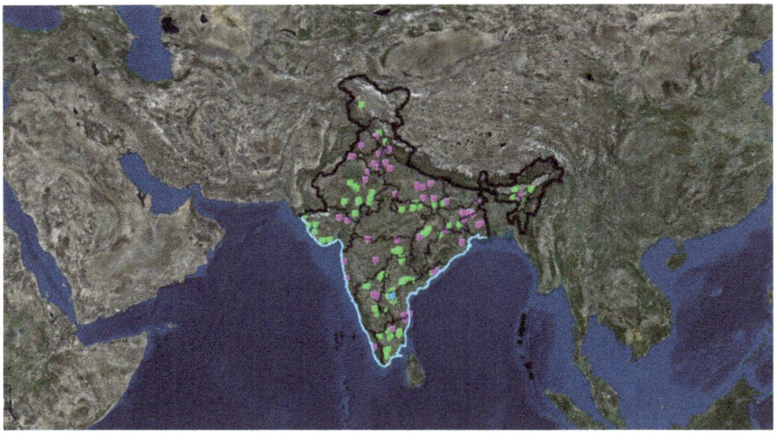

Fig. 1. Location of major integrated cement plants and grinding units

Step 2

In the second step, the location of all the major thermal power plants was plotted in the GIS map. The data of all major thermal power plants were collected from the Report on Fly Ash Generation at Coal/Lignite based Thermal Power Station and its Utilization in the Country for the Year 2014–2015 [3] prepared by the Central Electricity Authority, New Delhi on behalf of the Ministry of Power. A total of 145 thermal power stations have been plotted based on the available data. The location of the thermal power plants was also cross checked with the actual location. However no captive thermal power plants were plotted due to lack of any authentic published data.

Figure 2 shows the location of the thermal power plants marked as red dots. The locations can also be blown up to find the actual plants. Detailed information of the thermal power plant is given in drop down lists, including the amount of fly ash produced.

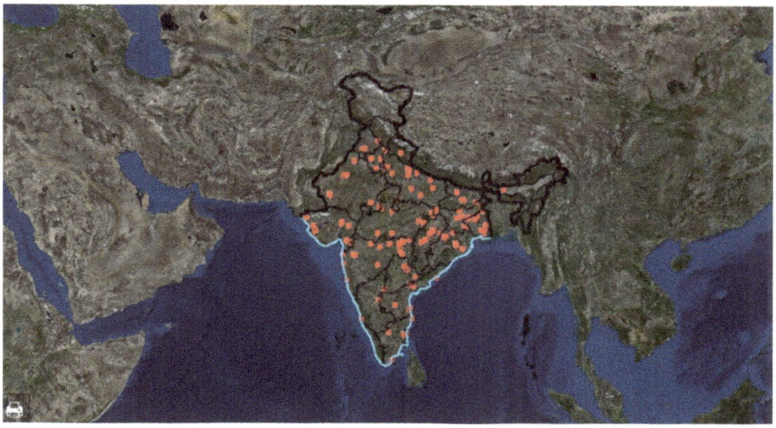

Fig. 2. Location of major thermal power plants in India (CEA 2015)

Step 3

In the third step for establishing the DSS tool, the location of all the major china clay areas was also plotted in the same GIS map. The data of the china clay availability locations were taken from Indian Minerals Yearbook 2012 published by the Indian Bureau of Mines, under the Ministry of Mines, Government of India in February 2014 [4]. During this period, there were a total of 71 reporting china clay mines. The china clay areas plotted were based on the district wise data for occurrence of china clay total

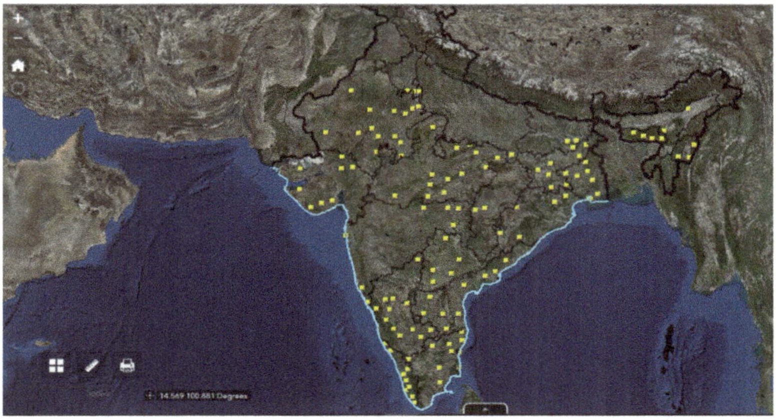

Fig. 3. Location of china clay mines (Indian Minerals Yearbook 2012)

resources as reported in the Mineral Yearbook. These were the most authentic reported data available.

Figure 3 shows the district wise china clay locations. It might be possible that detailed zooming-in will not reveal any china clay mines since it is still in the virgin state and not commercially exploited. Besides the mines plotted there are other occurrences of china clay also. These are under reserve or being prospected category.

Once the three attributes were plotted in the database, overlaying of all the attributes were done by switching on and off various layers. This helped to establish the business feasibility and raw material availability to enable informed decision making.

Figure 4 given below shows the aggregated database of the three attributes of cement plants, grinding units, thermal power plants and china clay mines plotted together.

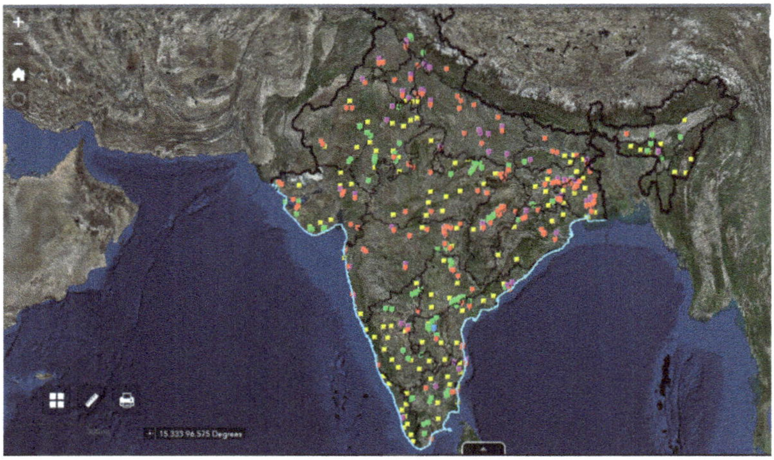

Fig. 4. Aggregated database of cement and thermal power plants and china clay mines

When the aggregated database was analyzed, it was found that there were many regions where fly ash is not available or cost of fly ash will be quite high due to longer transportation leads. These are the regions where the Limestone Calcined Clay Cement or LC^3 will be extremely feasible for cement production.

7 Conclusions and Outlook

The Resource Mapping – Decision Support System is a tool to enable decision making to locate a new LC^3 production unit or facility augmentation of existing unit based on china clay availability and suitability for use. This tool is not an absolute one and will vary with time. Based on the resource mapping system, it can be concluded that kaolinitic clay is abundant across the country. Contrary to the view that abundant fly ash is available in and around cement plants, the map proves that in certain areas it might be beneficial to produce Limestone Calcined clay cement (LC^3) rather than transporting fly ash from far off distance based on production economics. Thus, the GIS map is expected to support the primary objective in understanding the availability, quality and suitability of china clays. It will also help in cement companies finding a suitable location for new ventures based on raw material availability.

References

1. Technology Roadmap: Low Carbon technology for the Indian Cement Industry, WBCSD (2012)
2. Antoni, M., Rossen, J., Martirena, F., Scrivener, K.: Cement substitution by a combination of metakaolin and limestone. Cem. Concr. Res. **42**, 1579–1589 (2012)
3. Report on fly ash generation at coal/lignite based thermal power stations and its utilization in the country for the year 2014–2015, Central Electricity Authority, Ministry of Power, October 2015
4. Indian Minerals Yearbook 2012 – Part III: Mineral Reviews, Kaolin, Ball Clay, Other Clays and Shale, Indian Bureau of Mines, 51st edn. (2014)

Chloride Transport Behavior of LC^3 Binders

H. Maraghechi(✉), F. Avet, and K. Scrivener

Laboratory of Construction Materials (LMC), Ecole Polytechnique Fédérale de Lausanne, Lausanne, Switzerland

Abstract. Corrosion of metal reinforcement is the most important durability concern of concrete infrastructures. In this research, chloride transport in mortars prepared using ternary blends of OPC-calcined clay-limestone (LC^3) as the cementitious materials is assessed and compared to a reference OPC system. Three clays with varying kaolinite content (17%, 50% and 80%) were employed. Through XRD analysis, transformation of carboaluminate phases - the stable form of AFm in LC^3 binders - to Friedels salt, as a consequence of exposure to NaCl solution was observed. In addition, porosity of LC^3 binders, with kaolinite content of 50% and higher was found to be significantly refined, which helps reduction of the depth of chloride penetration. The findings support excellent behavior of LC^3 cement, prepared using low-grade clays, in corrosive environments.

1 Introduction

The use of calcined kaolinite clays as a supplementary cementitious material, SCM, dates back to 1932 in San Francisco, USA and since the 1970's Brazil has a constant production of calcined clay [1]. Despite excellent behavior, a limiting factor against global use of calcined clays has been the high price of high-grade clays. However, in the last few years, low-grade calcined clays, which is obtainable with lower price and is abundant in large quantities worldwide - much more than conventional SCMs - has been considered as a very promising SCM. It has also been proven that simultaneous utilization of calcined kaolinite clays and limestone, in preparation of ternary cementitious blends (known as LC3), with up to 50% clinker replacement, provides tremendous environmental, economic and technical benefits to concrete structures [1]. Various aspects of LC3 cements and concrete, such as mechanical properties, hydration and phase assemblage have been studied in the past few years [2, 3]. Reactivity of a clay source, and hence the mechanical performance of concrete, is found to be mainly and directly related to the kaolinite content of the clay (an indication of a clay grade) [4]. This study presents resistivity of LC3 mortars, prepared with kaolinite clays with varying kaolinite contents against chloride transport. In addition, porosity of LC3 binders and phase assemblage after exposure to NaCl solution are studied and compared to a 100% OPC binder.

F. Martirena et al. (eds.), *Calcined Clays for Sustainable Concrete*, RILEM Bookseries 16,
https://doi.org/10.1007/978-94-024-1207-9_49

2 Materials and Methods

Three different clays with kaolinite contents of 17, 50 and 80 wt.% were used; where the rest of the clays contained non-reactive impurities, mainly as quartz. Clay materials were calcined at around 800°C for 1 h in a high temperature furnace (Borel FP1100) before use. A type I cement, with the main clinker phases of C3S = 69%, C2S = 5%, C3A = 7% and C4AF = 15%, and a limestone powder, obtained from Omya with > 98% purity, were also used. Ternary blends of OPC-calcined clay-limestone (LC^3), were prepared with 50 wt.% clinker, 5 wt.% gypsum, 30 wt.% calcined clay and 15 wt.% limestone. Table 1 shows characteristics of the cementitious materials in this study.

Table 1. Properties of three clays, limestone and OPC used in this study.

			Calcined kaolinite clay		
	OPC	Limestone	K-80	K-50	K-17
Kaolinite [wt.%] *	–	–	79.4	50.3	17
Dv_{50} [μm]	8.4	7.2	5.3	10.9	5.8
BET surface [m^2/g]	0.9	1.8	15.3	45.7	18.7

Paste samples were prepared with w/solids ratio of 0.5, cured for 1 d in sealed cylindrical containers (D = 2 cm). After that, the paste specimens were submerged in a small amount of water for further curing with minimum leaching. After 28 d of curing, 2 mm thick slices were cut from the paste samples and submerged in either water or 1 M NaCl solution for a duration of 2 months. Phase assemblage of the pastes were assessed using XRD analysis. A PANalytical X'Pert Pro MPD diffractometer in a θ–θ configuration using CuKα source ($\lambda = 1.54$ Å) with a fixed divergence slit size of 0.5° was employed. MIP analysis was used to assess porosity distribution of OPC as well as the three LC3 pastes after 28 d of curing.

The three LC3 blends, and a reference OPC cement were used in preparation of mortar mixtures, with w/solid = 0.5, where the mass of sand was 3 times of the cementitious powders. Cylindrical mortar samples were cast in 11 cm diameter and 30 cm height molds. After 28 d of curing in moist room with > 95% RH, mortar samples were cut into two halves, sealed using epoxy resin in all surface except the saw cut face, and submerged in 0.3 M NaCl solution at 23 C. After 1 and 2 years of exposure, profiles of total chloride content were obtained following ASTM C 1543 method.

3 Results

Figure 1a shows the XRD patterns of the OPC and the three LC3 binders. The main difference in phase assemblage between OPC and LC3 is formation of mono-carboaluminate (Mc) and Hemi-carboaluminate (Hc) phases as the stable AFm phase. This is due to the presence of limestone, as it is also thermodynamically recognized that limestone reacts with aluminates to form carboaluminate AFm phase [4, 5]. For OPC, the presence of carbonate impurities may also trigger formation of carboaluminates, as it appears to be the case in this study. In presence of calcined clay, a synergetic effect is observed, where due to the presence of more aluminum, more carboaluminate phase form. After exposure to 1 M NaCl solution, AFm phases transforms into Fridel's salt (dashed lines in Fig. 1a). It should be noted that mono sulfoaluminate forms in OPC system, which due to its low crystallinity, it could not be detected very easily in XRD patterns. However, formation of Fridel's salt in OPC paste is also clearly seen in XRD patterns.

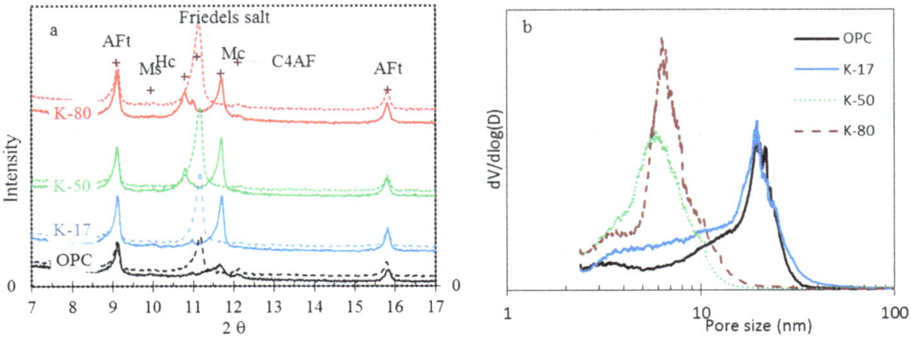

Fig. 1. (a) XRD patterns of OPC and LC3 pastes, cured for 28 d, and exposed to 1 M NaCl solution (dashed line) or water (solid line) for 2 months; (b) analysis of porosity of OPC and LC3 binders

As seen in Fig. 1b, the critical pore size entry for LC3 pastes with kaolinite contents of 50% and 80%, has significantly reduced. This helps with decreasing the depth of chloride transported through LC3 mortars, as it is shown in Fig. 2. Less than 1 cm is the

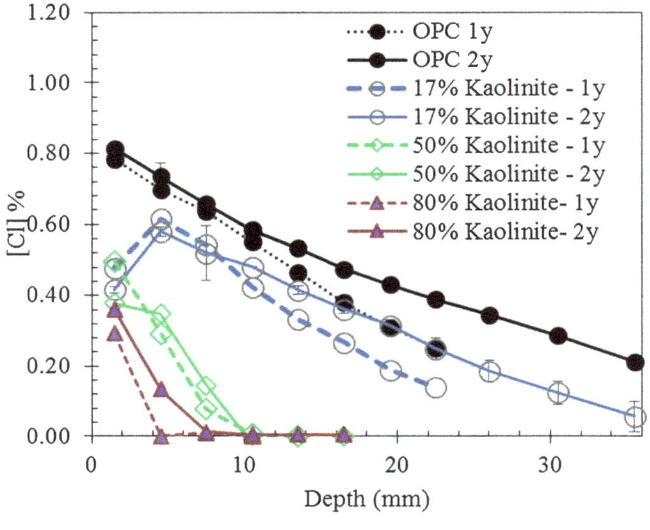

Fig. 2. Chloride profiles in OPC and LC3 mortars after exposure to NaCl solution in a ponding experiment for 1 and 2 years

depth of chloride diffused in LC3 mortars with 50% and 80% kaolinite content. This is 4–5 times less than the depth of diffused chloride in OPC mortar.

It should be noted that kaolinite clays with as low as 50% kaolinite content can be found in large quantities worldwide as the over- or under- burden from existing quarrying operations for high quality clays, being used in ceramic industry. Exploitation of these reserves represents an enormous potential to increase the global supply of SCMs [1].

References

1. Scrivener, K., John, V.M., Gartner, E.M.: Eco-efficient cements: Potential economically viable solutions for a low-CO2 cement-based materials industry. UNEP (United Nations Environment Program, December 2016
2. Antoni, M., Rossen, J., Martirena, F., Scrivener, K.: Cement substitution by a combination of metakaolin and limestone. Cem. Concr. Res. **42**, 1579–1589 (2012)
3. Avet, F., Snellings, R., Diaz, A.A., Haha, M.B., Scrivener, K.: Development of a new rapid, relevant and reliable (R3) test method to evaluate the pozzolanic reactivity of calcined kaolinitic clays. Cem. Concr. Res. **85**, 1–11 (2016)
4. Matschei, T., Lothenbach, B., Glasser, F.P.: The role of calcium carbonate in cement hydration. Cem. Concr. Res. **37**, 551–558 (2007)
5. Lothenbach, B., Le Saout, G., Gallucci, E., Scrivener, K.: Influence of limestone on the hydration of Portland cements. Cem. Concr. Res. **38**, 848–860 (2008)

Blended Cements with Calcined Illitic Clay: Workability and Hydration

Guillermina Marchetti[1], Jaroslav Pokorny[2], Alejandra Tironi[1], Mónica A. Trezza[1], Viviana F. Rahhal[1], Zbyšek Pavlík[2], Robert Černý[2], and Edgardo F. Irassar[1(✉)]

[1] Facultad de Ingeniería, CIFICEN (UNCPBA- CICPBA-CONICET), B7400JWI Olavarría, Argentina
[2] Faculty of Civil Engineering, Czech Technical University, Prague, Czech Republic

Abstract. In this paper, the paste workability and hydration progress of blended cements containing different calcined illitic clays were studied. For blended cements with different replacement percentages, the particle packing, the water film thickness (WFT) and the flow spread was modelled and measured. Results indicate that blended cement with ground illitic calcined clays maintain or reduce the packing, so the flow spread of blended cement pastes decrease when the replacement percentage increases. For blended cements with 25% of calcined illitic clay, the early hydration was described by the calorimetric curve, and later the hydration products were analysed by XRD and TG analysis and the pore size distribution (MIP) at 2, 7, 28 and 91 days. Finally, the performance of blended cements was evaluated by the compressive strength. For blended cements, the hydration products are similar to that corresponding to ordinary Portland Cement (OPC) and it also produce a pore size refinement that improve the compressive strength at later age.

1 Introduction

Low grade calcined kaolinitic clays as SCM to OPC is widespread recommendation to reduce of CO_2 emission. When this type of calcined clay is incorporated to blended cements, several simultaneous effects occur such as filler and dilution effects and the pozzolanic reactions that improve the cement based material. The main shortcoming of this SCM is the large water demand [1].

Among the natural clays, illitic clays have a high activation temperature and its pozzolanic activity is slow. However, it is the most abundant type of clay in several regions of the world and therefore their potential as SCM in cement based materials contributing to reduce of CO_2 emission. Previous studies [2, 3] show that calcined illitic clays have low water demand and its pozzolanic contribution occurs at 28 days. However, this type of calcined clays has a reduction on early compressive strength due to dilution effect on the blended cement.

The aim of this paper is to understand the contribution of calcined illitic clay on the packing, flowability and hydration process of blended cement to associate it with the strength development.

F. Martirena et al. (eds.), *Calcined Clays for Sustainable Concrete*, RILEM Bookseries 16,
https://doi.org/10.1007/978-94-024-1207-9_50

2 Materials and Procedures

An Ordinary Portland Cement (OPC) and two ground calcined illitic clays (R and O) were used. Their solid densities are 3.13, 2.63 and 2.72, respectively; and the chemical composition was reported in other paper [4]. Natural illitic clays were calcined at 950 °C and ground until the retained on 45 μm-sieve was lower than 12%. Pozzolanic activity for both calcined illitics clays was positive after 14 days according to Frattini test. The particle size distributions (PSD) was measured by laser diffraction particle diffraction particle size analyzer (Malvern Mastersize 2000) and the results are given in Fig. 1. The median particle size (d50) was 19.65, 8.76 and 12.74 μm; whereas the specific surface area measured by Blaine test method was 354, 552 and 507 kg/m^2 for OPC, R and O, respectively.

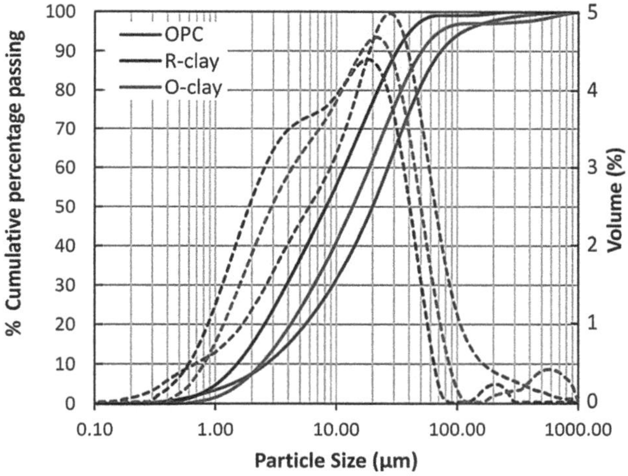

Fig. 1. Particle size distribution of materials

The ***packing density (φ)*** of cementitious materials (OPC, R and O) was measured using the wet packing method [5], the packing densities of blended cement pastes with 0 and 35% by mass was predicted by the Compressible Packing Model (CPM) [6] and the eater film thickness (WFT) was calculated at different w/cm (0.5 to 1.5 by volume). Complete details of method and prediction model were recently published [7].

Basically, φ was determined as the maximum solid concentration of the solid particles obtained when varying the water content from insufficient to more than sufficient to fill the voids. The WFT has the physical meaning of being the average thickness of water films coating the solid particles [8, 9] and it can be calculated as the ratio of water excess (excess water to solid volume ratio) to fill the voids of paste and the specific surface area of the cementitious materials in the paste. For plain (0%) and blended cement pastes (15, 25 and 35%); the flow spread was measured at a w/cm ratio of 0.5 by mass (1.5 by volume) using the mini-slump cone test [10].

The rate of ***heat released*** during hydration was measured in conduction calorimeter on plain and 25% blended cement pastes (w/cm = 0.50 by mass) under isothermal conditions at 20°C during 48 h. The ***flow*** (ASTM C 230) and the ***compressive strength*** (EN 196-1) was obtained on mortar (1:3; w/cm = 0.50 by mass) and prisms (40 × 40 × 160 mm) cured during 24 h in the molds and later immersed in water until 2, 7, 28 and 90 days.

To study the progress of hydration and the pore size distribution, blended cement pastes were prepared using a *w/cm* of 0.5 and cured in sealed plastic bags at 20 °C during 2, 7, 28 and 90 days. At this time, fragment of paste sample were carefully ground to particle size lower than 45 μm and the crystalline hydration phases were identified by ***XRD analysis*** and the mass loss with ***TG-DSC***. XRD was performed on X′Pert Philips PW 3710 diffractometer (CuKα radiation at 40 kV and 20 mA) and DSC-TG was performed on Labsys Evo (25 to 1000 °C; heating rate 5 °C min–1; inert atmosphere). ***Pore size distribution*** was measured on mercury porosimetry principle using device Pascal 140 and Pascal 440 (Thermo Scientific) measuring the 4–100 μm and the 3–4000 nm pore radium. Before the measurement, the samples were dried at vacuum drier.

3 Result and Discussion

The ***packing density*** (ϕ) for OPC, R-Clay and O-Clay was 0.636, 0.608 and 0.576, respectively. For blended cement pastes, ϕ predicted with the CPM is shown in the Fig. 2a. For R-clay, the incorporation of 15% causes some increase of ϕ (0.637) and it decreases for large replacements because the lower ϕ of SCM becoming dominant. For O-clay, the ϕ decreases for all replacements. This may be attributed to O-clay has lower packing density than that OPC and its particle size distribution is similar, so the filling effect is null when it replaces the OPC.

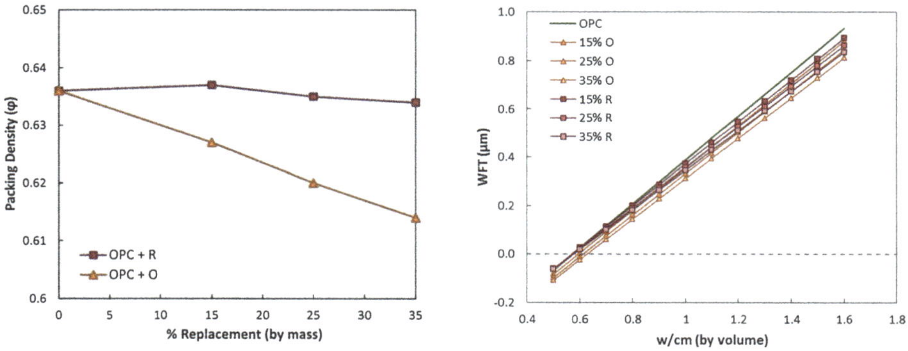

Fig. 2. Blended cement pastes (a) Packing density vs replacement and (b) WFT vs w/cm (by volume)

The addition of fine SCM affects the packing density but it also increases the total surface area of the solid particles and thus for the same amount of excess water, the average thickness of the water films coating the solid particles decreases [11].

These two opposite effects are measured by the WFT as function of the w/cm by volume (Fig. 2b). For low w/cm, WFT values are negative indicating that the amount of water in the cement paste was not sufficient to fill the interparticle voids. The WFT is a linear function of the w/cm. Comparing with OPC pastes with those containing illitic clays, it can be seen that the WFT decreases with the incorporation of both R and O-clay for all replacements. It can be explained because the addition of illitic clays decreases the ϕ and simultaneously increases the surface area of blended cements. Hence, the addition of R and O-clays to fill the voids between cement grains has a negative net effect.

Figure 3a shows the flow spread of blended cement pastes against the percentage of clay replacement. The flow spread values decrease when both R and O-clays are incorporated in blended cement. This is in agreement with the results obtained from WFT showing that the incorporation of these clays does not have a filling effect, therefore the amount of excess water for lubricating the solid particles to provide flowability decreases. Comparing with OPC mortar flow (Fig. 3b), the flow spread increases (6.6%) for 25% of replacement of R-clay and it decreases (5.6%) for O-clays. On contrary to occurred in paste, the volume of voids in the standard sand is constant and it should be filled by paste. The low relative density of calcined clays (2.63 and 2.72 for R and O) increases the volume of paste in ~ 1.9%. The mortar flow is greater for R-clay due to the less change in the packing density at 25% replacement (Fig. 2a).

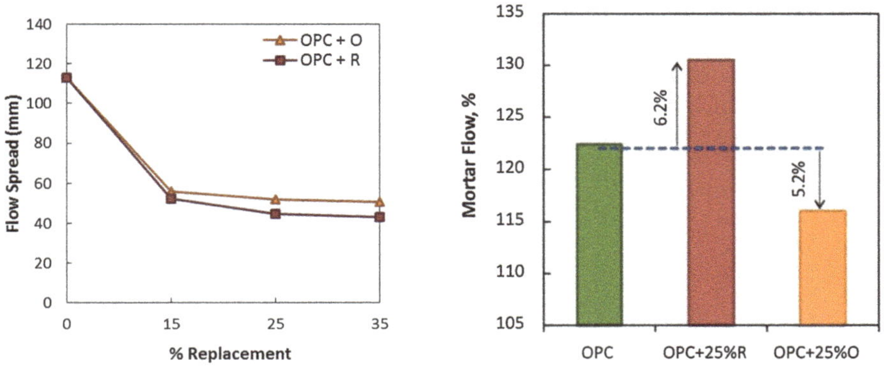

Fig. 3. (a) Flow spread of blended cement pastes (b) Flow of standard mortar (ASTM C230)

Figure 4 shows the curve of rate of heat evolution for OPC and both blended cements. It can be observed that the dormant period is shorter for blended cements. After the dormant period, the slope of the curve for O-clay is similar to OPC, while it was lower for R- clay. The intensity of second peak is lower for both blended cement at the same occurrence time (~ 14 h). The third peak is not revealed due to the low C_3A

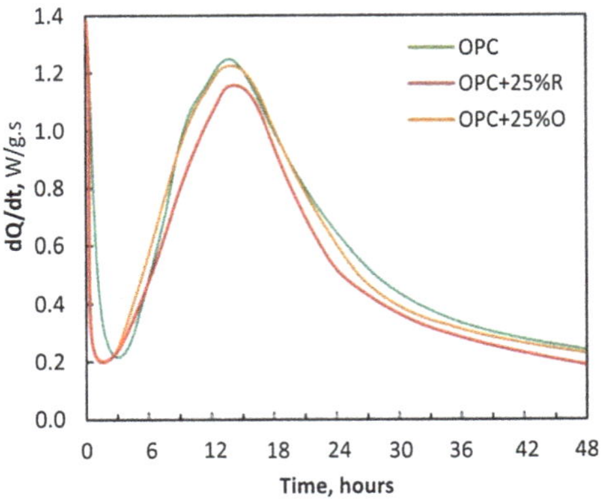

Fig. 4. Isothermal calorimetric curve for blended cement

of OPC. Finally, the descending branch during the deceleration period has the same shape for blended cement copying the OPC curve.

Figure 5 illustrates the XRD patterns for OPC and blended cements. At 2 days, the OPC reveals ettringite (E), calcium hydroxide (CH), calcite (CC) and un-hydrated cement phases (C_4AF and C_2S). For blended cements, the hydrated products detected are E and CH as occurred for OPC. Quartz (Q) is a clay-associated mineral. At 7 days, the formation of AFm (hemicarboaluminate-Hc) is detected for all cements. For blended cements with R-clay, the intensity of CH peaks decreases at 28 days, but it has high intensity for blended cement with O-clay. The AFm transformation begins and the monocarbo-aluminate formation (Mc) is insipient coexisting with Hc. Finally, the Mc is the predominant AFm phase and the CH-peak intensity is very high for O-clay indicating its limited pozzolanic reaction at 90 days. This observation is corroborated by the amount of CH determined by TG at 28 days: XX%; XX% and XX% for OPC, R—clay and O-clay; respectively.

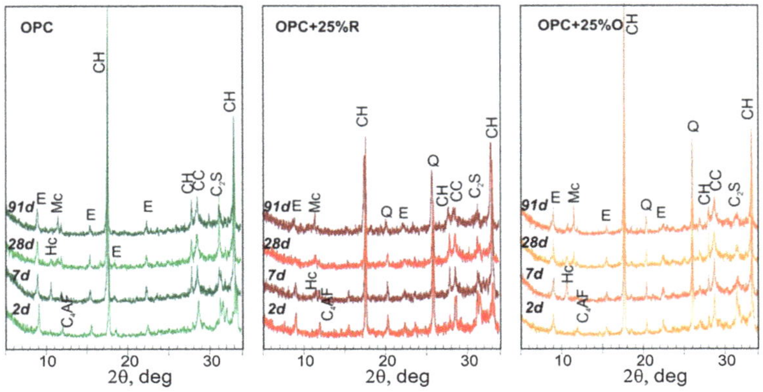

Fig. 5. XRD-pattern for pastes OPC; 25%R and 25%O

Figure 6 shows the relative volume of pore for selected ranges of pore diameter. For all cements, the volume of macropores (100–1 μm) is approximately constant; while the volume of 1–0.1 μm pore range is drastically reduced with the age. For blended cements, the dilution effect cause a large volume for this pore range at all ages. The progress of pozzolanic reaction increases the volume of finer pores (0.1–0.01 μm) attaining to large values than the corresponding to OPC after 7 days. Volume of very finer pores (0.001–0.01 μm) also increases for blended cement at 28 days.

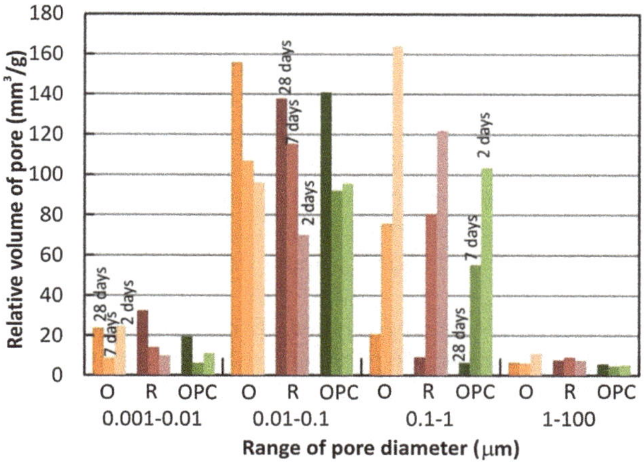

Fig. 6. Relative volume of pore by diameter range

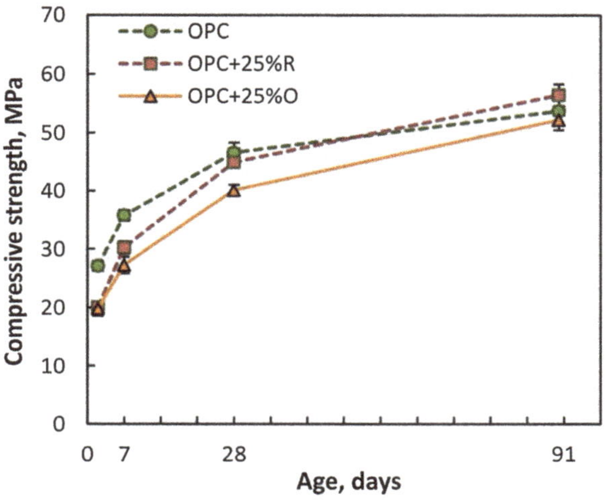

Fig. 7. Compressive strength of blended cements

The compressive strength (CS) development of standard mortar is showing in Fig. 7. Both mortars with calcined clays have ~ 25% reduction of CS at 2 days. Then, CS of R-clay increases attaining to similar strength to that of plain OPC mortar (0.96) and the final strength is greater than that control mortar (1.05). For O-clay, the CS-development occurs with low strength-gain rate (0.76 and 0.86 at 7 and 28 days, respectively), but CS reaches to similar value to OPC at 90 days (0.97).

4 Conclusions

Based on the results of this study, the following conclusions can be drawn:

- The initial flow spread of cement paste is reduced by addition of illitic calcined clays (15 to 35%) due to the loss of packing and the increase of the surface area in the blended pastes when the replacement percentage increases. This observation indicates that PSD for these additions will be improved. It does not have direct correlation with the mortar flow because the good PSD of standard sand.
- The incorporation of illitic clays reduces the heat released during cement hydration up to 48 h without significant changes in the time of occurrence of main peaks of hydration.
- Calcined clays behave as filler and dilution effect is preponderant at early ages but when the hydration progress, its pozzolanic activity consumes partially the CH causing a reduction of coarse porosity contributing to enhance the later compressive strength of mortar. The hydration product assemblage is similar to the OPC-paste and the AFm phase evolution is from Hc to Mc.

References

1. Sabir, B., Wild, S., Bai, J.: Metakaolin and calcined clays as pozzolans for concrete: a review. Cem. Concr. Comp. **23**(6), 441–454 (2001). doi:10.1016/S0958-9465(00)00092-5
2. He, C., Osbaeck, B., Makovicky, E.: Pozzolanic reactions of six principal clay minerals: Activation, reactivity assessments and technological. Cem. Concr. Res. **25**(8), 1691–1702 (1995). doi:10.1016/0008-8846(95)00165-4
3. Lemma, R., Irassar, E.F., Rahhal, V.: Calcined illitic clays as portland cement replacements. In: Calcined Clays for Sustainable Concrete. RILEM Bookseries, vol. 10. Springer (2015). doi:10.1007/978-94-017-9939-3_3
4. Cordoba, G., Rossetti, A., Falcone, D., Irassar, E.F.: Sulfate and alkali-silica performance of blended cements containing illitic calcined clays. In: Second Calcined Clay for Sustainable Concrete Conference (2017)
5. Wong, H.H.C., Kwan, A.K.H.: Packing density of cementitious materials: part 1 measurement using a wet packing method. Mater. Struct. **41**(4), 689–701 (2008). doi:10.1617/s11527-007-9274-5
6. De Larrard, F.: Concrete Mixture Proportioning: A Scientific Approach. E&FN Spon, London (1999)

7. Marchetti, M.G., Rahhal, V.F., Irassar, E.F.: Influence of packing density and water film thickness on early-age properties of cement pastes with limestone filler and metakaolin. Mater. Struct. **50**(2), 1–11 (2016). doi:10.1617/s11527-016-0979-1
8. Fung, W.W.S., Kwan, A.K.H.: Role of water film thickness in rheology of CSF mortar. Cem. Concr. Comp. **32**(4), 255–264 (2010). doi:10.1016/j.cemconcomp.2010.01.005
9. Li, L.G., Kwan, A.K.H.: Mortar design based on water film thickness. Constr. Build. Mater. **25**(5), 2381–2390 (2008). doi:10.1016/j.conbuildmat.2010.11.038
10. Kantro, D.L.: Influence of water-reducing admixtures on properties of cement paste: a miniature slump test. Cem. Concr. Aggr. **2**(2), 95–102 (1980). doi:10.1520/CCA10190J
11. Kwan, A.K.H., Chen, J.J.: Roles of packing density and water film thickness in rheology and strength of cement paste. J. Adv. Concr. Tech. **10**, 332–344 (2012). doi:10.3151/jact.10.332

Low Carbon Cement LC^3 in Cuba: Ways to Achieve a Sustainable Growth of Cement Production in Emerging Economies

Fernando Martirena[1(✉)] and Karen Scrivener[2]

[1] CIDEM, Central University "Marta Abreu" of Las Villas (UCLV), Carretera a Camajuanì km 5 1/2, Santa Clara, Villa Clara, Cuba
fmartirena@ecosur.org

[2] LMC, Ecole Polytechnique Federal de Lausanne, Lausanne, Switzerland

Abstract. Through a collaborative work established between CIDEM and the Laboratory of Construction Materials at EPFL, a new cementitious system based on a synergetic combination of calcined clay and limestone as Portland's clinker replacement was developed. The novelty of the system was that despite the low clinker content (50%) the resulting cement reached similar performance to pure Portland cement. The Cuban cement industry has collaborated with the introduction and take up of LC^3 since early 2011, and it has gathered experiences, which could be useful for other cement makers around the globe. This paper discusses the roadmap followed by the technical team to introduce LC^3as a mainstream product. The main areas of engagement were: (i) identification of reserves of suitable clay for the production of LC^3, (ii) Realization of industrial trials for the manufacture of LC^3, (iii) assessment of economic and environmental feasibility of the production, and (iv) formulation of new standards which cover the formulation of the new cement as well as its application in concrete. Results of the work of the multi-disciplinary team dealing with the introduction of LC^3 shall be presented, including the formulation of the material with local raw materials, considerations about the process, and performance and durability of concrete.

1 Introduction

Cuba has entered a process of economic reforms since 2009, and a boost on overall demand of cement for the forthcoming years is foreseen. Cuban cement consumption has historically been following the same trend of production output, since demand exceeds the supply by far. Forecasted demand based on the cement group's estimations would be in the order of 18, 15 and 10% growth rate by the subperiods 2016–2020, 2020–2025, 2026–2030, respectively [1].

To meet the demand, a short-term strategy should be devised. Increasing the availability of clinker is a long term and expensive path. An alternative could be increasing the amount of blended cements based on a modest increase of clinker production. Cuba's average clinker factor is in the range of 0.7–0.95, thus there is a huge potential for improvement [2].

F. Martirena et al. (eds.), *Calcined Clays for Sustainable Concrete*, RILEM Bookseries 16,
https://doi.org/10.1007/978-94-024-1207-9_51

Through a collaborative work established between CIDEM and the Laboratory of Construction Materials at EPFL, a new cementitious system based on a synergetic combination of calcined clay and limestone as Portland's clinker replacement was developed. The novelty of the system was that despite the low clinker content (50%) the resulting cement reached similar performance to pure Portland cement. [3] This paper discusses the roadmap followed by the technical team to introduce LC^3 as a mainstream product.

2 Identification of Suitable Clay Deposits

Selection of clay deposits was made based on its geological characteristics and its potential reactivity. The main criteria were preliminary identification of clay mineral type 1:1, and sufficient availability of resources. Five clay deposits were surveyed and representative samples from the clay deposits were studied for reactivity at lab conditions. Table 1 presents the main tests carried out to assess reactivity, as well as the kaolinite content of the clay and the potential reserves. With the exception of clay deposit "Cayo Guam" the rest are linked to the cement plant Siguaney [4].

Table 1. Clay deposits studied and experimental program carried out

Clay deposits	K %	Reserves (M ton)	XRF/XRD	TG/DTA	Strength	Soluble Al
Pontezuela	40–50	0.5–1	x	x	x	x
La Loma	40–50	1–2	x	x	x	x
Loma Sur	45–50	0.5–1	x	x	x	x
Yaguajay	40–45	4–6	x	x	x	x
Cayo Guam	60–70	>5	x	x	x	x

Pontezuela was used for a preliminary and exploratory industrial trial to assess the properties of the ternary cement LC^3 [2], and Yaguajay was selected as the clay deposit to use for commercial production, mainly because of the proximity to the cement plant and the volume of reserves.

3 Realization of Industrial Trials

The Cuban cement industry decided to make an industrial trial at cement factory Siguaney, which was carried out in 2013. [4] The target ternary cement should have clinker content around 50%. The trial included the calcination of 110 tonnes of clay from the clay deposit Pontezuela; mixing and homogenizing of the calcined material with limestone in a 2:1 ratio; and co-grinding of the synergetic materials with clinker and gypsum by using a ball mill with a double chamber grinding system.

The total production of LC^3 cement was 130 tonnnes. The material was characterized following the protocol established for blended cements in Cuban standards. Results are presented in Table 2. Different kinds of concrete and mortar applications were made successfully with the 1:1 substitution of Portland cement [2].

Table 2. Results of physical and mechanical test of the industrial low carbon cement

Material	Retained 4900 Sieve (%)	Consistency (%)	Setting time		Volume Stability (mm)	Compressive strength (Mpa)		
			Initial (min)	Final (hr.)		3d	7d	28d
LC3	12.0	25.0	135	2.9	0.3	11.0	17.5	30.3

Trial productions of several types of concrete were made using the LC3 produced at the industrial trial. Among several applications, 25 MPa precast concrete elements was cast. Table 3 presents the results of this concrete, where Portland cement has been replaced on a 1:1 basis by LC3.

Table 3. Results of compressive strength of concrete made for prefabricated elements with OPC and LCC

Material	Cement consumption (kg/m^3)	Average compressive strength at 3d (MPa)	Average compressive strength at 7d (MPa)	Average compressive strength at 28d (MPa)
LCC	360	–	21.0	31.4
OPC	360	20.4	–	33.2

4 Economic and Environmental Analysis

The production of LC^3 in Cuba was studied aided with the Life Cycle Assessment tool (LCA). The analysis was carried out for various technological levels of investment.

1. The first level (Pilot) considers no investment. The data used for energy and material consumption were measured during the first industrial trial [2].
2. The second technological level considers a low investment throughout retrofitting a wet cement kiln into a clay calciner, as a low CAPEX alternative to produce calcined clay.
3. The third technological level considers massive investment in Best Available Technology (BAT), with state of the art equipment for both the calcined clay and the clinker production.

The study shows that the LC^3 technology is an energy and cost efficient technology. Savings in term of greenhouse gases emissions as well as production and investment costs are significant. LC3 has therefore a great potential to provide a viable opportunity to meet an increase in cement demand with low CO2 released and low cost investment.

5 Standards

European Standards do only allow up to 65% of clinker for blended cements, and also the common practice in the industry is to limit the amount of limestone to 5%. [5] The ASTM standard C595/C595M–14, “Standard Specification for Blended Hydraulic

Cements", launched in 2016, provides a solution to accept lower clinker contents and higher amount of limestone. This standard describes a ternary blended cement consisting of an intimate and uniform blend produced either by intergrinding, by blending or by a combination of intergrinding and blending Portland cement or clinker with several SCMs. This cement is described as type IT. The type IT cement shall have a maximum limestone content of 15% by mass and is permitted to contain hydrated lime. All other ternary blends shall have a maximum pozzolan content of 40% by mass of the blended cement, and the total content of pozzolan, limestone and slag shall be less than 70% by mass of the blended cement. Further, the loss on ignition is limited to 10%, mainly due to the inclusion of a higher limestone content.

6 Conclusions

The interdisciplinary endeavour carried out by the technical team Switzerland-Cuba has proven that the innovative cementitious system can be a real alternative for developing countries with a rise on cement demand. The team has carefully followed all steps needed for the introduction of the technology, and commercial production is scheduled to begin in 2018, so the entire process took approximately 10 years for completion.

Acknowledgments. The authors would like to acknowledge Siguaney cement factory for the technical and material support. The authors would like to thanks to the whole team of the "Low Carbon Cement" Project for all the advisory and technical help during this paper creation.

References

1. Sánchez Berriel, S., Favier, A., Rosa Domínguez, E., Sánchez Machado, I.R., Heierli, U., Scrivener, K., Martirena Hernández, F., Habert, G.: Assessing the environmental and economic potential of limestone calcined clay cement in Cuba. J. Clean. Prod. **124**, 361–369 (2016). doi: 10.1016/j.jclepro.2016.02.125
2. Vizcaíno-Andrés, L., Sánchez-Berriel, S., Damas-Carrera, S., Pérez-Hernández, A., Scrivener, K., Fernando, M.-H.: Industrial trial to produce a low clinker, low carbon cement. Mater. Construcción **65** (2015). doi:10.3989/mc.2015.00614
3. Antoni, M., Rossen, J., Martirena, F., Scrivener, K.: Cement substitution by a combination of metakaolin and limestone. Cem. Concr. Res. **42**, 1579–1589 (2012). doi:10.1016/j.cemconres. 2012.09.006
4. Almenares, R.S., Vizcaíno, L.M., Damas, S., Mathieu, A., Alujas, A., Martirena, F.: Industrial calcination of kaolinitic clays to make reactive pozzolans. Case Stud. Constr. Mater. **6**, 225–232 (2017). doi:10.1016/j.cscm.2017.03.005
5. Hooton, R.D.: Cement and concrete research current developments and future needs in standards for cementitious materials. Cem. Concr. Res. **78**, 165–177 (2015). doi:10.1016/j.cemconres.2015.05.022

Studies on the Influence of Limestone-Calcined Clay Blend on the Hydration of Cement

G. Mishra(✉), A. Emmanuel, and S. Bishnoi

Indian Institute of Technology Delhi, New Delhi, India

Abstract. The conventional use of binary cement is changing towards a composite cementitious system which allows two or more supplementary materials in combination. In this way, present study focuses on the preparation of limestone-calcined clay blend and introduction into cement system at variable replacement level to study their influence on the properties of resulting concrete. The hydration behaviour of the ternary blended cement was carried out using heat of hydration and XRD analysis. Similarly, compressive strength was performed on mortar cubes at different hydration ages (1, 3, 7 and 28 days). From this study, it was observed that, the limestone in the presence of calcined clay results in the formation of additional hydrates. Furthermore, it is found that this ternary combination can significantly improve the substitution level of clinker with good mechanical characteristics.

Keywords: Limestone · Calcined clay · Strength · Hydration

1 Introduction

Cement is widely used material around the world. In developing countries, increasing rate of growth of population, demand for new infrastructures. Due to this situation, the cement industry needs to find the solution between increasing the cement production and without increasing the negative environmental impact. In the year 2014, India produced around 280 million tonnes of cement, making it the 2nd largest cement producer in the world, this figure is expected to double every 8 to 12 years during coming decades [1]. Cement production already contributes to approximately 7% of the CO_2 emissions [2]. In this regard, a promising solution has been found with the use of supplementary cementitious materials (SCMs), these materials are used to replace clinker in cement because of their pozzolanic reactivity. Since late 1990s, fly ash has played an important role in cement production with around 70% of the cement being produced in India containing 25% to 27% of fly ash [3]. While the Indian standards allow up to 35% fly ash content in Portland Pozzolana Cements (PPC) [4]. The quality and local availability of fly ash limits their higher replacement levels. Other than these other SCMs such as ground granulated blast furnace slag (GGBS), silica fume (SF) and metakaolin (MK) are being used in construction practices. From the reported literature, it can be seen that these have a positive influence on strength, chloride resistance and other durability properties, instead there are some other factors such as cost, availability, high water demand, lower/higher replacement level, limit their use in construction sectors.

F. Martirena et al. (eds.), *Calcined Clays for Sustainable Concrete*, RILEM Bookseries 16,
https://doi.org/10.1007/978-94-024-1207-9_52

In this regards Limestone-calcined clay pozzoloan offers an economically feasible alternative to overcome the adversity with other SCMs and to reduce the environmental impact of construction as well [5]. The advantage of limestone-calcined clay (LC^2) pozzolan with respect to the other pozzolanic materials is that it can be produced using materials that are abundantly available at production costs lower than cements and help in a substantial reduction in CO_2 emissions [6]. In the present paper, this Limestone-calcined clay is prepared and introduced to cement matrix to observe their influence on hydration kinetics and strength development.

2 Experimental Programme

2.1 Materials and Methods

In the present study, 43 grade OPC was used confirming from IS 8112:1996. To prepare LC^2, kaolinite clay was calcined at 800 °C to obtained calcined clay and mixed with low grade limestone in the proportion of 2:1 (Calcined clay: limestone). Further, this LC^2 was introduced in cement system at different replacements such as 10%, 20% and 30%. Mortar cubes were cast according to consistency of the blends and kept at 27° ± 2 °C for 24 h. Afterwards, the cubes were cured under water till the day of testing. The compressive strength was measured at the different age such as 1, 3, 7 and 28 days as per IS 4031.

For heat of hydration measurement, Isothermal calorimeter was used and paste samples were prepared at fixed water to cement ratio i.e. 0.45. Heat release was observed till 72 h from the moment of addition of water.

For XRD analysis, cement paste with LC^2 at different replacement 10% to 30% were prepared in a plastic tube and sliced at different hydration age. Further these slices were used to obtain XRD curves.

3 Results and Discussion

3.1 Heat of Hydration

From the heat of hydration curve shown in Fig. 1, it is observed that the rate of heat evolved with the replacement of clinker by LC2 is higher and the appearance of first peak (corresponding to silicate phases) is slightly shifted to left side of the curve indicating acceleration in hydration mechanism. Further, from the cumulative heat curve it is significantly observed that the total heat evolved is higher than plain cement paste.

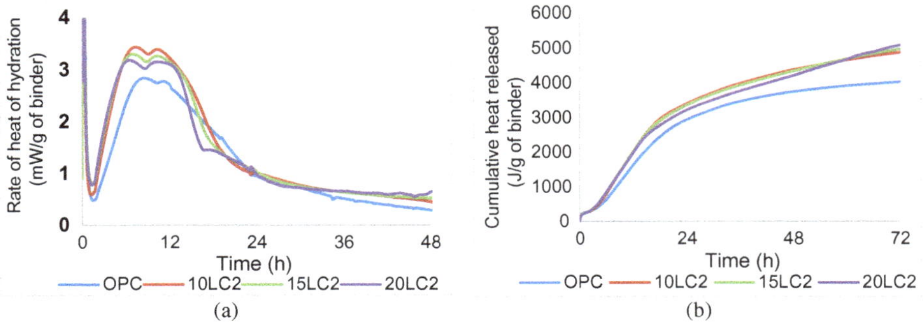

Fig. 1. (a) Rate of heat of hydration curve, (b) Cumulative heat evolution

3.2 Compressive Strength

The strength test results of the LC^2 blends specimens for different replacement ratios are shown in Fig. 2 at 1, 3, 7 and 28 days. From the results, it is observed that the early age strength (at 1 and 3 days) of LC2 blends is lower than normal cement mortar. This may be due to the variable water content in each replacement. In LC^2 blends, the requirement of water is higher than normal cement mortar, therefore the casting was done as per their consistency, which may lead to decrease in strength at early age. A hydration proceeds till 28 days the strength is significantly increased and observed higher than normal cement mortar. This increment in strength can be inferred due to the pozzolanic reaction, wherein calcium hydroxide is consumed by reactive silica from Calcined clay and formation of carbo-aluminate phases.

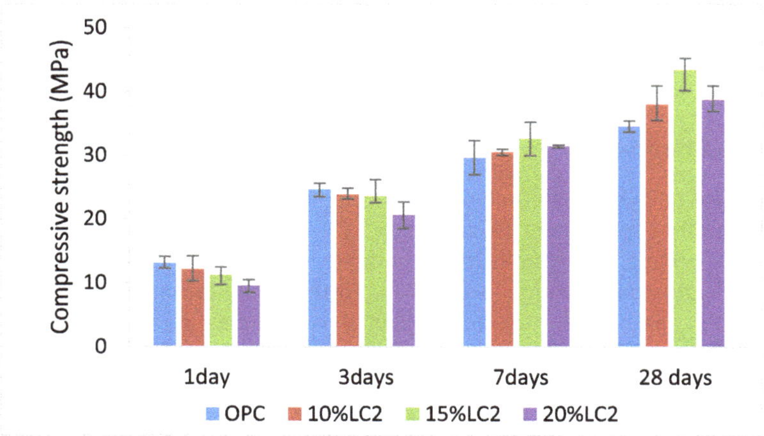

Fig. 2. Compressive strength at different replacement of LC^2

3.3 XRD Analysis

XRD profile of LC^2 blends at 1 and 28 days are shown in Fig. 3. At 28 days, the peak at ~10.7 (2θ) and ~11.6 (2θ) corresponds to hemi and mono carboaluminates, respectively, which appear in LC^2 blends, whereas in normal cement these peaks are not present. At the same time intensity of calcium hydroxide (CH) peak at ~18 (2θ) is reducing gradually indicating the consumption of CH during pozzolanic reaction, which may be responsible for increase in strength at later ages.

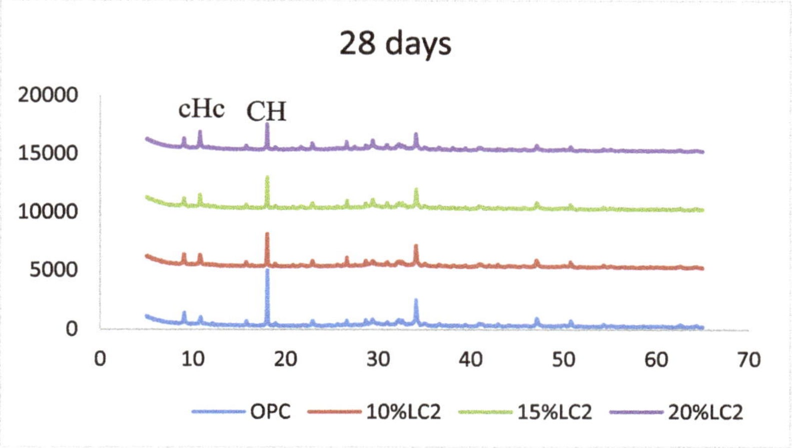

Fig. 3. XRD profile of LC2 blends at 28 days of hydration

4 Conclusion

From the experimental results, it can be concluded that the limestone-calcined clay blend can be used as replacement of cement to improve hydration kinetic and strength of the cement system.

Acknowledgements. The authors would like to acknowledge Swiss Agency for Development and Cooperation (SDC) for providing the financial support for the Limestone Calcined Clay Cement project in India.

References

1. Planning Commission: Interim report of the expert group on low carbon strategies for inclusive growth. Government of India (2011)
2. Cement Sustainability Initiative, Technology Roadmap: Low-carbon technology for the Indian cement industry. World Business Council for Sustainable Development and International Energy Agency (2013)
3. Cement Manufacturers Association: Basic data. Cement Manufacturers Association (2012)

4. Bureau of Indian Standards: IS1489: 1991 – Portland Pozzolana Cement – Specification, Part I: Fly Ash Based. Government of India (1991)
5. Bishnoi, S., Maity, S., Mallik, A., Joseph, S., Krishnan, S.: Pilot scale manufacture of limestone calcined clay cement: the Indian experience. Indian Concr. J. **88**(6), 22–28 (2014)
6. Krishnan, S., Emmanuel, A.C., Bishnoi, S.: Feasibility of ternary cements as potential green cements in India. In: 14th NCB International Seminar on Cement and Building Materials, New Delhi, India (2015)

Thermal Resistivity of Chemically Activated Calcined Clays-Based Cements

Marangu J. Mwiti[1(✉)], Thiong'o J. Karanja[1], and Wachira J. Muthengia[2]

[1] Department of Chemistry, Kenyatta University, Nairobi, Kenya
jmarangu2011@gmail.com
[2] Department of Physical Sciences, University of Embu, Embu, Kenya

Abstract. The study investigated the effects of selected potential chemical activators on thermal resistivity of calcined clay based cement mortars. 0.5 M Na_2SO_4 and 0.5 M NaOH were used as activator solutions. The chemical composition of sampled clays was determined by use of X-Ray Florescence (XRF) technique. Clays were incinerated at a temperature of 800 °C for 4 h. The calcined clays obtained were blended with OPC at replacement level of 35 percent by mass of the OPC to make the test cement labeled PCC35. The PCC35 mortar prisms measuring 40 mmx40mmx160mm were cast with activator solutions and cured in water. Compressive strength was determined at the 28th day of curing. As a control, OPC and PCC35 were similarly investigated without activator solutions. The 28 day cured mortars were exposed to a temperature of 700 °C for 2 h then cooled in water to room temperature and their compressive strengths determined. Chemically activated PCC35 and non-activated PCC35 exhibited lower loss in weight than OPC after exposure to the elevated temperatures. Chemically activated PCC35 and non-activated PCC35 exhibited higher residual compressive strength than OPC after exposure to the said temperatures. Na_2SO_4 activated mortars showed higher thermal resistance than NaOH activated mortars. Generally, chemically activated PCC35 exhibited the highest thermal resistance compared to non-activated PCC35 and commercial OPC mortars.

Keywords: Activators · Blended cements · Calcined clay · Cement · Compressive strength · Thermal resistance

1 Introduction

The durability of cement based structures is of great concern in construction industry globally. The structural integrity of cement based materials is guaranteed at ambient temperatures [1]. Hazardous fires are reported to destroy many cement based structures in the world annually resulting to enormous financial losses [2]. Emergence of domestic fires and electrical faults majorly contribute to hazardous fires that affect these structures. Hazardous fires expose these materials to elevated temperatures that are deleterious. High temperatures affect both physical and chemical aspects of the concrete / mortar matrix due to thermal effects on pore water and hydration products [3]. The service life of cement based structures thus reduced by hazardous fires.

F. Martirena et al. (eds.), *Calcined Clays for Sustainable Concrete*, RILEM Bookseries 16,
https://doi.org/10.1007/978-94-024-1207-9_53

Production of Portland cement results in significant environmental damage due to raw material acquisition and production of carbon dioxide [4]. Carbon dioxide is the main greenhouse gas mainly responsible for global warming and climate change [4]. There is an increasing demand for eco-friendly cement in the world. Production of Portland cement is also an energy intensive process mainly due to the fact that temperatures in excess of 1300 °C are requires during clinkerisation [5]. This is mainly fuelled by petroleum or coal. The high energy demand during production of Portland cement makes it unaffordable especially in developing countries Kenya included [6].

Pozzolanic materials have been used to partly replace Portland cement to produce blended cements since antiquity [7]. Blended cements are commonly used in construction due to their high ultimate compressive strength and high durability of the cement structure as a result of the pozzolanic reactions increasing the amount of calcium silicate hydrates (CSH) while diminishing $Ca(OH)_2$ [8]. Hazardous fires and exposure to high temperature environments such in hydrothermal wells compromise both strength and durability of cement based structures [9]. Pouring of water on the structures on fire is the most common practice of cooling the cement based structures in case hazardous fires occur [10]. There is need therefore to develop high temperature resistant cementious materials.

The pioneering research conducted by [6] reported that Kenyan selected clay-Portland cement blends met the EAS 148-1(2000) when 35 percent of calcined clay was used as pozzolana in blended cement. The test cement used in this study was essentially blended cement prepared by mixing of Portland cement with 35 percent of calcined clay. This paper presents the research findings of the said cement on its resistance to high temperatures when mixed with Na_2SO_4 and NaOH as potential chemical activators.

2 Materials and Methods

2.1 Materials

Ordinary Portland cement (OPC 42.5 N) and standard sand used in this work was supplied by East Africa Portland Cement Company – Athi River, Kenya. Raw clays were sampled from three different places within Ugweri region (longitude 37 ^{0} 34′ 19.47″ E, latitude 0^{0} 25′ 20.44″ S) in Embu County- Kenya. In each place, three clay samples were obtained from a depth of 3 feet. Clays sampled from a given place were mixed mechanically to obtain a homogenous mixture and stored in labelled polythene bags.

2.2 Methods

2.2.1 Calcination of Clay

The sampled clay was dried to a constant weight at 105 °C to a constant weight in an oven. 6 kg of the dried clay, in a platinum clay trough, was placed in an electrical muffle furnace and allowed to heat for four hours at 800 °C. The resultant clay was cooled grounded in a laboratory ball mill until 95 percent of the calcined clay particles passed below 45 μm BS sieves.

2.2.2 Preparation of PCC35

350 g of calcined clay and 650 g of OPC were mechanically mixed in an automatic mixer for one hour to make 1 kg of PCC35 cement.

2.2.3 Chemical Analysis

X-ray Fluorescence (XRF) technique was used for the chemical analysis of the sampled clay. The samples were analyzed using a sequential X-ray spectrophotometer model number PW4025 in accordance with the instruction manual. The results were presented as a percentage of the oxides.

Loss on ignition of the cement sample was determined in accordance with EAS 148-2:2015. 1.0 ± 0.05 g of clay sample was placed in a crucible and covered with a lid. The covered crucible was transferred in the electric furnace controlled at 950 ± 25 °C for 1 h. The crucible was allowed to cool at room temperature in the desiccator. Loss on ignition was expressed in percentage as the difference between the original mass of the sample and the final mass.

2.2.4 Mixing Casting and Curing

Mortar preparation and curing was done in accordance to KS EAS 148–1:2000. However slight modifications were adopted. First, a w/c ratio of 0.55 was used to ensure a workable mix was achieved. Secondly, 0.5 M Na_2SO_4 and 0.5 M NaOH were also used in place of water.

2.2.5 Thermal Resistivity

The 28 day cured mortar prisms measuring 40 mmx40mmx160mm in size were weighed at the saturated surface in dry condition (W_1). Their compressive strength was determined and recorded as C_{S1}. The specimens were introduced in an electric furnace and heated at 700 °C for two hours with furnace temperature increment of 10 °C per min. After two hours of heating, the mortars were cooled to room temperature for two hours using cold water. Subsequently, the dry weight of the specimens was taken as W_2 and the residual compressive strength of the specimens was determined (C_{S2}). Finally, loss in weight was calculated as $(W_1 - W_2)/ W_1$ and the strength loss of the specimens due to the thermal effect was determined as $(C_{S1} - C_{S2})/C_{S1}$; where, C_{S1} is the compressive strength of the mortar before heating and C_{S2} compressive strength of the mortar after heating.

3 Results and Discussion

3.1 Chemical Analysis

The chemical analysis of clays is given in Table 1.

Table 1. Chemical composition of the sampled clays

Oxide	SiO_2	Al_2O_3	Fe_2O_3	CaO	MgO	K_2O	Na_2O	Others	L.O.I
% Composition	55.26	15.35	11.43	0.54	2.13	4.7	3.19	4.4	3.0

EAS 148-5-2000; prescribes that the sum of SiO_2, Al_2O_3 and Fe_2O_3 should be above 70 per cent for the material to qualify for use as pozzolana. The sum of SiO_2, Al_2O_3 and Fe_2O_3 in the clay sample was 82.04 per cent. Alumina and silica react with calcium hydroxide produced during hydration of Portland cement to form secondary cementious materials. This reaction is referred to as pozzolana reaction. It is given in Eqs. 1 and 2.

$$2SiO_2 + 3Ca(OH)_2 + 5H_2O \rightarrow 3CaO.2SiO_2.8H_2O \quad (1)$$

$$Al_2O_3 + 4Ca(OH)_2 + 9H_2O \rightarrow 4CaO.Al_2O_3.13H_2O \quad (2)$$

3.2 Compressive Strength at 28th Day of Curing

The compressive test results are given in Fig. 1.

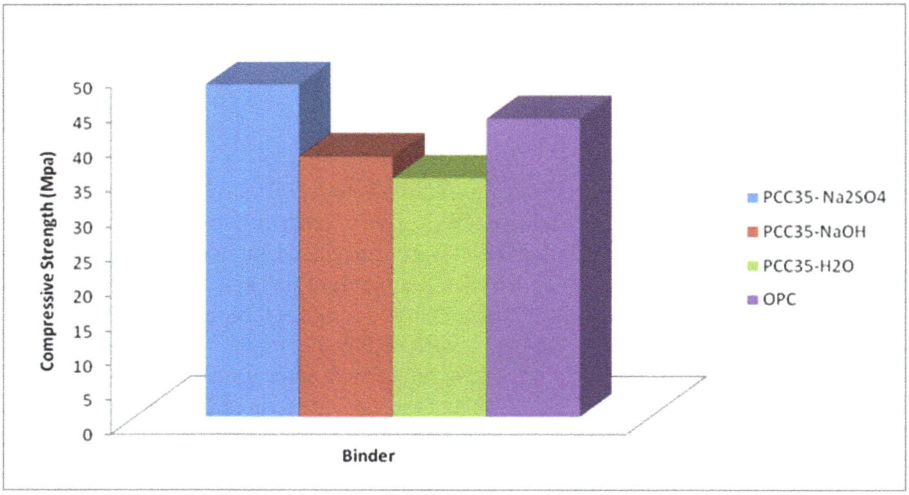

Fig. 1. Compressive strength of test cements on verses binder type on the twenty eighth of curing

It was observed that mortars with chemical activators (PCC35-Na_2SO_4 and PCC35-NaOH) exhibited higher compressive strength than the non-activated mortars (PCC35-H_20). This could be attributed to the fact that both Na_2SO_4 and NaOH accelerate the pozzolana reaction in blended cement leading to high compressive strength [11–13]. The lower compressive strength exhibited by non-activated PCC35 is possibly due to presence of large amount of unactivated pozzolana in PCC35 cement matrix due to the absence of chemical activators.

PCC35-Na_2SO_4 exhibited higher compressive strength than PCC35-NaOH. This can be attributed to the different modes of activation namely alkali and sulfate activation. Sulphate activation is based on the ability of sulfates to react with aluminium oxide in the glass phase of pozzolana to form ettrigite [14, 15]. Presence of ettrigite has been found to contribute to strength at early ages [14] More ettringite formation results in a significant solid volume increase hence forming a less porous structure and subsequently leads to higher early strength.

During alkali activation, the pH of pore water in cement matrix is greatly raised [16]. High pH has been reported to increase the dissolution of pozzolana materials hence improving pozzolanic reaction [13, 17–20]. This subsequently increases the early compressive strength of cement mortars. Shi and Day [14] examined the pozzolanic reaction mechanisms of fly ash in the presence of $CaCl_2$ and Na_2SO_4. The authors Shi and Day [14] reported that addition of Na_2SO_4 to fly ash cement increases the alkalinity of the solution and the dissolution of fly ash during initial stages, and accelerates the pozzolanic reaction. High 28-day compressive strength is normally achieved with improved pozzolana reaction.

3.3 Thermal Resistivity

Thermal resistivity was expressed as the change in weight/compressive strength of the mortars after exposure to 700 °C. Thermal resistivity of the mortars is presented by Fig. 2.

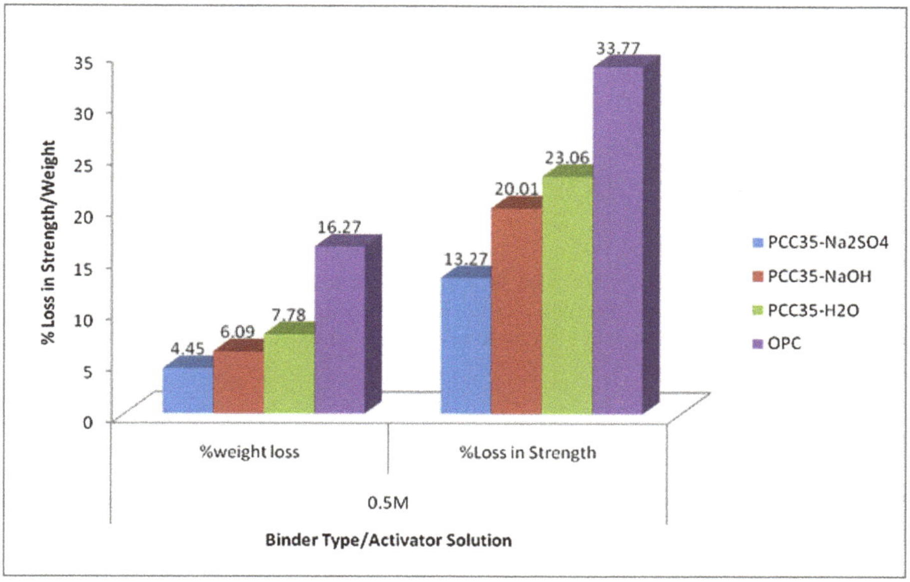

Fig. 2. Binder type/activator concentration verses percentage loss in compressive strength/ weight

Loss in weight and compressive strength was observed in all the cement categories after exposure to the elevated temperatures. This could be attributed to the decomposition of hydration products [10]. High temperatures have a significant influence on the thermal deformation, cracking, spalling and compressive strength losses [1]. High temperatures lead to the decomposition of CSH [3]. Decomposition of CSH due to elevated temperatures compromise the strength of cement based materials [9, 21].This could have led to the decrease in compressive strength of the mortars.

Hydrated mortars contain pore water that is, free, adsorbed or chemically combined in the mortar [21]. At elevated temperatures, the free and/or adsorbed water as well as combined water from CSH and CASH is removed [9, 21, 22]. This normally occurs at temperature range of 100–300 °C and leads to the collapse of the CSH gel that is mainly responsible for the strength of cement based materials the decreases in weight and compressive strength of the mortars [21]. Based on our study, a temperature of 700 °C was therefore sufficient to cause dehdroxylation leading to weight and strength loss in the specimens. $Ca(OH)_2$ in cement mortars has also been reported to undergo decomposition forming CaO and CO_2 at temperatures between 400–550 °C [22]. This is could have also led to the loss in weight of the mortars at high temperature observed in this study.

PCC35-H_2O exhibited higher loss in weight and compressive strength compared to and PCC35-Na_2SO_4. This could be attributed to the fact that PCC35-Na_2SO_4 exhibited higher compressive strength than PCC35-H_2O after 28 days of curing. High compressive strength offers buffering capacity to cement mortars when exposed to elevated temperatures [23]. Chemically activated mortars have been found to exhibit high resistance to high temperatures in other related studies [23].

4 Conclusion

Chemically activated cement mortars (PCC35-Na_2SO_4 and PCC35-NaOH) exhibit higher thermal resistance compared to non activated cement mortars (PCC35-H_2O and OPC) mortars.

Acknowledgements. The authors wish to acknowledge the African Development Bank in Collaboration with Kenyatta University for funding this research. We also wish to thank East Africa Portland Cement Company – Athi River, Kenya for allowing us to use their Cement Laboratory for this research.

References

1. Amin, M.S., El-Gamal, S.M.A., Hashem, F.S.: Fire resistance and mechanical properties of carbon nanotubes clay bricks wastes (Homra) composites cement. Constr. Build. Mater. **98**, 237–249 (2015)
2. Martin, A., Pastor, J.Y., Palomo, A., Fernández Jiménez, A.: Mechanical behaviour at high temperature of alkali-activated aluminosilicates (geopolymers). Constr. Build. Mater. **93**, 1188–1196 (2015)

3. Mousa, M.I.: Effect of elevated temperature on the properties of silica fume and recycled rubber-filled high strength concretes (RHSC). HBRC J. **13**(1), 1–7 (2017)
4. Torgal, F.P. (ed.): Eco-Efficient Concrete. Woodhead Publishing, Cambridge (2013)
5. Aïtcin, P.-C.: Cements of yesterday and today: concrete of tomorrow. Cem. Concr. Res. **30**(9), 1349–1359 (2000)
6. Marangu, J.M., Muthengia, J., Wa-Thiong'o, J.: Performance of potential pozzolanic cement in chloride media (2014)
7. Shi, C., Jiménez, A.F., Palomo, A.: New cements for the 21st century: the pursuit of an alternative to Portland cement. Cem. Concr. Res. **41**(7), 750–763 (2011)
8. Meddah, M.S.: Durability performance and engineering properties of shale and volcanic ashes concretes. Constr. Build. Mater. **79**, 73–82 (2015)
9. Sugama, T., Pyatina, T., Gill, S.: Thermal shock-resistant cement. Geotherm. Resour. Counc. Trans. **36**, 445–451 (2012)
10. Cree, D., Green, M., Noumowé, A.: Residual strength of concrete containing recycled materials after exposure to fire: a review. Constr. Build. Mater. **45**, 208–223 (2013)
11. Shi, C., Day, R.L.: Acceleration of strength gain of lime-pozzolan cements by thermal activation. Cem. Concr. Res. **23**(4), 824–832 (1993)
12. Shi, C., Day, R.L.: Acceleration of the reactivity of fly ash by chemical activation. Cem. Concr. Res. **25**(1), 15–21 (1995)
13. Rakhimova, N.R., Rakhimov, R.Z.: Alkali-activated cements and mortars based on blast furnace slag and red clay brick waste. Mater. Des. **85**, 324–331 (2015)
14. Shi, C., Day, R.L.: Comparison of different methods for enhancing reactivity of pozzolans. Cem. Concr. Res. **31**(5), 813–818 (2001)
15. Sassoni, E., Mazzotti, C., Pagliai, G.: Comparison between experimental methods for evaluating the compressive strength of existing masonry buildings. Constr. Build. Mater. **68**, 206–219 (2014)
16. Palomo, A., Grutzeck, M.W., Blanco, M.T.: Alkali-activated fly ashes: a cement for the future. Cem. Concr. Res. **29**(8), 1323–1329 (1999)
17. Fernández-Jiménez, A., Puertas, F.: Alkali-activated slag cements: kinetic studies. Cem. Concr. Res. **27**(3), 359–368 (1997)
18. A review on alkaline activation new analytical perspectives.doc
19. Shi, C., Day, R.L.: Chemical activation of blended cements made with lime and natural pozzolans. Cem. Concr. Res. **23**(6), 1389–1396 (1993)
20. Hanjitsuwan, S., Hunpratub, S., Thongbai, P., Maensiri, S., Sata, V., Chindaprasirt, P.: Effects of NaOH concentrations on physical and electrical properties of high calcium fly ash geopolymer paste. Cem. Concr. Compos. **45**, 9–14 (2014)
21. Amer, A.A., El-Sokkary, T.M., Abdullah, N.I.: Thermal durability of OPC pastes admixed with nano iron oxide. HBRC J. **11**(2), 299–305 (2015)
22. Trník, A., Scheinherrová, L., Kulovaná, T., Reiterman, P., Vejmelková, E., Černý, R.: Thermal analysis of high-performance mortar containing burnt clay shale as a partial Portland cement replacement in the temperature range up to 1000 & #xB0;C: thermal analysis of mortar containing burnt clay shale. Fire Mater. **41**(1), 54–64 (2017)
23. Karim, M., Hossain, M., Khan, M., Zain, M., Jamil, M., Lai, F.: On the utilization of pozzolanic wastes as an alternative resource of cement. Materials **7**(12), 7809–7827 (2014)

Promising Early Age Evaluations of Fly Ash - Calcined Marl - OPC Ternary Cement

Serina Ng[(✉)] and Tone Østnor

SINTEF Building and Infrastructure, Richard Birkelandsvei 3, 7034 Trondheim, Norway

Abstract. The search for new supplementary cementing materials (SCM) creates opportunities to using new materials such as calcined clays and marl. However, such changes require a transition period, where introduction of new SCMs are most successful when they can be coupled and employed with the existing materials.

For this purpose, a series of ternary mixes employing fly ash (F), calcined marl (C) and ordinary Portland cement (OPC) was evaluated against mono- and binary mixes of the components. It was found that due to the complementary water demand of fly ash and calcined marl, forming pastes made of ternary blends of FC-OPC leads to better rheology than pure OPC pastes, particularly when a F:C ratio of 1:1 was present, even up to 60% replacement of OPC. Any increase in C results in a decrease in flowability, while increasing the F proportion resulted in higher flow, with eventual bleeding. This was coupled with a synergetic improvement in the early age of hydration of the ternary mixes as compared to mono- or binary mixes.

The results here indicate that ternary mix of fly ash-calcined marl-OPC can pave the way for future cementing systems, where it can be easily utilized even in the transition stage and allow the successful implementation of this material.

1 Introduction

As one of the main CO_2 emitter globally [1], cement production is constantly scrutinized to enable greener and more environmentally friendly binders. One of the many approaches in tackling this is to increase the supplementary cementing materials (SCMs) content in cements. Increasing SCM content in blended cements result in a direct proportionate reduction in CO_2 emission, making them a popular choice in the cement industry.

The most widespread and bulk volume SCM employed is fly ash (F) generated from coal-fired power plants. However, due to its inherent CO_2 emission during production and eventual decrease in amount from the reduction of coal firing, fly ash may not be sustainable for the construction market in the long run. A rising star as SCM, on the other hand is calcined clay. Employment of this material gave very favorable mechanical properties and durability, even when contaminated calcined clays, e.g. calcined marl (C) were employed [2]. A drawback of this material however is the reduction in early age workability of such blended binder systems [3], impairing the commercialization value

F. Martirena et al. (eds.), *Calcined Clays for Sustainable Concrete*, RILEM Bookseries 16,
https://doi.org/10.1007/978-94-024-1207-9_54

of this material. In contrast, we have previously shown that fly ash improves the initial flow of OPC [4].

The purpose here is thus to highlight the feasibility of improving rheology of calcined marl loaded binder system by means of a ternary blend – fly ash, calcined marl and OPC. The early age rheological behaviors of such binder systems are explored and analyzed in correlations to their heat of hydrations.

2 Materials and Methods

An OPC and F supplied by Norcem AS (Brevik, Norway), and a smectite rich (~ 50%) C from Saint-Gobain Weber (Oslo, Norway) were employed. Detailed chemical compositions of the materials can be found in previous investigations [3, 4]. All materials were utilised as per obtained. Dry powder were manually blended before wetting to produce the 16 binder mixes (Table 1).

Table 1. Formulation of dry mixes for investigation

Amount of SCM in blends [wt%]						
0	Ordinary Portland cement (OPC)					
20	F20	C5F15	C10F10		C15F5	C20
40	F40	C10F30	C20F20		C30F10	C40
60	F60		C30F30			C60
100	Calcined marl (C100)			Fly ash (F100)		

*Number next to F and C represent the wt% of SCM in the blend

All cement pastes were prepared at a low w/b of 0.36. Dry powder was added to water and mixed under high shear for 1 min, let stand for 5 min and a final high shear mixing of 1 min to avoid false setting. Rheological measurements were performed with a Physica MCR 300 rheometer (Anton Paar, Graz/Austria) equipped with parallel plate geometry. The Bingham viscosity (μ_1 and μ_2), dynamic yield point (τ_d), and flow resistance (FR_2) were measured. Calorimetric investigation was conducted using an isothermal TAM Air calorimeter (TA Instrument, New Castle/USA) up to 24 h. Water demand was determined by a modification of the centrifugal consolidation method, proposed by Miller [5]. More details on the experimental procedures can be found in previous investigations [3].

3 Results and Discussion

3.1 Workability of Pastes

Analysis of the rheology of reference samples OPC, C100 and F100 revealed that the flow of these three materials was very different. No measurement was possible with C100 (indicated by the extremely high values of 500 Pa in Fig. 1). On the other hand, at an initial τ_d of 269 Pa and FR_2 of 40,100 Nm/m^3, OPC paste was ~ 10 times more resistant to flow (τ_d) than FA pastes, whereas the Bingham viscosity was similar for both samples ($\mu_2 = 0.19$ Pa•s). The differences in rheological values may be attributed to the water affinity of these materials: C > OPC > F. Binary blends demonstrated an almost additive rheological property from each material. When C was employed, no flow was measureable for C40 and C60 pastes, while initial τ_d, μ_2 and FR_2 were 375 Pa, 0.49 Pa·s and 56,300 Nm/m^3 respectively for the C20 paste. The results indicated that both the flow and rate of flow are affected by addition of C. On the other hand, F20, F40 and F60 displayed intermediate rheological properties between that of OPC and F; increase F, decrease FR_2 and τ_d. Unlike binary blends with C, μ_2 remained consistent for the F rich pastes. Preparation of ternary blends generated samples with an initial yield stress that is additive of that from F, C and OPC, with the exception of C15F5 (Fig. 1). On the other hand, Bingham viscosity of ternary blends were more in the range of C-OPC, indicating that in ternary blends, the rate of flow is dominated by C. Workability of ternary mix with 1:1 F:C displayed similar dynamic yield stress as OPC, whereas Bingham viscosity was higher for SCM20 and SCM40.

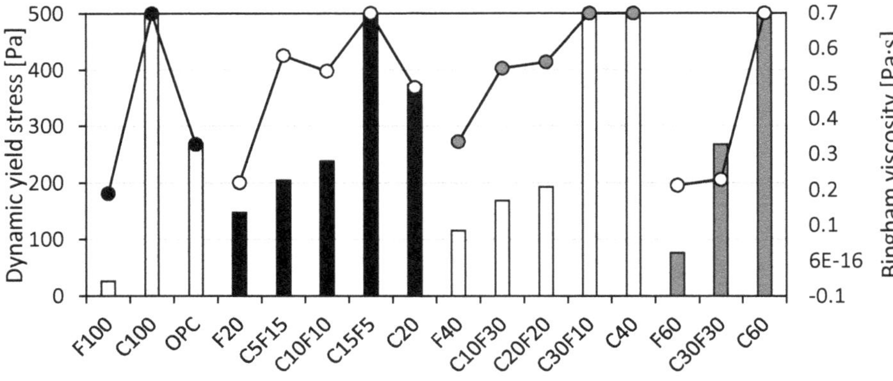

Fig. 1. Initial dynamic yield stress (bars) and Bingham viscosity of pastes (line), w/b = 0.36.

3.2 Heat of Hydration

Figure 2 demonstrates the cumulative heat evolved during the first 24 h after initial wetting. The cumulative heat evolved decreased with increase amount of SCM. However, at all replacement levels, the cumulative heat evolved was the highest for the ternary blends, particularly the F:C = 1:1, followed by binary blends with C and finally when F was added. The higher early age heat evolved from ternary mix indicated that combination of F and C may result in a synergetic reaction, which was absent in binary mixes.

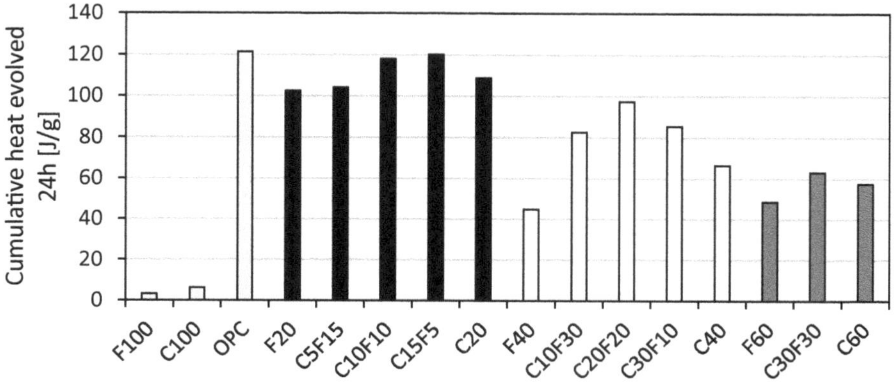

Fig. 2. Cumulative heat evolved of pastes in first 24 h, w/b = 0.36.

3.3 Water Demand

Figure 3 displays the packing density of samples prepared via centrifugal consolidation. The flow behaviour of the ternary mix can be attributed to the high packing density of 0.649 for F, driven by the pure physical surface adsorption, and low packing density of 0.519 for calcined marl where H-bonding of water molecule to the clay layered surfaces dominate [6] respectively. The OPC on the other hand, displayed a moderate packing density at 0.546. The complementing water affinity of these two SCMs make them suitable as combination to improve rheology of the paste, while maximising OPC replacement by SCMs.

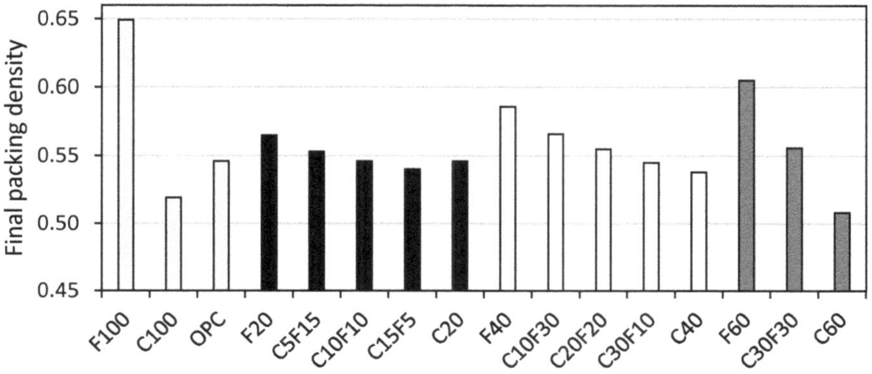

Fig. 3. Packing density of pastes, w/b = 0.36.

It is of interest to note that the variation in packing density from high F to high C blends was not correlated to the heat of hydration, but to the dynamic yield stress. This indicated that initial flow of these paste may potential be a function of physical attribute than chemical hydration.

4 Conclusion

This investigation is a first step towards fulfilling the search for new supplementary cementing materials, where calcined clay is in focus. Complementary employment of calcined clay and fly ash in ternary blends demonstrated that rheological properties similar to ordinary Portland cement was achievable even in the absence of superplasticizers, attributed to the availability of free water to facilitate flow of the paste. The rheological property was decoupled from the heat of hydration, where in fact, higher heat of reaction was observed when ternary mixes was employed as compared to binary mixes. The results here indicated that ternary mix of fly ash-calcined marl-OPC can pave the way for future cementing systems, where desirable rheological and eventual mechanical properties can be achieved. Further investigation including evaluation of the mechanical properties, durability etc. would be needed to eventually demonstrate the commercial value and feasibility of this material.

References

1. Cement Industry Energy and CO_2 Performance – Getting the numbers right, World Business Council for Sustain. Develop, The cement sustainability initiative (2011)
2. Østnor, T., Justnes, H.: Durability of mortar with calcined marl as supplementary cementing material. Adv. Cem. Res. **26**, 344–352 (2014)
3. Ng, S., Justnes, H.: Influence of plasticizers on the rheology and early heat of hydration of blended cements with high content of calcined marl. Cem. Concr. Compos. **60**, 123–134 (2015)
4. Ng, S., Justnes, H.: Influence of plasticizers on the rheology and early heat of hydration of blended cements with high content of fly ash. Cem. Concr. Compos **65**, 41–54 (2016)
5. Miller, K.T., Melant, R.M., Zukoski, C.F.: Comparison of the compressive yield response of aggregate suspensions: pressure filtration, centrifugation, and osmotic consolidation. J. Am. Ceram. Soc. **79**, 2545–2556 (1996)
6. Manning, D.: Handbook of Clay Science (Developments in Clay Science). Elsevier's Science & Technology, vol. 1 (2007). Bergaya, F., Theng, B.K.G., Lagaly, G. (eds.)

Applicability of Lime Reactivity Strength Potential Test for the Reactivity Study of Limestone Calcined Clay Cement

Anuj Parashar(✉), Vineet Shah, and Shashank Bishnoi

Indian Institute of Technology Delhi, New Delhi, India

Abstract. Limestone calcined clay cement (LC^3) allows clinker replacement up to 50% but it can also be designed for other replacement levels based on the quality of raw materials and required properties. In this study, LC^3 was prepared with different proportions using clinker replacement levels of 50%. Two different types of calcined clay and three different types of carbonates were used as supplementary cementitious materials (SCMs) for preparing LC^3. The compressive strength of LC^3 cement mortar cubes were checked for the age of 28 days. The pozzolanic strength potential of limestone and calcined clay (LC^2) blends were tested with the lime reactivity test as per Indian Standard 1727. The lime reactivity test showed the highest reactivity for blends comprising clay and limestone in the proportion of 2:1. Similar result was observed in the case of LC^3 mortar strength. The 28 days cement mortar strength results were correlated with the lime reactivity strength potential test and good correlation was observed. On the basis of results, it was concluded that the lime reactivity strength potential test could be directly applicable to the reactivity study of LC^3 even at varying proportions of limestone and calcined clay.

Keywords: LC^3 · SCMs · Lime reactivity

1 Introduction

Limestone calcined clay cement (LC^3) is one of the most promising ternary blended cement providing benefits of better strength, durability and environmental sustainability even at 50% of clinker replacement levels when compared with traditional ordinary portland cement [1, 2]. The economic benefit of LC^3 is already discussed in details [3]. In LC^3, the synergy can be defined with the help of reactivity of soluble aluminates from clay with calcium carbonate for producing mono-carboaluminates (Mc) and hemi-carboaluminates [1]. Additional benefits of using limestone are improved workability and better mixing.

But, before adding any SCMs to cement, their reactivity need to be studied as per the standard test methods. SCMs, when used as a blend in ternary cements, need to be checked even more carefully because of the synergy effect. At the same time, the quality of SCMs is always questioned because of their non-homogeneity. Due to this fact, the quality of any one SCM can play a major role in the overall performance of such blended cement. Surprisingly, no standardised technique is available for checking

F. Martirena et al. (eds.), *Calcined Clays for Sustainable Concrete*, RILEM Bookseries 16,
https://doi.org/10.1007/978-94-024-1207-9_55

the reactivity potential of blended SCMs. The existing methods were designed a long time ago for one particular type of SCM and are hardly checked for their applicability on such blended SCMs.

For this reason, this study will look forward to the applicability of Indian Standard (IS) lime reactivity test as per IS-1727 [4] over the limestone calcined clay (LC^2) blended pozzolana. This method is one of the classical approaches for looking at the reactivity of SCMs. This method is already tested for pond ash, calcined clay, rice husk ash etc. [5–7]. In this method, standard mortar cubes of 5 cm prepared and conditioned according to the specifications need to be tested for compressive strength at the age of 10 days [4, 6]. This method is easy to perform, requires less efforts, can be applicable over a wide range of pozzolanic material and result is very simple to understand.

The proportions of LC^2 used in the study were selected on the basis of literature [1, 2]. To check the robustness of this method, the proportions of clay and LS were varied. Additionally, different SCMs graded as low grade and high grade were procured. The blended pozzolana (LC^2) were tested in the presence of cement at 50% replacement levels of clinker (LC^3-50). The reactivity of LC^2 pozzolana were tested with the help of lime reactivity test followed by IS 1727. Finally, the results from both the tests were correlated.

2 Materials and Methods

One good quality calcined clay (K1) procured from metakaolin industry and one medium grade kaolinitic clay (K2) calcined in rotatory kiln was selected. For the preparation of LC^2 blends, three different sources of carbonates LS1, LS2 and LS3 were ground in the lab for the final blending. The LS1, LS2 and LS3 was crushed to almost the same fineness levels and later the calcined clay and LS were blended with the help of one lab scale ball mill. During this process, the steel balls of size, 12.5 mm diameter were added to avoid agglomeration in the ratio of 3:1 powder to balls. The time for blending was fixed for 30 min. The blends produced by blending K1, K2 with LS1, LS2 and LS3 are reported in Table 2. The clinker procured from one of the cement company in India was ground in the lab with 3% of gypsum in it. The grinding of cement was targeted to produce a 43 grade OPC as per Indian Standard. The raw materials were characterised with the help of XRF, XRD, and laser diffraction for the particle size distribution. The properties of calcined clays and 3 types of LS are presented in Table 1. The Particle size distribution of all the materials is plotted in Fig. 1.

From the XRD data (not presented), some amount of kaolinite was found to be un-calcined in both clays. In the case of LS1, high quantity of calcium carbonate and small impurity as quartz was observed. LS2 was pure dolomite and that is why no other phases were traced except magnesium carbonates. One low carbonate content siliceous limestone LS3 intentionally selected. In LS3, high amount of quartz, kaolinite, muscovite and calcite were measured. One quartz (Q) was also used to prepared one blend of Q+LS. This step was performed to study the effect of LS on the reactivity study.

Table 1. Chemical and physical properties of raw materials used for LC^2 blends preparation

S. No.	Constituent %		K1	K2	LS1	LS2	LS3
1	Loss on ignition		3.03	4.24	36.96	46.09	25.9
2	Silica SiO_2		53.53	52.70	11.02	0.75	25.84
3	Iron Fe_2O_3		2.03	4.00	1.55	0.64	4.80
4	Aluminium Al_2O_3		40.03	38.14	2.53	1.02	10.12
5	Calcium CaO		0.08	0.09	44.24	29.28	31.95
6	Magnesium MgO		0.02	0.02	1.96	21.26	1.05
7	Sulphate SO_3		0.12	0.10	–	–	0.04
8	Sodium Na_2O		0.24	0.18	0.50	–	0.29
9	Potassium K_2O		0.07	0.17	0.28	–	0.82
10	Specific Gravity		2.64	2.60	2.64	2.88	2.67
11	PSD (μ)	d(10)	1.89	1.69	1.73	1.99	2.34
		d(50)	5.86	5.39	11.1	11.1	12.6
		d(90)	21.9	29.7	56.1	49.8	45.1

Table 2. Ratio of raw materials in different LC2 blends

Blends	Clay	LS	Ratio
B1	K1	LS1	2:1
B2	K1	LS1	1:1
B3	K1	LS 1	1:2
B4	K1	LS2	2:1
B5	K1	LS3	2:1
B6	K2	LS1	2:1
B7	Q	LS 1	2:1

For the final testing, the blended pozzolana (LC^2) was tested in the presence of cement as per IS 4031 (part 6) [8]. For the reactivity part, the strength potential of lime-pozzolana mortar cubes prepared and tested as per IS 1727 were checked.

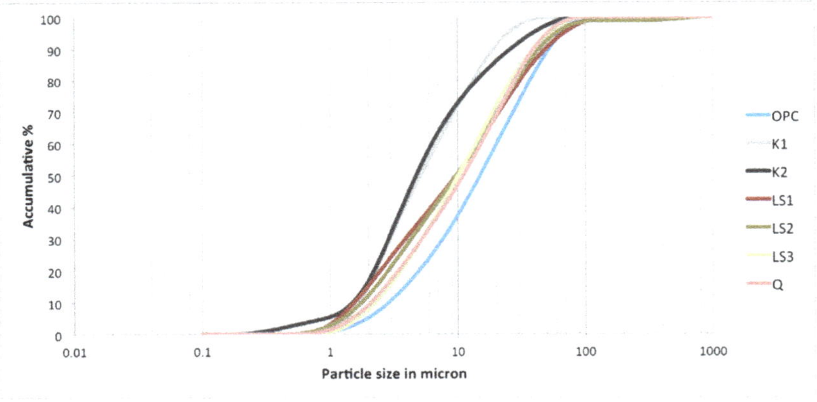

Fig. 1. Particle size distribution of raw materials

3 Methods

3.1 Cement Mortar Compressive Strength

The standard cement mortar cubes of 7.06 cm as per IS 4031 (part 6) [8] were used for this study. The mix design is reported in Table 3. For the ease of comparing the strength of different mixes, fix w/c ratio of 0.45 was selected for all the mixes. The cubes were stored under the temperate conditioned of 27 ± 2° C for first 24 h. The samples were water cured till the age of 28 days. A total of 3 cubes were tested for each mix and the mean value with error bars are plotted in Fig. 2.

Table 3. Mix design for cement mortar strength test

Mix type	LC^3 50%
Water	202.5
Sand	1350
Clinker + gypsum	243
Blend	207

From the results, it was observed that the maximum strength was delivered by B1 blend containing LC^2 in 2:1 ratio. Furthermore, there were only slight variations in strength after changing the source of carbonate.

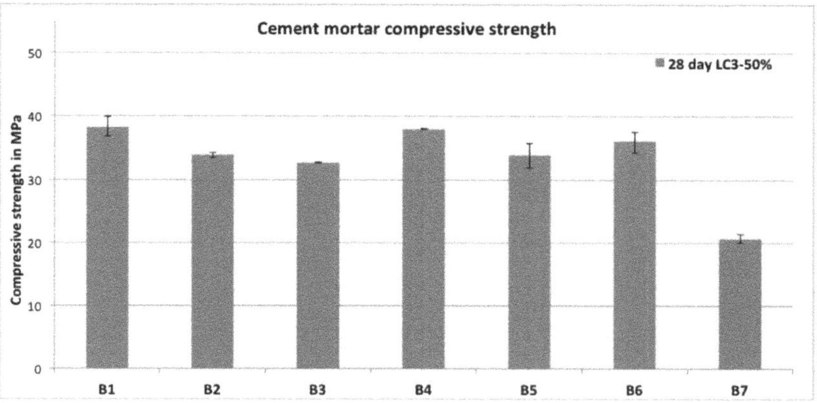

Fig. 2. Compressive strength results for cement mortar cubes at the age of 28 days

3.2 Lime Reactivity

The LC^2 blends were tested as per the lime reactivity test. The quantity of LC^2 and required flow was checked as per the standard protocols. After finalising the workable water demand, the mortar casting was done with the fresh mix using cubes of 5 cm. The lime-mortar cubes were stored in 27 ± 2° C for first 48 h. After 48 h., the cubes were demoulded gently and then transferred to the chamber with the temperature of 50 ± 2° C and humidity of 90 ± 5% for another 8 days. Finally, the cubes were tested under compression at the loading rate of 35 $kg/cm^2/min$ [8].

The result from this testing is reported in Fig. 3. It was found that the compressive strength reduced considerably by reducing the clay content in the blends. By changing the source of carbonates, slightly decrease in strength was observed which might be

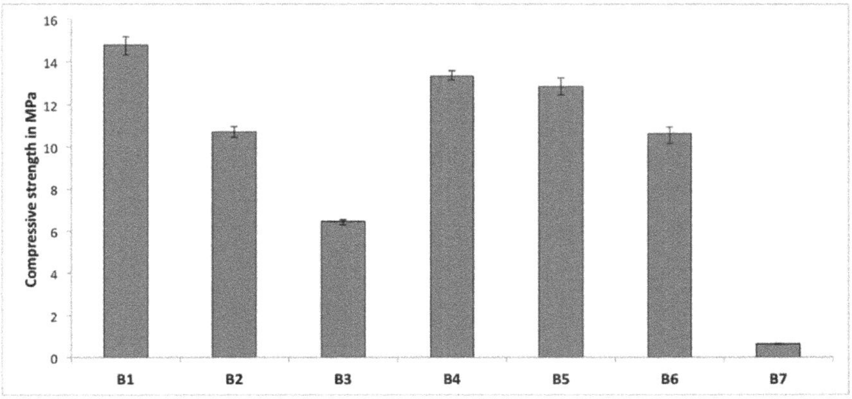

Fig. 3. Lime reactivity strength test results at the age of 10 days

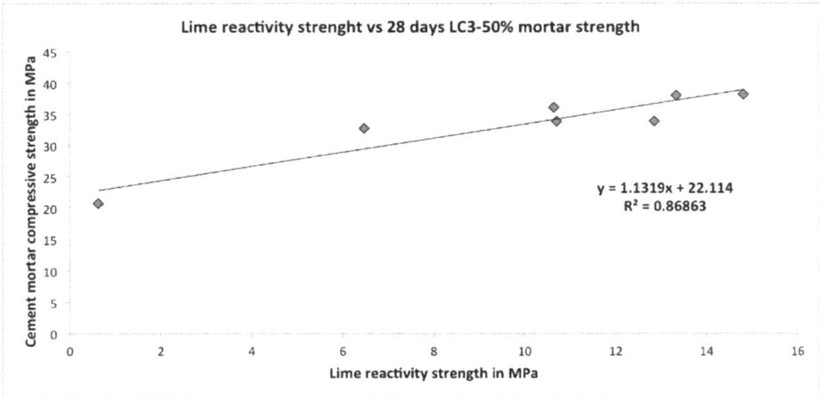

Fig. 4. Correlation of compressive strength of 28 days cement mortar cubes having 50% LC2 with lime reactivity

due to the synergy reaction in between Al and Ca. Blend B7 shows very little strength and was also expected to behave like this. The B7 blend gives strength of 0.6 MPa only.

Finally, the applicability of lime reactivity test over LC^2 pozzolana was checked by simply correlating the 28 days compressive strength of cement mortar cubes and the strength data of lime reactivity test. The correlation of 0.868 for LC^3 50% reported in Fig. 4 was found to be promising for the applicability of this test over the blended pozzolana for LC^3.

4 Conclusions

The applicability of the existing standardised method as per IS 1727 to blends of SCMs used in ternary cements was tested. The test was originally designed for individual SCMs and tried first time for the LC^2 blended pozzolana. The compressive strength result of lime reactivity shows considerable change in strength with the change of the ratio of clay and LS. The correlation of lime reactivity test with the cement mortar strength test was found to be good even after varying the proportions of SCMs. The results indicated that lime reactivity test can be used for the reactivity study of LC^2 pozzolana.

References

1. Scrivener, K.L.: Options for the future of cement. Indian Concr. J **88**(7), 11–21 (2014)
2. Antoni, M., Rossen, J., Martirena, F., Scrivener, K.: Cement and concrete research cement substitution by a combination of metakaolin and limestone. Cem. Concr. Res. **42**, 1579–1589 (2012)
3. Joseph, S., Bishnoi, S., Maity, S.: An economic analysis of the production of limestone calcined clay cement in India. Indian Concr. J. 1–8 (2014)
4. Bureau of Indian Standard (BIS): Methods of test for pozzolanicn materials, IS 1727 (1967)
5. Ranganath, R.V., Bhattacharjee, B., Krishnamoorthy, S.: Influence of size fraction of ponded ash on its pozzolanic activity. Cem. Concr. Res. **28**, 749–761 (1998)
6. Bahurudeen, B., Wani, K., Basit, M.B., Santhanam, M.: Assessment of pozzolanic performance of sugarcane bagasse ash. J. Mater. Civil Eng. 28(2) (2016)
7. Luxan, M.P., Rojas, M.I.S., Frias, M.: Investigations of fly ash- calcium hydroxide reaction. Cem. Concr. Res. **19**, 69–80 (1989)
8. Bureau of Indian Standard (BIS): Methods of physical tests for hydraulic cement, IS 4031 (part 6) (2003)

Limestone and Calcined Clay Blended Cement Used as Low-Cost Binder to Reduce Heat Production and Potential for Delayed Ettringite Formation

Jerry M. Paris(✉) and Christopher C. Ferraro

Engineering School of Sustainable Infrastructure and Environment, University of Florida, Gainesville, FL 32611, USA

Abstract. The use of mass concrete for the construction of infrastructure has become more prevalent in recent history. However, the use of Portland cement as a binder material for massive concrete structures is both highly expensive and can come with durability issues such as delayed ettringite formation resulting from heat produced during hydration. The use of limestone calcined clay cement binder offers a low-cost alternative to mitigate potential problems borne from construction of mass concrete. This paper presents research in which limestone and calcined clay were used as partial replacement of Portland cement to create low heat, low cost binder for mass concrete members. The specimens were evaluated for the potential for delayed ettringite formation; results were confirmed with scanning electron microscopy. Isothermal conduction calorimetry was performed at temperatures of 23 °C, 70 °C, and 85 °C which showed a reduction in heat production in paste amended with limestone calcined clay cement (LC^3). The results of this study indicate that the incorporation LC^3 binder of calcined clay and limestone provides higher alumina and lower sulfate which contributes to the mitigation of deleterious ettringite formation. Several conclusions were drawn from this study: LC^3 can be used to create more financially viable materials for developing areas to create massive concrete structures and the concrete produced with LC^3 as the concrete has a lower potential for delayed ettringite formation due to both lowered heat and chemical composition.

1 Introduction

The development of the infrastructure typically involves the construction of structural components made of massive concrete elements. In order to encourage the use of large structural elements for the construction of the built infrastructure, the use of sustainable, cost-effective materials is necessary for the continued use of Portland cement in construction. Mass concrete members composed of traditional Portland cement concrete typically experience a large temperature rise, which can cause durability issues associated with alkali-silica reactivity (ASR) and delayed ettringite formation (DEF) [1, 2]. The replacement of Portland cement in traditional concrete with a mixture of Portland cement, limestone, gypsum, and calcined clay is more economical and reduces the heat of hydration. The reductions which take place within the Portland cement system which

F. Martirena et al. (eds.), *Calcined Clays for Sustainable Concrete*, RILEM Bookseries 16,
https://doi.org/10.1007/978-94-024-1207-9_56

has been amended with clay provides several benefits which reduce the potential for DEF and ASR.

- Lower heat of hydration results in a lower ultimate temperature
- The resultant products of hydration of Portland cement produce a more stable cementitious product than concrete made with Portland cement alone

Delayed ettringite formation is a chemical reaction borne from ettringite deteriorating at 70 °C, becoming more soluble, and later recrystallizing as ettringite [3]. The recrystallization phase results in the formation of long crystals, causes internal pressure, and ultimately leads to cracking. The problems borne from DEF may take years to evolve and result in the cracking of concrete; additionally, this chemical reaction is often accompanied by alkali-silica reactivity [4–6]. Historically, determining the DEF potential for a particular cement or concrete mixture has been tedious due to the time required to perform laboratory experiments (normally taking several years before expansion is observed in laboratory specimens). In an effort to expedite the laboratory experimentation portion of DEF evaluation, two modifications to the currently accepted Kelham method and Fu method have been developed, which result in much faster determination [7, 8]. In addition to measuring the changes due to the modifications of the methods, it was observed that the replacement of Portland cement with limestone, calcined clay cement (LC^3) successfully mitigated DEF at early ages.

2 Materials and Methods

2.1 Materials

The Portland cement used for this research was an ASTM C150 Type I/II cement from Jacksonville, Florida, USA. Powdered limestone and kaolin clay were obtained from Georgia, USA. High purity, food-grade gypsum was obtained from Yonkers, New York, USA. X-ray fluorescent spectroscopy was performed on the cement to determine elemental oxide composition; a summary of this evaluation is in Table 1.

Table 1. Oxide composition of type I/II cement by XRF.

Analyte	Weight, %	Analyte	Weight, %
CaO	64.10	TiO_2	0.22
SiO_2	19.81	P_2O_5	0.13
Al_2O_3	4.79	M_2O_3	0.06
Fe_2O_3	3.86	SrO	0.13
MgO	1.02	Cr_2O_3	0.02
SO_3	2.66	ZnO	0.11
Na_2O	0.16	Na_2O_{eq}	0.35
K_2O	0.30		

Pretreatment of the material included oven drying at 100 °C, then crushing, and sieving the clay material; denoted as raw clay. After pretreatment, two different calcination temperatures of 650 °C and 800 °C to create two different treatment groups. The thermal treatment consisted of increasing the temperature at 5 °C/min to 650 °C or 800 °C, respectively with a 120-minute dwell time, followed by a decreasing ramp of 5 °C/min until ambient conditions were achieved.

X-ray diffraction was performed on the three clays (raw, 650 °C, and 800 °C) to evaluate the conversion of raw kaolin clay to metakaolin using the different thermal treatments. The XRD diffraction patterns for the clays are presented in Fig. 1 which shows that the kaolinite clay was mostly decomposed by the thermal treatment at 650 °C with further conversion to metakaolin at 800 °C; this result is in agreement with literature [9]. However, the thermal treatments were not aggressive enough to decompose mica impurities, which are still present in the XRD patterns for the thermally treated clays.

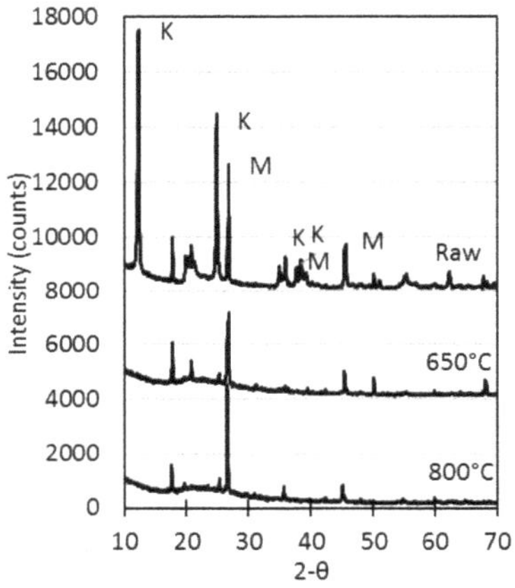

Fig. 1. XRD pattern for the raw kaolin clay and two calicined clays; peaks attributed to kaolin are denoted with 'K' and mica impurities are noted by 'M'.

Particle size distribution was determined for all of the cementitious materials used in the study by laser light diffraction using ethanol dispersion. The median particle size was 28 μm, 37 μm, and 48 μm for the raw clay, 650 °C clay, and 800 °C clay, respectively. The results of the particle size distribution are shown in Fig. 2.

Table 2 presents the mortar mix designs used to create the specimens for each curing regimen. These mixes follow the mix design as specified in the accelerated alkali-silica reactivity test method ASTM C1260 [4. 10]. This method was used to verify that the aggregates utilized for this research were not reactive.

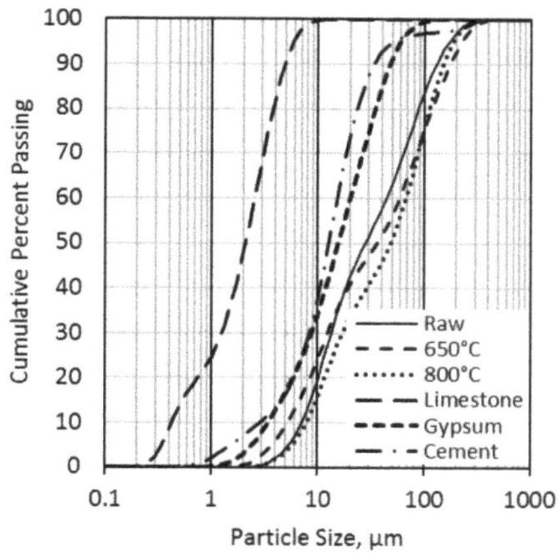

Fig. 2. Particle size distribution for the cementitious materials used for this research.

Table 2. Mortar mix designs used for DEF testing; values are in kg/m^3.

Mix	Control	650 °C	800 °C
Type I/II cement	607	295	295
Calcined clay	–	177	177
Limestone	–	88	88
Gypsum	–	30	30
Water	285	277	277
Sand	1366	1327	1327

3 Methods

To determine the potential for delayed ettringite formation, several researchers have proposed accelerated curing methods of mortars that induce ettringite formation after several months or years. The two most notable being the Kelham Method and the Fu Method [7, 8]. Kelham prescribed curing mortar bars at ambient conditions for four hours, then raising the temperature and humidity to 95 °C and 95% RH at 20 °C/min, holding these conditions for 12 h, then bringing the specimens back to ambient conditions and storing in limewater. Measurements of the bars are taken at the end of the curing period as a datum and periodically for years [7]. Fu prescribed a more aggressive curing regimen, which involved curing the mortar bars at ambient conditions for one hour, bringing the specimens to 95 °C and 95% RH over one hour, then holding at these conditions for 12 h, taking the specimens to ambient conditions over four hours, holding at ambient conditions for six hours, placing into a 85 °C oven for 24 h, then placing in a limewater bath with periodic measurements [8]. The final curing method used as a comparison involved curing specimens at ambient conditions for 24 h, then storing in limewater at ambient temperature.

For this evaluation, researchers created a new curing method to expedite expansion using an even more aggressive curing regimen. This method involved curing at ambient conditions for four hours, ramping to 95 °C and 95% RH over four hours, soaking in this environment for 24 h, then bringing to ambient conditions and storing in limewater.

4 Results

The evaluation procedures investigated for this research began by comparing curing methods to determine potential for delayed ettringite formation. For the control mix, four curing methods were investigated as described in the Methods section. The results of nearly 12 months of expansion data for the control mix are presented below in Fig. 3; it is evident that the proposed "modified" curing method accelerates the expansion due to delayed ettringite formation. Alkali-silica reactivity was investigated, but the results are not presented for brevity; the aggregates used were found to be not reactive. The expansion seen in the modified method is greatly accelerated compared to the Kelham or Fu method; failure criteria for this test method was assumed to be 0.10% in line with ASTM C1260. The same criterion was used to define dimensional instability (0.10% expansion) and therefore, is unacceptable for the use of concrete and structural applications.

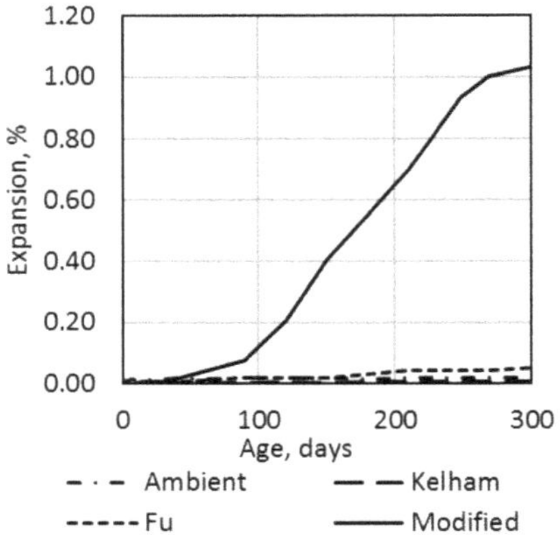

Fig. 3. Summary of expansion data for the control mortar using different curing regimens.

From these results of control mortar tested with various curing methods, Fig. 3, the modified method accelerated expansive reactions and was used for all future testing. For the full-sized mortar bars, curing under the modified method, the LC^3 cements showed 46% and 27% of the expansion experienced by the control mixes at 65 days for the 650 °C and 800 °C mixes, respectively.

5 Conclusions

From the results of the experiments performed, the following conclusions could be drawn:

- The modified method provides results regarding the potential for delayed ettringite formation much faster than traditionally accepted methods.
- LC^3 cement can be used to reduce the potential for delayed ettringite expansion as evidenced by reduced expansion using the aggressive modified curing method.
- The reduced reactivity of LC^3 cements provides less heat of hydration, which further reduces potential for delayed ettringite formation in structural mass concrete.
- Along with lowered costs, these positive benefits for mass concrete provides a benefit to developing countries who are in need of economical, durable structures for creating infrastructure.

References

1. Gajda, J.: Mass concrete for buildings and bridges. Portland Cement Association 34 (2007)
2. Ferraro, C.: Determination of test methods for the prediction of the behavior of mass concrete. Dissertation, University of Florida, 277 p (2009)
3. Taylor, H.: Cement Chemistry, 2nd edn., pp. 374–378. Thomas Telford Publishing, London (2004)
4. Shayan, A., Quick, G.W.: Microscopic features of cracked and uncracked concrete railway sleepers. Mater. J. **89**, 348–361 (1992)
5. Shayan, A., Quick, G.W.: Alkali aggregate reaction in concrete railway sleepers from Finland. In: Proceedings 16th International Conference Cement Microscopy, ICMA, Duncanville, TX, USA, pp. 69–79 (1994)
6. Diamond, S.: Delayed ettringite formation — processes and problems. Cem. Concr. Compos. **18**, 205–215 (1996). doi:10.1016/0958-9465(96)00017-0
7. Kelham, S.: Delayed ettringite formation the effect of cement composition and fineness on expansion associated with delayed ettringite formation. Cem. Concr. Compos. **18**, 171–179 (1996). doi:10.1016/0958-9465(95)00013-5
8. Fu, Y.: Delayed ettringite formation in Portland cement products, University of Ottawa, Report 220 (1996)
9. Wan, Q., Rao, F., Song, S.: Reexamining calcination of kaolinite for the synthesis of metakaolin geopolymers - roles of dehydroxylation and recrystallization. J. Non-Cryst. Solids **460**, 74–80 (2017). doi:10.1016/j.jnoncrysol.2017.01.024
10. ASTM C1260: Standard Test Method for Potential Alkali Reactivity of Aggregates (Mortar-Bar Method), ASTM International, West Conshohocken, PA, 5 (2014)

Hydrate Phase Assemblages in Calcium Sulfoaluminate – Metakaolin – Limestone Blends

M. Pedersen[1(✉)], B. Lothenbac[2], F. Winnefeld[2], and J. Skibsted[1]

[1] Department of Chemistry and iNANO, Aarhus University, Langelandsgade 140, 8000 Aarhus C, Denmark

[2] Laboratory for Concrete and Construction Chemistry, EMPA, Swiss Federal Laboratories for Materials Science and Technology, Überlandstrasse 129, 8600 Dübendorf, Switzerland

Abstract. The combination of a commercial calcium sulfoaluminate (CSA) cement with metakaolin (MK) and limestone (LS) as supplementary cementitious materials (SCMs) is investigated for a CSA replacement level of 20 wt%. In addition to a pure CSA cement, paste samples have been prepared for three blends with MK to LS ratios of MK/(MK + LS) = 0, 0.5 and 1. All blends used a molar gypsum/ye'elimite ratio of 0.5 and a water to binder ratio of w/b = 0.7. The hydration for the four series has been followed from 1 day to 182 days and the hydrate phase assemblages have been identified using XRD, TGA as well as ^{27}Al and ^{29}Si NMR and compared with predictions from thermodynamic modelling. The results show almost full degrees of reaction for ye'elimite in all blends and that the degrees of reaction for belite and MK have an important impact on the hydrate phases and their quantities formed in the ternary systems.

1 Introduction

Calcium sulfoaluminate (CSA) cement currently attracts significant attention as an alternative to ordinary Portland cement (OPC), as it with its principal phase of ye'elimite ($Ca_4Al_6O_{12}SO_4$) can be produced with lower energy consumption and lower embodied CO_2 emission [1]. Moreover, a further reduction in CO_2 footprint may be achieved by partly replacement of the CSA cement with supplementary cementitious materials (SCMs) such as slags, fly ashes and calcined clays. Calcined clays represent an important type of SCMs as a result of their world-wide abundance in large amounts and since they may exhibit rather high pozzolanic reactivities in cementitious systems. For example, degrees of reaction above 90% can be achieved for metakaolin (MK) in Portland cement blends incorporating up to 30 wt% MK [2]. Furthermore, in ternary blends of Portland cement, calcined clays and limestone (LS), a synergetic effect of the two SCMs on the compressive strength can be obtained for cement substitution degrees up to 45 wt% [3, 4]. Limestone provides nucleation sites for the formation and growth of the calcium-silicate-hydrate (C-S-H) phase, and the synergetic effect is primarily associated with its partial reaction, which results in the formation of calcium monocarboaluminate hydrate ($Ca_4Al_2(OH)_{12}CO_3{\cdot}5H_2O$) [5], utilizing aluminate from the calcined clays. Ye'elimite of the CSA cement reacts with water and forms monosulfate ($Ca_4Al_2(OH)_{12}SO_4{\cdot}6H_2O$)

F. Martirena et al. (eds.), *Calcined Clays for Sustainable Concrete*, RILEM Bookseries 16,
https://doi.org/10.1007/978-94-024-1207-9_57

and $Al(OH)_3$ whereas ettringite ($Ca_6Al_2(OH)_{12}(SO_4)_3{\cdot}26H_2O$) and $Al(OH)_3$ forms if an additional sulfate source (e.g. gypsum) is added to the system. The high aluminum content in CSA cement, provided by ye'elimite, may be utilized to form additional amounts of AFm phases such as hemi- and monocarbonate aluminate hydrate by addition of small amounts of limestone, which may potentially lead to improved mechanical properties [6].

The present work is our first account on an investigation of ternary CSA cement – MK – LS blends with the aim of exploring the potential of calcined clays in CSA cement blends and any synergetic effect between calcined clay and limestone. The first task is to establish the hydrate phase assemblages which has been accomplished by investigations of four different blends. A challenge for the CSA cement – MK – LS system is the availability of CaO which drives the pozzolanic reactions for MK and LS. Thus, a CSA cement replacement degree of only 20 wt% has been used in the present work. Although, the hydration has been followed from one day to 182 days, this paper presents only the data obtained after 91 days.

2 Materials and Methods

A commercial CSA cement, with the main phases ye'elimite (62.7 wt%), belite (β-Ca_2SiO_4, 10.7 wt%), fluorellestadite ($Ca_5(SiO_4)_{1.5}(SO_4)_{1.5}F$, 8.2 wt%) and bredigite ($Ca_7Mg(SiO_4)_4$, 5.7 wt%) was used along with a commercial LS and chemical-grade gypsum. MK was prepared from calcination of kaolinite (Imerys Performance Minerals, UK) at 560 °C for 20 h. Three CSA – MK – LS blends were prepared with a total substitution degree of 20 wt% and different MK/(MK + LS) ratios (Table 1). In addition, a pure CSA blend was prepared as reference. All blends used a molar sulfate/ye'elimite ratio of 0.5 and a water/binder ratio of 0.7. Pastes were prepared and hydrated for 91 days before characterization.

Table 1. Composition (wt%) of the four blends investigated in this work

	CSA	MK0	MK0.5	MK1
CSA cement	93.3	74.7	74.7	74.7
Gypsum	6.7	5.3	5.3	5.3
Metakaolin	–	–	10	20
Limestone	–	20	10	–

The X-ray diffraction patterns were obtained with a Panalytical X'pert Pro diffractometer in the 2θ-range of 5–90°, using CoK_α radiation. The phases were assigned using the X'Pert Highscore Plus software. The ^{27}Al MAS NMR spectra were recorded on a Varian Direct-Drive VNMR-600 (14.09 T) spectrometer using a homebuilt CP/MAS NMR probe for 4 mm o.d. zirconia (PSZ) rotors and a spinning speed of $\nu_R = 13.0$ kHz. The ^{29}Si MAS spectra were acquired on a Bruker Avance III HD 9.4 T spectrometer using a homebuilt CP/MAS NMR probe for 7 mm o.d. PSZ rotors and a spinning speed of $\nu_R = 6.0$ kHz. Thermodynamic modelling was carried out using the Gibbs free energy minimization program, GEMS 3.3 [7].

3 Results and Discussion

The crystalline hydrate phases are identified by powder XRD, as illustrated by the patterns shown in Fig. 1 for the four blends hydrated for 91 days. Ettringite is the main hydrate phase for all blends, resulting from the reaction of ye'elimite with gypsum and water. In addition, strätlingite ($2CaO{\cdot}Al_2O_3{\cdot}SiO_2{\cdot}8H_2O$) is observed in all samples except for the MK0.5 blend, which may indicate a lower degree of belite reaction in this system. Hemicarbonate aluminate hydrate ($(Ca_4Al_2(OH)_{12}(OH)(CO_3)_{0.5}{\cdot}5H_2O)$) is detected for the two LS-containing blends whereas monosulfate is detected in the hydrated pure CSA cement. Ye'elimite is not observed in any of the XRD patterns whereas reflexions from belite C_2S, fluorellestadite and bredigite can be identified in all samples, demonstrating an incomplete hydration for these phases from the CSA cement. Moreover, significant amounts of LS are still present in the MK0 and MK0.5 blends.

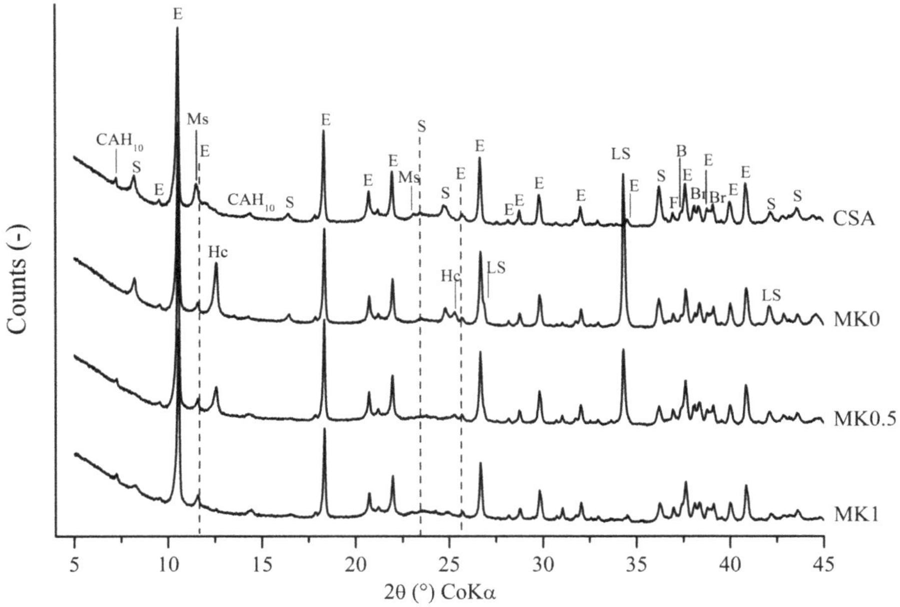

Fig. 1. X-ray diffractograms of the four blends after 91 days of hydration. CAH10 = calcium aluminate hydrate (CaO·Al2O3·10H2O), S = strätlingite, E = ettringite, Hc = hemicarbonate, Ms = monosulfate, LS = limestone, F = fluorellestadite, B = belite, Br = br

The XRD measurements are complemented by ^{27}Al and ^{29}Si MAS NMR spectra (Fig. 2), which provide additional information about the less-crystalline and amorphous phases in the hydrated systems. The ^{29}Si NMR spectra reveal that all samples contain unreacted belite (−71.3 ppm), consistent with the XRD analysis. For the pure CSA cement, the degree of belite reaction after 91 days is estimated to 40% from the observed intensities for the anhydrous and hydrated CSA cement. The principal silicate-containing hydration product is strätlingite, as seen by the resonance at −85.5 ppm, which is observed in all samples. The formation of a C-S-H phase would result in peaks in the range −79 to −85 ppm and intensity in this region is observed for the pure CSA cement and most clearly for the CSA - LS blend (MK0), which may indicate that LS promotes formation of C-S-H, as observed for Portland cement – limestone blends. Unreacted MK (−80 to −120 ppm) is observed for the MK0.5 and MK1 blends, however, the degree of reaction for MK has not been estimated from the present spectra since the broad resonance from MK overlaps with the peaks from strätlingite.

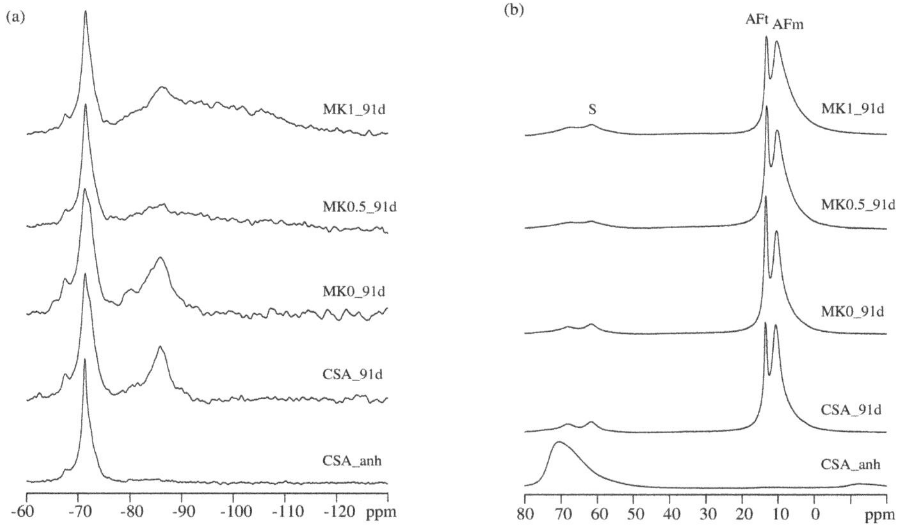

Fig. 2. (a) 29Si MAS NMR and (b) 27Al MAS NMR spectra of the anhydrous CSA cement and the four blends hydrated for 91 days of hydration. S indicates the Al(4) peak from strätlingite.

The ^{27}Al MAS NMR spectra show that almost all of the ye'elimite has reacted for all blends. Strätlingite is present in all samples, which is clearly identified by its resonance at 61 ppm from the tetrahedral Al sites in its structure [2]. Most intensity is observed in the spectral region for octahedral aluminum (20–0 ppm) which only allows distinction of resonances from AFt (ettringite) and the AFm phases (monosulfate, monocarbonate, hemicarbonate and strätlingite). The largest amount of ettringite is found in the two LS-containing blends, in accordance with that the formation of hemi- and monocarbonate stabilizes ettringite. An improved resolution of the individual AFm phases can potentially be achieved at very high magnetic fields, as indicated by a ^{27}Al NMR study of a model CSA cement at 22.3 T [8].

Thermodynamic calculations have been performed for the four blends (Fig. 3), assuming the following degrees of hydration: ye'elimite – 100%, belite −40%, fluorellestadite – 50%, bredigite – 50% and MK −70%. Silicious hydrogarnet, kaolinite, gibbsite and thaumasite were suppressed in the equilibrium calculations due to their slow hydration kinetics [9, 10]. For the pure CSA cement, strätlingite, ettringite and monosulfate are predicted as the main hydration products, in accord with the experiments.

Microcrystalline $Al(OH)_3$ is predicted to form as a result of ye'elimite hydration and it increases with increasing MK content. For the MK0 blend, the calculations predict that LS leads to the formation of monocarbonate at the expense of the monosulfate, however, only hemicarbonate is observed in the XRD patterns. This deviation has previously been explained by faster formation kinetics of hemicarbonate compared to monocarbonate [9]. The presence of a C-S-H phase is virtually only predicted for the MK1 blend, most probably as a result of its high silicate content, whereas indications of a C-S-H phase is only seen for the CSA and MK0 blends in the ^{29}Si NMR spectra. Improved agreement between the experimental and predicted phases is expected if the principal anhydrous phases of the original blends can be determined with good precision.

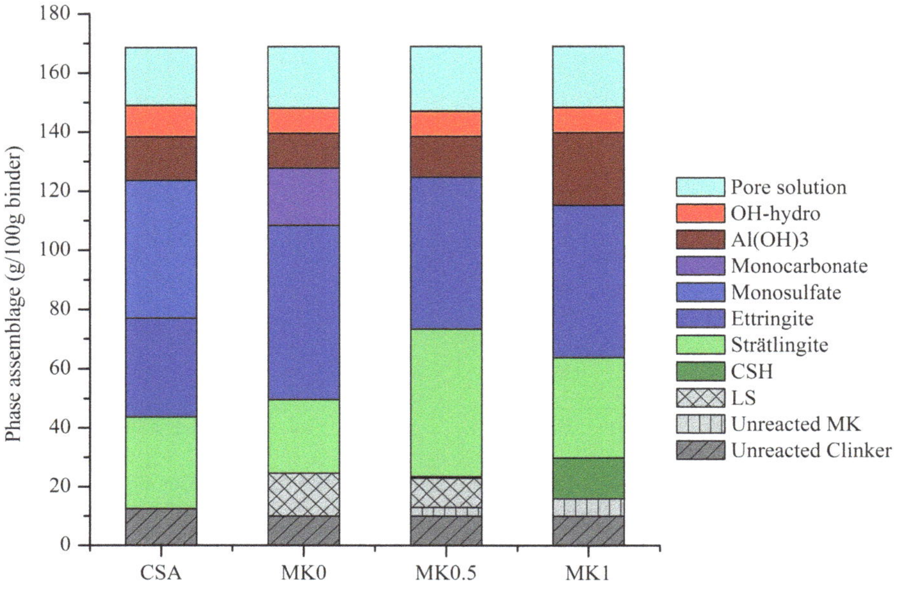

Fig. 3. Calculated phase assemblages for the four blends using GEMS and the assumptions given in the text.

4 Conclusions

Ye'elimite exhibits nearly full reaction for all studied blends after 91 days of hydration, the primary hydration products being strätlingite, ettringite and different AFm phases. The presence of limestone stabilizes ettringite, most likely by the formation of hemi- or monocarbonate aluminate hydrate. A faster reaction of metakaolin compared to belite may slow down the hydration for belite and thereby result in the release of less CaO to drive the pozzolanic reactions. The thermodynamic calculations for the present systems are sensitive to the degree of reactions for the principal phases, implying that improved data on hydration kinetics and degrees of reactions should be obtained for these phases experimentally.

References

1. Juenger, M.C.G., Winnefeld, F., Provis, J.L., Ideker, J.H.: Advances in alternative cementitious binders. Cem. Concr. Res. **41**, 1232–1243 (2011)
2. Dai, Z., Tran, T.T., Skibsted, J.: Aluminum incorporation in the C-S-H phase of white Portland cement – metakaolin blends studied by ^{27}Al and ^{29}Si MAS NMR spectroscopy. J. Am. Ceram. Soc. **97**, 2662–2671 (2014)
3. Steenberg, M., Herfort, D., Poulsen, S., Skibsted, J., Damtoft, J.: Composite cement based on Portland cement clinker, limestone and calcined clay. In: 13th International Congress of the Chemistry of Cement, Madrid, p. 97 (2011)
4. Antoni, M., Rossen, J., Martirena, F., Scrivener, K.: Cement substitution by a combination of meta-kaolin and limestone. Cem. Concr. Res. **42**, 1579–1589 (2012)
5. De Weerdt, K., Ben Haha, M., Le Saout, G., Kjellsen, K.O., Justnes, H., Lothenbach, B.: Hydration mechanisms of ternary Portland cements containing limestone powder and fly ash. Cem. Concr. Res. **41**, 279–291 (2011)
6. Martin, L.H.J., Winnefeld, F., Müller, C.J., Lothenbach, B.: Contribution of limestone to the hydration of calcium sulfoaluminate cement. Cem. Concr. Comp. **62**, 204–211 (2015)
7. Kulik, D.A., Wagner, T., Dmytrieva, S.V., Kosakowski, G., Hingerl, F.F., Chudnenko, K.V., Berner, U.R.: GEM-Selektor geochemical modeling package: revised algorithm and GEMS3 K numerical kernel for coupled simulation codes. Comput. Geosci. **17**, 1–24 (2013)
8. Skibsted, J., Pedersen, M.T., Holzinger, J.: Resolution of the two Al sites in ettringite by ^{27}Al MAS and MQMAS NMR at very high magnetic field (22.3 T). J. Phys. Chem. C **121**, 4011–4017 (2017)
9. Winnefeld, F., Lothenbach, B.: Phase equilibria in the system $Ca_4Al_6O_{12}SO_4$ – Ca_2SiO_4 – $CaSO_4$ – H_2O referring to the hydration of calcium sulfoaluminate cements. RILEM Techn. Lett., 10–16 (2016)
10. Lothenbach, B.: Thermodynamic equilibrium calculations in cementitious systems. Mater. Struct. **43**, 1413–1433 (2010)

Influence Grinding Procedure, Limestone Content and PSD of Components on Properties of Clinker-Calcined Clay-Limestone Cements Produced by Intergrinding

A. Pérez[1](✉), A. Favier[2], K. Scrivener[2], and F. Martirena[3]

[1] CIDC, Havana, Cuba
[2] LMC, EPFL, Lausane, Switzerland
[3] CIDEM, UCLV, Santa Clara, Cuba

Abstract. This paper looks at the study of intergrinding for the production of ternary cement based on clinker, calcined clay, limestone and gypsum with 50% of clinker substitution (LC^3). The impact of grinding time on clinker, limestone and calcined clay PSD, and how this parameter influences the overall performance of the ternary cement is assessed. Laboratory cement blends were produced by grinding all components in a batch laboratory mill. Industrial cements produced through intergrinding in a continuous ball mill were used for comparison. Three fractions were identified: d<7 μm, 7 μm < d < 40 μm and d< 40 μm, for each of the cements studied and the amount of each component were assessed. Fresh and hardened state properties of blends were tested. Results indicate that in intergrinding most of clinker remains at the medium fraction, and further grinding cannot improve clinker fineness due to fine calcined clay muffle clinker fineness gaining. PSD of limestone and calcined clay is wider than clinker PSD, with a high amount of each material on fine fraction, having a strong impact on rheology. A change in calcined clay/limestone ratio from 2:1 to 1:1 improves clinker grinding and rheology but has a negative impact on strengths due to the less proportion of calcined clay that impact negatively on the pozzolanic reaction.

1 Introduction

Concrete is the most widely used material on earth; It is strong, durable and relatively inexpensive. The cement industry is responsible for about 5–8% of the global man-made CO_2 emissions. On average 0.8–0.9 tons CO_2 is emitted for the production of 1 ton of cement. During the production of cement clinker, calcium carbonate from limestone decomposes to calcium oxide, liberating CO_2, it represents about 50–60% of the total amount of CO_2 emitted during cement production [1].

Blended cements are replacing part of the clinker by mineral additions. It causes not only a reduction in the consumption fossil fuel and in the CO_2 emissions, but can also contribute to better concrete properties in both fresh and hardened state [2], due to pozzolanic reaction and/or filler effect [1]. Metakaolin (MK) ($Al_2Si_2O_7$) is a highly reactive pozzolan produced through the calcination of clays rich in kaolinite mineral and is widely used in order to produce blended cements reducing clinker amount. But

F. Martirena et al. (eds.), *Calcined Clays for Sustainable Concrete*, RILEM Bookseries 16,
https://doi.org/10.1007/978-94-024-1207-9_58

the cost of this material, its availability and the use of this material by other industries restrict its use as pozzolan [1].

Medium purity kaolinite clay has proven to be a good alternative to MK in this system; that expands the opportunity to use low grade clays that are rejected by MK consumer industries and ceramic industries. Low grade clays are also distributed along a wide area that includes the developing countries, made this cements an alternative to accelerate its infrastructure development. LC3 is included in this cement class.

The synergy of the combined addition of calcined clays and limestone has been set in evidence by comparing with quartz filler with very similar particle size distribution, bringing a significant enhancement of the mechanical properties [1]. Limestone reacts with the alumina in presence of calcium hydroxide and favours the production of carboaluminates reaction products. This allows increasing the clinker substitution to 50%.

Besides components reactivity, PSD of each one plays an important role on final performance of blended cements. Contrary to OPC, fineness, specific surface and PSD of multi-component cement are not a consequence of one material; during intergrinding the different component interact with one another due to the difference in grindability [3]. During intergrinding, the PSD of the more grindable component becomes finer and wider and the PSD of the harder to grind one becomes coarser and narrower [2]. For LC3 clinker is the harder component. That means that for normal cement fineness and grinding time, clinker will be the coarser material on the blend probably due to the presence of relatively soft calcined clay particles that shields harder clinker particles from being ground. Harder and coarser clinker particles also abrade the softer ones increasing its fineness [2]. Has been observed also that the agglomeration of the finer particles upon continuous grinding leading to a sudden decrease in Blaine fineness. Due to this, clinker hydration is less effective and early strength most be affected. Wide limestone PSD led to decreasing water demand and improves the workability.

Calcined clay had a high specific surface due to its surface structure. Previous results of BET and Blaine tests show that total specific surface of LC^3 is driven by specific surface of the calcined clay fraction [1] increasing the water demand. A compromise between limestone and calcined clay fineness is needed; by minimizing the void space between the cement particles, the packing can be optimized. That improves rheology reducing water demand. It was of great interest to investigate how intergrinding affects LC^3 component PSD and how this parameter affects cement performance on fresh and hardened state in order to adjust final LC^3 fineness. The influence of calcined clay/ limestone ratio was also investigated.

2 Materials and Methods

Clinker (CK) was produced at Siguaney cement factory in Cuba. Calcium sulfate (GS) and limestone (LS) originated also from Siguaney are used in the manufacture of plain Portland cement.

The clay originated from Pontezuela is classified as medium grade kaolinite clay, with an average content of kaolinite of 48.6%, measured by thermogravimetric analysis

(Vizcaino et al. 2015). The clay was calcined in one of the clinker kilns of Siguaney Cement factory at 750 °C and the calcined clay (CC) was obtained.

In order to determine the impact of the calcined clay/limestone, two different blended cements were produced for this study. 50.0% of CK, 28.0% of CC, 14.4% of LS, 6.8% of GS for a CC/LS ratio of 2:1 (2 1), and 50% of CK, 21.6 of CC, 21.6 of LS, 6.8% of GS for a CC/LS ratio of 1:1 (1 1). A reference Portland cement was also produced using the same CK and GS.

Laboratory grinding was carried out using an ellipsoidal KHD laboratory mill with 50 L volume and 15 L of charge volume. Reference Portland cement and reference blended cements were grounded before the test. Samples were extracted at regular intervals and their fineness measured, in order to determine the grinding kinetics and adjust the grinding time to obtain the requested fineness values for each blend. Using these results, three different blends were grounded to a range of fineness between 91% and 98%.

Industrial grinding was carried out using a continuous semindustrial grinding plant with all the industrial process present in a real industrial plant. The mill used has three chambers, 570 L of total volume and 170 L of charge volume with a production regime of 80 L/h. One blend for each calcined clay/limestone ratio and a reference were produced.

The chemical composition of all materials was characterized by X-ray Fluorescence (XRF) (APC Solutions, Denges, Switzerland) using a Bruker AXS S4 Explorer spectrophotometer operating at a power of 1 kW and equipped with a Rh X-ray source. Results are showed in Table 1.

Table 1. Oxidic composition of all materials studied

Oxides	Calcined clay	Gypsum	Clinker	Limestone
SiO_2	48.4	4.6	18.8	4.1
Al_2O_3	29.5	1.4	3.2	1.1
Fe_2O_3	16.5	1.4	4.7	0.9
CaO	0.4	36.1	68.0	51.8
MgO	1.0	2.3	1.3	0.8
SO_3	1.1	38.6	0.2	0.5
Na_2O	–	–	0.5	–
K_2O	0.7	0.4	1.1	0.1
TiO_2	0.6	0.2	0.2	0.1
LOI	1.5	14.8	1.5	40.6

Cement fineness was determined by the material percent passing through the sieve with an aperture of 90 μm. Mortar prisms 40 x 40 x 160 mm were made with water to cement ratio of 0.5 according to EN 196-1, demoulded after 24 h and cured by immersion in tap water in a storage tank at 25 ± 2 °C. Specific surface was determined by air permeability test (Blaine) according to ASTM C204.

For fraction separation, air filtering was done using a Multi Plex 100 Almine air classifier. Air separation was carried out with two different air speed 3000 m^{-1} and 9000 m^{-1}

and two corresponding flows respectively 52 Nm^3/h and 46 Nm^3/h. these parameters were calibrated to separate the blend in 3 fractions, fine (less than 7 μm), medium (between 7 and 40 μm) and coarse (more than 40 μm). Each fraction was weighted to know the relative amount of each one.

Rietveld adjustment was used to quantify the phases in each fraction. XRD analysis was carried out in a Panalytical X'Pert Pro MPD diffractometer from 3° to 70° with ¼ slit for 35 min. Raw materials were used to determine all the phases. The blends without separation were used to calibrate the metakaolin content inside the blends.

Workability was tested using a minislump test in pastes. Cement blend pastes were prepared with a water/binder ratio of 0.5, mixed in a laboratory mixer at 200 rpm for 2 min. After a stop of 3 min was mixed again for 1 min and poured in a minicone with 56 mm height, 19 mm of upper diameter and 38 mm of lower diameter. The minicone was pulled out vertically, and paste spread was measured.

3 Results and Discussions

Figure 1 shows the development of fineness and Blaine for each cement produced on laboratory conditions. At 35 min, the LC^3 mixes are finer ground than OPC. From 40 min LC^3 2:1 and 1:1 achieve the same fineness. At this range of fineness Blaine stabilizes for LC3 2:1, while still at 60 min, Blaine for LC^3 1:1 keeps growing. Blaine for LC^3 2:1 is higher than Blaine for LC^3 1:1. For fineness higher than 95% further grinding increases energy consumption without a major impact on fineness. It is need a compromise between grinding time and reactivity of the system in order to save energy.

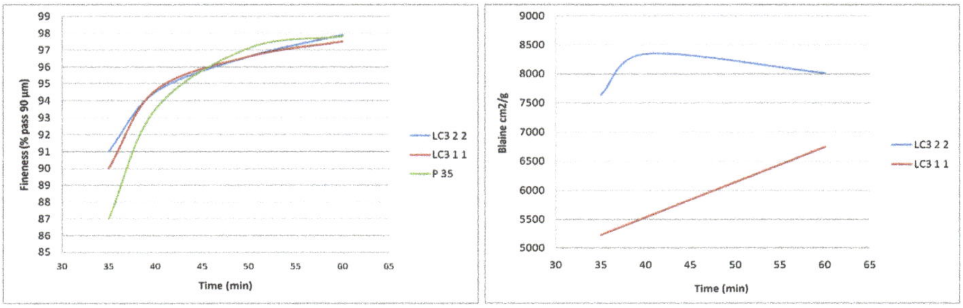

Fig. 1. Fineness and specific surface vs. grinding time

Figure 2 shows the evolution of the amount of each component in all the fractions studied during the grinding process in the range of fineness defined before for both LC^3 cement produced and the results for the industrial trials. In both cases, clinker fineness at fraction d < 7 μm is not increased by further grinding. Further grinding on LC^3 2:1 does not have any major impact on clinker fineness at all fractions. Further grinding on LC^3 1:1 increases clinker content at fraction 7 μm < d < 40 μm while decreases at fraction d > 40 μm. The more amount of Calcined clay in 2:1 blends increase the shielding process on the mill balls and clinker particles surface produced by the

Calcined clay fine particles. This process muffles the interaction between the mill balls and clinker particles stopping the gain of fineness. For 1:1 blends, the smaller amounts of Calcined clay reduce this effect.

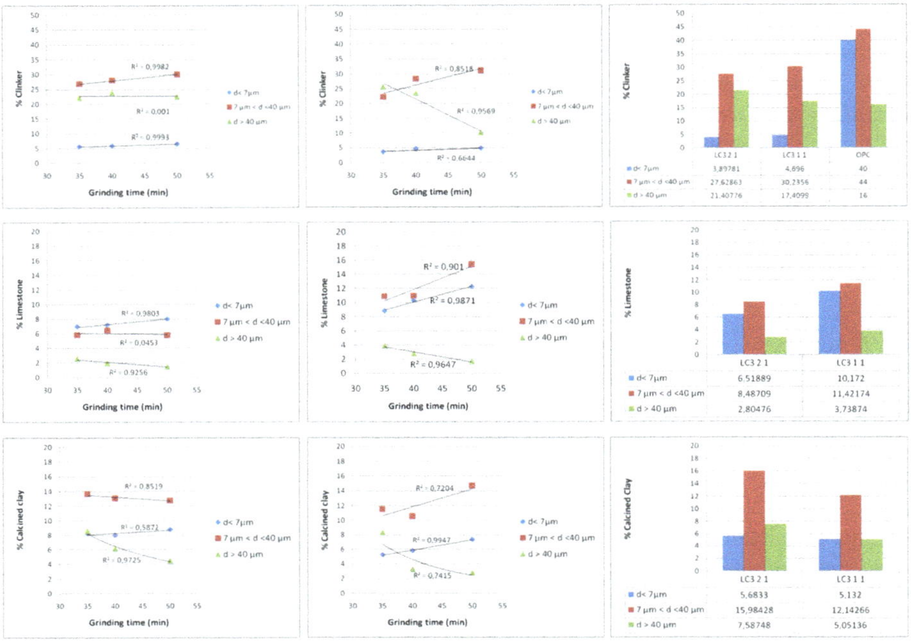

Fig. 2. Evolution of components fineness (laboratory) for LC^3 2:1, LC^3 1:1 and PSD (industrial)

As designed the total amount of limestone in LC^3 1:1 is higher than LC^3 2:1. Further grinding on both blends decreases the amount of limestone at d> 40 μm fraction and the total amount for both is similar. There is a slow increase of fineness at fraction d < 7 μm for LC^3 2:1 and no major changes at fraction 7 μm < d < 40 μm. That is also explained by the shielding process. For LC3 1:1, further grinding increases fineness for the finer fractions and the amount of material in both cases is higher than for LC^3 2:1 (almost the double).

For both formulations further grinding decreases the amount of coarse calcined clay. For LC^3 2:1 further grinding has no important impact on the remaining fractions. The major increase for LC^3 1:1 is in the d < 7 μm fraction. That can be explained by the effect of more limestone and less calcined clay present on the grinding balls surface.

In general terms, for industrial blends, LC^3 2:1 corresponds with laboratory cement in a range from 35 min to 40 min of grinding time while LC^3 1:1 corresponds with laboratory cement in a range from 40 min to 45 min.

Figure 3 shows the results of compressive strength for all the blends studied and the reference Portland cement. For LC^3 2:1 increasing fineness increases compressive strength. According to the fact that there is no mayor change on clinker fineness, the increase of limestone fineness d < 7 μm most increase the filler effect increasing strengths. For LC^3 1:1, the increase of clinker fineness increases strength at all ages. Finer LC^3 1:1 has early strength similar to the best LC^3 2:1 (similar amount of reactive clinker), however strength at 28 days is slightly lower because of the less calcined clay/ limestone ratio (less synergy). Compressive strength results for industrial blends correspond to its similar laboratory blends. For LC^3 2:1 less amount of clinker and calcined clay in d < 7 μm, compared with laboratory blends, impact negatively on compressive strength at 7 days. Strengths for finer 2:1 blend and 2:1 industrial blend are similar to reference OPC.

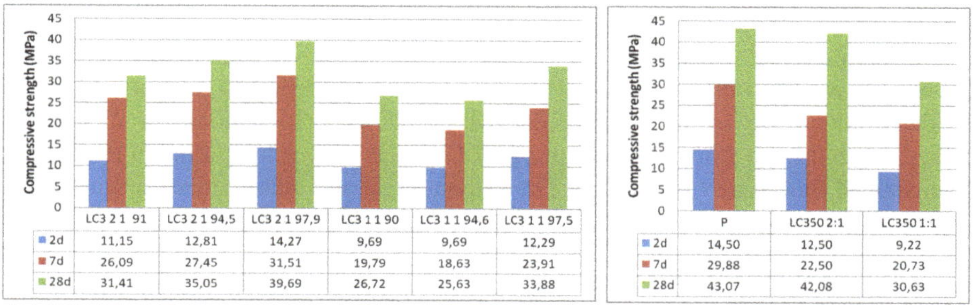

Fig. 3. Compressive strength vs. Fineness for laboratory and industrial blends

Figure 4 shows the results of specific surface measured by Blaine and rheology measured by minislump. Increasing the amount of limestone rheology is improved. In general terms, reducing calcined clay content reduce specific surface, water demand and improves rheology.

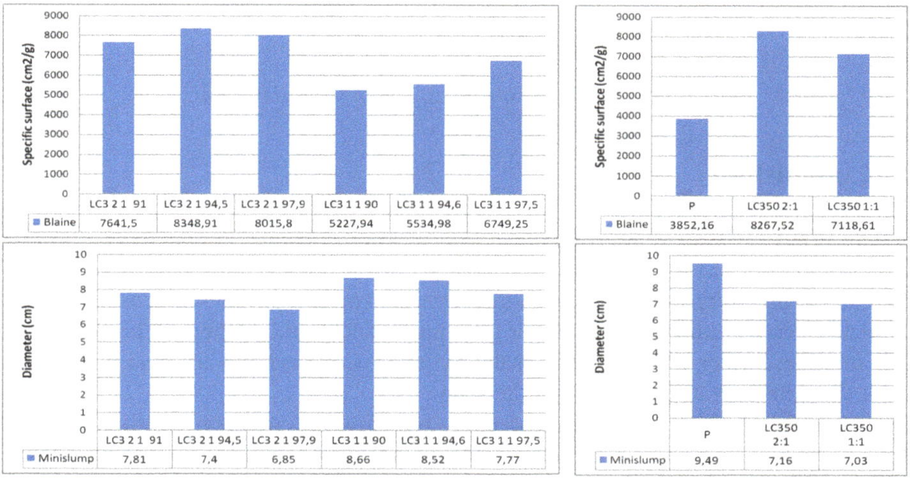

Fig. 4. Specific surface and rheology of cement pastes

On laboratory blends, for LC^3 2:1, rheology decreases with the increase of fineness. For LC^3 1:1 rheology is affected by the increase of finer clinker and finer calcined clay, rising specific surface, but the general more amount of limestone, even at finer fractions, reduce the specific surface improving rheology due to a better PSD of the system (worst LC^3 1:1 rheological behavior, similar to best LC^3 2:1 behavior). The results for industrial blends correspond to its similar laboratory blends. For LC^3 2:1 the more amount of calcined clay in the 7 μm < d < 40 μm fraction, compared with LC^3 2:1 blends from laboratory, increases specific surface rising water demand and affecting rheology. Due to de absence of calcined clay, reference OPC has less specific surface and a better rheological behavior.

Pozzolanic effect of calcined clay is also corroborated by Isothermal calorimetry. Figure 5 shows the normalized heat flow/g clinker for industrials cements. For both blended cements curves are very similar. More calcined clay amount in all fraction studied, especially at 7 μm < d < 40 μm for LC^3 2:1 improves hydration more than the increasing of limestone for LC^3 1:1. The sulphating is also different, according to the different amount of alumina due to the change on calcined clay/limestone ratio.

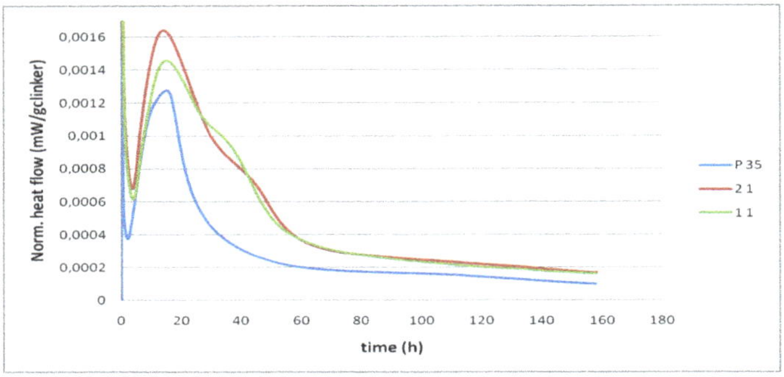

Fig. 5. Isothermal calorimetry results for industrial cements

4 Conclusions

By intergrinding is possible to produce LC3 blended cements. Augmenting grinding time compressive strength is improved but water demand rises. Difference between raw materials grindability and morphology affect the gain of fineness for hard ones. Presence of calcined clay muffles proper clinker grinding, reducing the amount of clinker in the finer fraction, further grinding don't make a change. Calcined clay fineness is the main factor that affects water demand and rheology.

Changing calcined clay/limestone ratio, reducing the amount of calcined clay, clinker grinding is improved. By the other hand is possible to find a better compromise between strength and rheology. Increasing limestone amount reduces water demand while is possible to maintain high compressive strength. Using grinding aids (grinding intensifiers) will be possible to improve clinker grinding.

Acknowledgements. Special thanks are given to Aurelie Favier (LMC) for the Rietveld analysis.

References

1. Vizcaino, L., Antoni, M., Alujas, A., Martirena, F., Scrivener, K.: Effects of particle size distribution and intergrinding in blend of cement, calcined clays and limestone. In: Twin International Conferences 2nd Civil Engineering and 5th Concrete Future (2013)
2. De Weerdt, K.: Separate grinding versus intergrindig. State of the art, Sintef Report (2007)
3. Tsivilis, S.T., Moutsatsou, A.: Contribution to the problems arising from grinding of multicomponent cements. Cem. Concr. Res. **22**, 95–102 (1992)

Durability of Concretes Made with Calcined Clay Composite Cements

Roland Pierkes(✉), Simone E. Schulze, and Jörg Rickert

Cement Chemistry Department, Research Institute of the Cement Industry, VDZ GGmbH, Tannenstrasse 2, 40476 Duesseldorf, Germany

Abstract. To prove the applicability of cement with calcined clays as main constituent in concrete the performance of such composite cements, especially with regard to durability, has to be demonstrated. Composite cements with different calcined clays were used in conventional concrete mixtures and tested for their resistance to carbonation, chloride migration, frost and frost-deicing salts. Attention had to be paid to the workability of the concretes which could be affected strongly by the increases in the water demand of cements with calcined clays. Adjustments with superplasticisers could become necessary. The results of the durability tests showed that the performance of such binder systems is comparable to cements with other substitution material like limestone, slag or fly ash.

1 Introduction

Within two research projects VDZ was able to show the performance of several calcined impure clays as cement main constituent. The used kaolinitic, illitic and chloritic clays show low ceramic qualities and represent typical clays of cement plant quarries. Suitable calcination conditions in laboratory scale led to appropriate pozzolanic reactive compounds (silica and alumina), resulting in acceptable strength developments when mixed to cements with a clinker substitution of 20 and 40 mass% [1]. Further investigations showed that there is some potential for optimization in the adjustment of clinker, calcined clay and sulphate carrier [2, 3]. Details on clay compositions, calcination and cement properties could also be found in [2, 3].

To prove the applicability in concrete the performance of such composite cements, especially with regard to durability, has to be demonstrated. Composite cements with different calcined clays were produced on laboratory scale, again using a substitution rate of 20 and 40 mass% (CEM II/A-Q, CEM IV/A-Q). Concrete mixtures as used in tests for European technical approval (ETA) were produced with these cements and tested for their resistance to carbonation, chloride migration, frost and frost-deicing salts.

F. Martirena et al. (eds.), *Calcined Clays for Sustainable Concrete*, RILEM Bookseries 16,
https://doi.org/10.1007/978-94-024-1207-9_59

2 Test Methods

Table 1 show the concrete compositions and respective test standards used for the durability tests, and the range of compressive strengths obtained after 28 days of hydration of concretes using calcined clay composite cements.

Table 1. Concrete test parameters and results of concrete compressive strength

Parameter	Concrete type A	Concrete type B
Composition	Cement content 300 kg/m^3; w/c-ratio 0.60; aggregate grading ≈ B16: no air entraining agent (AEA)	Cement content 320 kg/m^3; w/c-ratio 0.50; aggregate grading ≈ AB16; AEA, air voids = 4.5-5.5 vol.%
Compressive strength	DIN EN 12390-3	DIN EN 12390-3
Resistance to carbonation	DIN EN 12390-10	–
Resistance to chloride migration	chloride migration test[a)]	–
Resistance to frost	cube frost (EN CEN/TS 12390-9)	–
Resistance to frost-deicing salts	–	CDF test (EN CEN/TS 12390-9)
Compressive strength range CEM II/A-Q (at 28 days)	37 - 52 MPa (requirement acc. EN 206-1: ≥ C 25/30, **passed**)	38 – 50 MPa (requirement acc. EN 206-1: ≥ C 30/37, **passed**)
Compressive strength range CEM IV/A-Q (at 28 days)	27 - 45 MPa (requirement acc. EN 206-1: (≥ C 25/30, **passed**)	28 MPa (requirement acc. EN 206-1: ≥ C 30/37, **not passed**)

(a) following NT Build 492 (11.99) [4]; - = not determined

3 Results

Table 1 shows that concretes with useful compressive strengths up to C 35/45 acc. EN 206-1 could be achieved. Only the air entrained concrete with CEM IV/A-Q (40 mass% calcined clay) did not reach the strength required by EN 206 for such an exposure.

The workability of the concretes was strongly influenced by the type of calcined clay and by the calcination temperature. Kaolinitic clays, calcined at temperatures of about 600–700°C to reach a high pozzolanic reactivity, show a very high specific surface and water demand. Concretes using this clay type in cement show a loss in workability which sometimes must be compensated with a superplasticiser, which itself could strongly be absorbed by the calcined clay compound [5]. Illitic clays which need higher calcination temperatures up to sintering conditions did not show this effect. In single cases the concrete workability could even be slightly improved.

A crucial aspect of concrete durability is the resistance against carbonation, which might be critical for cements with extended use of pozzolanic main constituents. In Fig. 1 the range obtained for carbonation behaviour of concrete mixtures with and without calcined clays is shown. As expected, the carbonation resistance is somewhat lower compared to CEM I (OPC), but is not leaving the carbonation range of blast furnace cements or portland limestone cements. Further improvement will be achieved with some extended curing.

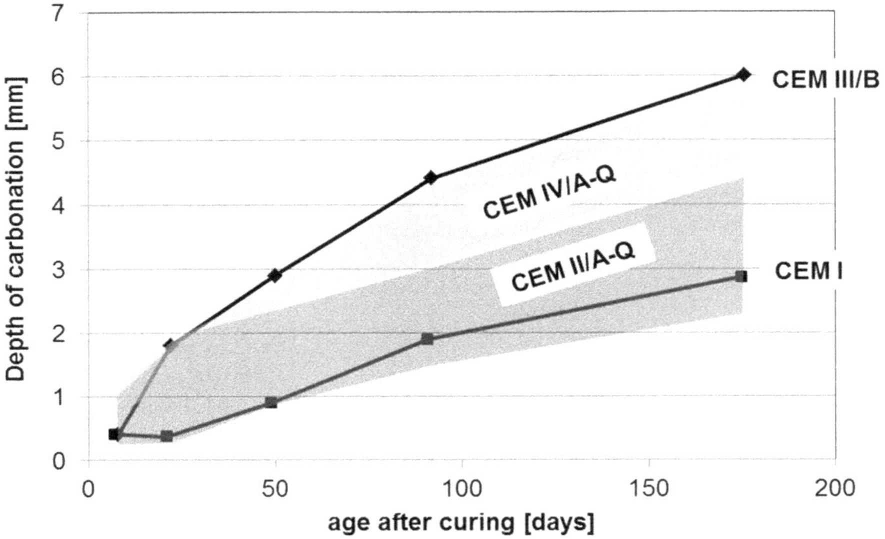

Fig. 1. Ranges in depth of carbonation of concretes with CEM II/A-Q (20 mass% calcined clay) and CEM IV/A-Q (40 mass% calcined clay), related to typical CEM I and blast furnace cement

As known from experience with e.g. fly ash cements, pozzolanic cement main constituents could decrease the chloride migration coefficient in concrete. Figure 2 shows that this improvement is not reached of all calcined clays concretes. Kaolinitic clays reach comparably low coefficients already after 35 days (Mix A, Mix B), predominantly illitic clays show a more delayed pozzolanic effect with a strong decrease of the coefficient at 98 days (Mix C, Mix D). Less reactive calcined clays with higher impurities of quartz and carbonates (Mix E, Mix F) do not reduce the migration coefficient due to a higher porosity of the concrete microstructure.

For a broad use of calcined clays composite cements in Europe a suitable frost resistance of respective concretes will be required. Figure 3 displays the results of the so-called "cube test", described in EN CEN/TS 12390-9. In this test, concrete cubes are subjected to 100 frost-thaw cycles, and a loss in weight of up to 10 mass% is accepted to pass the test ("high frost resistance"). All CEM II/A-Q tested in the research project felt below a weight loss of 2 mass%, indicating a very suitable frost resistance. Even the tested CEM IV/A-Q did not exceed a weight loss of 5 mass% and passed this test very successfully.

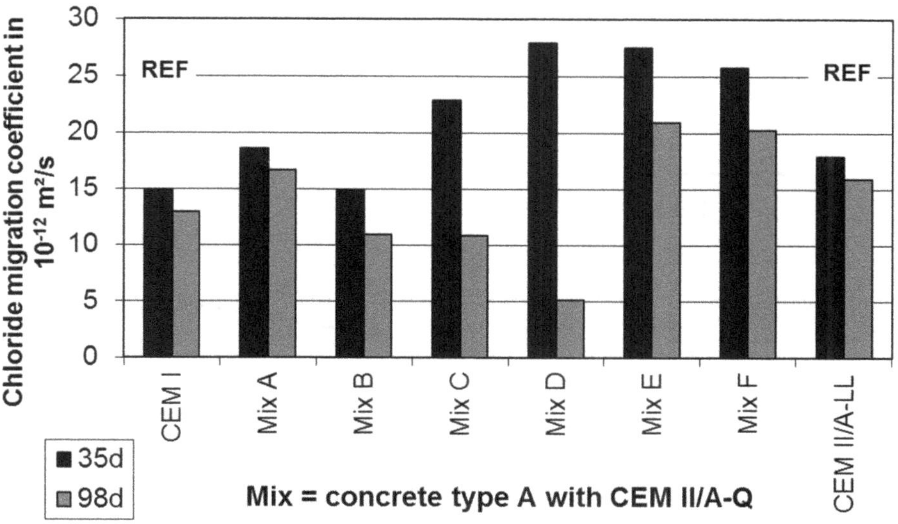

Fig. 2. Chloride migration coefficient of concretes with different CEM II/A-Q (20 mass% of calcined clay), related to typical CEM I and portland limestone cement results

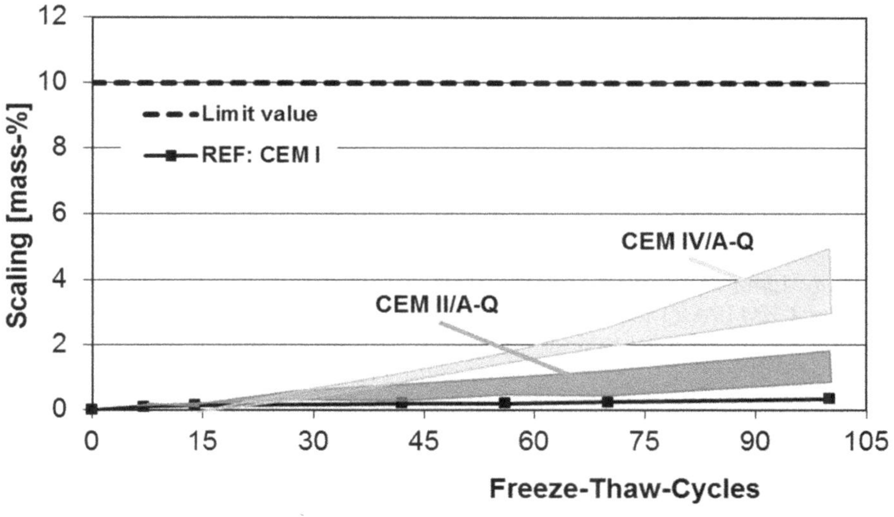

Fig. 3. Scaling ranges of concretes with CEM II/A-Q and CEM IV/A-Q in cube test (acc. EN CEN/TS 12390-9), compared to CEM I concrete and limit value

To prove the frost-deicing salt resistance of a concrete, acc. EN CEN/TS 12390-9 an additional test is necessary, called "CDF test". The concretes were immersed in salt solution and exposed to 28 frost-thaw cycles. To pass the test, a scaling of less than 1500 g/m^2 concrete surface is necessary. Figure 4 shows the results of several calcined clay concretes subjected to the CDF test. All concretes tested showed a scaling in a range similar to the experiences gained by concretes with Portland cements or blast furnace cements and pass the criteria of a scaling of less than 1500 g/m^2.

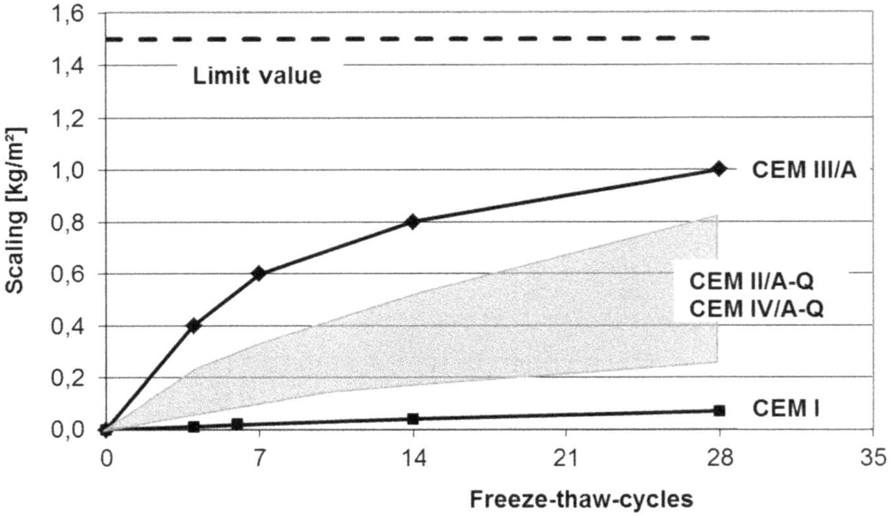

Fig. 4. Scaling range of concretes (type B) with CEM II/A-Q and CEM IV/A-Q in CDF test (acc. EN CEN/TS 12390-9), compared to CEM I and CEM III/A concrete and the limit value

4 Conclusions

A precondition for the application of composite cements containing calcined clays in the European market is to prove the cement performance in concrete, especially with respect to durability. Within a research project VDZ tested a couple of concretes containing CEM II/A-Q and CEM IV/A-Q laboratory cements on durability aspects like resistance to carbonation, chloride migration and frost as well as frost-deicing salt resistance.

The results showed that the performance of such systems is comparable to cements with other substitution compounds like limestone, blast furnace slag or fly ash. As the pozzolanicity of calcined clays which do not mainly consist of pure kaoline gets more effective with higher hydration age, a further increase of the concrete performance is expected with ongoing hydration time.

To ensure an appropriate workability of the concretes which could be affected by strong increases in the water demand of specific types of calcined clays in the cement, corrective measures using plasticisers could be required.

References

1. Schulze, S.E., Rickert, J.: Pozzolanic activity of calcined clays. In: Twelfth International Conference on Recent Advances in Concrete Technology and Sustainability Issues (Prag 30.10.–02.11.2012). Farmington Hills: American Concrete Institute, ACI, (ACI Publication SP-289) (2012)
2. Pierkes, R., Schulze, S.E., Rickert, J.: Optimization of cements with calcined clays as supplementary cementitious materials. In: Scrivener, K., Favier, Aurélie (eds.) Hrsg Calcined Clays for Sustainable Concrete: Proceedings of the 1st International Conference on Calcined Clays for Sustainable Concrete, Lausanne, 23–25 June 2015
3. Schulze, S.E., Pierkes, R., Rickert, J.: Optimization of cements with calcined clays as supplementary cementitious materials. In: China Building Materials Academy (Hrsg): 14th International Congress on the Chemistry of Cement, ICCC, Beijing, 13–16 October 2015
4. NT Build 492 11.99. Concrete, Mortar and Cement-Based Repair Materials: Chloride Migration Coefficient from Non-Steady-State Migration Experiments
5. Herrmann, J., Rickert, J.: Interactions between cements with calcined clay and superplasticizers. In: Malhotra, V.M., Gupta, P.R, Holland, T.C. (eds.) 11th International Conference on Superplasticizers and Other Chemical Admixtures in Concrete, 12–15 July 2015, Ottawa, Canada, Proceedings. Ottawa: American Concrete Institute, ACI (ACI Special Publication SP 302 (2015)

Alkali-Activation of Calcined Clays – Past, Present and Future

John L. Provis(✉)

Department of Materials Science and Engineering, University of Sheffield, Sheffield S1 3JD, UK

Abstract. The combination of a calcined clay with an alkali silicate or hydroxide solution has been identified since the 1920s to yield potentially useful materials. More recently these have become termed 'geopolymers', and have been popularised under that name. This paper briefly summarises some of the earlier history of alkali-calcined clay binders and related materials including synthetic zeolites, exploring some of the reasons underlying the more recent broadening of interest in this research field, and identifying some of the future opportunities that arise through the use of these materials. These cements may particularly be capable of offering very good technical performance and cost-effectiveness in a variety of applications, with an environmental emissions footprint lower than that of competing materials.

1 Introduction

The term 'geopolymer' was introduced by Davidovits in the 1970s to describe binders based on the reaction between calcined clays and a source of alkalinity, most commonly an aqueous alkali metal silicate or hydroxide solution [1–3]. However, this combination of reactants had been described in the technical literature some time prior to the work of Davidovits, in some cases yielding the monolithic alkali aluminosilicate gels which characterise geopolymer materials, but in other cases focused on the synthesis of crystalline zeolites. Some of this work will be reviewed briefly here, and the potential for further future development of alkali-activated binders based on calcined clays will be noted. The details of the chemistry of the alkali aluminosilicate gels which are formed by such processes have been described in detail in the technical literature [4–8]. The scope of this paper is limited to calcium-free binding systems, although lime-pozzolan cements based on calcined clays are certainly both well-known and of significant technological value.

It is also worthwhile to mention that some workers have used the combination of phosphoric acid and calcined clays to produce what they describe as an 'acid geopolymer' [9–11]; although potentially of some technical interest due to their potential for rapid strength gain, the high cost of phosphoric acid limits the use of such cements on a very large scale in construction and they will not be the focus of further discussion in this brief paper.

F. Martirena et al. (eds.), *Calcined Clays for Sustainable Concrete*, RILEM Bookseries 16,
https://doi.org/10.1007/978-94-024-1207-9_60

2 The First Steps and Further Development

The earliest identified study of the reaction of an alkali silicate with calcined clay (kaolinite, in this instance) was published by the U.S. National Bureau of Standards in 1920 [12], where this combination of materials was included in a set of tests aimed at identifying the preferred cement for use in joining spark plug wires to porcelain insulator bodies. It was reported that "*When brought into contact with a strongly alkaline substance such as sodium silicate, calcined kaolin reacts quite rapidly to form a friable, porous mass*", with "*little strength*" [12]. Fortunately, the development of materials based on similar chemical reaction processes, for this and other applications, did not stop at this rather unpromising beginning. The addition of lard was found to improve the properties of these cement sufficiently to enable their successful use in spark plugs [13], while applications for similar materials in construction were also developed, for example in waterproof structural wallboard [14] and surfacing granules [15].

The longest-running research and development programme related to alkali-activated binder materials is the group in Kiev, Ukraine, which was initiated by V.D. Glukhovsky and is now continued in the institute which bears his name. An early landmark publication from that group was a 1957 book entitled *Gruntosilikaty* (Soil Silicates) [16]. This book describes binder formulations based on combinations of metallurgical slags, clay soils such as loess, brown clay, and loam, and alkaline solutions containing NaOH, silicates, carbonates and/or fluorides. Subsequent work by the same group led to the development of a broad range of alkali-activated binder systems, mainly based on metallurgical slags, but with systems based on calcined clays particularly highlighted as having potential in the immobilisation of radioactive wastes [17].

The reaction of calcined clays with alkaline solutions was also studied in the context of zeolite synthesis, with the work of the group led by Barrer [18–20] being of particular importance in understanding the alkali-hydrothermal reactions of metakaolin. Although the purpose of that work was to produce crystalline zeolites rather than monolithic binder phases, the insight developed into the conditions which lead to the formation of numerous different types of zeolitic crystal has proven essential in understanding the nanostructures of the binders which are formed through analogous reactions under non-hydrothermal conditions [21].

In this context, Davidovits in the early 1970s began to develop alkali-aluminosilicate cements based on metakaolin as an inorganic, fireproof alternative to organic polymers [22], and subsequently introduced 'geopolymer' terminology as a descriptor for this class of materials [1–3]. His work in promoting and popularising geopolymer cements has led to a high degree of international prominence for the materials since the 1980s, and the term geopolymer has become applied to ever broader groups of materials beyond its original definition. Nonetheless, applying the original sense of a geopolymer as a mineral binder based on alkali-aluminosilicate chemistry, with physical characteristics resembling to some degree those of an organic thermoset, research in this area has since led to many valuable products for both cement-like and ceramic-like applications.

3 The Present and the Future; Development Opportunities and Needs

Innovation in the construction materials industry is currently being driven by environmental pressures, specifically the need to reduce sector-wide CO_2 emissions [23]. It has been identified that calcined clays offer probably the greatest scope among all materials to be used on a gigaton per annum scale in place of Portland cement [24]; their incorporation into blends with Portland cement is certainly a key avenue by which these materials will add value in the global built environment, but the production of alkali-activated clay-based cements is certainly of strong technological and societal interest in areas where the necessary resources are available [25]. Significant commercial advances have recently been made in this regard in the UK [26, 27] and elsewhere, but there is a clear need for further advancement in both the design and testing of clay-based alkali-activated binders to enable scale-up and deployment to continue at pace, to exercise the full potential of these materials. This will also feed into ongoing standardisation efforts (e.g. [28]), where the availability of validated testing methods will enable the confident application of performance-based routes to specification.

In terms of materials design, one of the greatest strengths of alkali-activation as a route to cement production is the ability to match activators to precursors in a controlled way, to manipulate and optimise material properties in a very flexible way. The activator that usually gives the best results in conjunction with a calcined clay is an aqueous alkali metal silicate (also called 'waterglass'), which is a commercial product usually sold for applications in detergent manufacture and other fields where the total mass of material used is orders of magnitude below the demand for construction materials. The ability to produce and/or source activators at reasonable cost and sufficient volume, and with an acceptable ecological footprint to enable the 'green' credentials of the material to be attractive, lies at the heart of the quest for industrial uptake of these cements, and there are rapid and ongoing developments in this area worldwide.

In terms of materials testing, it has been identified that the durability tests which give reliable results for materials based on Portland cement are often less representative when applied to non-Portland cement binders. The alkali-activated materials based on calcined clays are among the groups of materials for which improved testing methods are needed; the chemistry of these binders is very different from that of Portland cement (much more so than many other alkali-activated materials which may contain higher levels of calcium), and so many of the durability tests designed to determine the performance of calcium-rich binders may not provide realistic insight into the behaviour of a material that is so different in chemistry. Much ongoing work, including through a RILEM Technical Committee and a European Federation of Corrosion Task Group, is aimed at remedying this mismatch and providing guidance for specifiers and testing laboratories regarding how these materials can realistically be tested to provide an understanding of in-service performance.

4 Concluding Remarks

Alkali-activation of calcined clays has been trialled for almost 100 years as a means of producing cementitious products, for specialty and/or bulk applications. The current drive towards sustainable development in the construction industry has generated significant impetus for the deployment of these materials, and the technical and non-technical hurdles which have historically impeded their uptake are being eroded or removed as field experience is gained, standards and specifications are put in place, and improved methods of designing, producing and utilising these materials are developed.

Acknowledgements. The research of the author into geopolymer and alkali-activated cements is funded by the UK's Engineering and Physical Sciences Research Council (EP/M003272/1 and EP/P013171/1), and by the European Research Council (FP7 StG #335928)

References

1. Davidovits, J.: Solid-phase synthesis of a mineral blockpolymer by low temperature polycondensation of alumino-silicate polymers: Na-poly(sialate) or Na-PS and characteristics. In: IUPAC Symposium on Long-Term Properties of Polymers and Polymeric Materials, Stockholm, Sweden (1976)
2. Davidovits, J.: The need to create a new technical language for the transfer of basic scientific information. In: Gibb, J.M., Nicolay, D. (eds.) Transfer and Exploitation of Scientific and Technical Information, EUR 7716, pp. 316–320. Commission of the European Communities, Luxembourg (1982)
3. Davidovits, J., Orlinski, J.: Proceedings of Geopolymer 1988 - First European Conference on Soft Mineralurgy, Compeigne, France, 369 pp. (1988)
4. Duxson, P., Provis, J.L., Lukey, G.C., Separovic, F., van Deventer, J.S.J.: 29Si NMR study of structural ordering in aluminosilicate geopolymer gels. Langmuir **21**, 3028–3036 (2005)
5. Provis, J.L., Bernal, S.A.: Geopolymers and related alkali-activated materials. Annu. Rev. Mater. Res. **44**, 299–327 (2014)
6. Provis, J.L., Palomo, A., Shi, C.: Advances in understanding alkali-activated materials. Cem. Concr. Res. **78A**, 110–125 (2015)
7. Pouhet, R., Cyr, M.: Formulation and performance of flash metakaolin geopolymer concretes. Constr. Build. Mater. **120**, 150–160 (2016)
8. Palomo, A., Glasser, F.P.: Chemically-bonded cementitious materials based on metakaolin. Br. Ceram. Trans. J. **91**, 107–112 (1992)
9. Liu, L.-P., Cui, X.-M., Qiu, S.-H., Yu, J.-L., Zhang, L.: Preparation of phosphoric acid-based porous geopolymers. Appl. Clay Sci. **50**, 600–603 (2010)
10. Lassinantti Gualtieri, M., Romagnoli, M., Pollastri, S., Gualtieri, A.F.: Inorganic polymers from laterite using activation with phosphoric acid and alkaline sodium silicate solution: Mechanical and microstructural properties. Cem. Concr. Res. **67**, 259–270 (2015)
11. Perera, D.S., Hanna, J.V., Davis, J., Blackford, M.G., Latella, B.A., Sasaki, Y., Vance, E.R.: Relative strengths of phosphoric acid-reacted and alkali-reacted metakaolin materials. J. Mater. Sci. **43**, 6562–6566 (2008)
12. Staley, H.F.: Cements for Spark-Plug Electrodes (Technologic Papers of the National Bureau of Standards, #155), Government Printing Office, Washington DC, 10 pp. (1920)

13. Schwartzwalder, K., Ortman, C.D.: Sodium silicate type cement (U.S. Patent 2,793,956) (1957)
14. Parry, R.E.: Structural unit and method of manufacture (U.S. Patent 2,549,516) (1946)
15. Morrow, G.W., Sackrison, N.B.: Mineral surfacing granules containing calcined clay (U.S. Patent 3,169,075) (1965)
16. Glukhovsky, V.D.: Gruntosilikaty (Soil Silicates). Gosstroyizdat, Kiev (1959)
17. Krivenko, P.V.: Alkaline cements. In: Proceedings of the First International Conference on Alkaline Cements and Concretes, Kiev, Ukraine, pp. 11–129 (1994)
18. Barrer, R.M., Mainwaring, D.E.: Chemistry of soil minerals. Part XI. Hydrothermal transformations of metakaolinite in potassium hydroxide, J. Chem. Soc.-Dalton. Trans., 1254–1265 (1972)
19. Barrer, R.M., Mainwaring, D.E.: Chemistry of soil minerals. Part XIII. Reactions of metakaolinite with single and mixed bases. J. Chem. Soc.-Dalton. Trans., 2534–2546 (1972)
20. Barrer, R.M., Beaumont, R., Colella, C.: Chemistry of soil minerals. Part XIV. Action of some basic solutions on metakaolinite and kaolinite. J. Chem. Soc.-Dalton. Trans., 934–941 (1974)
21. Provis, J.L., Lukey, G.C., van Deventer, J.S.J.: Do geopolymers actually contain nanocrystalline zeolites? - A reexamination of existing results. Chem. Mater. **17**, 3075–3085 (2005)
22. Davidovits, J.: Geopolymer Chemistry and Applications. Institut Géopolymère, Saint-Quentin (2008)
23. Scrivener, K.L., John, V.M., Gartner, E.M.: Eco-efficient cements: Potential, economically viable solutions for a low-CO_2. Cement-based materials industry, Paris, 50 pp. (2016)
24. Scrivener, K.L.: Options for the future of cement. Indian Concr. J. **88**, 11–21 (2014)
25. Provis, J.L.: Alkali-activated materials. Cem. Concr. Res. (2017, in press). doi:10.1016/j.cemconres.2017.1002.1009
26. McIntosh, A., Lawther, S.E.M., Kwasny, J., Soutsos, M.N., Cleland, D., Nanukuttan, S.: Selection and characterisation of geological materials for use as geopolymer precursors. Adv. Appl. Ceram. **114**, 378–385 (2015)
27. Lawther, S.E.M., McIntosh, J.A., Nanukuttan, S., Provis, J.L., Soutsos, M.N., Jose, D.: Understanding the microstructure of alternative binder systems – banahCEM, a metakaolin based geopolymer. In: Civil Engineering Research in Ireland 2016, Galway (2016)
28. British Standards Institute, BSI PAS 8820:2016, Construction materials – Alkali-activated cementitious material and concrete – Specification, London, UK (2016)

Influence of a Calcined Clay and the Temperature on the Hydration of an Oilwell Cement

Juan Alberto Ribalta[1(✉)], Adrián Alujas Díaz[2], and José Fernando Martirena[3]

[1] Departamento de Licenciatura en Química, Universidad Central "Marta Abreu" de Las Villas, 54830 Santa Clara, Cuba
juanrq@uclv.edu.cu
[2] Centro de Estudios de Química Aplicada, Universidad Central "Marta Abreu" de Las Villas, 54830 Santa Clara, Cuba
[3] Centro de Investigación y Desarrollo de Estructuras y Materiales, Universidad Central de Las Villas, 54830 Santa Clara, Cuba

Abstract. The influence of a calcined clay on the hydration of an oilwell cement was studied in slurries. The replacement grades of the cement (by weight of cement) were 5, 10 and 20% of the calcined clay. The slurries were prepared at a density of 1.80 $g{\cdot}cm^{-3}$ and hydrated in sealed containers at 30 and 60 °C during 24 h. A reference slurry was prepared where the cement was substituted by 5% of a zeolitic tuff. The water–to–solid ratio was 0.52 for each slurry. From the TGA and pozzolanic reactivity tests were shown that the calcined clay was more reactive than the zeolitic tuff. The isothermal calorimetry tests showed that the induction periods of the slurries were shorter when the cement was replaced by the calcined clay at both temperatures and they decreased with the increment of the temperature and the replacement level by the mineral addition, what could impact negatively the thickening time in real conditions. The MIP assays evidenced that the influence of the kind and the quantity of the mineral addition on the porosity of the hardened slurries is complex but in general it was observed a refinement of porosity and less total porosity when the slurries were hydrated at 60 °C. The XRD experiments showed that the phase assemblages when the slurries were cured at 60 °C were more complex and with more crystalline phases presents, which could have influences on the performance of the hardened slurries in real well conditions.

1 Introduction

Portland cement is by far the most important oilwell binding material [1]. It is widely used for primary cementing, one of the most important and crucial operations performed in an oil and/or gas well [2]. However, Portland cement cannot fulfill by itself with all the requirements that are needed for each situation [3]. Among the most demanding conditions the Portland cement can meet during a cementing operation are

F. Martirena et al. (eds.), *Calcined Clays for Sustainable Concrete*, RILEM Bookseries 16,
https://doi.org/10.1007/978-94-024-1207-9_61

high temperatures. As temperature increases with the depth of the wellbore, oil well cements are subjected to wide ranges of temperature. It is important to study the effects of these different curing conditions on hydration of oil well cements. In Cuba the higher attained temperature in a primary cementing operation is about 60 °C. A wide variety of additives are used to modify the properties of cement slurries to let them to be used in different oilwell conditions. Specifically, extenders are employed to reduce slurry density and/or increase slurry yield. Among them the pozzolans are perhaps the most important group of cement extenders. The most common pozzolanic extenders are diatomaceous earths, fly ashes and silica [1]. Others pozzolans have been studied as extender for oilwell cementing, such as metakaolin [4], sugarcane biomass waste [5], blast furnace slag [6] and rice husk ash [7]. In Cuba, zeolites are the most important national pozzolans used as extender in primary oilwell cementig, fundamentally that of from the deposit San Andrés, but its use is limited to a 5% by weight of cement (BWOC), because of a higher replacement level induce too much free fluid, i.e. sedimentation. (M. L. Morgado 2016, pers.comm., 15 January).

Clays with low and moderate kaolin contents are promissory pozzolanic extenders to be used during oilwell cement, overall because they are widely spread and easily activated by calcination. Several recent works have demonstrated that clays with low and moderate kaolin content can be used as pozzolanic materials when they are adequate activated [8–11]. The necessity to increase the replacement of the Portland cement in slurries for oilwell cementing is the main motivation of this work, overall taking into account that to the best of this work authors′ knowledge there are not many studies about the hydration of Portland cement in presence of low or moderate metakaolin content calcined clays in slurries submitted at temperatures higher than room temperature.

2 Materials and Experimental Procedures

The clay and the zeolitic tuff were collected, respectively, from the deposits Pontezuela and San Andrés, Cuba. The clay was stationary calcined during 1 h at 850 °C. The zeolite tuff was used as received. The cement was a Cuban Portland cement produced at the Siguaney plant. The chemical compositions of the materials were determined by X – ray fluorescence. The results are presented in Table 1.

Table 1. Summary of the chemical compositions of the materials used in these work (expressed in weight percent).

	Al_2O_3	SiO_2	Fe_2O_3	CaO	MgO	Na_2O	K_2O	SO_3	LOI*
Pont850[£]	46.80	28.97	19.93	0.08	0.772	0.090	0.614	0.048	1.35
ZSA[¥]	11.73	67.46	1.98	2.833	1.75	1.325	1.446	0.004	10.8
PC[§]	4.52	18.53	5.51	61.13	1.110	0.224	0.433	4.215	2.22

* Loss on ignition

[£] Calcined clay from Pontezuela

[¥] Zeolite from San Andrés.

[§] Portland cement.

The phase identification and quantification for the clay were made by X ray diffraction. The phase compositions of the clay, and that of its clayey fraction, as well as that of the zeolitic tuff are presented in Table 2.

Table 2. Phase compositions of the materials used in these work (expressed in weight percent).

	Kaolinite	2:1 Clay	Total of clayey phases	Quartz	Goethite
Pont*	45.7	38.7	84.4	7.5	8.1
	Kaolinite	Montmorillonite	Montmorillonite – Illite	Illite	
Clayey fraction of Pont*	65.8	20.0	8.0	6.2	
	Clinoptilolite	Mordenite	Total of zeolites	Montmorillonite	Calcite
ZSA**	50	28	78	11	< 1

* Refers to the clay from Pontezuela without calcining
** Refers to the zeolite from San Andrés (The phase composition was that of reported by [12])

The particle size distribution was determined by laser diffraction on a device Malvern MasterSizer type S. The suspensions were prepared with the aid of ultrasound and using the same dispersant to prepare the slurries. The percentiles characterizing each particle size distribution (D_{10}, D_{50} and D_{90}) are shown in the Table 3.

Table 3. D_{10}, D_{50} and D_{90} for the calcined clay from Pontezuela (Pont850) and the zeolitic tuff from San Andrés (ZSA).

	Maximum diameter (μm)	
Percentile	Pont850	ZSA
D_{10}	1.36	0.28
D_{50}	5.08	10.98
D_{90}	30.20	40.70

The slurries were prepared at a density of 1.80 $g \cdot cm^{-3}$ and the mix design is shown in Table 4. The pozzolanic reactivities of both mineral additives were estimated by R3 protocol [13].

After mixing the slurries were cured in sealed plastic containers at 30 and 60 °C during 24 h. Fresh slices from these specimens were taken for X – ray diffraction. Other slices were taken, their hydration were stopped using isopropanol and then crushed for thermogravimetric analysis.

The calorimetric assays on the slurries were done at 30 and 60 °C during 24 h in a TAM AIR calorimeter. The pore structures were studied by MIP at a Thermoscience (Pascal 140/440) device. The samples for porosimetry were taken from the same

Table 4. Compositions of the slurries.

$\%_{Extender}$* (BWOC)	$\%_{Free\text{-}fluid\text{-}control}$ (BWOC)	$\%_{Dispersant}$ (BWOC)	$\%_{Fluid\text{-}loss\text{-}control}$ (BWOC)	$\%_{Retarder}$ ($L{\cdot}sk^{-1}$)	$\%_{Antifoam}$ ($L{\cdot}sk^{-1}$)	w/s	w/c
5	0.50	0.20	0.60	1.00	0.07	0.52	0.54
10							0.58
20							0.63

* Calcined clay from Pontezuela and zeolitic powder from San Andrés. The zeolitic powder was employed as a reference and only at 5%

specimens cured at 30 and 60 °C and their hydration stopped with isopropanol. TGA curves were obtained at a device TGA/SDTA851[e] from Metter Toledo, the gas purge was N_2. The samples for thermogravimetry were prepared similarly to those for porosimetry.

3 Presentation and Discussion of the Main Results

The calorimetric curves when the mineral additives were subjected to pozzolanic the reactivity assay are presented in Fig. 1

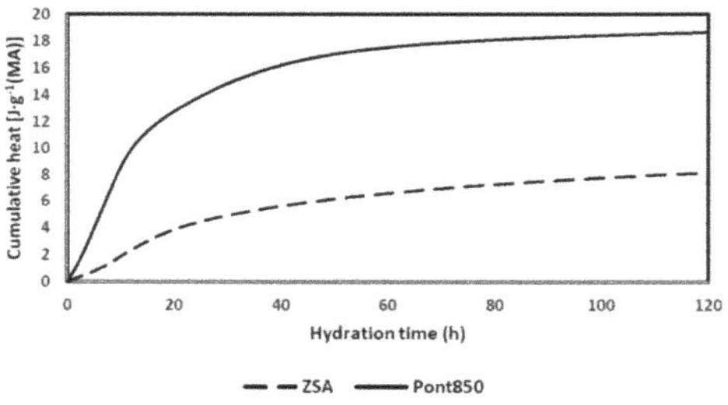

Fig. 1. Results from pozzolanic reactivity assay. MA: Mineral additive

As it is possible to see the calcined clay is more reactive than the zeolitic tuff from the pozzolanic point of view. That is because, as it is shown in Table 1 the calcined clay has more alumina containing phases than zeolite [14], it is finer (Table 3) and the metakaolin it contains makes it more reactive [15].

Table 5 shows the bound water and $Ca(OH)_2$ contents in the slurries hydrated at 30 and 60 °C during 24 h. As it is possible to see, the bound water, when the cement is partially substituted at 5% by the mineral additives, is lower when calcined clay is used in the slurries hydrated at 30 °C but the situation is the contrary in the slurries hydrated at 60 °C. For the slurries with calcined clay the bound water contents increase from the 5 to 10% of substitution but decrease when the substitution level is 20%, at both temperatures.

Table 5. Influence of the kind and the quantity of mineral additive and temperature on bound water and $Ca(OH)_2$ contents

	PC-ZSA_5%	PC-Pont850_5%	PC-Pont850_10%	PC-Pont850_20%
Slurries cured at 30 °C				
Bound water (g)	6.39	5.86	7.34	6.67
$Ca(OH)_2$ [$g \cdot g^{-1}$(PC)]	0.10	0.09	0.09	0.07
Slurries cured at 60 °C				
Bound water (g)	8.04	8.32	8.83	8.03
$Ca(OH)_2$ [$g \cdot g^{-1}$(PC)]	0.13	0.10	0.06	0.03

The Portlandite content in the slurries is always lower when the calcined clay is the mineral additive than when the zeolite is, in keeping with the higher pozzolanic reactivity of the former. Besides, the $Ca(OH)_2$ content decreases when the content of both pozzolans in the slurries increases at the temperatures of the experiments due to the higher temperature speeds up the onset of the pozzolanic reactions [16].

The following pictures shows the calorimetric curves during the first 24 h of hydration of the slurries at 30 and 60 °C. As it is shown at 30 °C the induction time is not so influenced by the kind or the replacement level when the PC is partially substituted until 10% by the mineral additives, but when the calcined clay from Pontezuela is used at 20% the dormant period is lower. When the curing temperature is increased to 60 °C the induction periods are shorter than when the slurries are hydrated at 30 °C [≈ 10 – 12 h on average (at 30 °C) vs ≈ 2.5–5 h (at 60 °C)]. The higher temperature speeds up the hydration rate of both the PC and the mineral additives, mainly that of the calcined clay due to its higher Al_2O_3 containing phases [16]. That situation can negatively affect the thickening time in real conditions (Fig. 2).

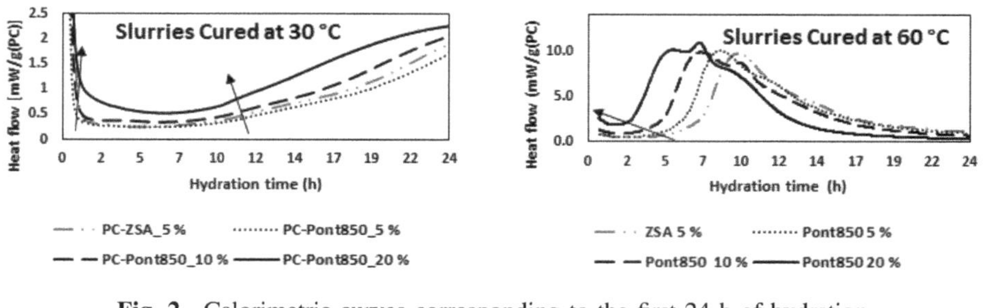

Fig. 2. Calorimetric curves corresponding to the first 24 h of hydration

Table 6 shows the total percolated volumes in the slurries hydrated at 30 and 60 °C during 24 h. As it can be observed the behavior is complex but this parameter is always lower when the slurries are cured at 60 °C. That could be associated to the more quantity of bound water found when the slurries are cured at 60 °C.

Table 6. Influence of the kind and the quantity of mineral additive and temperature on total percolated volumes.

	PC-ZSA_5%	PC-Pont850_5%	PC-Pont850_10%	PC-Pont850_20%
Slurries cured at 30 °C				
Total percolated pore volume (%)	38.78	36.33	33.15	42.03
Slurries cured at 60 °C				
Total percolated pore volume (%)	29.03	29.29	28.51	28.39

4 Conclusions

- The calcined clay from Pontezuela, being finer and with higher quantity of Al_2O_3 containing phases has more pozzolanic reactivity than the zeolite from Pontezuela what provokes that bound water contents be normally higher and the quantities of $Ca(OH)_2$ be slower in Portland cement slurries cured at both 30 and 60 °C.
- The induction periods are slower when the Portland cement is partially substituted by calcined clay from Pontezuela than when the zeolite is used and these decreases are higher when the content of calcined clay and the curing temperature are increased.
- The total percolated porosity in the slurries studied is lower when they are submitted to higher temperature which speeds up both the PC hydration rate and the onset and the rate of the pozzolanic reactions.

References

1. Nelson, E.B., Guillot, D.: Well Cementing. 2da edn. Sugar Land, Texas: Schlumberger Educational Services (2006)
2. Mangadlao, J.D., Cao, P., Advincula, R.C.: Smart cements and cement additives for oil and gas operations. J. Petrol. Sci. Eng. **129**, 63–76 (2015)
3. Bu, Y., et al.: Properties of oil well cement with high dosage of metakaolin. Constr. Build. Mater. **112**, 39–48 (2016)
4. Dos-Anjos, M., et al.: Cement slurry with silica flour and metakaolin for cementation of oil-wells subject to high temperature. J. Mater. Sci. Eng. **4**(12), 1–5 (2010). (Serial No. 37)
5. Anjos, M.A.S., Martinelli, A.E., Melo, D.M.A.: Effect of sugarcane biomass waste in cement slurries submitted to high temperature and pressure. Mater. Sci. Eng. A **529**, 49–54 (2011)
6. Sánchez, R., Palacios, M., Puertas, F.: Cementos petroleros con adición de escoria de horno alto. Características y propiedades. Materiales de Construcción **61**(302), 185–211 (2011)
7. Soares, L.W.O., et al.: The effect of rice husk ash as pozzolan in addition to cement Portland class G for oilwell cementing. J. Petrol. Sci. Eng. **131**, 80–85 (2015)
8. Díaz, A.A., et al.: Pozzolanic reactivity of low grade kaolinitic clays: Influence of calcination temperature and impact of calcination products on OPC hydration. Appl. Clay Sci. **108**, 94–101 (2015)

9. Danner, T.: Reactivity of calcined clays, in Department of Natural Sciences and Engineering. Doctorate thesis, Norwegian University of Science and Technology: Trondheim, Norway (2013)
10. Fernández, R., Martirena, F., Scrivener, K.L.: The origin of the pozzolanic activity of calcined clay minerals: a comparison between kaolinite, illite and montmorillonite. Cem. Concr. Res. **41**, 113–122 (2011)
11. Díaz, A.A.: Obtención de un material puzolánico de alta reactividad a partir de la activación térmica de una fracción arcillosa multicomponente, in Departamento de Ingeniería Civil. Doctorate thesis, Universidad Central "Marta Abreu" de Las Villas: Santa Clara, p. 106 (2010)
12. Hervé, D.M., Garrido, M.V.: Comparación de dos muestras de zeolita en la adsorción de humedad y remoción de olores. INFOMIN **4**(1), 21–31 (2012)
13. Snellings, R., Scrivener, K.L.: Rapid screening tests for supplementary cementitious materials: past and future. Materials and Structures (2015)
14. Talero, R., Rahhal, V.: Calorimetric comparison of Portland cements containing silica fume and metakaolin. J. Therm. Anal. Calorim. **96**, 383–393 (2009)
15. Valipour, M., et al.: Comparing a natural pozzolan, zeolite, to metakaolin and silica fume in terms of their effect on the durability characteristics of concrete: A laboratory study. Constr. Build. Mater. **41**, 879–888 (2013)
16. Deschner, F., et al.: Effect of temperature on the hydration of Portland cement blended with siliceous fly ash. Cem. Concr. Res. **52**, 169–181 (2013)

Influence of the Kind of Mineral Addition and the Seawater on the Hydration of a Portland Cement

Juan Alberto Ribalta[1(✉)], Leidys Laura Pérez[1], and Adrián Alujas Díaz[2]

[1] Departamento de Licenciatura en Química, Universidad Central "Marta Abreu" de Las Villas, 54830 Santa Clara, Cuba
juanrq@uclv.edu.cu

[2] Centro de Estudios de Química Aplicada, Universidad Central "Marta Abreu" de Las Villas, 54830 Santa Clara, Cuba

Abstract. The influence of two pozzolanic additions, a calcined clay and a zeolitic tuff, on the hydration of a Portland cement was studied in pastes prepared with fresh water and substitute seawater. The pastes contained 0, 5, 10, 20 and 30% of each pozzolanic addition and were prepared with a water–to–solid ratio of 0.50. The hydration kinetic was evaluated by isothermal calorimetry at 30 °C and during 72 h. The bound chloride contents in the pastes were also quantified. The results of the experiments indicated that there was not a correlation between the kind of mixing water and the induction periods but these periods increased with the percent of substitution of the Portland cement and the increments were higher with the finest pozzolan and with the higher content of alumina containing phases (calcined clay). It was observed that during the acceleration periods the mineral additions increased the reaction rates and the release of heat during the principal peaks and the increments were higher in pastes with the calcined clay and in those prepared with artificial seawater. The total cumulative heats were higher in the systems where the Portland cement was substituted by the calcined clay (with higher content of alumina) and in the pastes that were prepared with substitute seawater. The capacity to bind chloride was related to the content of Al_2O_3 y Fe_2O_3 in the mineral addition and it was higher in pastes with calcined clay (with higher content of Fe_2O_3 and Al_2O_3 containing phases) but the zeolitic tuff was more effective to bind chlorides, in spite of having less quantity of phases providing Fe_2O_3 and Al_2O_3.

1 Introduction

The Portland cement is among the most used materials and with higher production level in the world. However, because of its production on a massive scale, it accounts for approximately 5% of worldwide man-made CO_2 emissions [1]. One of the approaches to lower the environmental impact of cement production focus on partial substitution of clinker by supplementary cementitious materials (SCMs). Several authors have studied, by isothermal calorimetry, the influence of calcined clays and zeolites on the hydration of Portland cements in pastes prepared with fresh water [2–4]. They found that this kind of pozzolanic additions accelerate the hydration reactions during the main maximum

F. Martirena et al. (eds.), *Calcined Clays for Sustainable Concrete*, RILEM Bookseries 16,
https://doi.org/10.1007/978-94-024-1207-9_62

and the released heat were higher when the content of alumina in the reactive phases increased. In general, all these works were done in pastes prepared with fresh water. During the construction of marine structures in coastal areas, the substitution of fresh water by seawater will reduce the cost of transportation and improve the work efficiency. Portland cement is the most frequently kind of cement used during oil well cementing. Some mineral additions are used as additives during well cementing jobs, like bentonite, silica fume, zeolite, diatomeceous earth, etc. Seawater is used extensively for mixing cement slurries at offshore locations. It contains up to 2.5 wt% NaCl, resulting in acceleration [5]. However, the use of seawater may modify the cement hydration [6] and the properties and microstructure of concrete [7, 8] and accelerate the corrosion of the casing in the wellbore. Cuba is trying to expand the areas for drilling and constructing oil wells to depth seas. It is inevitable, in those cases, to use seawater to prepare the slurries for the cementing. It is also necessary to find new mineral additives that be compatible with these conditions.

For that reason, the aim of this work is to determine how the hydration of a Portland cement is affected in the presence, concomitantly, of seawater, a calcined clay or a zeolitic tuff. To the best of our knowledge the hydration of Portland cement in pastes where it was partially substituted by calcined clay or zeolitic material and prepared with sea water has been little studied.

2 Materials and Experimental Procedures

The clay and the zeolitic tuff were collected, respectively, from the deposits Pontezuela and San Andrés, Cuba. The clay was stationary calcined during 1 h at 850 °C. The zeolite tuff was used as received. The cement was a Cuban Portland cement produced at the Siguaney plant. The chemical compositions of the clay and its calcination product were determined by X – ray fluorescence at a spectrometer Bruker AXS S4 operated at 1 kW of power with a Rh cathode. The chemical compositions of the calcined clay, the zeolite and the Portland cement are presented in Table 1.

Table 1. Summary of the chemical composition of the materials used in these work (expressed in weight percent)

	Al_2O_3	SiO_2	Fe_2O_3	CaO	MgO	Na_2O	K_2O	SO_3	CaO free	LOI*	IR**
Pont850[£]	45.69	28.02	21.48	0.15	0.83	0.15	0.54	–	–	–	–
ZSA[¥]	10.48	69.50	1.48	2.59	0.82	1.61	1.34	–	–	10.94	–
PC[§]	4.84	19.29	4.80	61.56	1.29	0.32	0.62	3.25	1.27	3.01	0.99

* Loss on ignition
** Insoluble residue
[£] Calcined clay from Pontezuela
[¥] Zeolite from San Andrés. (The chemical composition was the reported by [10])
[§] Portland cement. (The chemical composition was supplied by the factory)

The phase identification and quantification of clay was made with a Panalytical Xpert Pro MPD diffractometer using CuKα radiation, a divergence slit size of 0.5°. The sample was scanned between 4 and 70° 2θ, the step size and time per step were 0.017° 2θ and 80 s, respectively. The phase compositions of the clay, and that of its clayey fraction, as well as that of the zeolitic tuff, are presented in Table 2.

Table 2. Phase compositions of the materials used in these work (expressed in weight percent)

	Kaolinite	2:1 Clay	Total of clayey phases	Quartz	Goethite
Pont*	45.7	38.7	84.4	7.5	8.1
	Kaolinite	Montmorillonite	Montmorillonite – Illite	Illite	
Clayey fraction of Pont*	65.8	20.0	8.0	6.2	
	Clinoptilolite	Mordenite	Total of zeolites	Montmorillonite	Calcite
ZSA**	50	28	78	11	< 1

* It refers to the clay from Pontezuela without calcining
** It refers to the zeolite from San Andrés (The phase composition was that of reported by [10])

The particle size distribution was determined by laser diffraction on a device Malvern MasterSizer type S. The percentiles characterizing each particle size distribution (D_{10}, D_{50} and D_{90}) are shown in the Table 3. The artificial seawater was prepared following the standard ASTM D1141 – 98 (Reapproved 2003) *"Standard practice for the preparation of substitute ocean water"*. The calorimetry experiments were carried out with a TAM AIR 3238 calorimeter from Thermometric and calibrated at 600 mw. All the calorimetry experiments were made at 30 °C during 72 h. The pastes contained 0, 5, 10, 20 and 30 of each mineral addition and the w/s ratio was 0.5 for both kind of water (fresh water and artificial seawater).

Table 3. D_{10}, D_{50} and D_{90} for the calcined clay from Pontezuela (Pont850) and the zeolitic tuff from San Andrés (ZSA).

Percentile	Maximum diameter (μm)	
	Pont850	ZSA
D_{10}	0.34	1.83
D_{50}	3.01	11.66
D_{90}	26.80	35.26

The pastes to quantify their capacities to bind chlorides were prepared at 0.5 of artificial seawater – to – solid ratio, mixed using the same procedure and mineral additions substitution levels of the calorimetric experiments. They were cured during 30 days at 30 °C in a water bath. The water – soluble chloride contents were determined following the procedures described in ASTM C1218/C1218 M – 15 *"Standard Test*

Method for Water-Soluble Chloride in Mortar and Concrete" and ASTM C114–15 *"Standard Test Methods for Chemical Analysis of Hydraulic Cement"*.

3 Presentation and Discussion of the Main Results

3.1 Hydration on the Pastes

As it is shown in Fig. 1 when the Portland cement is partially substituted by the addition minerals the induction periods are always higher than in the paste with only Portland cement, in the mixtures prepared with distilled water or artificial seawater. Beside, except when the Portland cement is partially replaced by 10% of mineral additions in pastes prepared with distilled water and 5% in that of prepared with artificial seawater, the observed induction periods in the mixtures where the Portland cement was partially substituted by calcined clay (the finest and with higher containing alumina phases of both mineral additions) are higher than that of prepared with zeolitic powder, in pastes mixed prepared with both distilled water and artificial seawater.

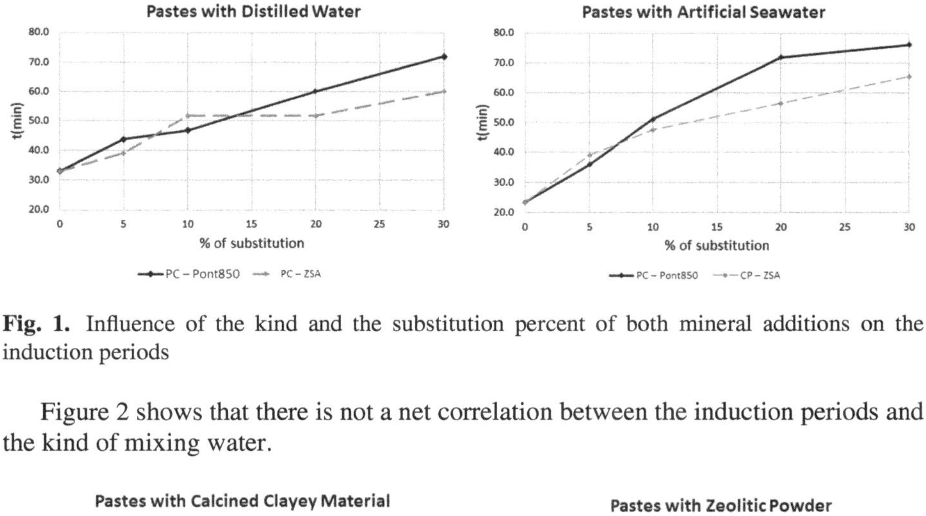

Fig. 1. Influence of the kind and the substitution percent of both mineral additions on the induction periods

Figure 2 shows that there is not a net correlation between the induction periods and the kind of mixing water.

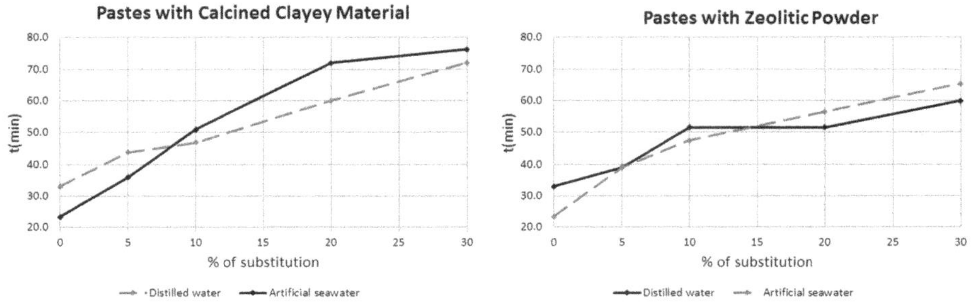

Fig. 2. Influence of the kind of water on the induction periods

Figures 3 and 4 display that there is not either a clear relation between the kind of mineral and the reaction rates (slopes) and between the kind of mixing water and the reaction rates (slopes) during the acceleration periods but the reactions rates during this stage are always higher in pastes with mineral additions and increase with the replacement level in pastes prepared with both distilled water or artificial seawater.

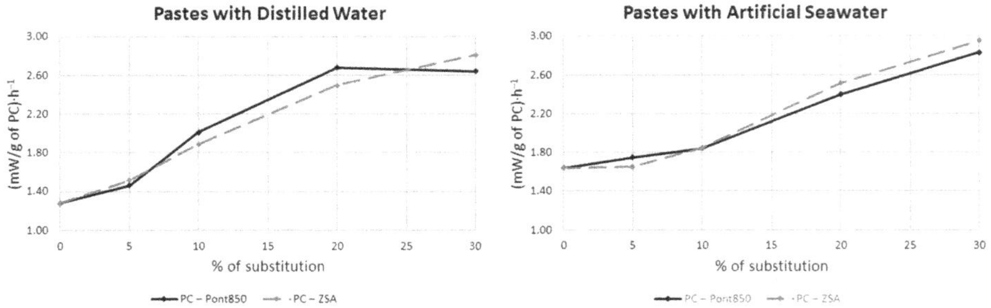

Fig. 3. Influence of the kind and the substitution percent of both mineral additions on the acceleration periods

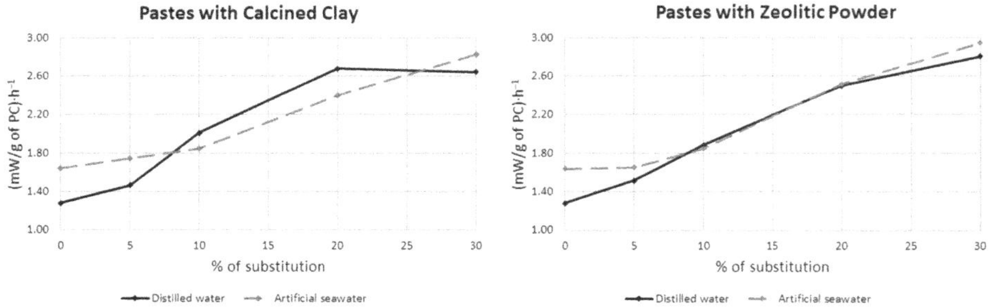

Fig. 4. Influence of the kind of water on the acceleration periods

4 Chloride Binding

Figure 5 shows the capacity of the system to bind chloride increases with the content of mineral addition in the pastes and it is always higher when the Portland cement is partially replaced by calcined clay (with more phases providing Al_2O_3 and Fe_2O_3), compared to that of where the cement was partially substituted by the zeolitic tuff.

In spite of the capacities of binding chloride are higher in pastes with calcined clay, as it is shown in the following figure the calcined clay, containing more Al_2O_3 and Fe_2O_3, is less effective fixing chloride ions on these experiment conditions (Fig. 6).

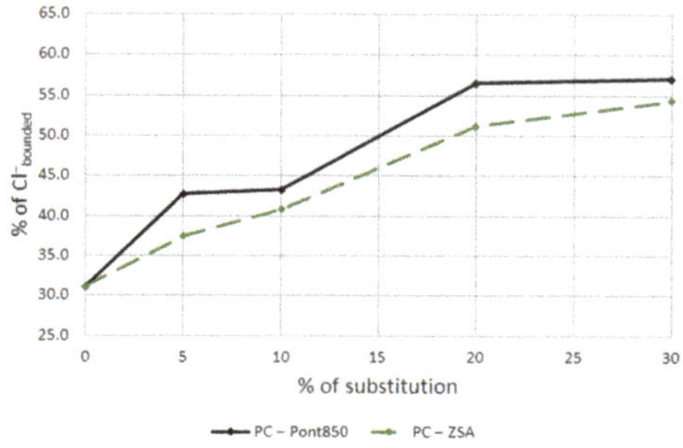

Fig. 5. Influence of the kind and the substitution percent of both mineral additions on the capacities of pastes to bind chloride ions

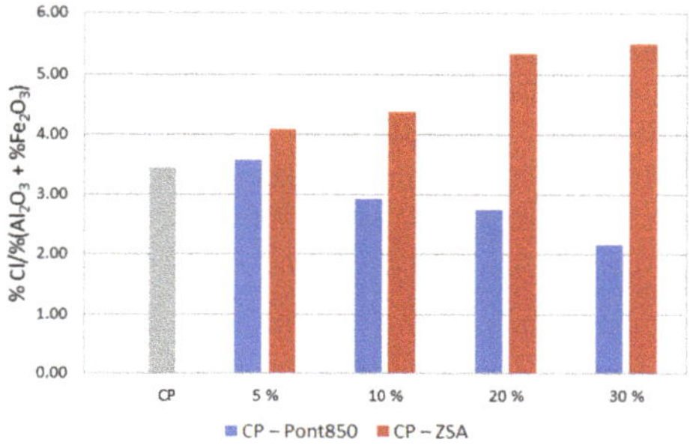

Fig. 6. Effectiveness of both mineral additions on the capacities of pastes to bind chloride ions

5 Conclusions

- The presence of calcined clay from Pontezuela and zeolitic powder from San Andrés, essentially inert during the early hydration provoke increments in the induction periods and the increments were higher, generally, in pastes where the Portland cement was partially replaced by calcined clay.
- The kind of water does not exert a great influence during either the induction and the acceleration periods from the hydration rates point of view.
- The capacity to bind chlorides is directly related to the content of phases providing Al_2O_3 and Fe_2O_3 but it seems the effectiveness does not depend of these contents exclusively.

References

1. Gartner, E., Hirao, H.: A review of alternative approaches to the reduction of CO_2 emissions associated with the manufacture of the binder phase in concrete. Cem. Concr. Res. **78**((Part A)), 126–142 (2015)
2. Danner, T.: Reactivity of Calcined Clays, in Department of Natural Sciences and Engineering. Norwegian University of Science and Technology: Trondheim, Norway (2013)

3. Tydlitát, V., Černý, R., Zákoutský, J.: Early-stage hydration heat development in blended cements containing natural zeolite studied by isothermal calorimetry. Thermochim. Acta **582**, 53–58 (2014)
4. Snellings, R., Mertens, G., Elsen, J.: Calorimetric evolution of the early pozzolanic reaction of natural zeolites. J. Therm. Anal. Calorim. **101**, 97–105 (2010)
5. Nelson, E.B., Guillot, D.: Well Cementing. 2da ed. Sugar Land, Texas: Schlumberger Educational Services (2006)
6. Pang, X., Boul, P., Jimenez, W.C.: Isothermal calorimetry study of the effect of chloride accelerators on the hydration kinetics of oil well cement. Constr. Build. Mater. **77**, 260–269 (2015)
7. Li, Q., et al.: Effect of metakaolin addition and seawater mixing on the properties and hydration of concrete. Appl. Clay Sci. **115**, 51–60 (2015)
8. Galan, I., Perron, L., Glasser, F.P.: Impact of chloride-rich environments on cement paste mineralogy. Cem. Concr. Res. **68**, 174–183 (2015)
9. Thomas, M.D.A., et al.: The effect of supplementary cementitious materials on chloride binding in hardened cement paste. Cem. Concr. Res. **42**, 1–7 (2012)
10. Hervé, D.M., Garrido, M.V.: Comparación de dos muestras de zeolita en la adsorción de humedad y remoción de olores. INFOMIN **4**(1), 21–31 (2012)
11. Sabir, B.B., Wild, S., Bai, J.: Metakaolin and calcinded clays as pozzolans for concrete: a review. Cement Concr. Compos. **23**, 441–454 (2001)

Standardization Strategy of Low Carbon Cement in Cuba. Case Study for "Siguaney" Cement Factory

D. Rocha[1(✉)], R. Almenares[2], S. Sanchez[3], A. Alujas[4], and F. Martirena[1]

[1] CIDEM, Universidad Central de Las Villas, Santa Clara, Villa Clara, Cuba
drocha@uclv.cu
[2] Instituto Superior Minero Metalúrgico de Moa, Moa, Holguín, Cuba
[3] Facultad de Economía, Universidad Central de Las Villas, Santa Clara, Villa Clara, Cuba
[4] CQA, Universidad Central de Las Villas, Santa Clara, Villa Clara, Cuba

Abstract. Nowadays, the Cement Cuban Industry has installed a productive capacity of 2.8 MMt of clinker per year, for a use of the 40% of this productive capacity for different reasons. The reported production in 2014 was 1.8 MMt. The current demand of cement exceeds the production, indicating the necessity of new investments in the cement sector to increase the productive capacity and the production but this is impossible in short-term. The projection of this demand in the period 2015–2020 will imply an annual increase of 18% due to the new age of the economic reorganization we are currently facing in our country and the increase of constructive activity. Today, the LC^3 production constitutes a viable alternative to reduce this problem. The Cuban Cement Industry has shown a huge interest on the development and implementation of LC^3. With that purpose, it is necessary to work on the introduction of this cement in the Cuban standard.

Since 2015, CIDEM has representation in the CTN No. 22 "Cement" and the CTN No. 37 "Concrete and Mortars" portraying an active participation. At present, our research is based on three points:

1. Development of a Technical Instructive about the determination clay potential to be use as pozzolan.
2. Confection of the new standard about the specifications of ternary blended cement.
3. Application of the NC 120:2014 "Hydraulic concrete — Specifications" for the development of concrete with LCC for the different levels of atmospheric and chemical aggressiveness.

The main goal of this work is to develop an Implementation Strategy for LC^3 in Cuba. This strategy links the standardization results to environmental and cost results including the clay deposit study, durability test and the experience acquired during several industrial trials. Siguaney Cement Factory has been selected to put into practice this strategy.

Keywords: Low carbon cement · Cuban standard · Cement Cuban Industry · Concrete exposure site

F. Martirena et al. (eds.), *Calcined Clays for Sustainable Concrete*, RILEM Bookseries 16,
https://doi.org/10.1007/978-94-024-1207-9_63

1 Introduction

With the increasing demand of the binders in our country, the Cement Industry has shown a great interest for raising the production of cements with active additions, which represents the increase of production with regard to the Clinker consumption, environmental improvement referred to CO_2 emissions and production costs. Although historically, this kind of cement has been produced in Cuba, mainly using natural pozzolans substituting the clinker with less than 20%, this production has always been around 25% of the cement total production. Nowadays, we can confirm that the world tendency is very different as the production of cements with active additions is around 72% of the total production.

This research paper focuses on the study of the National and Foreign Standards that allow the creation of our own standards with the purpose of implementing the production of ternary cements in our country and the elaboration of an implementation strategy with the main results of the different research lines of this project.

2 Development

2.1 Pozzolan Standard

Nowadays, Cuba relies on a specifications standard for the use of pozzolans, in which the slate and the calcined clays class N are included. However, it is considered that due to the inexperience in the use of calcined clays as MCS in our country, this standard document needs complementary information derived from the studies carried out by Cuban scientists on a group of deposits of kaolin clays in the country.

From these investigations results [1, 2], and the discussion of the relative results to the characterization of raw materials and to the pozzolanic reactivity of calcination products, we can conclude that the most important parameter in the pozzolanic reactivity determination of the calcination products of a material is its clay minerals content, specifically the ones belonging to kaolin group, in contents with no less than 40% in mass, expressed as content of equivalent kaolin (% KE). Furthermore, limit criteria have also been established for the preliminary selection of the clay material [3].

As a result of the work of the CIDEM in conjunction with the CTN 22, a proposal of Technical Instructive has been carried out for the determination of the calcined clays potentialities as a source of SCM, which all interested industries and institutions will have access to [4].

2.2 Ternary Blended Cement Standard

A ***Cuban Standard Project: Ternary Blended Cement. Specification***, regarding the use of Low Carbon Cements was approved in the Standardization Technical Committee N° 22 “Cement”. This Standard is the result of more than one year of work together with the Technical Committee N° 22. For this purpose, many international and Cuban standards about the cement, blended cement and pozzolan use were analysed. The scientific results and experiences of Low Carbon Cement Project in Cuba, Switzerland and India during the last five years were incorporated to this standard development work.

Finally, the work team decided to make a new standard about the use of Ternary Blended Cement, based in the technical specifications of ***NC 96:2011 "Cement with active additions — Specifications"*** and ***ASTM C595/C595 M – 16 "Standard Specification for Blended Hydraulic Cements"*** as our country doesn't have a standard for this type of cement.

The main technical specifications of this Standard are:

Terminology

Ternary blended cement: A blended hydraulic cement consisting of Portland cement with either a combination of two different pozzolans, slag and a pozzolan, a pozzolan and a limestone, or a slag and a limestone [5].

Manufacture

Ternary Blended Cement: Ternary blended cement shall be a hydraulic cement consisting of an intimate and uniform blend produced either by intergrinding, by blending, or a combination of intergrinding, and blending Portland cement clinker or Portland cement with (1) two different pozzolans, (2) slag and a pozzolan, (3) ***a pozzolan and a limestone***, or 4) a slag and a limestone. Ternary blended cement shall have a maximum pozzolan content of 40% by mass of the blended cement, a maximum limestone content of 15% by mass of the blended cement, and the total content of pozzolan, limestone, and slag shall be less than 70% by mass of the blended cement [5].

Classification and designation

The product is classified by two grades of quality, in accordance with addition percent and the compressive strength to 28 days, expressed in MPa (kgf/cm^2).

The product is designed with the T letter of ternary and the initials of the additions' names, separated by a hyphen of the numeric value of compressive strength to 28 days [6].

In case of limestone utilized as addition, it should comply with the next requirements:

1. The CaCO3 content, calculated from calcium oxide, more than 75% in mass.
2. The Methylene blue index, max: 1.2 g/100 g.

In case of using pozzolans as addition, the same should comply with the requirements established in the Cuban Standard NC TS 528:2013 "Hydraulic Cement-Pozzolans-Specifications"

Blended Hydraulic Cement:

1. Ternary Cement 25 (Addition of calcined clay - limestone between 35% and 50%, recommended ratio 2:1): **TAC-25 Cement**
2. Ternary Cement 35 (Addition of calcined clay - limestone between 35% and 50%, recommended ratio 2:1): **TAC-35 Cement**

3 Requirements and Test Methods

See Table 1.

Table 1. Requirements and test methods for the Ternary Blended Cement.

Index	Requirements	MU	TAC-25	TAC-35	Test method
Physicals	Time of initial setting, Vicat test, min	min	45	45	NC 524
	Time of final setting, max	h	12	12	
	Volume stability for Le Chatelier, max	mm	10	10	NC 504
Mechanics	Compressive strength, min 3 días	MPa		17	NC 506
	7 días		17	25	
	28 días		25	35	
Chemicals	Loss on ignition, max	%	10	10	NC EN 196-2
	Magnesium oxide (MgO), max	%	5	5	
	Sulfate reported as SO3, max	%	3,5	3,5	

4 Concrete Standard

Nowadays in Cuba, there is an excellent standard of specifications for hydraulic concrete focused on the durability by performance in function of the aggressiveness conditions of the environment where the material is implemented. This standard is also supported by important investigations of Cuban scientists and up-to-date bibliography about this topic by the main world authors.

From the study of the standard NC 120:2014 "Hydraulic Concrete. Specifications", unique standard of specifications for hydraulic concretes of Portland cement in our country, our centre has been in charge of constructing several concrete Exposure sites aimed at assessing the constructed concretes in function of the atmospheric and chemical aggressiveness according to Table 1 of this Cuban Standard. These concretes' durability is evaluated each year by means of the extraction of core and the implementation of a test protocol to which the tests and the specifications have been incorporated as it is posted in the Table 11 of the NC 120:2014 [7].

5 Strategy of Implementation. Siguaney Case Study

5.1 General Information About Siguaney Cement Plant. Strategy's Structure

The Siguaney cement plant, mainly devoted to the production of grey cement and the only one producing white cement in Cuba, is located in Siguaney town in Sancti Spiritus province. Since the beginning of this investigation, the management of this industry have cooperated with the CIDEM with the purpose of implementing the LC^3 cement production. In 2013, an industrial trial of 130 tons of LC^3 was carried out and nowadays, another trial is being prepared. According to its location (central region of Cuba) it could even turn into a calcined clays supplier to other plants of the country. Due to the aforementioned, we have drawn an implementation strategy based on providing the Cuban Cement Industry with all the relevant information to enable them to take the decision of using this technology. This information will be provided with technical reports by each of the work lines of this project.

5.2 Raw Materials Report

After the deposit site being identified and studied by CIDEM and Empresa Geominera del Centro, the Group of Cement (GECEM) makes a request to the National Geologic Service of the exploration Bamburanao Deposit for LC3, since the use of these clays constitutes an alternative for the increasing demand of the cement in Cuba. These works were financed by the government budget with a value of $379 000.

The place for these works is located at the municipality of Yaguajay, province of Sancti Spiritus, which is situated at 300 m to the South of this City, with road access and energy sources at 200 m to the East. The study methodology applied was the Instructions Manual of the Geologic Activity. Code I.A.G. 03, which regulates the utilization of Geologic Itineraries Method, documents and sampling, Hydrogeological studies, Topographic works, utilization of mines works and perforation and cores sampling. Lab Works and Industrial Sample are to be processed at Siguaney Cement Plant and the calculation of the measured resources superior to 5MNt [3].

5.3 Durability Studies Report

By means of the application of the Torrent Method, it was confirmed that the concretes elaborated with LC^3 have a minor air permeability compared to the ones elaborated with Portland Cement in very high aggressiveness sites. The concretes of the Portland Cement P-35 have a superior carbonation depth compared to the one produced with LC^3 in very high aggressiveness sites (Exposure Site Cayo Santa Maria). The Chloride transport in concretes with LC^3 is inferior than in Portland Cement Concretes, due to the refinement of the pores structure and superior tortuosity because of the pozzolanic reaction. The concrete with LC^3 present superior indexes of electric resistivity than Portland Cement Concretes [8].

5.4 Report About Costs

With regard to the economy field, a partnership work has been developed in conjunction with the Siguaney Cement Plant to elaborating cost sheets, the estimation of the possible sale prices for the calcined clay, demand estimation and feasibility studies.

Figure 1 presents results obtained when the best available level of technology is used. The initial construction of the cement plant (CAPEX) which represents the major costs for the production of one ton of cement and not so much the operation costs for the fuel and raw materials (OPEX). Furthermore, for both the OPC and the LC^3, environmental and monetary costs are mainly conditioned by the clinkerization phase. LC^3 outperforms OPC in all phases.

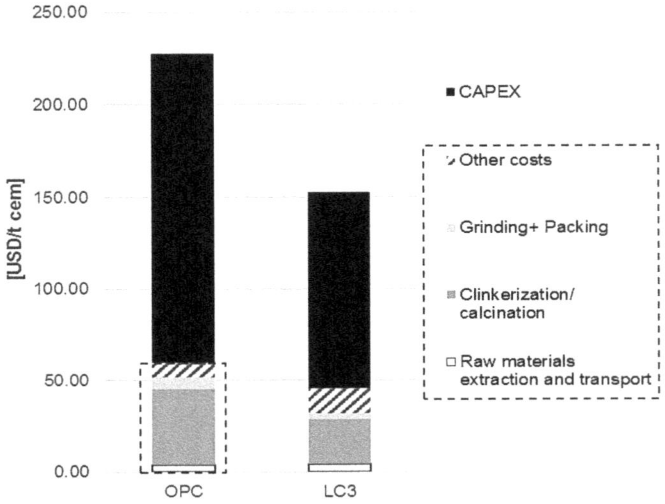

Fig. 1. Capital and operational expenditures LC[3] vs. OPC (BAT scenario) [9]

Main results in investment assessment shows that the best alternative to produce LC[3] in Cuba is, in the short term, to install retroffited calciners. In the long-term flash calciners could be installed in order to gain efficiency.

5.5 Environmental Studies Report

To assess the environmental impact of LC[3], a Life Cycle Assessment is being carried out comparing cements OPC, PPC and LC[3]. Many midpoint and endpoint where assessed. Due to its importance, the energy consumption is analysed in detail (Energy demand), expressed in Megajoules (MJ) and the Global Warming Potential (GWP) for a projection period of 100 years expressed in kilograms of CO_2 equivalent to (kgs CO2 eq) and calculated by the methodology IPCC2013 [10].

Sánchez Berriel et al. (2016) compare the environmental impact for OPC, blended cement PPC and LC[3] for three different technical levels: Pilot, Industrial and BAT. LC[3] cement always produces around 30% savings. Furthermore, it is noticed that the worse LC[3] cement made in the pilot industrial trial is better than the best OPC that can be produced with the BAT. Emissions reduction are between 25-35% respect PP-25 and P-35 respectively.

6 Conclusions

- Cuba counts with all the standards documentation to LC[3] production and use.
- A clay deposit with measured resources superior to 5MNt it´s available near to Siguaney factory.

- As result of durability studies to several atmospheric aggressiveness zones it´s been proven that LC3 can be used as a cement of general use, but it best performance occurs in very high aggressiveness environments (coast distance < 500 m).
- LC3 introduction has a significant impact reducing costs in approximately 10% and emissions between 25–35%.

References

1. Vizcaíno, L.M.: Cemento de bajo carbono a partir del sistema cementicio ternario clínquer - arcilla calcinada - caliza, Tesis Doctoral, Universidad Central « Marta Abreu » de Las Villas (2014)
2. He, C., Osbaeck, B., Makovicky, E.: Pozzolanic reactions of six principal clay minerals: activation, reactivity assessments and technological effects. Cem. Concr. Res. **25**, 1691–1702 (1995). doi:10.1016/0008-8846(95)00165-4
3. Almenares, R.: Potencialidades de arcillas caoliníticas cubanas para la obtención de materiales cementicios suplementarios, Tesis Doctoral, Universidad Central « Marta Abreu » de Las Villas (2016)
4. NC TS 528: Puzolanas. Especificaciones, Cuba (2013)
5. ASTM-C-595: Standard Specification for Blended Hydraulic Cements. Estados Unidos (2016)
6. NC 96: Cemento con adición activa. Especificaciones, Cuba (2011)
7. NC 120: Hormigón Hidráulico. Especificaciones, Cuba (2014)
8. Torrent, R.: Non-destructive air-permeability measurement: From gas-flow modelling to improved testing. In: The 2nd International Conference of Microdurability, 7 April 2012 Amsterdam, The Netherlands (2012)
9. Sánchez Berriel, S., et al.: Assessing the environmental and economic potential of limestone calcined clay cement in Cuba. J. Clean. Prod. **124**, 361–369 (2016)
10. IPCC: Climate change 2014: synthesis report (2015)

Antibacterial Metakaolin-Based Geopolymer Cement

Jose-Carlos Rubio-Avalos(✉)

Innovation Materials Laboratory, Faculty of Civil Engineering, Universidad Michoacana de San Nicolás de Hidalgo, 58040 Morelia, Michoacán, Mexico
j.c.rubio.avalos@gmail.com

Abstract. Geopolymer cement can be used as a *ceramic matrix carrier* for different applications such as antibacterial materials. In this work, an antibacterial geopolymer cement has been synthesized by loading the organic compound 5-chloro-2-(2, 4-Dichlorophenoxy) phenol also known as "Triclosan" into a geopolymer metakaolin and nano-silica based cement matrix. The antibacterial efficiency of Triclosan-geopolymer cement against *Escherichia Coli (Gram-negative bacteria) and Staphylococcus Aureus (Gram-positive bacteria)* was evaluated using the Halo method; results show that triclosan based Geopolymer cement are effective for preventing the reproduction of the *E. coli and S. aureus* bacteria. From results, a "room-temperature" antibacterial metakaolin based geopolymer cement for buildings with health and environmental protection can be obtained.

1 Introduction

Antibacterial materials are of great interest for building materials with health and environmental protection. Cleaner Environments and free bacteria surfaces are ideal for hospitals, kitchens, bathrooms, etc. to assist in minimizing the spreading of disease and reducing the prevalence of illness. Nosocomial infections have been increasing in recent decades. Two billion people worldwide carry *Staphylococcus aureus (S. aureus)* and up to 53 million of these carry the resistant form, methicillin-resistant S. *aureus* (MRSA). In 2005, in the USA alone, S. *aureus*-related diseases were responsible for nearly 500,000 hospitalisations and 11,000 deaths. In the European Union nosocomial infections are estimated to cost 5.5 billion of euros per year. Additionally, walls, floors and other surfaces provide an easily accessible platform for pathogenic bacterial to settle and proliferate and microbial activity can have an important impact on the durability of building materials [1, 2]. Antibacterial agents have been included in different matrices such as polymers, glass, Ordinary Portland Cement (OPC), zeolites, and some minerals [3, 4, 5, 6].

On the other hand, geopolymers or mineral polymers are amorphous to nano-crystalline aluminosilicates materials and can also be viewed as the amorphous equivalent of certain synthetic zeolites having similar composition although the distinctive zeolitic structure is absent which make them amorphous to X-rays [6, 8]. They are composed of cross-linked $[AlO_4]^-$ and $[SiO_4]^-$ tetrahedral with charge balancing

F. Martirena et al. (eds.), *Calcined Clays for Sustainable Concrete*, RILEM Bookseries 16,
https://doi.org/10.1007/978-94-024-1207-9_64

K^+ or Na^+ ions [9]. Geopolymers are obtained using alkaline solutions mixtures with precursors such as: metakaolin, fly-ash, blast furnace slag, etc. Polycondensation and solidification can occur between 20 °C and 90 °C. Geopolymers are heat resistant when are compared with organic resins and also have the potential capability to incorporate and immobilize a variety of radioactive waste ions including Sr^+ and Cs^+ ions [9–12]. Additionally, geopolymers have free-ions (cations K^+ or Na^+) increasing its water affinity and water matrix absorption. This means that water can be diffused into the ceramic body but retain waste or benefic ions. Using these geopolymer properties, some authors have proposed some geopolymer adsorbents based on fly-ash for dye removal from aqueous solution and it was found that the adsorption capacity depends on the preparation conditions and the metal: silica/alumina ratio [13]. Geopolymers are also considered as a green ceramic material when they are compared with traditional ceramics including the highly produced Ordinary Portland Cement (OPC). Additionally, it is also known that the production of geopolymer cements requires around 60% less energy, results in three times less CO_2 emissions compared to Ordinary Portland Cement (OPC), so they are in line with the principles of sustainable development [14].

The aims of this study were the synthesis of Triclosan- metakaolin based geopolymer cement, their antibacterial efficiency evaluation against *Escherichia Coli (Gram-negative bacteria) and Staphylococcus aureus (Gram-positive bacteria)* and its determination to their ability to inhibit its growth on the antibacterial-geopolymer cement surfaces for health and environmental building products applications.

2 Experimental Procedure

2.1 Starting Materials and Sample Preparation

The starting materials to synthesize the antibacterial meta-kaolin based geopolymer cement were:

1. Potassium hydroxide pellets (J.T. Baker Chemical Analytical Grade).
2. Dehydroxylated kaolinite (Metamax, Meta-Kaolinite, Engelhard Corp. USA).
3. Amorphous Nano-silica with a particle size distribution between 7 to 20 nm, Average d = 12 nm (AEROSIL® 200 is a hydrophilic fumed silica with a specific surface area of 200 m2/g. Evonik Industries AG. Germany).
4. Deionized water.

A geopolymer cement paste was obtained by using a potassium silicate solution prepared by dissolving nano-silica and KOH pellets in deionized water to make a solution with a molar ratio of $K_20:SiO_2 = 0.25$ and $H_2O:K_2O = 14$. The solution was allowed to mature under stirring for at least one day in order to dissolved the nano-silica completely, after that metakaolin was mixed into the potassium silicate solution. The metakaolin used was the necessary to complete a molar ratio SiO_2: Al_2O_3 of 4.

To achieve antibacterial properties with this nano-silica and metakaolin geopolymer paste, different batches with only 0.5% and 1.5% (w/w) of organic compound (5-chloro-2-(2, 4-dichlorophenoxy)phenol) commercially known as "triclosan" (Millikan S.A. Mexico) were prepared. Additionally for reference, a standard nano-silica and metakaolin based geopolymer cement paste was synthesized. Finally, the standard and antibacterial pastes were cast into a petri dish of 50 mm in diameter and 10 mm in thickness. All Samples were cured for 4 h at ambient temperature (20 ° C) followed by 24 h at 65 °C, after cooling standard and antibacterial nano-silica and meta-kaolin based geopolymer cements were obtained for evaluation.

2.2 Antibacterial Test Method

The Halo Method was used to determine the antibacterial efficiency. It is an established method to provide a semi-quantitative analysis [4]. A disk of each sample of standard and antibacterial meta-kaolin based geopolymer cements of 50 mm in diameter and 10 mm in thickness were first placed on an *Escherichia Coli ATCC25922 (Gram-negative bacteria)* culture spreading on a medium. After 18 to 24 h at 35 °C of incubation the areas surrounding the nano-silica and metakaolin based geopolymer cements disk shows the bacteria activity. After the *E. Coli* evaluation all samples were sterilized in autoclave at 110 °C and placed on a *Staphylococcus aureus ATCC 25923 (Gram-positive bacteria)* culture spreading on a medium and incubated 18 to 24 h at 35 °C the activity bacteria evaluation was performed, same sample of each batch were compared with gram-negative and gram-positive bacteria.

3 Results

3.1 Activity Evaluation Against E. Coli and S. Aureus Using Triclosan as Antibacterial Agent

In Figs. 1 and 2, a standard metakaolin based geopolymer cement disk can be observed, as well as their activity evaluation against *E.* Coli and *S.* aureus bacteria. The result is clear and obvious, there is no bacteria inhibition.

On the other hand, organic antibacterial compounds such as Triclosan, also called Triclocarban, are used in a wide range of consumer products, including soaps, deodorants, cosmetics and in the field of materials is typical its used in organic polymers; however, since its nature is organic can not be used as antibacterial agent in traditional ceramic/cement manufacturing where manufacturing temperature rises 1250 °C, since geopolymer cements are low temperature and room temperature cements, it was found interesting its antibacterial activity evaluation using triclosan as antibacterial agent showing a high effectiveness against Gram-positive and Gram-negative bacteria as it can be probed by the Halo method results (Figs. 3, 4, 5 and 6).

Fig. 1. Activity evaluation against E.Coli using a standard metakaolin based geopolymer cement without E. Coli inhibition.

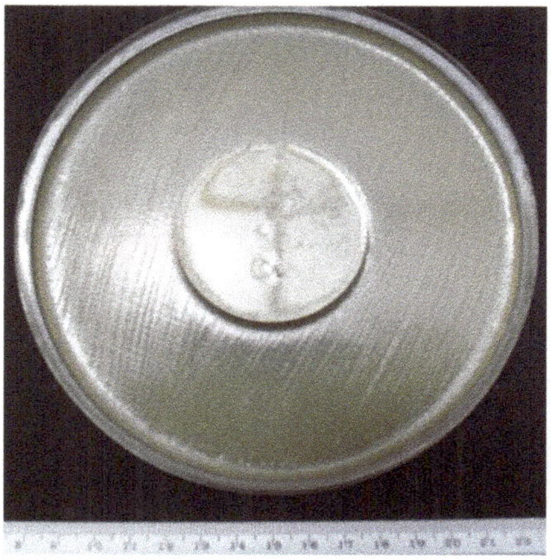

Fig. 2. *S. aureus* activity evaluation of a standard metakaolin based geopolymer cement disk without bacteria inhibition.

Fig. 3. A geopolymer cement disk with 0.5% (w/w) of Triclosan against *E. Coli* and a 90 mm Halo was developed on it

Fig. 4. A geopolymer cement disk with 1.5% (w/w) of Triclosan against E. Coli. A 90 mm Halo was developed on it.

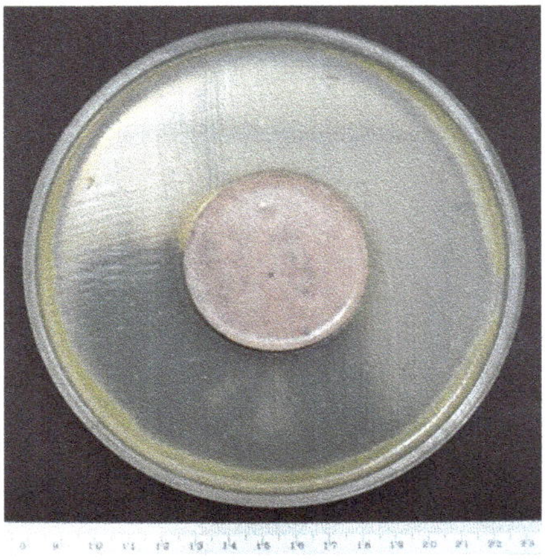

Fig. 5. Activity evaluation against S. aureus of a geopolymer cement disk with 0.5% (w/w) of Triclosán. A 116 mm Halo was developed on it.

Fig. 6. Activity evaluation against S. aureus of a geopolymer cement disk with 1.5% (w/w) of Triclosán. A 116 mm Halo was also developed on it.

4 Conclusion

From the *in vitro* susceptibility tests of antibacterial metakaolin based geopolymer cement against pathogenic microorganisms such as *E. coli* and *S. aureus* have been shown that these geopolymer cement matrices are effective "carriers", with similar behavior to that reported in literature for zeolites. By this reason, antibacterial "triclosan compound" can be successfully used in geopolymers cements for bacteria inhibition against Gram-positive and Gram-negative bacteria. Since the organic compound is encapsulated into a geopolymer cement a certain U.V. protection is achieved; however, further research considering longer periods of time evaluation needs to be done to verify the antibacterial efficiency.

Acknowledgments. The author would like to thank to the microbiologist research team of SIALATO S.A. de C.V. and CEDIMI S.A. de C.V: M.R. Rubio-Avalos, M.P. Rubio-Avalos, B. Ballesteros-Silva, L.E. Mercado-Bravo for their assistance in sample preparation and characterization of such samples by the techniques used in the present work. J.C. Rubio-Avalos thanks to Coordinación de la Investigación Científica de la Universidad Michoacana de San Nicolás de Hidalgo (grant 12.11) for financial support.

References

1. Hanus, M.J., Harris, A.T.: Nanotechnology innovations for the construction industry. Progress Mater. Sci. **58**, 1056–1102 (2013). Elsevier
2. Galayrde, C., Silva, M.R., Warscheid, T.: Microbial impact on building materials: an overview. Mater. Struct. Mater. et Constructions **36**, 342–352 (2003). Rilem
3. Bal, B.T., Yilmaz, H., Aydin, C.: Antibacterial and antifungal properties of polyether impression materials. J. Oral Sci. **49**(4), 265–270 (2007)
4. Ge, J., Shi, X., Cai, M.: A novel biodegradable antimicrobial PU foam from wattle tannin. J. Appl. Poly. Sci. **90**, 2756–2763 (2003). Wiley Periodicals, Inc
5. Li, B., Shuhui, Yu., Hwang, J.Y.: Antibacterial vermiculite nano-material. J. Miner. Mater. Charact. Eng. **1**(1), 61–68 (2002)
6. Zhang, Y., Zhong, S., Zhang, M., Lin, Y.: Antibacterial activity of silver-loaded zeolite a prepared by a fast microwave-loading method. J. Mater Sci. Springer Science **44**, 457–462 (2009)
7. Duxson, P., Provis, J.L., et. al.: Understanding the relationship between geopolymer composition, microstructure, and mechanical properties. Colloids Surf. A Physicochem Eng. Aspects **269**, 47–58 (2005). Elsevier B.V
8. Van Jaarsveld, J.G.S., Van Deventer, J.S.J., Lorenzen, L.: The potential use of geopolimeric materials to immobilise toxic metals: Part I. Theory Appl. Miner. Eng. **10**(7), 659–669 (1997). Pergamon, Elsevier Science Ltd
9. Perera, D.S., et al.: Fe speciation in geopolymers with Si/Al molar ratio of $\sim$ 2. J. Eur. Ceram. Soc. **27**, 2697–2703 (2007). Elsevier Ltd
10. Khalil, M.Y., Merz, E.: Immobilization of intermediate-level wastes in geopolymers. J. Nucl. Mater. **211**, 141–148 (1994)
11. Zosin, A.P., Priimak, T.I., Avsaragov, K.B.: Geopolymer materials based on magnesia-iron slags for normalization and storage of radioactive wastes. Atom. Energy **85**, 510–514 (1998)
12. Perera, D.S., Vance, E.R., Aly, Z., Davis, J., Nicholson, C.L.: Immobilization of Cs and Sr in geopolymers with Si/Al $\sim$ 2. Environmental issues and waste management technologies in the nuclear industries XI. Ceram. Trans. **176**, 91–96 (2006)
13. Li, L., Wang, S., Zhu, Z.: Geopolymeric adsorbents from fly ash for dye removal from aqueous solution. J. Colloid Interface Sci. **300**, 52–59 (2006)
14. Komnitsas, K., Zaharaki, D.: Geopolymerisation: a review and prospects for the minerals industry. Miner. Eng. **20**, 1261–1277 (2007)

Identification of Reactive Sites in Calcined Kaolinite and Montmorillonite from a Combination of Chemical Methods and Solid-State NMR Spectroscopy

C. Ruiz-Santaquiteria(✉) and J. Skibsted

Department of Chemistry and Interdisciplinary Nanoscience Center (INANO), Aarhus University, Langelandsgade 140, 8000 Aarhus C, Denmark

Abstract. Kaolinite and montmorillonite have been examined at heating temperatures from 500–1100 °C. For each sample, the degrees of dissolution in acidic and basic media have been determined, using a 1.0 vol.% HF solution and an 8.0 M NaOH solution, respectively. The solid residues from these experiments are analyzed along with the calcined starting materials by ^{27}Al and ^{29}Si MAS NMR. A comparison of these spectra, before and after dissolution, enables a clear differentiation of the silicon and aluminium environments that are present in each sample, providing direct information about the aluminate and silicate species which are dissolved under basic and acidic conditions, i.e., identification of the active sites in calcined clays. Moreover, this procedure facilitates the structural assignment of the different silicon environments observed by ^{29}Si NMR, shedding light on the dehydroxylation process and on the structural changes that occur for kaolinite and montmorillonite upon heat treatment.

1 Introduction

The use of calcined clays as supplementary cementitious materials (SCMs) may provide a valuable contribution to the reduction in CO_2 emissions associated with cement production since calcined clays can be produced at significantly lower temperatures and do not involve a decarbonation reaction. Thermal activation of clay minerals involves dehydroxylation of the aluminate and silicate sheets which introduces a progressive degree of disorder in the heated material. ^{27}Al and ^{29}Si MAS NMR spectroscopy have the advantage that amorphous and crystalline phases are detected in an equal and unambiguous manner and thus, these tools are ideal in studies of the thermal activation of both pure and impure clay minerals [1–3]. In addition, ^{27}Al and ^{29}Si MAS NMR can provide the degrees of reaction for the principal cement and calcined clay phases which can be utilized in thermodynamic modeling of the hydrate phase assemblages in these blends [4]. Moreover, valuable structural information is also achieved about the calcium-silicate-hydrate phase (C-S-H) [5], formed as the principal hydration product in Portland cement – calcined clay blends.

A common approach to evaluate the reactivity of SCMs is the Chapelle's test which probes the pozzolanic reactivity by measuring the amount consumed $Ca(OH)_2$ in $Ca(OH)_2$ – SCM suspensions. For these experiments, additional information can be

F. Martirena et al. (eds.), *Calcined Clays for Sustainable Concrete*, RILEM Bookseries 16,
https://doi.org/10.1007/978-94-024-1207-9_65

achieved by studying the solid residues by ^{29}Si and ^{27}Al MAS NMR, which allows identification of the formed hydration products and the remains of SCM in a semi-quantitative manner. In addition to this approach, the present work employs a combination of chemical methods and solid-state NMR to identify the reactive species in clay minerals calcined at different temperatures. Although both kaolinite and montmorillonite have been examined in this work, the paper presents only data observed for kaolinite.

2 Experimental

Premium-grade kaolinite (Imerys Minerals, UK), containing a small impurity of quartz (~2.6 wt%), has been heat-treated in air for 2 h at temperatures between 500 °C and 1100 °C in intervals of 50°C. For each heated sample, one gram was subjected to chemical attack in 100.0 mL HF (1.0 vol.%) at room temperature under stirring for five hours. The same experimental procedure was carried out for a basic medium using 8.0 M NaOH for selected samples, however, the temperature for these experiments was set at 80 $\pm$ 2 °C to further facilitate dissolution. After the exposure, the liquid and solid residue was separated by filtration and the residue was rinsed with distilled water until a neutral pH was reached. The mass of the solubilized clay was calculated from the weight difference between the starting material and the dried residue. The Si/Al ratio of the dissolved phase in HF was determined from the Si and Al contents in the filtrate, obtained with a Spectro Arcos ICP-AES instrument. The ^{29}Si MAS NMR spectra were recorded on a Varian INOVA-300 spectrometer using a home-built CP/MAS probe for 5 mm o.d. zirconia (PSZ) rotors and a spinning speed of $\nu_R = 10.0$ kHz. The ^{27}Al MAS NMR spectra were recorded on a Varian Direct-Drive VNMR-600 spectrometer, using a homebuilt CP/MAS probe for 4 mm o.d. zirconia rotors and a spinning speed of $\nu_R = 13.0$ kHz.

3 Results and Discussion

The fraction of dissolved material for the heated kaolinite samples in the 1.0 vol.% HF solution is shown in Fig. 1a as a function of the heating temperature. The fraction of dissolved material increases with temperature until 550 °C where nearly complete dehydroxylation of kaolinite is expected to have occurred. Only minor variations are observed in the range 500 – 900 °C, in accordance with the presence of the metastable metakaolin phase, and it decreases from 900–1100 °C, the latter associated with the transformation of metakaolin into mullite and a spinel-type phase. The decrease in reactivity from 900–1100 °C is accompanied by an increase of the Si/Al molar ratio for the dissolved phase (Fig. 1b), which is ascribed to the formation of the mullite and aluminate-rich spinel-type phases. These phases exhibit a higher degree of structural order than metakaolin and thus, the dissolution of aluminate species from these phases is expected to be lower. Kaolinite has a Si/Al molar ratio of 1.0 whereas slightly lower ratios are observed for the dissolved phase of kaolinite from 500–900 °C. This indicates a preferential dissolution of aluminate species and thereby a slightly incongruent dissolution of silicon and aluminium.

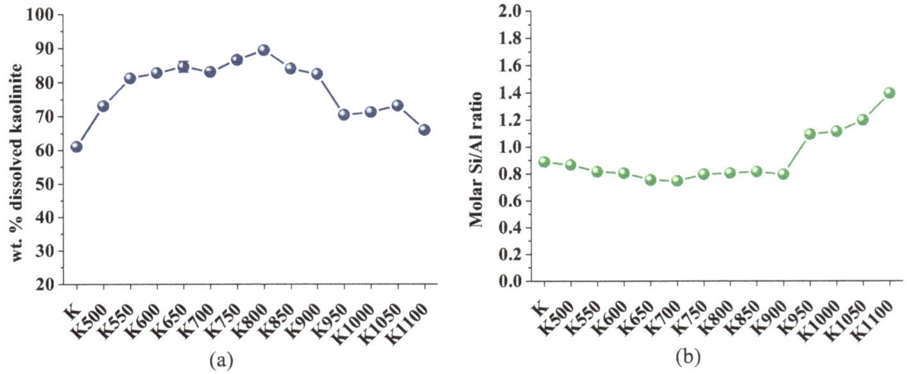

Fig. 1. (a) Fraction (wt%) of dissolved material in the 1.0 vol.% HF solution as a function of the heating temperature. (b) The Si/Al molar ratio of the dissolved phase determined by ICP-AES analyses of the filtrates.

^{29}Si MAS NMR spectra of selected heated kaolinite samples and their corresponding residues from the HF experiments are shown in Fig. 2. The deconvolutions of the individual spectra have been performed using the same number of resonances and only minor variations of their chemical shifts and linewidths. Comparison of the spectra and deconvolutions for the heat-treated samples and their residues reveals that the different silicon sites in the clay material exhibit different reactivities. The contribution to the spectra from metakaolin is simulated by four resonances (Fig. 2) located at −82, −84, −94.5 and −101 ppm and by three resonances at lower chemical shifts than quartz at −108, −115 and −121 ppm. The chemical shifts of the latter three peaks strongly suggest that they originate from Q^4(0Al) silicon sites. From the deconvolutions of the ^{29}Si MAS NMR spectra of the residues, it is apparent that the three Q^4(0Al) peaks give a major contribution to the spectra, demonstrating that these fully condensed SiO_4 units are less reactive than the other components of the metakaolin structure. Moreover, for the samples heated at 950 °C and above, a narrow resonance around −112 ppm appears in the spectra which reflect a partial crystallization of silica and thereby the formation of a less-reactive phase, in agreement with the data in Fig. 2.

^{27}Al MAS NMR spectra of the samples heated at 550 and 800 °C and their residues from the HF solutions are shown in Fig. 3. The spectra contain overlapping centerbands from Al in tetrahedral, fivefold and octahedral coordination. The peak from five-fold coordinated Al (35 ppm) is almost absent in the spectra of the residues, demonstrating that the pentahedral Al sites are highly reactive. In addition, the residues contain at least two different AlO_4 sites, which may also be present in the samples before acid attack, and a rather narrow AlO_6 resonance that may arise from hydroxylated surface sites.

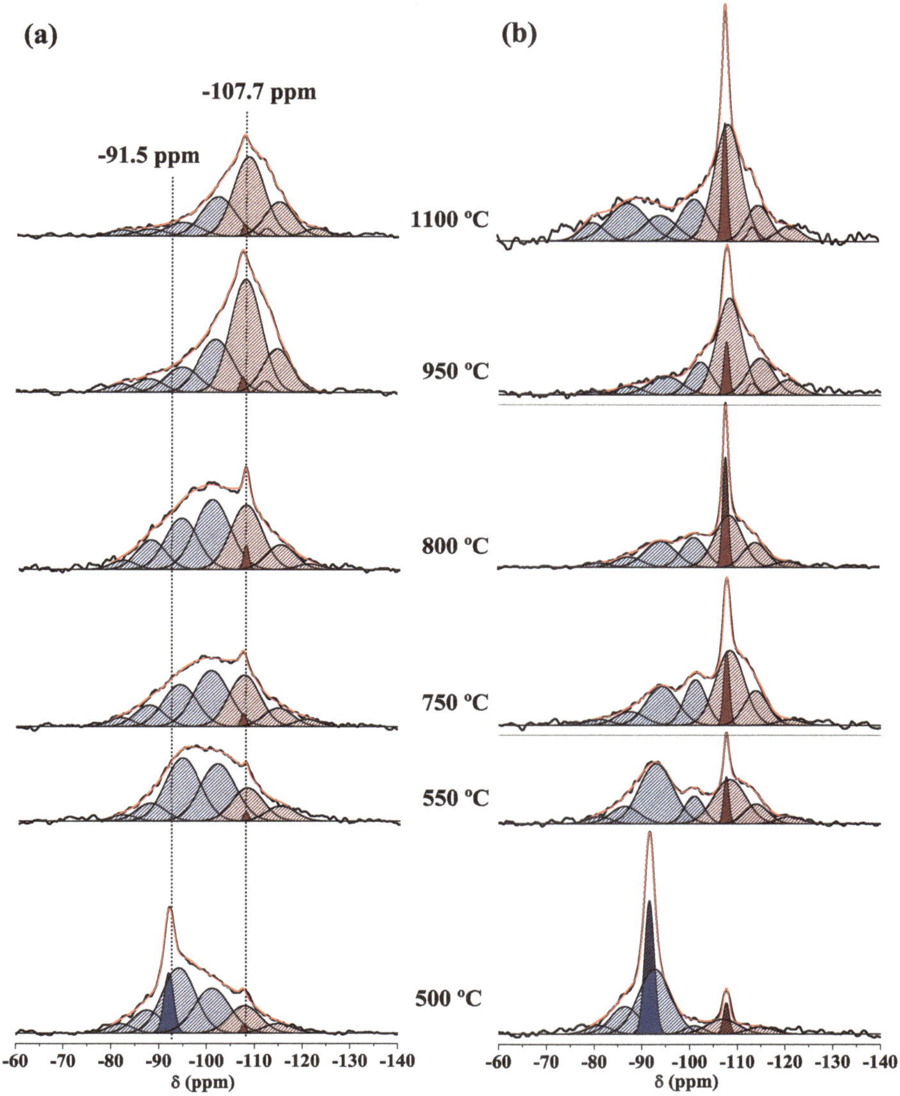

Fig. 2. ^{29}Si MAS NMR spectra (7.1 T, ν_R = 10.0 kHz) of (a) selected heat-treated kaolinite samples and (b) their corresponding residues from the HF acid-attack experiments. The quartz impurity is observed at −107.7 ppm.

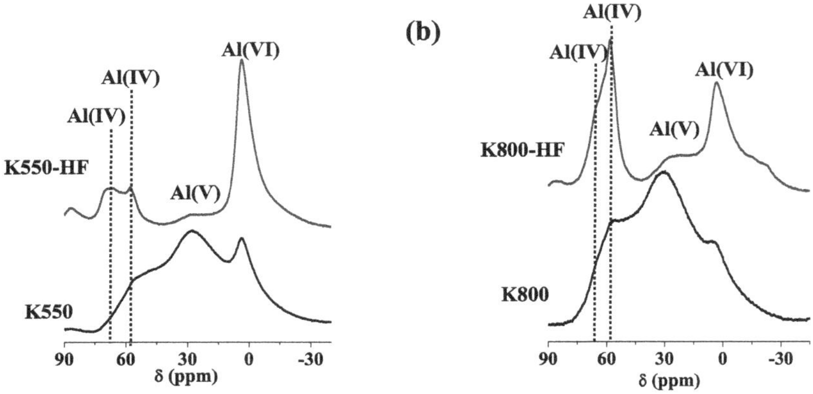

Fig. 3. ^{27}Al MAS NMR spectra (14.1 T, $\nu_R = 13.0$ kHz) of kaolinite heated at (a) 550 °C and (b) 800 °C and of the corresponding residues from the HF solution (upper row spectra).

4 Conclusions

The combination of chemical procedures and solid-state MAS NMR represents a valuable approach to gain new information about the nanostructure and reactivity of heat-treated kaolinite samples. Comparison of the ^{29}Si NMR spectra for the heated clays and their residues allow identification of different silicate environments in metakaolin and characterization of the chemical reactivity for these silicate species. ^{27}Al MAS NMR spectra of the heated samples and their residues have shown that the fivefold coordinated Al sites in metakaolin is the most reactive Al sites and that almost a full degree of reaction is observed for these sites in the chemical attack experiments.

Acknowledgements. The Danish Council for Strategic Research is acknowledged for support to the LowECEM project (No 11-116724).

References

1. Garg, N., Skibsted, J.: Thermal activation of a pure montmorillonite clay and its reactivity in cementitious materials. J. Phys. Chem. C **118**, 11464–11477 (2014)
2. Garg, N., Skibsted, J.: Pozzolanic reactivity of a calcined interstratified illite/smectite (70/30) clay. Cem. Concr. Res. **79**, 101–111 (2016)
3. Hollanders, S., Adriaens, R., Skibsted, J., Cizer, Ö., Elsen, J.: Pozzolanic reactivity of pure calcined clays. Appl. Clay Sci. **132–133**, 552–560 (2016)
4. Kunther, W., Dai, Z., Skibsted, J.: Thermodynamic modeling of hydrated white Portland cement – metakaolin – limestone blends utilizing hydration kinetics from ^{29}Si MAS NMR spectroscopy. Cem. Concr. Res. **86**, 29–41 (2016)
5. Dai, Z., Tran, T.T., Skibsted, J.: Aluminum incorporation in the C-S-H phase of white Portland cement – metakaolin blends studied by ^{27}Al and ^{29}Si MAS NMR spectroscopy. J. Am. Ceram. Soc. **97**, 2662–2671 (2014)

The Decay of the Historical Site of Malecon in Havana, Cuba: Salt Crystallization Damage at Repair Interfaces

A.M. Aguilar Sanchez[1(✉)], F. Caruso[2], F. Girardet[3], F. Martirena[4], T. Wangler[1], and R.J. Flatt[1]

[1] ETH Zurich Institute for Building Materials, Zurich, Switzerland
[2] Department of Archaeology, Conservation and History, University of Oslo, Oslo, Norway
[3] Rino Sarl, Blonay, Switzerland
[4] CIDEM Facultad de Construcciones, Universidad Central de las Villas, Santa Clara, Cuba

Abstract. The buildings of the Malecon, the historical section of Havana along the coastline, suffer accelerated degradation due to the aggressive environment. The primary damage mechanism at play is the crystallization of sodium chloride at repair mortar interfaces, as shown in petrographic and SEM analysis. This preferential precipitation leads to crystallization pressures that reduce the adherence of the lime based mortars from the substrate. Environmental monitoring shows that the daily relative humidity fluctuates around the deliquescence point of sodium chloride, exacerbating the problem through continual cycles of dissolution and recrystallization. The potential for an alternative repair material such as limestone calcined clay cement based mortar is discussed.

1 Introduction

The city of Havana, founded in 1519, is the largest city in the Caribbean with a population of 2 million people. Most of its buildings with historical value have been preserved, which prompted UNESCO to declare Old Havana and its Fortification System as a World Heritage Site [1].

However, nearly 50% of the existing buildings located along the coastal line known as "Malecon" are in critical condition. Malecon's construction was started May 1901 and finally completed in 1959. It is considered a symbol of the city, a major attraction for tourism and an important part of the identity of the Cuban people.

The buildings in Malecon area are subjected to very aggressive environmental conditions because of the proximity to the sea and extreme weather events, such as hurricanes and intense rains. In November 2013, more than a hundred buildings collapsed during a week of heavy rain in the Historical Center due to extreme corrosion of the reinforcements. The renders that cover the structures and walls are damaged and, in large areas, they are totally lost [2].

Since 1994, the Cuban institution in charge of the restoration works (OHC, Oficina del Historiador de la Ciudad de la Habana), in collaboration with other national and international institutions, has made several restoration attempts to repair the renders by using locally developed lime-based mortars and also imported expensive repair

F. Martirena et al. (eds.), *Calcined Clays for Sustainable Concrete*, RILEM Bookseries 16,
https://doi.org/10.1007/978-94-024-1207-9_66

materials. The repair mortars applied on stone, brick or concrete walls have had little or no success. The consequences are bad performance of the repair mortars just a few months the interventions, with implications for costs, as seen in Fig. 1.

Fig. 1. Building Malecon 251 (left) Main substrates of brick masonry and reinforced concrete. Restoration work: waterproofing of roof, repair of renders on the facade and columns. Repair of renders was done with MAXREST and lime-based mortar. (center, right)

In this paper, we aim to understand the nature of the damage mechanisms affecting the brick and stone masonry as well as the repair mortars in Malecon, and to identify potential solutions. Chemical and mineralogical characterization of the original substrates and repairs have therefore been carried out, and onsite monitoring as well to give an idea of site conditions that could lead to degradation. The use of alternative, locally available repair mortars with potentially helpful properties for durability is also discussed.

2 Materials and Methods

The assessment of current state of conservation and the observation of the main degradation patterns in representative reference buildings, as selected by the Havana restoration office experts (OHC), was the initial focus of the project. Criteria for the choice of the subject buildings were type of substrate: stone – brick masonry and the type of render used. Selected representative results are shown in Fig. 2.

To understand the real conditions to which buildings are exposed, sensors to measure temperature and relative humidity in the atmosphere and at the surface of the wall were installed in the building Malecon 21, on the surface of a stone column covered by original lime mortar.

Samples of different substrates such as stone and bricks, as well as original mortars and further repair mortars were taken from the reference buildings. Samples from the salt efflorescence were taken to determine composition of the main salts.

Petrography and SEM analysis were carried out in the stones, bricks and repair mortars by using an optical microscope Leica M750 P and SEM FEI Quanta 200 3D.

Fig. 2. Building Malecon 21. Stone column covered by highly porous lime mortar, intensely disintegrated on the surface, likely due to salt crystallization. Buildings have had no restoration intervention since 1910.

3 Results and Discussion

3.1 Description of Original Building Materials and Most Used Repair Mortars

The historic buildings are primarily composed of reinforced concrete, bricks and ashlars of fossiliferous limestone. Original renders and joint mortars are constituted by highly porous lime mortars with quartz and feldspar sand.

There are two most commonly used recipes of repair mortar that have been used: first, to repair the original renders, and then for the successive interventions. The first formulation is a lime mortar elaborated with common commercial calcium hydroxide. The second repair mortar consists of a mix of calcium hydroxide and 5% of white cement. For both repair mortars, a mix of calcareous and siliceous washed sand from La Victoria quarry was used [3].

Besides the previously mentioned repair mortars, which are the most used ones, some reference buildings have been repaired with three different ready-to-use commercial repair mortars used to fill fissures and holes in stone facades and concrete substrates, respectively.

3.2 Degradation Patterns

The preliminary assessment of the damage in the buildings Malecon 21, 251, 259, 307, 313 showed that salt crystallization and chloride induced corrosion appear to be the main problems affecting the renders and the substrates. In this paper, we will focus on the damage caused by salt crystallization, a known and widely studied damage phenomenon in the field of historic conservation [4–6]. Detachment of paint and mortars from the substrates is a very common problem. The most representative damage patterns associated to salt crystallization that can be observed are granular disintegration on the surfaces of stone, bricks and mortars. Alveolar formation on stone, renders and bricks is also observed extensively.

3.3 Microscopy of Materials

The microscopic study of the original stone and original mortars revealed the presence of fossiliferous dolomitic limestone. The original mortars are a lime-based mortar with calcareous sand; sodium chloride is contained in the porosity and crack as observed in Fig. 3. It was observed that sodium chloride crystals preferentially precipitate at the interface between substrate and repair mortar as well as at the interface between two generations of repair mortars. This can be seen in the Fig. 4.

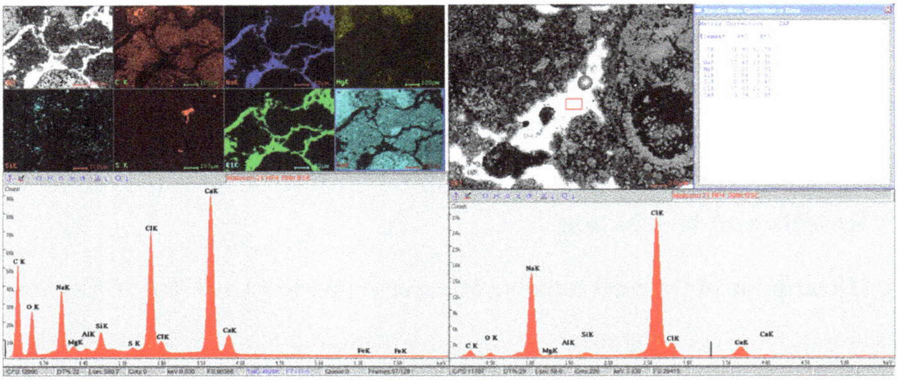

Fig. 3. Compositional mapping performed on a lime mortar on stone substrate. Sample HP-4, Building Malecon 21. Sodium chloride is located within the cracks of the mortar matrix.

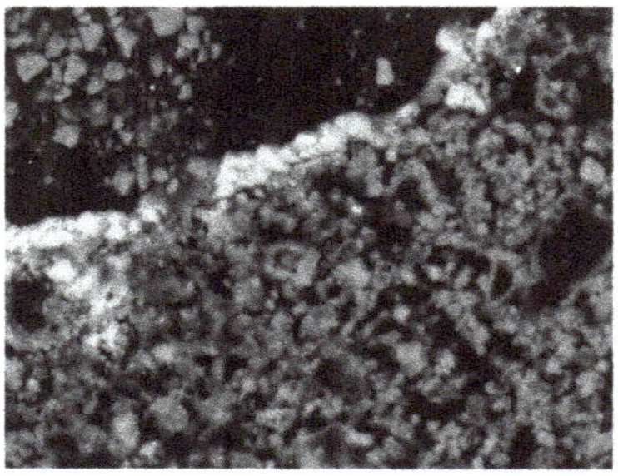

Fig. 4. Sodium chloride crystals (white) at the interface between the original lime mortar and the ready-mix MAXREST repair mortar.

Preliminary measurement of environmental variables of Havana show high relative humidity (60–90%) and high temperature (25–34 °C) throughout the entire year (Fig. 5). The relative humidity oscillates between 60 and 90% during the day, which sandwiches the NaCl deliquescence RH and consequently causes damage-inducing dissolution and recrystallization [7].

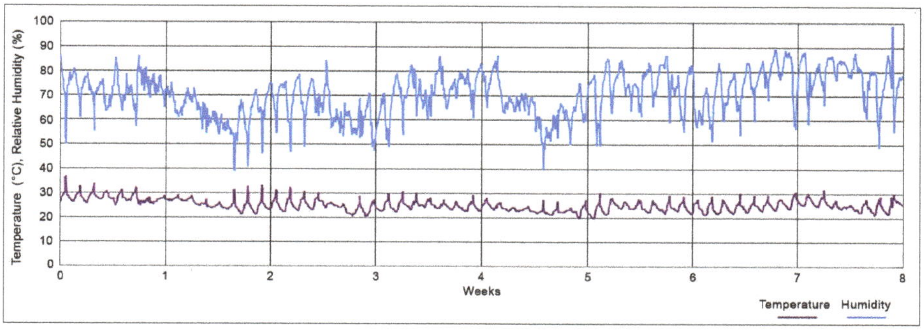

Fig. 5. Temperature and relative humidity data, obtained by the sensor located in building Malecon 21. These data were obtained from September-October 2013. The mean temperature is 28 °C.

Sea salt spray is one of the main salt sources. Additionally, several buildings were built on places where the level of underground water is close to the surface, and the geological structure of the soil supporting the structure (calcareous rock with many caverns caused by leaching) favors the presence of high moisture content and capillary rise in the existing buildings. Besides sodium chloride, the presence of $CaSO_4$ and $MgSO_4$ was confirmed by X ray diffraction and EDX, and is probably associated with the proximity of the sulfur-rich exhaust gases of a nearby power plant.

4 Conclusions

The extreme conditions of the environment in Malecon, the co-existence of diverse materials, and the insufficient characterization of old materials with regard to composition and porosity, can complicate the further understanding of the causes of the degradation.

Crystallization by sodium chloride is very likely the main degradation mechanism acting in Havana, Malecon. The salt is coming primarily from sea spray, and is found to be saturated within the substrate materials. Most damage is occurring from salt precipitating at the interface of the repair material and the substrate.

Future work will focus on the crystallization dynamics based on material properties and conditions, as well as examining alternative repair materials with potentially better adherence strength, such as limestone calcined clay cement based repair mortars based on locally available materials. These mortars offer potential resistance to chlorides due to the formation of Friedel's salt, which also serve as a chloride sink in the protection of any reinforcement that may be present.

Acknowledgments. The authors acknowledge the contribution of Arq Mrs Sofia Martinez Guerra, from Oficina del Historiador de la ciudad de la Habana (OHC), for helpful discussions and access to job sites.

References

1. Old Havana and Its Fortification System - UNESCO World Heritage Centre. http://whc.unesco.org/en/list/204. Accessed 24 Oct 2014
2. Juventud Rebelde. http://www.juventudrebelde.cu/cuba/2013-11-30/lluvias-intensas-en-occidente
3. Cleofas Buajasan, Maria: Manual de Reparacion y Mantenimiento de Edificaciones Del Centro Historico de La Habana, Oficina del Historiador de la Ciudad Habana (2013)
4. Tsui, N., Flatt, R.J., Scherer, G.W.: Crystallization damage by sodium sulfate. Jour. Cult. Her. **4**, 109–115 (2003)
5. Goudie, A.S., Viles, H.A.: Salt Weathering Hazards. Wiley, London (1997)
6. Evans, I.S.: Salt crystallization and rock weathering: a review. Rev. Géomorphologie Dyn. **19**, 153–177 (1969)
7. Lubelli, B.A.: Sodium chloride damage to porous building materials. Ph.D. thesis (2006)

Introducing Low Carbon Cement in Cuba - A Life Cycle Sustainability Assessment Study

S. Sánchez Berriel[1(✉)], Y. Ruiz[2], I.R. Sánchez[1], J.F. Martirena[3], E. Rosa[4], and G. Habert[5]

[1] Faculty of Economy, Central University "Marta Abreu" of Las Villas (UCLV), Carretera a Camajuanì km 51/2, Santa Clara, Villa Clara, Cuba
ssanchez@uclv.edu.cu

[2] Faculty of Constructions, Central University "Marta Abreu" of Las Villas (UCLV), Santa Clara, Cuba

[3] Center for Research and Development of Structures and Materials (CIDem), UCLV, Santa Clara, Cuba

[4] Faculty of Chemistry and Farmacy, Central University "Marta Abreu" of Las Villas (UCLV), Santa Clara, Cuba

[5] Chair of Sustainable Construction, ETH Zurich, 8093 Zurich, Switzerland

Abstract. The main goal of the paper is to carry out the first implementation of a Life Cycle Sustainability Assessment (LCSA) in Cuba throughout a case of study related with the introduction of Low Carbon Cement in the country. First, an adapted model to assess economic, social and environmental impacts is developed according to cement sector conditions in the island. Secondly, the "adapted LCSA" is carried out comprising a life cycle assessment (LCA), an economic life cycle analysis (EcLCA) and a social-LCA (S-LCA). The environmental assessment includes impacts on climate change, particulate matter emission, fossil fuel depletion among others. The systems explored include all processes from cradle to industry 'gate'. The cements assessed were Ordinary Portland Cement (OPC), Blended Cement (PPC) and Low Carbon Cement (LC^3). For the EcLCA and S-LCA, indicators adjusted to Cuban conditions are proposed and assessed, using midpoint and endpoint levels to match with LCA methodology.

Economically, conventional Portland cement presents the highest costs per ton, followed by PPC and LC^3. In the S-LCA, LC^3 introduction reduces sensitive social impacts in relation with different stakeholders: workers, local community and society. Environmentally, main impact, at global scale, is climate change and, at local level, particulate matter emission. Integrating the results from LCA, S-LCA and EcLCA shows a significant reduction on the climate and social impacts per cost when producing LC^3. Finally, results show that LC^3 introduction is the best option to meet sustainability goals of the cement industry in Cuba and LCSA can be considered a valuable tool to support decision-making processes oriented to sustainability.

F. Martirena et al. (eds.), *Calcined Clays for Sustainable Concrete*, RILEM Bookseries 16,
https://doi.org/10.1007/978-94-024-1207-9_67

1 Introduction

Cement demand has increased exponentially in recent years, reaching 4600 million tons in 2015 [1]. World Bussiness Council for Sustainable Development (WBCSD) and International Energy Agency (IEA) in their predictions for 2050 indicate that in a scenario of high demand the increase in global production will reach 6000 million tons of cement, where 90% will correspond to emerging or developing countries. If current production conditions are maintained, this will lead to increased CO_2 emissions from the sector.

The proposal of low carbon cement (LC^3) introduction in Cuban industry allows to maintain the normalized resistances that allow the substitution of P-35 (OPC), both in mortars and concretes [2, 3]; increase productive capacities with a low level of investment and reduce CO_2 and other GHG emissions by approximately 30% [4].

Life Cycle Sustainability Assessment (LCSA) is one of the most modern tools applicable to integrated assessment of investment impacts and sustainability-oriented programs [5]. Therefore, the main goal of this paper is to assess the economic-social and environmental impacts of the introduction of low-carbon cement into the Cuban cement industry through a LCSA.

2 Method

According to UNEP/SETAC (2011) the Life Cycle Sustainability Assessment considers all stages of the product and service life cycle the complete study of its production and value chain. Kloepffer (2008) proposed the LCSA structure through a conceptual formula [6]. Later, UNEP has published several documents that serve as a methodological guide for the discussion, discussion and implementation of this tool analysis [7].

In Eq. 1 (proposed by Kloepffer), the LCSA is conceptually defined:

$$\text{LCSA} = \text{LCA} + \text{LCC} + \text{S - LCA} \quad (1)$$

Where LCSA stands for Life Cycle Sustainability Assessment, LCC is the Life Cycle Costing, LCAis the Life Cycle Assessment (environmental) and S-LCA the Social Life Cycle Assessment.

The environmental life cycle assessment (environmental LCA) looks at the potential impacts of products and services in the environment. ISO standards for LCA guide these studies in four phases: (i) definition of objectives and scope; (ii) inventory analysis; (iii) impact assessment; (iv) interpretation of results; with close interrelation between the phases.

Life Cycle Costing (LCC) and financial analysis are combined with the economic analysis of the life cycle from the methodological guide proposed by Neugebauer et al. (2016), understanding that the economic analysis comprises a number of variables that allow holistic evaluation of the impacts of an activity or service beyond the cost category [8, 9].

Social Life Cycle Assessment (S-LCA) were published by UNEP/SETAC in 2009, then in 2013, methodological sheets for impact categories and indicators by stakeholder groups were published [10]. In this investigation, the subcategories and indicators are adjusted to Cuban conditions to be evaluated. From the UNEP proposal (2013) 14 indicators are identified for evaluation.

3 Results

The central objective is to evaluate the environmental, economic and social impacts of the introduction of low carbon cement in Cuba comparing LC^3 with cements currently produced P-35 and PP-25.

The technical study carried out focuses on the production and transportation of raw materials from cement to obtaining a ton at the factory door (from cradle to door), while the analysis of the economic, social and environmental impacts extends to its use. The functional unit used in the study is 1 ton of cement.

The comparative life cycle analysis of the P-35, PP-25 and LC^3 cements is performed to evaluate 11 intermediate categories and 3 final categories of the ReCiPe methodology. All of the results are shown in Fig. 1, noting that the LC^3 has the least impact in 8 categories of the midpoint categories. The calculations are performed using the professional software Simapro vs-8.0.3.14 and the Windows tool Microsoft Excel 2013, for the treatment of some data.

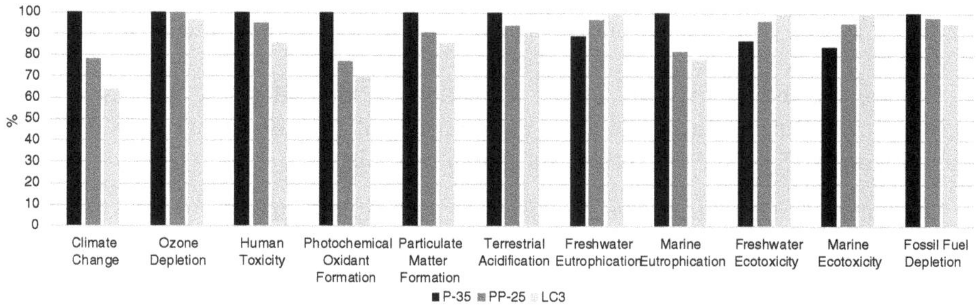

Fig. 1. Comparison of environmental profiles expressed in percentage (ReCiPe midpoint)

The analysis of the energy consumption shows that with the introduction of LC^3 can reduce in approximately 900 MJ the energy consumption per ton of cement produced. The main savings are obtained in the processes of clinkering, extraction of fuels and grinding.

The formation of particulate matter (dust) is generated in the cement industry by kilns, raw material mills, clinker coolers and cement mills. In the obtained results, the superiority of the LC^3 is visible in the different technological levels. Cement grinding is the process that causes more dust due to the technical state of the mills.

From the endpoint categories, the most affected category is human health, due to the damage caused by the gases emitted in the production process related to CO_2 and particulate matter. The P-35 causes greater damage to human health and the ecosystem.

Through the LCC, the cost composition of each cement is analyzed, as shown in Fig. 2. The main costs are given by the exploitation and transportation of energy resources, raw materials and equipment depreciation. Labor costs represent 15% of the total cost of production.

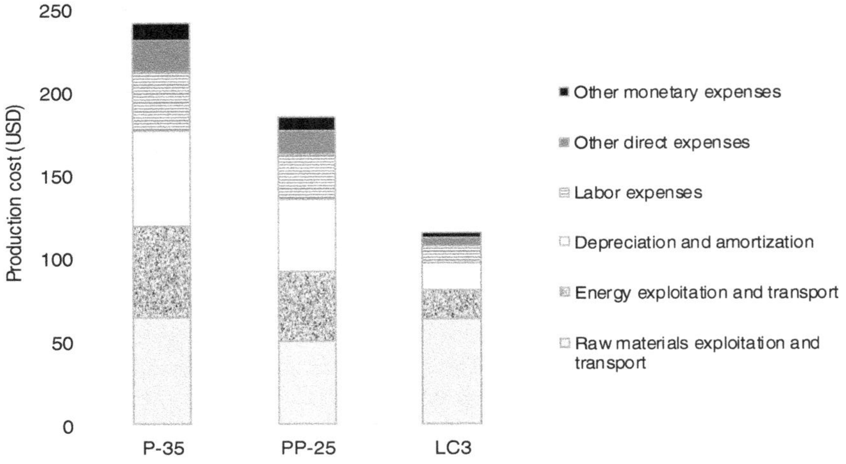

Fig. 2. Cost structure of cement production in Siguaney. Pilot scenario

To evaluate the profitability of the technological alternatives, 2 alternatives are compared with Business as Usual using a capital cost of 12%, discount rate of 35% and time horizon of 15 years. The results are shown in Table 1.

- Alternative 1: This alternative proposes the strategy of introducing the LC^3 in the Cuban cement industry as the partial substitution of traditional cements from the conversion of kilns to calciners. Four calciners with capacity of 300 thousand tons/ year of calcined clay are estimated.
- Alternative 2: Under this alternative, calcination of the clay should be done through flash calciners. Flash technology must be imported. The same amount of calciners is estimated as in alternative 1.

Table 1. Financial results for each of the investment alternatives

Alternatives	Indicators		
	NPV (MPesos)	IRR (%)	Payback period (Years)
LC^3_Retrofitted calciners	$227.34	58%	3 años, 81/3 mes
LC^3_ Flash calciners	$123.53	33%	5 años

The internal rate of return and the payback period show that the conversion of kilns to calciners is the best alternative in the short-term but in the long-term efficient flash calciners could be introduced.

The assessment of social impacts is carried out mainly by assessing the potential for change in the selected indicators. When possible, quantitative analysis is performed. The following levels are proposed to evaluate the potential for change: a: Negligible, B. Minor, C. Moderate and D. Significant. The results of the evaluation show that 79% of the impacts present moderate or significant change potential (Table 2).

Table 2. Potential of change (PC) of proposed social indicators

Sub-categories	PC	Indicators	PC
1.4 h of Work	A	No. of hours worked/t cement	A
1.7 Health and Safety	D	*Incidence of diseases attributable to cement production/worker*	D
3.3 Cultural Heritage	C	No. of buildings with patrimonial value restore/year	C
3.5 Local employment	B	Percentage of labor force contracted in the locality	A
		No. of jobs	B
3.7 Access to material resources	C	No. of infrastructure projects developed with access and benefit to the community	C
3.8 Safe and healthy living conditions	C	*Incidence of diseases attributable to cement production/community residents*	C
5.1 Public commitment to sustainability issues	C	Presence of publicly available documents as agreements on sustainability issues	C
		Implementation/signature of principles or other internationally reconciled codes of conduct	D
5.3 Contribution to economic development	C	% GDP relative to the construction sector	C
		Number of houses/year	D
		Changes in the purchasing power of the population	C
5.5 Technological development	C	Sectorial efforts for technological development	D
		Relationship with technology transfer programs or projects	C

Local employment is a significant variable. The municipality of Taguasco occupies 80% of the workforce of Siguaney and of this 60% corresponds to the town Siguaney, located very close to the factory. Although the generation of jobs is not declared with the introduction of the LC^3, this investment allows to keep open the factory that by its technical state has been in several occasions about to be closed.

The *percentage of diseases in the community attributable to cement production* is analyzed through information published in municipal statistical yearbooks on the incidence of acute respiratory diseases. It analyzes the localities of provinces in the central region where cement factories are located using as reference the average reported in the country.

In the province of Sancti Spíritus, there are several municipalities with a high incidence of respiratory diseases, among which the municipalities of Jatibonico and Yaguajay stand out, as there are no other factories that produce atmospheric pollution, which can be attributed to the high incidence of respiratory diseases to the influence of cement production in Siguaney, taking into account that the wind flows regularly to 34° northwest with an average speed of 1.7 meters per second (m/s)[1], where these localities are located.

4 Conclusions

The results of the Life Cycle Sustainability Assessment carried out show that the main impacts are associated with the reduction of greenhouse gas emissions and dust, which in turn influences the reduction of diseases associated with this type of production; the energy saving that is translated in the reduction of the cost of production and the consequent impact on the purchasing power of the population; the revitalization of the cement production in Cuba and in particular in Siguaney, which would allow to satisfy the domestic demand and to increase the exports of this good; in addition to the increase of productive capacities through small investments with high profitability and short recovery period.

Acknowledgments. The authors would like to acknowledge Siguaney cement factory for the technical and material support. The authors would like to thanks to the whole team of the "Low Carbon Cement" Project for all the advisory and technical help during this paper creation.

References

1. Scrivener, K.L., John, V.M., Gartner, Y.E.M.: Eco-efficient cements: potencial, economically viable solutions for low-CO2, cement-based materials industry, Paris (2016)
2. Antoni, M., Rossen, J., Martirena, F., Scrivener, Y.K.: Cement substitution by a combination of metakaolin and limestone. Cem. Concr. Res. **42**(12), 1579–1589 (2012)
3. Vizcaíno-Andrés, L.M., Sánchez-Berriel, S., Damas-Carrera, S., Pérez-Hernández, A., Scrivener, K.L., Martirena-Hernández, Y.J.F.: Industrial trial to produce a low clinker, low carbon cement. Mater. Construcción **65**(317), e045 (2015)
4. Sánchez Berriel, S., Favier, A., Rosa Domínguez, E., Sánchez Machado, I.R., Heierli, U., Scrivener, K., Scrivener, K., Martirena Hernández, F., Habert, Y.G.: Assessing the environmental and economic potential of limestone calcined clay cement in Cuba. J. Clean. Prod. **124**, 361–369 (2016)
5. UNEP/SETAC: Towards a Life Cycle Sustainability Assessment. Making informed choices on products (2011)

[1] According to data from the Meteorological Information Portal of Sancti Spíritus. http://www.cmpss.cu/index.php?id=emas. Accessed (20/03/2017).

6. Finkbeiner, M., Schau, E.M., Lehmann, A., Traverso, Y.M.: Towards life cycle sustainability assessment. Sustainability **2**, 3309–3322 (2010)
7. Valdivia, S., Ugaya, C.M.L., Hildenbrand, J., Traverso, M., Mazijn, B., Sonnemann, Y.G.: A UNEP/SETAC approach towards a life cycle sustainability assessment - our contribution to Rio +20. Int. J. LCA **18**, 1673–1685 (2013)
8. Swarr, T.E., Hunkeler, D., Klöpffer, W., Pesonen, H.L., Ciroth, A., Brent, A.C., Pagan, Y.R.: Environmental life-cycle costing: a code of practice. Int. J. LCA **16**, 389–391 (2011)
9. Neugebauer, S., Forin, S., Finkbeiner, Y.M.: From life cycle costing to economic life cycle assessment—introducing an economic impact pathway. Sustainability **8**, 1–23 (2016)
10. UNEP/SETAC: The Methodological Sheets for Sub - categories in Social Life Cycle Assessment (S-LCA) (2013)

Sulphate Optimization of Binders with Calcined Clay Using Isothermal Calorimetry

P. Sandberg[1,2(✉)] and S. Bishnoi[1,2]

[1] Calmetrix Inc, Holbrook, USA
[2] Indian Institute of Technology Delhi, Delhi, India

Abstract. The concept of using isothermal calorimetry for sulphate optimization of Portland cement was developed by Lerch more than 70 years ago. In this paper, we demonstrate a new calorimetry based approach for modern low clinker blended cements grounded in Lerch's concept and that can be used for both periodic sulphate optimization and continuous process control to ensure optimum performance during daily cement production. More frequent sulphate optimization, coupled with daily process control, is especially important for binders that rely on optimum aluminate hydration, such as the novel LC3 cement that utilizes aluminate phases from both clinker aluminates and calcined clay. In the LC3 cement, the sulphate component is the major phase controlling the early hydration, strength development and admixture compatibility of the alumina bearing phases. Moreover, the calorimetry based approach described herein can be substantially automated and does not require a traditional laboratory setting, nor air conditioning, since the calorimetry itself provides a laboratory environment in its temperature control chamber. The only manual labour required for this concept is the weighing of binder, water, calcium sulphate and optional admixtures. The mixing, data collection, calibration, evaluation of the sulphate optimum and subsequent process control can all be automated, such as in Calmetrix Inc's software suite for cement research and development.

1 Introduction

We previously discussed the basics of Lerch's approach to calorimetry based sulphate optimization of modern cements including limestone and LC3 cements [1]. One major feature of this approach is the use of heat of hydration when performing a sulphate addition dosage ramp on the target cement rather than the more labor intensive and less precise method based on compressive strength development. It has been shown that for a given cement, there is a close correlation between compressive strength and heat of hydration from 1 to 7 days of curing at room temperature [2]. Hence, we can determine the optimum sulphate content by testing cement paste according to ASTM C1702 at the desired temperature, and use Lerch's approach to sulphate depletion [3]. Optimum sulphate content corresponds to the sample with the highest heat of hydration. Once the optimum sulphate content has been established, we determine the elapsed time from the maximum of the main Alite hydration peak to the onset of the aluminate-induced sulphate depletion peak for the calorimetry curve corresponding to the optimum sulphate

F. Martirena et al. (eds.), *Calcined Clays for Sustainable Concrete*, RILEM Bookseries 16,
https://doi.org/10.1007/978-94-024-1207-9_68

content. This elapsed time from the maximum of the main peak to the onset of the sulphate depletion peak becomes the process control benchmark target to be used for continuous testing of the cement samples taken from the finished cement mill during production, with feedback to the sulphate dosing system.

2 Optimum Sulphate Content of LC3 Cement Using Isothermal Calorimetry

Figure 1 shows the results of a typical sulphate dosage ramp carried out for industrially produced LC3 cement containing only 50 mass % clinker, formulated to optimize the formation of calcium-carboaluminate hydrates. The test was carried out in a Calmetrix I-Cal 8000 isothermal calorimeter at 27 °C. All mixtures had 50 mass % clinker and approximately 30 mass % of calcined clay and 15% of limestone, respectively, with calcium sulphate dihydrate additions varying from 2 to 5 mass %, thereby resulting in total SO3 contents varying from 3.2% to 4.6%.

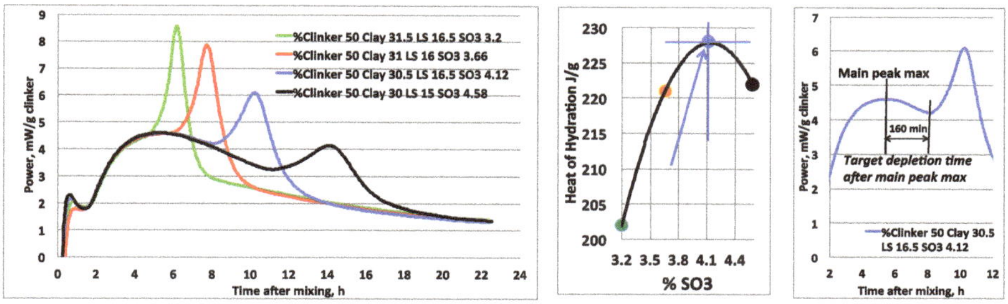

Fig. 1. Sulphate optimization dosage ramp on industrially produced LC3 cement. Left: Power. Middle: Heat of hydration after the end of the dormant period vs. total SO3 content. Right: Target delta time from maximum of main peak to onset of sulphate depletion peak

3 Automated Isothermal Calorimetry

Manual sample preparation and operation of the calorimeter remains an obstacle in most plant environments, which unlike central labs often lack personnel resources for the task. As an alternative to having an on-site operator trained in the use of an isothermal calorimeter, cement plants can also rely on specialized tools such as the automated mixer, Fig. 2, for its isothermal calorimeters. With the automated mixer, users only need to add the right amount of dry cement and calcium sulphate into the sample cup, as well as 50% water by weight of cement in the liquid compartment of the mixer. The mixer is then loaded into calorimeter and all the remaining steps required for operation of the calorimeter and data analysis, including stabilization, mixing, measurement and evaluation are carried out automatically by a software application. Such systems are not only efficient for not requiring any additional or trained laboratory personnel, but they also ensure the best repeatability and reliability of the measurements through very consistent mixing action and minimal thermal imbalance from the mixing. If necessary, any such imbalance can be corrected for by calibration whenever it is important to capture the very first reaction after mixing (i.e. effect of alkali modified aluminates, etc.).

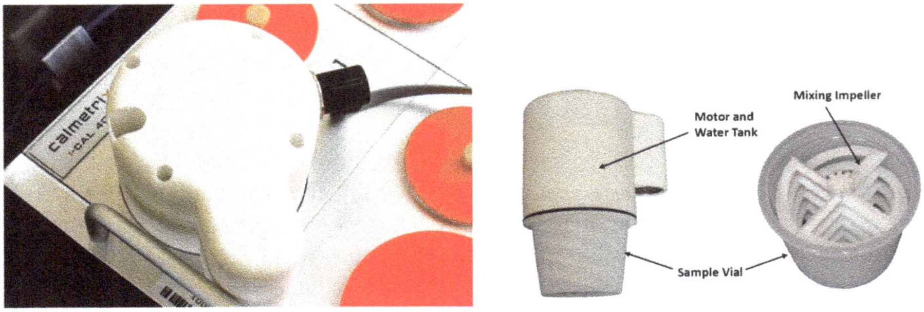

Fig. 2. Automated internal mixer for isothermal calorimetry SO_3 optimization and for daily process control, with software for automated sample mixing, data collection and interpretation

4 Novel Evaluation Methods of Isothermal Calorimetry Curves

The basic evaluation of calorimetry results for sulphate optimization and process control is generally simple and straightforward – simply measure the heat of hydration (HoH) after the end of the dormant period and select the maximum HoH at the desired age, which corresponds to the optimum SO3 content (and maximum compressive strength). In practice, the evaluation of the sulphate depletion as affected by the aluminate activity can be more complicated if the binder consists of multiple alumina bearing phases with different reactivity. This is the case not only for binders with calcined clay but also for many binders with granulated slag and some high calcium fly ashes with more reactive

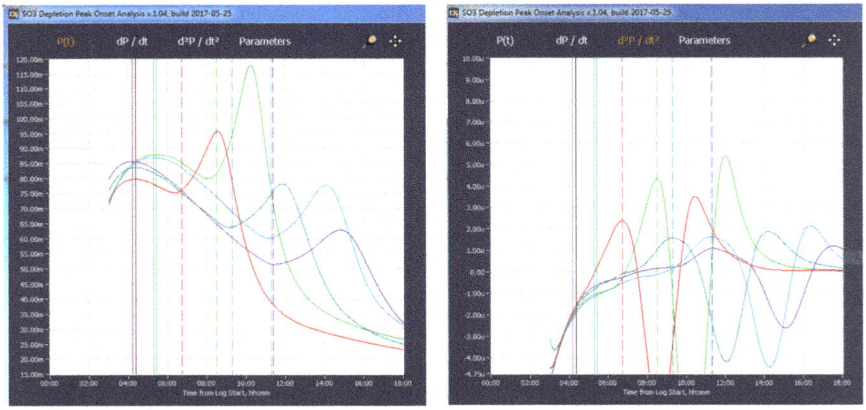

Fig. 3. Software for analysis of the sulphate depletion peak. Left: Power, with simple detection of the onset of the sulphate depletion peak using the first derivative. Right: Second derivative

aluminate phases than most low calcium fly ashes. In its simplest application, the optimization software developed by Calmetrix, and used for this paper, detects the onset of sulphate depletion using the first derivative. However, using the 2nd derivative also enables a more direct detection of the different reactions taking place during the sulphate depletion peak. Figure 3 shows an example using the same LC3 cement as in Fig. 1. By using the 2nd derivative, the different kinetics of the hydration taking place at sulphate depletion become more detectable.

5 Discussion

The sulphate optimization process of modern cements is complicated by the fact that parameters such as mill temperature, admixtures, secondary fuels, SCM's, curing temperature may have a profound effect on the optimum by impacting both the solubility of the sulphates and the reactivity of the aluminate bearing phases. Many producers prefer to perform the optimization in concrete to account for some of the more practical aspects of sulphate optimization such as propensity for false setting or early stiffening as a consequence of early sulphate imbalances in fresh concrete. However, it is difficult to perform systematic sulphate optimization in concrete for cement produced several days earlier and then feed the results back to the operation in an efficient manner. We therefore recommend that sulphate optimization be carried out in a more rapid and automated fashion, similar to other process control analysis tools like clinker analysis by XRF and XRD. Additional tests can then be performed if needed to further tweak the analysis and reflect certain concrete performance issues. For a typical cement production line, isothermal calorimetry is an ideal tool for sulphate optimization of finished cement, as it wholly captures the net balance in reactivity between the aluminate phases and the sulphate phases. The benefits from a continuous approach to sulphate optimization are economical and environmental, as one of the main benefits from the optimization is to ensure maximum strength, which in turns often allows for higher levels of clinker substitution and a commensurately lesser carbon footprint, while at the same time reducing the risk for admixture incompatibility. Since LC3 cement uses more reactive alumina compared to most traditional cements, it is very likely that a continuous approach to sulphate optimization will be even more beneficial for this novel and promising cement type.

6 Conclusion

Isothermal calorimetry has been used for sulphate optimization of Portland cement for more than 70 years. However, recent developments in hardware and software technology have enabled calorimetry to become much less costly and far more user friendly than in the past. Sulphate optimization using modern isothermal calorimetry is much more repeatable than traditional optimization using compressive strength, and the potential to make it essentially automated has now become a reality, thereby making it a suitable tool for process control. This paper shows that the calcined clay based LC3 cement is very sensitive to sulphate optimization, as expected given its rich content of reactive aluminates. We also present an approach to automating the sulphate optimization process as well as cement production process control using isothermal calorimetry with an automated mixer.

References

1. Bishnoi, S., Sandberg, P.: Sulfate optimization of cements using isothermal calorimetry. Cement International **14**, 70–77 (2016)
2. Frolich, L., Wadso, L., Sandberg, P.: Using isothermal calorimetry to predict one day mortar strengths. Cem. Concr. Res. **88**, 108–113 (2016)
3. Lerch, W.: The influence of gypsum on the hydration and properties of Portland cement pastes. In: Proceedings of the American Society for Testing Materials 46 (1946)

Reaction Kinetics of Basic Clay Components Present in Natural Mixed Clays

S. Scherb(✉), N. Beuntner, and K.-C. Thienel

University of the Federal Armed Forces, Munich, Germany

Abstract. The investigation reveals the reaction kinetics of three calcined phyllosilicates (metakaolin, metaillite and metamuscovite) in an alkaline solution which was prepared without the addition of clinker. The phyllosilicates were calcined at their individual optimal calcination temperature. Two test series without and with the addition of anhydrite were investigated. Calcite is present in all measurement series due to the impurity of portlandite. The reaction kinetics were investigated by means of isothermal calorimetry and in-situ x-ray diffraction (XRD). All measurement series show crystalline reaction products after 7 d in the absence of anhydrite. The addition of anhydrite leads to a first formation of ettringite within the first 13 h of reaction with all phyllosilicates tested. The water absorption capacity of phyllosilicates does not correlate with the specific surface area measure as BET-surface. Especially for metamuscovite, the high water absorption leads to a simultaneous formation of monosulfate and gypsum after 17 h. While metakaolin exhibits a significant concentration of dissolved alumina and silicon ions, the influences of metaillite and metamuscovite are less pronounced in that respect and exhibit reaction at a later stage. Nevertheless, it can be concluded that the reactivity of naturally occurring mixed layered clays cannot be reduced their metakaolin content but is due in addition to the contribution originating from other clays present in the mixtures. Metaillite and metamuscovite contribute even to the formation of hydrate phases at early ages of the reaction. Thus, the content and kind of phyllosilicates deserve more attention when used as supplementary cementitious material because of their high water absorption, their contribution to reactivity and their consumption of portlandite.

1 Introduction

Calcined clays represent an interesting perspective as supplementary cementitious materials (SCM). The use of naturally occurring clays as pozzolanic material becomes important due to their low material immanent CO_2 emission during calcination and their high global availability. The pozzolanic activity, primarily the amount and solubility rate of aluminum and silicon from calcined clays is influenced by the type and amount of the individual phyllosilicates, their structural order especially the degree of dehydroxylation after calcination and additional physical factors [1, 2]. Furthermore, the pozzolanic reaction mechanism differs depending on the Si/Al-ion content ratio of the calcined clay and the supply of ions into the pore solution. C-A-S-H and strätlingite are formed in alumina rich clay compositions. In the presence of sulfate and carbonate the

F. Martirena et al. (eds.), *Calcined Clays for Sustainable Concrete*, RILEM Bookseries 16,
https://doi.org/10.1007/978-94-024-1207-9_69

stability of AFm- and AFt-solid solutions are promoted. The availability of portlandite influences the kind and amount of the crystalline hydration products formed [3].

It is important to investigate the contribution of the individual components for a better understanding of the reaction mechanisms of a complex phase mixtures like natural clays. This research work identifies the reaction kinetics in cement free systems with and without sulfate being present.

2 Experimental Procedure

2.1 Characterization of Calcined Phyllosilicates

Three different preferably pure phyllosilicates were used. An industrial flash calcined metakaolin (containing quartz, anatase and phengite as impurities) was ready for use, while illite (containing kaolinite and calcite) and highly pure muscovite were calcined in a muffle oven for 30 min. The optimal calcination temperature of illite at 770 °C and muscovite at 800 °C was chosen due to their dehydroxylation temperatures determined by DTA and Si- and Al-ion solubility in alkaline solution after 20 h. Figure 1 shows the Al- and Si-ion solubility of metaillite and metamuscovite at different calcination temperatures and of metakaolin at 550 °C. The chemical composition was measured by inductively coupled plasma optical emission spectrometry (ICP-OES). Further the particle size distribution [4] (Fig. 2), the BET–specific surface area [5] and the water absorption capacity according to DIN 18132 [6] were measured (Table 2). Note that BET surface and water absorption do not have any correlation.

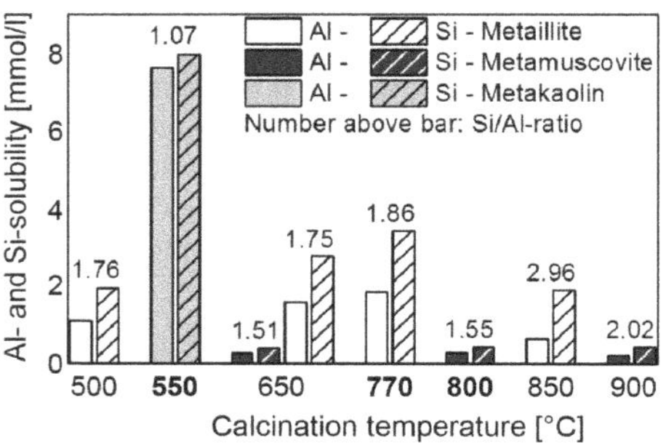

Fig. 1. Solubility of aluminum- and silicon-ions in alkaline solution

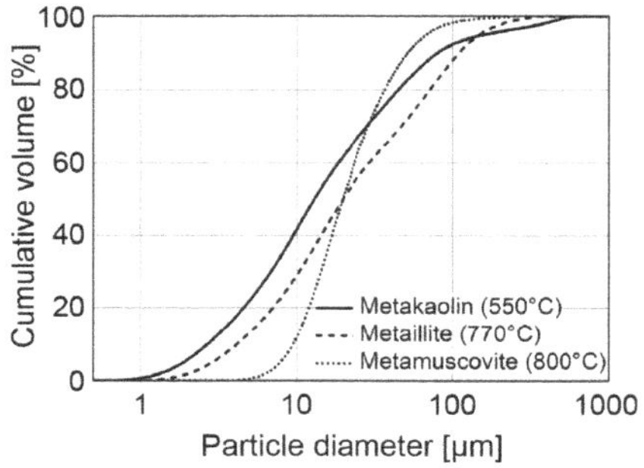

Fig. 2. Particle size distribution of the calcined phyllosilicates

Table 1. BET–specific surface area and water absorption capacity of the calcined phyllosilicates

	Metakaolin	Metaillite	Metamuscovite
BET [m^2/g]	17.76	82,38	10.89
Water absorption capacity [%]	77.0	76.4	154.5

Table 2. Mineralogical compositions (wt%) of the calcined phyllosilicates

Minerals	Metakaolin	Minerals	Metaillite	Minerals	Metamuscovite
Quartz	5.0	Illite	31.3	Muscovite	76.3
Anatase	0.6	Calcite	3.6		
Phengite	1.4	Lime	0.5		
		Portlandite	1.5		
X-ray amorphous	93.0	X-ray amorphous	63.1	X-ray amorphous	23.7

The mineralogical composition of the calcined clays was characterized by means of XRD (Table 2). Therefore, zincite was used as internal standard to quantify the x-ray amorphous content of the samples. While kaolinite lost its crystallinity completely, metaillite and metamuscovite still contain crystalline primary phyllosilicates after calcination procedure. The x-ray amorphous percentages contain the "meta-phases" of the phyllosilicates.

2.2 Equipment and Measurement Parameters

In situ XRD and isothermal calorimetry were used to investigate the reaction kinetics. For both measurements, the samples were stirred manually for one minute and transferred into appropriate sample crucibles which were closed for calorimetry and covered with a KAPTON film for XRD measurements. Isothermal calorimetry was performed with a TAM Air calorimeter at 25 °C for at least 70 h. For a good comparability of calorimetry and XRD the sample holder of the diffractometer was connected to a temperature device that allowed in situ XRD measurements at the same temperature. XRD was performed with a PANalytical Empyrean equipped with a primary Bragg-Brentano^{HD} monochromator and a PIXcel^{1D} linear detector. A diffractogram was taken every 15 min from 6 to 40° 2Θ at 40 kV and 40 mA up to 50 h of hydration. HighScore 4.2 was used as software to analyze the measurements.

Two test series were examined. The composition of the first series contained one calcined phyllosilicate each, calcium hydroxide in powder form and an alkaline solution with the ratio of 1:1:2. The second measurement series were conducted with a substitution of 10% by mass of the solid material by anhydrite. The alkaline solution to solid ratio was 1. The composition of the alkaline solution was 100 mmol/l NaOH and 500 mmol/l KOH. Calcite is due to the impurity of portlandite present in all measurement series.

3 Results and Discussion

3.1 Reaction Kinetics During Early Reaction Time

Figure 3 shows the heat flow of the two measurement series. The initial heat flow was not taken into account because of the external sample preparation for the measurements. While metakaolin and metaillite are homogeneous pastes after stirring the powder and the alkaline solution for one minute, metamuscovite has an earth-moist consistency because of its high water absorption capacity.

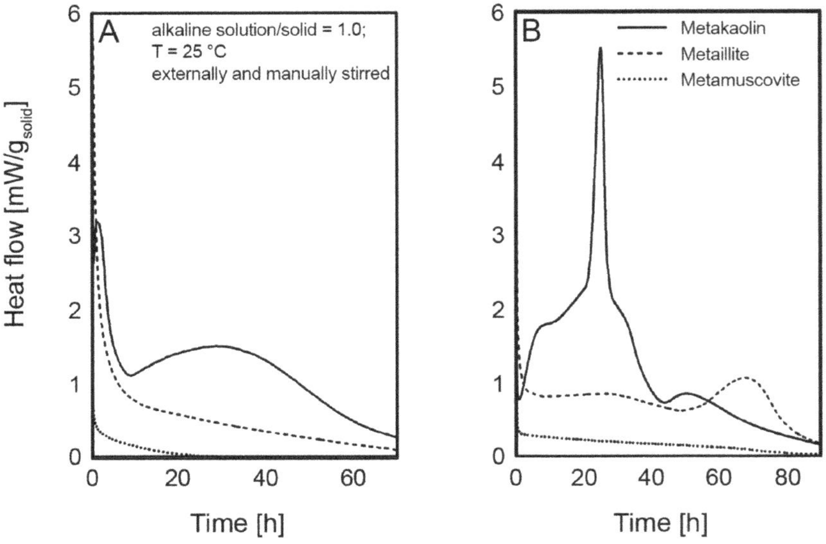

Fig. 3. Calorimetric measurements without anhydrite (A) and with anhydrite (B)

In the absence of anhydrite does Metakaolin cause two maxima at about 1 h (Q_{max} = 3.2 mW/g) and 30 h (Q_{max} = 1.5 mW/g) (Fig. 3A). The corresponding in situ XRD measurement does not show any formation of crystalline reaction products until 50 h (Fig. 5A). Only a dissolution of portlandite can be detected. Without anhydrite metaillite and metamuscovite do not have any maxima except for the initial heat. If anhydrite is added, the formation of ettringite can be detected in all measurements. For metakaolin, a first formation of ettringite can be observed after 3.5 h, for metaillite after 13 h and for metamuscovite after 12 h (Fig. 4). The complete dissolution of sulfates (anhydrite, gypsum) after 20 h leads to an increased ettringite formation in the mix containing metakaolin. The heat flow maximum at 25 h (Q_{max} = 5.5 mW/g) correlates well with the maximum content of ettringite, which remains constant after reaching the maximum. The heat flow of metaillite is relatively constant between 0.6 and 0.8 mW/g for the first 50 h of hydration and has its maximum at 68 h (Q_{max} = 1.1 mW/g). The visible dissolution of anhydrite starts after 3 h and lasts until 46 h. Gypsum can be detected after 15 h and is completely dissolved after 49 h again. Although the in situ XRD measurement just allows following the reaction kinetics until 50 h, it can be assumed that the maximal heat flow at 68 h correlates with the maximum content of ettringite comparable with the reaction of metakaolin. Metamuscovite exhibits a divergent reaction behavior. Following a slight ettringite formation a simultaneous crystallization of monosulfate, pyroaurite and gypsum takes place after 17 h (Fig. 4C). The high water absorption capacity seems to be the reason for a lower availability of water preventing continuously formation of ettringite.

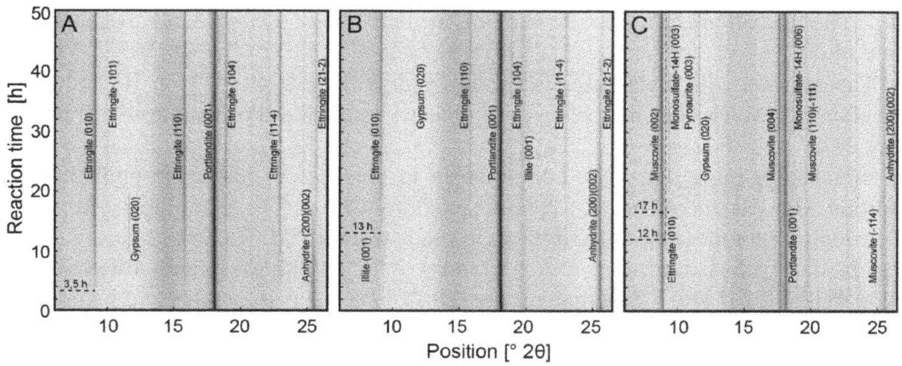

Fig. 4. In situ XRD measurement over 50 h from 6 to 26° 2Θ of metakaolin (A), metaillite (B) and metamuscovite (C) with the addition of anhydrite

3.2 Reaction Kinetics Until 28 Days

The XRD patterns of systems containing metakaolin and metaillite and prepared without the addition of anhydrite until 28 d are shown in Fig. 5. Hemicarboaluminate (AFm-Hc) and monocarboaluminate (AFm-Mc) are formed in both systems after 7 d. While portlandite is completely dissolved at 7 d with metakaolin, the dissolution process of portlandite needs until 28 d for metaillite. The higher calcite content in the measurement with metaillite leads to a preferential formation of AFm-Mc and AFm-Hc disappears after 28 d, while in the system with metakaolin both AFm-phases are still present after

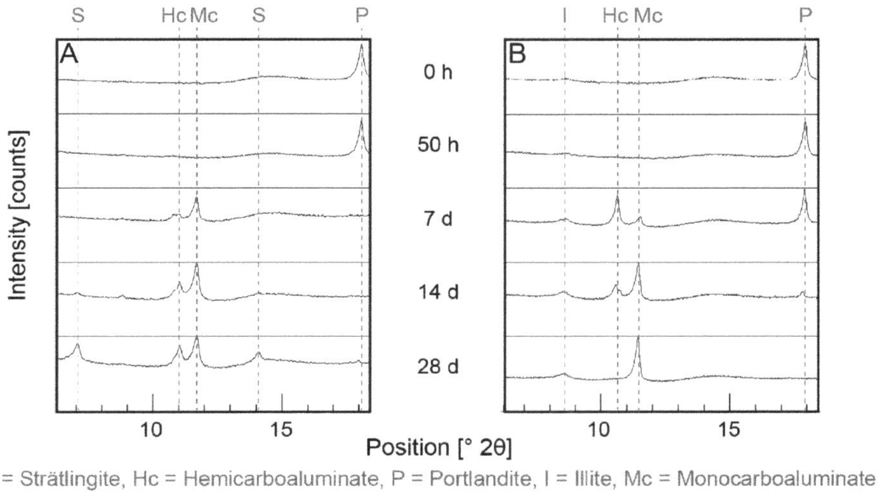

Fig. 5. XRD patterns of metakaolin (A) and metaillite (B) without anhydrite at 0 h, 50 h, 7 d, 14 d and 28 d

28 d. The formation of strätlingite can only be detected in the system with metakaolin after 14 d.

The reaction with metamuscovite shows a slight formation of AFm-Hc and AFm-Mc after 7 d, which does not change significantly until 28 d. Portlandite is still present after 28 d.

Metakaolin forms ettringite continuously in presence of anhydrite until 28 d. In addition AFm-Hc and AFm-Mc can be detected from 7 d on. For metaillite the formation of ettringite continues until 7 d and seems to stagnate until 28 d. The formation of kuzelite (hemisulfate) is visible after 7 d. Metamuscovite just shows small differences as compared to the in situ measurement after 28 d. It seems as if metamuscovite provides a small amount of water to form ettringite and lower the monosulfate content. In all measurements series undissolved portlandite is still present after 28 d.

In accordance with [7] strätlingite is formed in the absence of anhydrite in the system with metakaolin only. The Si/Al ratio of metaillite (1.86) and metamuscovite (1.55) is too high and offers insufficient aluminum to form strätlingite. If anhydrite is added, the formation of ettringite lowers the aluminum level in the pore solution and prevents the crystallization of strätlingite.

4 Conclusion

The investigation reveals the reaction kinetics of metakaolin, metaillite and metamuscovite in a clinker free alkaline solution:

- The reaction is slow in the absence of anhydrite and leads to the formation of crystalline reaction products (AFm-Hc, AFm-Mc) after 7 d.
- The addition of anhydrite leads to a first formation of ettringite with all phyllosilicates tested.
- The water absorption capacity of phyllosilicates does not correlate with the BET–specific surface area (Table 1). Especially for metamuscovite the high water absorption leads to a simultaneous formation of monosulfate and gypsum after 17 h.
- The contribution of phase mixtures like natural clay to pozzolanic reactivity cannot be reduced to its metakaolin content. Metaillite and metamuscovite contribute to the formation of hydrate phases even at early ages of the reaction as well even though on a much lower level than metakaolin.
- The content and kind of phyllosilicates deserve more attention due to their water absorption, their contribution to reaction kinetics and their consumption of portlandite when calcined clays are used as SCM.

References

1. Sabir, B.B., Wild, S., Bai, J.: Metakaolin and calcined clays as pozzolans for concrete: a review. Cement Concr. Compos. **23**, 441–454 (2001)
2. Tironi, A., Trezza, M.A., Scian, A.N., Irassar, E.F.: Kaolinitic calcined clays: Factors affecting its performance as pozzolans. Constr. Build. Mater. **28**, 276–281 (2012)

3. Antoni, M., Rossen, J., Martirena, F., Scrivener, K.: Cement substitution by a combination of metakaolin and limestone. Cem. Concr. Res. **42**, 1579–1589 (2012)
4. ISO 13320: Particle size analysis - Laser diffraction methods (2009)
5. DIN ISO 9277: Bestimmung der spezifischen Oberfläche von Feststoffen durch Gasadsorption nach dem BET-Verfahren (Determination of the specific surface area of solids by gas adsorption using the BET method), Beuth-Verlag (2003)
6. DIN 18132: Bestimmung des Wasseraufnahmevermögenes, Beuth-Verlag (2012)
7. De Silva, P.S., Glasser, F.P.: Hydration of cements based on metakaolin: thermochemistry. Adv. Cement Res. **3**, 167–177 (1990)

Colloid-Chemical Investigation of the Interaction Between PCE Superplasticizers and a Calcined Mixed Layer Clay

M. Schmid[1(✉)], N. Beuntner[2], K.-Ch. Thienel[2], and J. Plank[1]

[1] Chair of Construction Chemistry, Department of Chemistry, Technical University of Munich, Munich, Germany
[2] Institute for Construction Materials, University of the Federal Armed Forces, Munich, Germany

Abstract. This study investigates the dispersing performance of three different polycarboxylate superplasticizers (PCEs) and a cationic polymer in the suspension of a calcined clay holding a mixture of different meta phases. It was found that the anionic PCEs as well as the cationic polymer can disperse this clay, thus suggesting that this calcined clay exhibits particles with positive as well as negative surface charges. Relative to PCEs, their dispersion ability is highest when their anionicity is particularly high. Adsorption and zeta potential measurements of the PCEs further elucidated this interaction.

1 Introduction

The partial replacement of cement clinker by supplementary cementitious materials (SCMs) improves the sustainability of cement by reducing the carbon dioxide emissions originating from its production [1]. In this context, calcined clays present an attractive potential alternative due to their pozzolanic reactivity, their relatively low calcination temperatures and their abundant and widespread deposits [2, 3]. However, such substitution alters the chemical composition of the binder and consequently its behavior in combination with admixtures [4]. Here, the effectiveness of anionic and cationic superplasticizers to disperse a calcined clay sample (amorphous part 59%, mainly meta-kaolin and meta-mica) was studied. Furthermore, the kind of interaction of different superplasticizers with this calcined clay was assessed via adsorption and zeta potential measurements.

2 Materials and Methods

2.1 Calcined Clay Sample

A naturally occurring mixed layer clay known as Amaltheen from the black jura in Bavaria/Germany was used. This clay is particularly rich in mica and other clay minerals including kaolinite, illite and chlorite. Calcination of this clay was performed

F. Martirena et al. (eds.), *Calcined Clays for Sustainable Concrete*, RILEM Bookseries 16,
https://doi.org/10.1007/978-94-024-1207-9_70

at 750 °C in a rotary kiln and yielded a material with 59 wt.-% amorphous content [5]. The composition of the raw and the calcined clay as determined via quantitative XRD (Rietveld method) is shown in Table 1.

Table 1. Mineralogical composition of the raw and the calcined clay samples.

Mineral phases	Raw clay [wt.-%]	Calcined clay [wt.-%]
Kaolinite	25	0.0
Mica	30	3.8
Illite	11	4.4
Chlorite	6	0.0
Quartz	18	18.3
Feldspar	5	7.9
Carbonates	3	0.4
Sulfate	1	1.1
Pyrite	1	—
Secondary silicates	—	3.4
Hematite	—	0.6
Ore	—	0.8
X-ray amorphous	—	59.3
Total	100	100.0

2.2 Superplasticizer Samples

As anionic superplasticizers, two methacrylate (MPEG) ester based PCEs with molar ratios of 6:1 and 2:1 respectively between methacrylic acid and the MPEG methacrylate ester macromonomer and one isoprenol ether (IPEG)/acrylic acid based PCE were utilized. As cationic superplasticizer, poly(allylamine hydrochloride) (PAH), a linear, non-branched homopolymer, was used. The chemical structures of the superplasticizer samples and their denotation are shown in Fig. 1.

2.3 Dispersion via Slump Flow Test

Dispersing performance of the superplasticizer samples in clay suspensions was assessed via a modified mini slump test using a brass cylinder with a diameter of 3.0 cm and a height of 5.0 cm. The water-to-binder ratio was set to obtain an initial slump flow of 7.0 ± 0.5 cm without superplasticizer and was 0.70.

2.4 Adsorption of Superplasticizers

The adsorbed amounts of the superplasticizers on the calcined clay were determined by using the depletion method. The calcined clay was suspended in water holding the superplasticizer, then centrifuged and the amount of polymers consumed from the liquid phase due to surface adsorption on the clay was quantified from the total organic content in the centrifugate. Different dosages of the polymers were applied to obtain

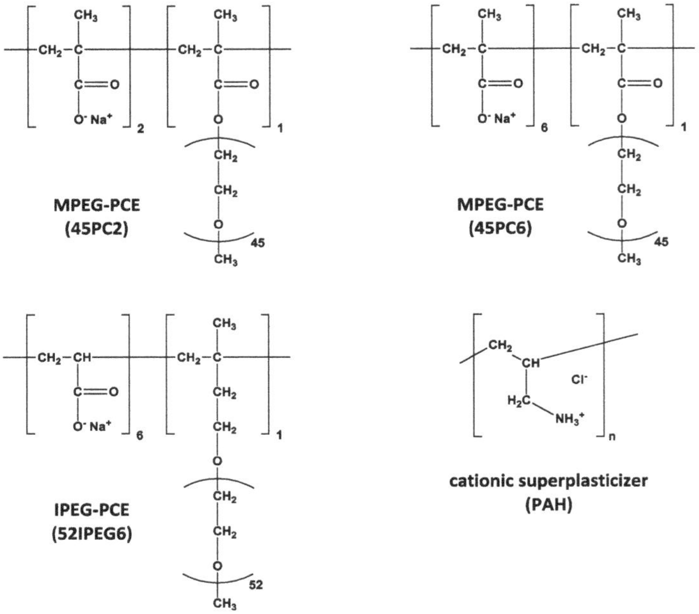

Fig. 1. Chemical structures of the polycarboxylate and the cationic superplasticizer samples.

adsorption isotherms. A High TOC II apparatus (Elementar, Hanau, Germany) was used to quantify the organic carbon. The adsorbed amounts were calculated by subtracting the concentration of superplasticizer found in the centrifugate from the initial concentration added to the suspension.

2.5 Zeta Potential of Clay Suspensions

The electrical surface charge (zeta potential) developed by the calcined clay when suspended in water was determined using a zeta potential instrument Model DT-1200 Electroacoustic Spectrometer (Dispersion Technology Inc., Bedford Hills, NY). For the determination of the zeta potential, aqueous solutions of the copolymers were stepwise titrated to the calcined clay suspended in water. The zeta potential of the slurry was recorded as a function of the polymer concentration.

3 Results and Discussion

3.1 Dispersing Performance

To assess the dispersing ability of the different superplasticizer samples relative to this calcined clay, mini slump tests were performed. The polymer dosages were determined at which a paste flow of 10 ± 0.5 cm was achieved. The result is shown in Fig. 2.

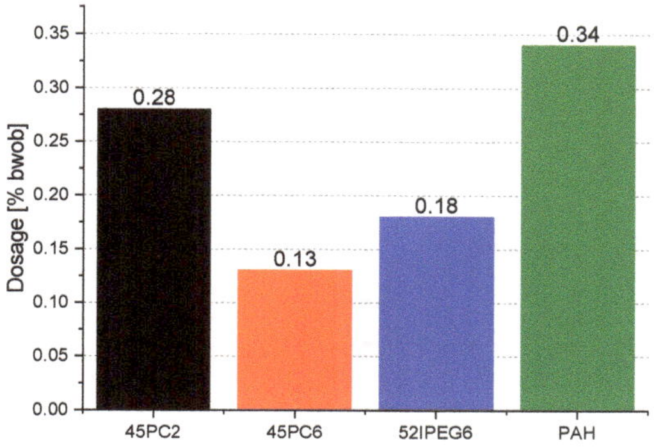

Fig. 2. Dosages of polycarboxylate and PAH superplasticizers required to obtain a target slump flow of 10 ± 0.5 cm for the suspension of the calcined clay.

The experiments revealed that the highly anionic MPEG-PCE sample 45PC6 required a significantly lower dosage compared to the less anionic MPEG-PCE 45PC2 (0.13% vs. 0.28% bwob). Moreover, the IPEG based PCE which because of its longer side chain possesses a slightly lower anionicity than MPEG-PCE 45PC6 consequently requires a slightly higher dosage than 45PC6 (0.18% vs. 0.13% bwob). The results suggest that the anionicity of the copolymers presents the key parameter regarding dispersing efficacy of superplasticizers for this calcined clay sample. It should be noted here that the superplasticizer dosages required for this calcined clay sample were very comparable to those found for a CEM I 52.5 R sample which was studied for comparison.

Furthermore, surprisingly it was found that even a cationic polymer (PAH) was quite effective in dispersing this clay. This result is highly important as it demonstrates that the calcined clay sample exhibits positively as well as negatively charged surface sites, apparently owed to the different chemical nature of the calcined meta phases.

3.2 Adsorption of PCEs

The dispersing effectiveness of superplasticizers on cement normally correlates with their adsorbed amount. To assess this interaction with the calcined clay, different dosages of the superplasticizers were added to the clay suspensions which were then centrifuged. Subsequently, the residual PCE content was determined and adsorption isotherms were developed. Figure 3 illustrates the results obtained for the superplasticizer samples.

All superplasticizer samples exhibit significant adsorbed amounts which correspond to Langmuir type isotherms, thus suggesting monolayer adsorption. For the PCE samples, the adsorption affinity as expressed by the saturated adsorbed amount of the polymers clearly follows the anionic character of the samples and is in perfect agreement with the results from the mini slump tests. In comparison, the cationic polymer PAH requires substantially higher dosages to reach the saturated adsorption. This effect might be owed to a denser packing of this linear, non-branched polymer on

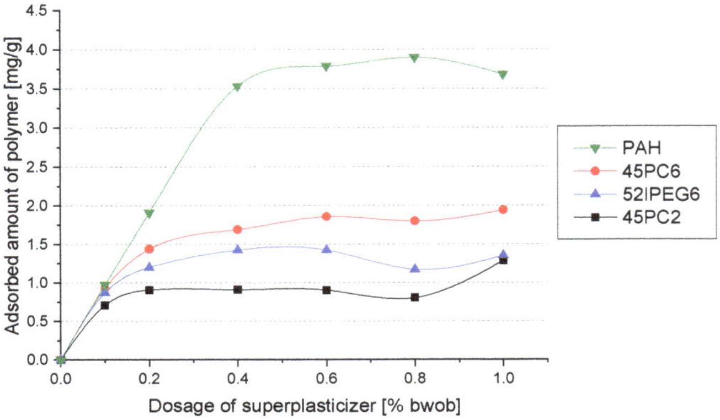

Fig. 3. Adsorption isotherms for the polycarboxylate and PAH superplasticizers on calcined clay dispersed in water

the clay surface, or to occurrence of significantly more negatively charged surface sites on this clay.

3.3 Zeta Potential Measurements

To further understand the colloidal properties of the clay suspension containing superplasticizer, their zeta potentials were measured under continuous titration of the superplasticizers to the slurry. The initial surface charge of the calcined clay dispersed in water was found to be slightly negative (−11 mV) at a pH value of 10.

As is shown in Fig. 4, for all copolymers a similar effect was observed. They move the zeta potential toward more positive values. In the case of the PCEs, the shear plane of the zeta potential is shifted to greater distance away from the surface due to the steric effect of the poly(ethylene glycol) side chains, thus producing zeta potentials which

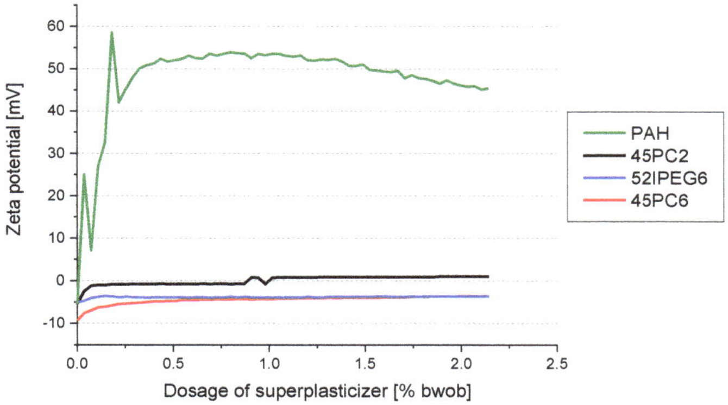

Fig. 4. Effect of PCE and PAH dosages on the zeta potential of the aqueous suspension of the calcined clay

approach the isoelectric point [6]. This way, the interaction between the PCEs and the clay and the steric effect of the PCEs are confirmed. For the positively charged PAH, the zeta potential increases significantly with dosage until it reaches a very high positive value of $\sim +50\,\mathrm{mV}$, suggesting strong electrostatic repulsion between the clay particles as dispersing mechanism.

4 Conclusions

The results demonstrate that conventional PCE superplasticizers can fluidify this specific calcined clay sample quite well. Interestingly, this calcined clay sample possesses a heterogeneous surface charge, similar to Portland cement. As such, both PCEs as well as cationic polymers can disperse this clay. The effectiveness of polycarboxylates is highest for samples possessing a particularly high anionicity. Consequently, this calcined clay is not inert towards PCEs, and the dosage of the PCEs needs to be adjusted when this calcined clay is used in a blended cement. Further investigations are currently underway to elucidate the interaction between the pure meta phases of calcined clay minerals and differently structured superplasticizers.

Acknowledgments. This research is financially supported by the Deutsche Forschungsgemeinschaft (DFG) grant no. PL 472/11-1 and TH 1383/3-1 entitled "Ecological and energetic optimization of concrete: Interaction of structurally divergent superplasticizers with calcined clays".

References

1. Schneider, M., Romer, M., Tschudin, M., Bolio, H.: Sustainable cement production - present and future. Cem. Concr. Res. **41**, 642–650 (2011). doi:10.1016/j.cemconres.2011.03.019
2. Tironi, A., Trezza, M.A., Scian, A.N., Irassar, E.F.: Assessment of pozzolanic activity of different calcined clays. Cement Concr. Compos. **37**, 319–327 (2013). doi:10.1016/j.cemconcomp.2013.01.002
3. Hollanders, S., Adriaens, R., Skibsted, J., Cizer, Ö., Elsen, J.: Pozzolanic reactivity of pure calcined clays. Appl. Clay Sci. **132–133**, 552–560 (2016). doi:10.1016/j.clay.2016.08.003
4. Plank, J., Sakai, E., Miao, C.W., Yu, C., Hong, J.X.: Chemical admixtures - Chemistry, applications and their impact on concrete microstructure and durability. Cem. Concr. Res. **78**, 81–99 (2015). doi:10.1016/j.cemconres.2015.05.016
5. Beuntner, N., Thienel, K.C.: Properties of calcined lias delta clay - technological effects, physical characteristics and reactivity in cement. Scrivener, K., Favier, A. (eds.) Calcined Clays for Sustainable Concrete. Proceedings for the 1st International Conference on Calcined Clays for Sustainable Concrete, Lausanne, Switzerland, RILEM Bookseries 10, pp. 43–50 (2015). doi:10.1007/978-94-017-9939-3
6. Plank, J., Hirsch, C.: Impact of zeta potential of early cement hydration phases on superplasticizer adsorption. Cem. Concr. Res. **37**, 537–542 (2007). doi:10.1016/j.cemconres.2007.01.007

Prediction of Carbonation Depth in Blended Systems

V. Shah(✉) and S. Bishnoi

Indian Institute of Technology Delhi, New Delhi, India

Abstract. The majority of cement available worldwide includes supplementary cementitious materials (SCMs) as a mineral additive to improve mechanical and durability properties of cement. The alkalinity of the concrete made using such cements is less as SCMs consume the calcium hydroxide produced on clinker hydration by the pozzolanic reaction. Quantity of calcium hydroxide available in the system is one of the major factors influencing carbonation resistance of the system. Numerous carbonation models are available which take the quantity of calcium hydroxide as input and predicts the carbonation depth. In blended systems, insignificant or very small amount of calcium hydroxide is present and hence prediction of carbonation depth using existing models becomes difficult. In this study, a new approach has been developed to predict carbonation depth in concrete made using blended cement based on total reserve alkalinity of the system along with other factors. The carbonation depth predicted by the above approach is in close relation with experimental data.

1 Introduction

Corrosion of reinforcement in concrete due to depassivation of steel bars induced by carbonation is one of the major causes of deterioration of concrete structures [1, 2]. Carbonation is the reaction of hydration phases of cement present in concrete with carbon dioxide present in the environment. The quantity of calcium hydroxide available in the system is one of the major factors influencing carbonation resistance of the system. As carbonation proceeds in concrete, the amount of calcium hydroxide and other calcium bearing phases reduce and extent of carbonation can be measured by measuring calcium hydroxide content. The majority of cement available worldwide includes supplementary cementitious materials (SCMs) as a mineral additive to improve mechanical and durability properties of cement. The amount of calcium hydroxide in such cements is lower due to the pozzolanic reaction of SCMs and dilution effect [3, 4]. In such cases, carbonation of other hydration phases becomes important. Because of the difference in molar solid volume of the hydration phases and the phases formed on carbonation, the microstructure of concrete is altered [5, 6]. The carbonation of calcium hydroxide results in an increase in molar solid volume, however, volume change occurring on carbonation of other hydration phases is not clearly established. Various literature has reported increase in porosity on carbonation of cement containing SCMs [7, 8].

Diffusion of carbon dioxide in concrete is slow owing to its low concentration in the atmosphere. To predict the long-term performance of concrete structures subjected

F. Martirena et al. (eds.), *Calcined Clays for Sustainable Concrete*, RILEM Bookseries 16,
https://doi.org/10.1007/978-94-024-1207-9_71

to carbonation in short time use of mathematical modelling is done to estimate the extent of carbonation. Most of the carbonation models available use amount of calcium hydroxide (CH) available to predict the carbonation [9–12]. Moreover, the existing carbonation model does not take into consideration the volume change occurring on carbonation of hydration phases other than calcium hydroxide.

In this study, an attempt has been made to develop a carbonation depth prediction model that can be functional to both Ordinary Portland Cement (OPC) and blended cements. The carbonation resistance of a concrete system is described in terms of reserve alkalinity of the system.

2 Materials and Methods

Two types of binder: Portland Pozzolana Cement (PPC) and Limestone calcined clay Cement (LC3) were studied. PPC was produced by replacing 30% clinker with fly ash and by interblending in a ball mill without balls to get a homogeneous mix. LC3 was produced by replacing 50% clinker with 31% calcined clay, 15% limestone and 4% gypsum. The details of the production of cement are given elsewhere [13]. Concrete prisms were cast at a water to cement ratio of 0.35. After casting, the concrete samples were cured under water for 120 days. Thereafter, the samples were preconditioned at 27 °C temperature and 60% relative humidity prior to exposure in carbon dioxide for 15 days. Half of the samples were exposed in 3% carbon dioxide concentration, 27 °C temperature and 60% relative humidity whereas remaining samples were placed in 3% carbon dioxide concentration, 27 °C temperature and 40% relative humidity. The carbonation depth was measured at the different age of exposure using phenolphthalein indicator. Cylindrical samples were cast to measure sorptivity according to guidelines provided in ASTM C1585.

Mercury Intrusion Porosimetry (MIP) and reserve alkalinity were measured on cement paste samples. The cement paste samples were also cast at the water to cement ratio of 0.35 and cured for 120 days. Thin slices of 2–3 mm were cut and kept in carbon dioxide exposure for 2 months at 3% carbon dioxide concentration, 27 °C temperature and 60% relative humidity. MIP measurements were done on non-carbonated samples i.e. after the end of the curing period and on carbonated samples. The reserve alkalinity of cement paste sample was measured by titrating the suspension of cement paste in the water against sulphuric acid.

3 Carbonation Mechanism and Implementation of Scheme

The rate of carbonation in concrete is simultaneously governed by the physical and chemical characteristics of concrete. The rate is governed physically by diffusion of carbon dioxide and chemically by the reserve alkalinity content of the concrete. One-dimensional numerical diffusion model based on Fick's second law is formulated by solving equations for the concentration of carbon dioxide, alkali and relative humidity numerically using finite difference method. Equation 1 shows the differential equation used to compute the change in carbon dioxide concentration with time [14, 15].

$$\frac{\partial C_{aq}}{\partial t} = D\frac{\partial^2 C_{aq}}{\partial x^2} + S \tag{1}$$

where $[CO_{2(aq)}]$ (mol/m^3) is the aqueous concentration of carbon dioxide, D(m^2/s) is the diffusion coefficient and S is the sink term. Similar, equations were used for computing alkali concentration and drying of concrete with time. Since calcium ions constitute the majority of alkalis in cement, the word alkalis and calcium are used interchangeably.

The carbonation reaction takes place in the solution form between the dissolved carbon dioxide and alkalis dissolved in pore solution. The rate of carbonation reaction can be described as follows:

$$r_N = HRTK\left[CO_{2(aq)}\right]\left[A_{(aq)}\right] \tag{2}$$

where $[CO_{2(aq)}]$ (mol/m^3) is the aqueous concentration of carbon dioxide, $[A_{(aq)}]$ (mol/m^3) is the aqueous concentration of the alkalis in pore solution, H is Henry's constant (mol/m^3 atm), R (m^3 atm/K mol) is the universal gas constant and K is the reaction rate constant. For simplicity, the neutralization reaction is considered to be instantaneous implying precipitation of calcium carbonate takes place right away [11, 16]. Due to precipitation of calcium carbonate, more carbon dioxide and alkalis can dissolve in the pore solution. The rate of dissolution of alkalis in pore solution decreases with time, probably due to the reduction in the overall amount of alkalis (because of precipitation of calcium carbonate) and formation of calcium carbonate layer on the surface of hydrates impeding further dissolution of it [17]. The dissolution of alkalis in pore solution was governed by Eq (3).

$$Rate\ of\ Dissolution : 10^{-5} * \frac{Present\ Reserve\ Alkalinity}{Initial\ Reserve\ Alkal\text{inty}} \tag{3}$$

The diffusion coefficient of carbon dioxide is a function of porosity, relative humidity and temperature. Papadakis et al. (1991) proposed an expression to compute the value of diffusion coefficient [14]:

$$D = 1.64 \times 10^{-6} \times (P)^{1.8} \times (1 - R.H.)^{2.2} \tag{4}$$

where P is the porosity of concrete and R.H. is the relative humidity. However, the author assumed uniform relative humidity throughout the concrete. Moreover, the pore structure of concrete is known to change along with porosity on carbonation. The new equation for diffusion coefficient of carbon dioxide based on Eq. 4 is proposed as follows:

$$D = 1.64 \times 10^{-6} \times (P_a)^a \times (1 - R.H.)^b \times P_s \tag{5}$$

where P_a is the air filled porosity and P_s is the pore size factor obtained by dividing the threshold pore diameter of the carbonated system to non-carbonated system. Air filled porosity value was obtained by multiplying the porosity obtained from MIP with (1-R.H.). The MIP was carried out on cement paste samples and the porosity value corresponding till 10 nm pore diameter was considered as the majority of diffusion of external ions in concrete takes place through capillary pores [18]. The value of diffusion coefficient changed with the change in relative humidity and porosity. The porosity value in the model for a particular system was varied between the porosity of non-carbonated and carbonated samples obtained from MIP. The rate of diffusion of alkalis due to concentration gradient within concrete was assumed to be constant and it was taken as 1×10^{-12} m^2/s [12]. The wetting diffusion coefficient was computed using rate of water absorption by concrete as described by Lockington et al. (1999) [19]. The drying diffusion coefficient value was obtained by dividing the wetting diffusion coefficient by a constant factor.

The differential equations were solved simultaneously considering one-dimensional space. At time $t = 0$, it was assumed that all the interior points of concrete, the carbon dioxide concentration is zero. Based on relative humidity at each point, the amount of alkali dissolved was computed. In a saturated pore solution of concrete 43.2 mol/m^3 of hydroxyl ions can dissolve [14]. Time step of 10 s was used considering the stability of the finite difference scheme. At each time step, the amount of alkalis consumed and corresponding change in porosity value was made. It was assumed that the porosity varied linearly from the non-carbonated value to carbonated value depending upon the initial and final reserve alkalinity value of the sample. Therefore, after each time step, the porosity value was updated. Similarly, the relative humidity value due to subsequent drying of concrete at each time step was also updated. After updating relative humidity value, the porosity value was again updated to get the air-filled porosity.

4 Results

The porosity of non-carbonated (NC) and carbonated (C) samples corresponding to pores greater than equal to 10 nm is given in Table 1. The reserve alkalinity values, drying diffusion coefficient and pore size factor is also given in Table 1.

Table 1. Compilation of experimental results

	Porosity		P_S	Reserve Alkalinity (mol/m^3)	Drying diffusion coefficient (m^2/s)
	NC	C			
PPC	6.94	11.36	5.38	1226	6.14E-11
LC3	12.15	14.33	5.36	1010	7.58E-11

Figures 1and 2 show the carbonation depth measured using phenolphthalein indicator test and predicted depth using model. All the results are not presented here due to space constraints; however, similar trends were observed for experimental and predicted depth for all the systems.

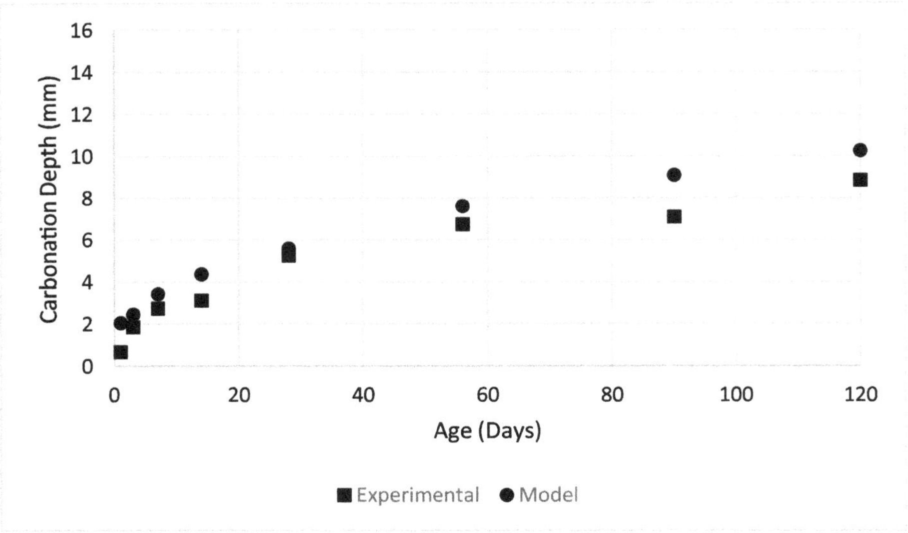

Fig. 1. Experimental and modelled carbonation depth of PPC sample exposed in 3% carbon dioxide concentration, 27 °C temperature and 60% relative humidity

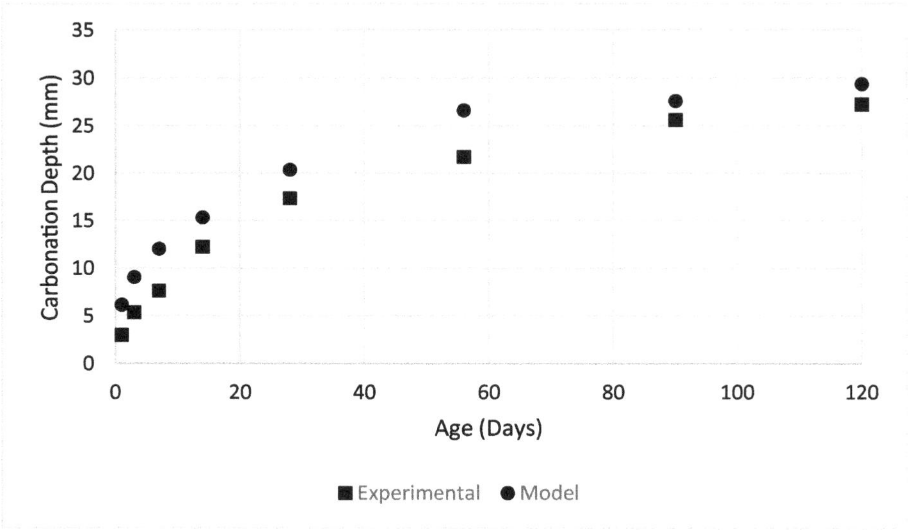

Fig. 2. Experimental and modelled carbonation depth of LC3 sample exposed in 3% carbon dioxide concentration, 27 °C temperature and 40% relative humidity

5 Conclusions

- The carbonation resistance of a cement can be described in terms of reserve alkalinity of the system.
- The model described in the paper is able to predict the carbonation depth irrespective of the type of cement and exposure conditions used for carbonation.

Acknowledgements. The authors would like to acknowledge Swiss Agency for Development and Cooperation (SDC) for providing the financial support for the Limestone Calcined Clay Cement project in India.

References

1. Song, H.W., Kwon, S.J., Byun, K.J., Park, C.K.: Predicting carbonation in early-aged cracked concrete. Cem. Concr. Res. **36**, 979–989 (2006). doi:10.1016/j.cemconres.2005.12.019
2. Bertolini, L., Elsener, B., Pedeferri, P., Polder, R.P.: Corrosion of Steel in Concrete Prevention, Diagnosis, Repair, 2nd edn. WILEY-VCH, Singapore (2013)
3. Wu, B., Ye, G.: Development of porosity of cement paste blended with supplementary cementitious materials after carbonation. In: 14th International Congress on the Chemistry of Cement, pp. 1–18 (2015). doi:10.1016/j.conbuildmat.2017.03.176
4. Ashraf, W.: Carbonation of cement-based materials: challenges and opportunities. Constr. Build. Mater. **120**, 558–570 (2016). doi:10.1016/j.conbuildmat.2016.05.080
5. Phung, Q.T., Maes, N., Jacques, D., Bruneel, E., Van Driessche, I., Ye, G., De Schutter, G.: Effect of limestone fillers on microstructure and permeability due to carbonation of cement pastes under controlled CO2 pressure conditions. Constr. Build. Mater. **82**, 376–390 (2015). doi:10.1016/j.conbuildmat.2015.02.093
6. Chang, C.F., Chen, J.W.: The experimental investigation of concrete carbonation depth. Cem. Concr. Res. **36**, 1760–1767 (2006). doi:10.1016/j.cemconres.2004.07.025
7. Gruyaert, E., Van Den Heede, P., De Belie, N.: Carbonation of slag concrete: effect of the cement replacement level and curing on the carbonation coefficient - effect of carbonation on the pore structure. Cem. Concr. Compos. **35**, 39–48 (2013). doi:10.1016/j.cemconcomp.2012.08.024
8. Hargis, C.W., Lothenbach, B., Müller, C.J., Winnefeld, F.: Carbonation of calcium sulfoaluminate mortars, Cem. Concr. Compos. **80** (2017). doi:10.1016/j.cemconcomp.2017.03.003
9. Khunthongkeaw, J., Tangtermsirikul, S.: Model for simulating carbonation of fly ash concrete. J. Mater. Civ. Eng. **17**, 570–578 (2005). doi:10.1061/(ASCE)0899-1561(2005)17:5(570)
10. Maekawa, K., Ishida, T., Kishi, T.: Multi-scale Modeling of Structural Concrete. Taylor & Francis, London and New York (2009)
11. Saetta, A.V., Schrefler, B.A., Vitaliani, R.V.: The carbonation of concrete and the mechanism of moisture, heat and carbon dioxide flow through porous materials. Cem. Concr. Res. **23**, 761–772 (1993)
12. Talukdar, S., Banthia, N., Grace, J.R.: Carbonation in concrete infrastructure in the context of global climate change – Part 1: experimental results and model development. Cem. Concr. Compos. **34**, 924–930 (2012). doi:10.1016/j.cemconcomp.2012.04.011

13. Emmanuel, A.C., Haldar, P., Maity, S., Bishnoi, S.: Second pilot production of limestone calcined clay cement in India: The experience. Indian Concr. J. **90**, 22–27 (2016)
14. Papadakis, V.G., Vayenas, C.G., Fardis, M.N.: Fundamental modelling and experimental investigation of concrete carbonation. Am. Concr. Inst. Mater. J. **88**, 363–373 (1991)
15. Isgor, O.B., Razaqpur, A.G.: Finite element modeling of coupled heat transfer, moisture transport and carbonation processes in concrete structures. Cem. Concr. Compos. **26**, 57–73 (2004). doi:10.1016/S0958-9465(02)00125-7
16. Peter, M.A., Muntean, A., Meier, S.A., Böhm, M.: Competition of several carbonation reactions in concrete: a parametric study. Cem. Concr. Res. **38**, 1385–1393 (2008). doi:10.1016/j.cemconres.2008.09.003
17. Šavija, B., Luković, M.: Carbonation of cement paste: understanding, challenges, and opportunities. Constr. Build. Mater. **117**, 285–301 (2016). doi:10.1016/j.conbuildmat.2016.04.138
18. Kumar, R., Bhattacharjee, B.: Porosity, pore size distribution and in situ strength of concrete. Cem. Concr. Res. **33**, 155–164 (2003). doi:10.1016/S0008-8846(02)00942-0
19. Lockington, D., Parlange, J.-Y., Dux, P.: Sorptivity and the estimation of water penetration into unsaturated concrete. Mater. Struct. **32**, 342–347 (1999). doi:10.1007/BF02479625

Autogenous Shrinkage and Creep of Limestone and Calcined Clay Based Binders

J. Ston(✉), A. Hilaire, and K. Scrivener

Laboratory of Construction Materials,
École Polytechnique Fédérale de Lausanne, Lausanne, Switzerland

Abstract. This study explored the delayed strains in limestone and calcined clay ternary blends. Autogenous shrinkage measurements are carried out over 2 months and compressive basic creep tests during 28 days after one month of curing. All tests are done using mixed with calcined clay at different metakaolin amounts or with variable mix designs. Results show that the presence of any type of clay, except pure metakaolin, has a similar impact on both autogenous shrinkage and basic creep. However, mix design seems to have an important contribution. Shrinkage rate is higher for blends than PC at 28 days, although reaching a similar amplitude at this age for most mixes. Creep amplitude and rate are reduced when using the blended systems.

1 Introduction

Comfort and modernity often come at the expense of the environmental resources and even the most basic rights, such as housing, come at a cost. Limitation of the environmental footprint and sustainability are among the long-term objectives of most industrial sectors, as they often come together with economic prosperity. Concrete is the most produced material by volume, and is therefore accounted for a non-negligible part of anthropic CO_2 emission. Most of the greenhouse gases are emitted during clinkerisation, originating from the chemical decomposition of the raw materials. The use of Supplementary Cementitious Materials (SCM) – often by-products of other industries – allows to decrease the clinker needed to obtain a binder, as well as reducing the cost and improving durability properties. This philosophy has for long been applied in the industry, as it allows to produce more binder using the same initial amount of clinker, hence being economically and environmentally effective. Nevertheless, most SCM come in limited amounts and cannot meet the increasing demand for building materials, especially in developing countries. But among these SCM, limestone and calcined clays have a considerable potential as both raw materials are widely available and require very simple processing to be used in cement fabrication. The use of limestone or calcined clay as SCM in binary blends has been studied by many [1–3], but the combination of both materials in high replacement fractions is only a recent approach [4]. Limestone on its own can be considered as an inexpensive, low reactivity filler but it also affects the phase assemblage by promoting the formation of a stable carboaluminate phase 6, especially in the presence of soluble aluminium. On the other hand, metakaolin – the reactive phase of calcined clayey soils – is known to be a very

F. Martirena et al. (eds.), *Calcined Clays for Sustainable Concrete*, RILEM Bookseries 16,
https://doi.org/10.1007/978-94-024-1207-9_72

efficient pozzolanic SCM. This material shows high reactivity and leads to porosity refinement and therefore is an interesting but expensive SCM for improving durability. Its effects on delayed strains is a decrease of the creep and shrinkage strains in sealed conditions [6–8]. Recent research showed that a combination of limestone and metakaolin could be used to replace a considerable fraction of clinker in a binder, whilst still obtaining high compressive strength. The main reason for this is the fast pozzolanic reaction of metakaolin and the extra reactivity of limestone thanks to the aluminates brought by the pozzolan. Moreover, studies demonstrated that using calcined kaolinic clays instead of pure metakaolin provided a very good source of metakaolin at a fraction of the price and were therefore suitable for use as SCM [9]. Limestone and calcined clay cements, referred as LC3 in this study, allow both high replacement fraction of clinker by SCM and mechanical strength comparable to that of plain cement (PC). The interest of such blends in regions with increasing cement demand is obvious; however, the long-term dimensional stability of such binders is as critical as their early age properties. This study explores the shrinkage and creep behaviour of limestone and calcined clay ternary blends in autogenous conditions. Autogenous shrinkage is triggered by self-desiccation of the microstructure, causing internal stresses, resulting in macroscopic shrinkage and cracking potential if restrained. Creep is a time-dependent phenomenon that is present during the entire concrete lifetime and can have some deleterious effects, such as loss of prestress or excessive deformation of a structure [10]. On the other hand, moderate creep can avoid early age cracking, hence the importance of a characterisation of this behaviour and its kinetics. The precise physical origin of these phenomena is not fully understood yet, but water, porosity and the number of amorphous phases play a major role.

2 Materials and Methods

Plain cement samples were used as reference and compared with LC3 mixes at various levels of replacement and clay purity. The same cement, a CEM I 42.5 N type, was used in all type of mixes. Calcined kaolinic clays of varying metakaolin content – from 45% to 95% in mass – were used as substituents together with Omya Durcal 5 limestone in the ternary blends. A minor addition of gypsum is required to control the reaction of the extra aluminium coming from the calcined clay. The chemical composition of the materials used is summarized on Tables 1 and 2 resumes the studied mix designs. Mixes are separated into two groups: the first group includes samples with an identical mix design, but varying calcined clay grade and the second group contains samples with variable mix design, but the same calcined clay (45% grade) for substitution. All pastes were prepared with a water to binder ratio of 0.4, including an addition of a PCE-based superplasticizer ranging from 0.4% to 1% depending the clay used.

Autogenous shrinkage was measured following the method proposed in ASTM 1698, using a silicon oil bath at 20 °C to maintain the temperature. Two corrugated tubes were prepared per mix and their strain was measured using a LVDT per sample, acquiring every 5 min. As complementary measurement to autogenous shrinkage, internal relative humidity was also measured for most of mixes. In this case,paste

Table 1. Chemical composition of the materials used for plain cement and ternary mixes, obtained by XRF analysis

Oxides	CEM I	CC 45%	CC 50%	CC 70%	CC 95%	D 5
CaO	63.6	0.1	1.3	0.8	0.0	55.0
SiO_2	19.3	53.5	44.9	48.1	52.0	0.1
Al_2O_3	5.7	34.9	32.3	35.1	43.8	0.0
Fe_2O_3	3.6	3.4	15.4	9.4	0.3	0.0
MgO	1.6	0.1	0.8	0.5	0.0	0.1
Na_2O	0.2	0.2	0.4	0.2	0.3	0.1
K_2O	1.2	0.2	0.2	0.1	0.1	0.0
SO_3	3.2	0.0	0.0	0.0	0.1	0.0

Table 2. Mix designs of the studied blends, in mass fraction

Label	CEM I	C. clay	Limest.	Gyps.	Water: binder
PC	100	–	–	–	0.4
LC3-50 (45%) 2:1	54	29	15	2	0.4
LC3-50 (50%) 2:1	54	29	15	2	0.4
LC3-50 (70%) 2:1	54	29	15	2	0.4
LC3-50 (95%) 2:1	54	29	15	2	0.4
LC3-50 (45%) 1:1	54	22	22	2	0.4
LC3-65 (45%) 2:1	70	19	10	1	0.4
LC3-65 (45%) 1:1	70	15	15	1	0.4

samples were coarsely crushed about 16 h after casting and placed in a container at 20 °C sealed with a RH sensor (Rotronic HygrologNT). Relative humidity was then measured automatically every 15 min for 28 days. The sensors were calibrated before and after the tests using salts in the range from 98% to 75% RH.

Paste samples for creep tests were cast into 25 × 25 × 80 mm prismatic moulds. 24 h after mixing, the samples were demoulded and wrapped in adhesive aluminium to ensure autogenous conditions. Samples were then cured at 20 °C during 28 days, then loaded at 7.3 MPa, which is about 10% of the compressive strength measured on cubes of 25 mm. A constant load is applied on the samples and the strain is automatically monitored by 3 LVDTs during 28 days. The acquisition speed is one reading every two seconds for the first ten minutes after loading, then one reading every two minutes. Two samples were tested for each mix and averaged. The total duration of shrinkage and creep measurements is therefore 56 days, including curing time.

3 Results

The autogenous shrinkage over time for the studied systems is presented on Figs. 1 and 2 for the two groups of mixes. All studied materials exhibit a moderate swelling during the first hours after setting time. In this study, this expansion is removed and shrinkage

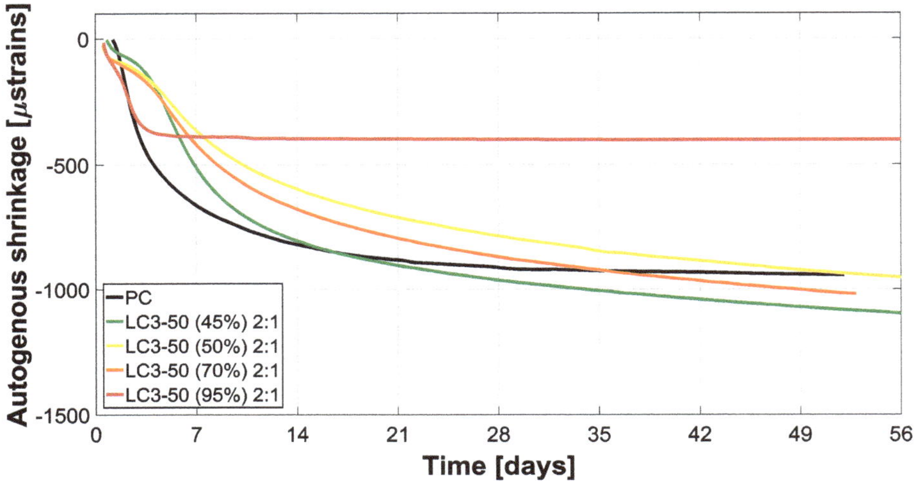

Fig. 1. Autogenous shrinkage during the first 2 months after casting for the group of mixes with varying clay purity. Initial expansion was removed.

strains were zeroed at the time of the local maximum after casting, in order to show only the shrinkage part. For the first group of samples, the main features are the fast stabilization of the LC3-50 (95%) 5 days after casting and the very similar shrinkage behaviour of the other ternary blends, regardless of the calcined clay used. In addition, shrinkage amplitude is quite similar between the PC reference mix and the LC3-50 (except the 95% grade) during the first month after casting, but still goes on after 28 days, when PC is stable. Regarding the second group of samples, the tendencies are the same, but shrinkage is actually more important for the LC3-65 series, even if they contain less clinker than the reference plain cement mix. A common feature to all the ternary blends is a delay of the initial shrinkage onset due to the dilution effect from the clinker substitution. As expected, the effect is somewhat more pronounced for the LC3-50 series compared to the LC3-65 series. The mass difference of the samples before and after the test is less than 0.5%, proving the suitable quality of the sealing. Figure 3 shows autogenous shrinkage along with internal relative humidity. For all samples, the initial drop in humidity corresponds to the shrinkage onset. At later age, RH stabilises whereas shrinkage goes on for the LC3-50 (45%), (50%) and (70%) samples.

Results on Fig. 4a show the evolution of basic creep compliance J versus loading time for the first group of studied paste samples. Compliance (or specific creep) is defined as the creep strain – in microstrains – divided by the applied stress. Creep strain is obtained by subtracting the autogenous shrinkage from the total creep strain. Data are plotted starting about 1 min after loading the sample, in order to remove the elastic deformation and the instability during loading. From these results, it is noticed that the substitution of cement with limestone and calcined clay causes a reduction in creep compliance. This decrease does not seem to be related to the clay grade as all the presented mixes display a similar behavior. The creep test results for the second group

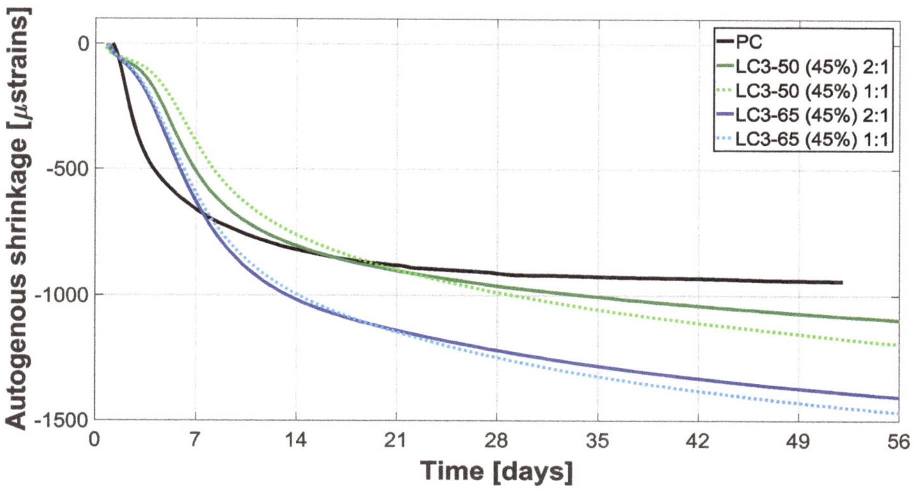

Fig. 2. Autogenous shrinkage during the first 2 months after casting for the group of mixes with varying mix design. Initial expansion was removed.

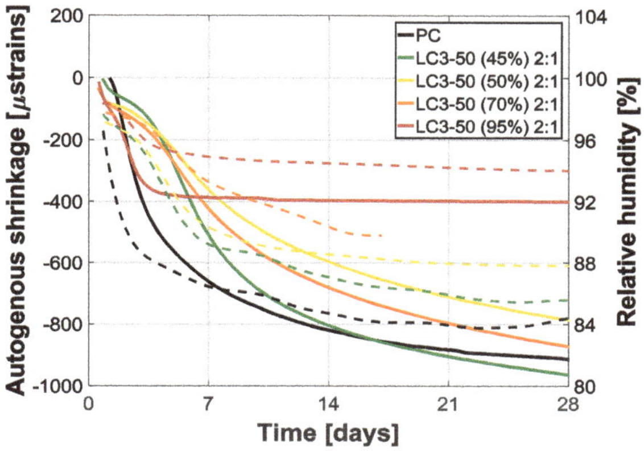

Fig. 3. Comparison of autogenous shrinkage (solid lines) and internal relative humidity (dashed lines) for the first month after casting

are showed on Fig. 4b. Contrarily to the autogenous shrinkage, a change in the calcined clay to limestone ratio has a noticeable effect on creep compliance. In this case, creep compliance decreases with increasing content of clay in the system.

It is useful to look at the kinetics of the creep strain to get an insight on the underlying mechanisms. To that end, the first time derivative of compliance J was computed using the general algorithm described in [11]. The resulting data is then plotted on a log-log graph, appearing as straight lines for the groups of studied samples on Fig. 5. The first derivative is then fitted using a simple logarithmic function with

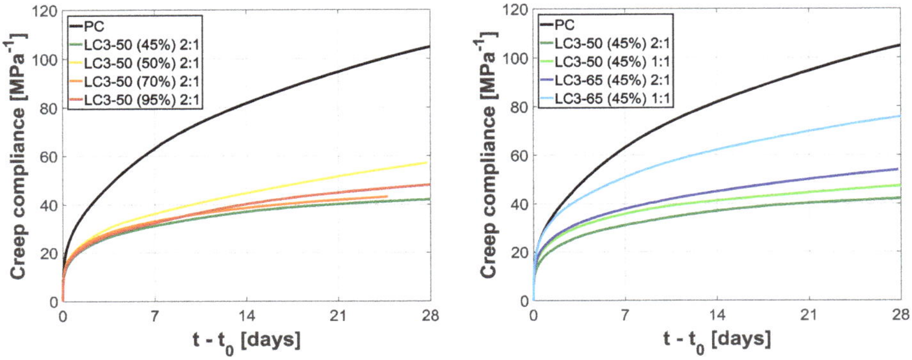

Fig. 4. Basic creep compliance over loading time. (a) First group of mixes with varying clay: after 28 days of curing, the effect of the substitution materials is similar regardless of their quality. (b) Second group of mixes with variable mix design: in this case the proportion and composition of the substituted cement fraction has an impact of creep compliance.

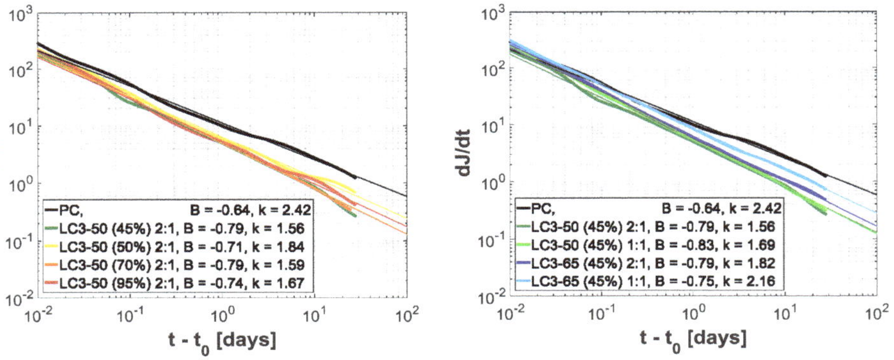

Fig. 5. Log-log plots of the first time derivative of compliance versus loading time. The caption contains the values of the fitting coefficients, using the least squares method. The slope B is similar among the ternary blends and lower than that of the PC sample.

only two coefficients, in the form of $\ln J(t - t_0) = B\,\ln(t - t_0) + k$. B is the slope of the line, therefore linked to the mechanisms causing the strain, and k is the offset, related to the history of the sample. The values are reported on the graphs as well. The coefficients B are similar among the LC3 samples and lower than that of PC, indicating that the creep is occurring at a slower rate in the ternary blends.

4 Discussion and Conclusion

In this study, autogenous shrinkage measurements were carried out over a two-month period using ternary blends of limestone and calcined clay. Results showed that, for a given mix design, the effect on shrinkage strain is similar regardless of the clay, except when using pure metakaolin instead of a calcined clay. The main effects are a delay of the onset of shrinkage due to a dilution effect and a noticeable strain rate still after 28 days, when a plain cement sample is stabilized. The amplitude of the autogenous shrinkage is comparable to that of a plain cement reference for the first 28 month. When changing the mix design but keeping the same materials, it was observed that a variation in the calcined clay to limestone ratio had a very minor influence on shrinkage, but that the total substitution level had an impact: total shrinkage was higher for the mixes containing 65% of cement than for those with 50% of cement. This feature could indicate the existence of an optimum mix design minimizing the amplitude of autogenous shrinkage. The peculiar behavior of the mix using pure metakaolin as one of the substituents could be due to its high surface are and fineness, a consequence of the flash-calcination process, producing a highly reactive and adsorbing material. The question of why the shrinkage rate is still considerable during the second month is still under investigation: measurement of the internal relative humidity of the materials showed no important drop in RH, which could be a driving force for shrinkage.

The results of the basic creep tests showed that any substitution of cement with limestone and calcined clay decreased the creep compliance of mature paste samples. Similar to autogenous shrinkage, the reduction is comparable for mixes with the same mix design but varying clay grade, including pure metakaolin. On the other hand, changes in the mix design – be substitution level of calcined clay to limestone ratio – have a measurable effect on creep compliance. The fact that there is an impact of the calcined clay to limestone ratio on creep compliance proves that the creep compliance reduction is not simply due to a decrease of the cement fraction in the binder.

Some questions remain open after this study, such as the reason causing some of the materials to shrink after one month of hydration, considering that the relative humidity is stable and that these same materials have a low creep compliance. In addition, the mechanism causing a creep reduction in the ternary blends is not elucidated yet. To that end, a more thorough characterization of the microstructure of the samples is necessary. Indeed, a straightforward explanation would be that all the samples have a similar microstructure and hydrates proportion, as they all have matured for 28 days. However, other contributions – such as a physical impact on viscoelasticity from the clay particles – have also to be investigated.

References

1. Ambroise, J., Maximilien, S., Pera, J.: Properties of metakaolin blended cements (1994)
2. Juenger, M.C.G., Siddique, R.: Recent advances in understanding the role of supplementary cementitious materials in concrete. Cem. Concr. Res. **78**, 71–80 (2015)

3. Sabir, B.B., Wild, S., Bai, J.: Metakaolin and calcined clays as pozzolans for concrete: a review. Cem. Concr. Compos. **23**(6), 441–454 (2001)
4. Antoni, M., Rossen, J., Martirena, F., Scrivener, K.: Cement substitution by a combination of metakaolin and limestone. Cem. Concr. Res. **42**(12), 1579–1589 (2012)
5. Matschei, T., Lothenbach, B., Glasser, F.P.: The role of calcium carbonate in cement hydration. Cem. Concr. Res. **37**(4), 551–558 (2007)
6. Brooks, J.J., MegatJohari, M.A.: Effect of metakaolin on creep and shrinkage of concrete. Cem. Concr. Compos. **23**(6), 495–502 (2001)
7. Gleize, P.J.P., Cyr, M., Escadeillas, G.: Effects of metakaolin on autogenous shrinkage of cement pastes. Cem. Concr. Compos. **29**(2), 80–87 (2007)
8. Khatib, J.M., Wild, S.: Pore size distribution of metakaolin paste. Cem. Concr. Res. **26**(10), 1545–1553 (1996)
9. Antoni, M.: Investigation of cement substitution by blends of calcined clays and limestone. École Polytechnique Fédérale de Lausanne (2013)
10. Ye, H.: Creep mechanisms of calcium–silicate–hydrate: an overview of recent advances and challenges. Int. J. Concr. Struct. Mater. **9**(4), 453–462 (2015)
11. Lubansky, A.S., Yeow, Y.L., Leong, Y.-K., Wickramasinghe, S.R., Han, B.: A general method of computing the derivative of experimental data. AIChE J. **52**(1), 323–332 (2006)

Hydration of Blended Cement with Halloysite Calcined Clay

Alejandra Tironi[1(✉)], Fernanda Cravero[2], Alberto N. Scian[2], and Edgardo F. Irassar[1]

[1] Facultad de Ingeniería, CIFICEN (CONICET-CICPBA-UNCPBA), Av. del Valle 5737, B7400JWI Olavarría, Argentina
atironi@fio.unicen.edu.ar

[2] Centro de Tecnología de Recursos Minerales y Cerámica CONICET La Plata – CICPBA, Cno. Centenario y 506, B1897ZCA Gonnet, Argentina

Abstract. The effects of calcined kaolinitic clays as supplementary cementitious materials (SCMs) on the performance of pastes and mortars have been well studied. Less attention has been paid to the thermal transformation of halloysite than that of kaolinite and its possibility to be used as SCMs. Halloysite and kaolinite have identical chemical composition, except that halloysite may have two molecules of H_2O, as interlayer water. The content of additional water in the interlayers of halloysite has a decisive influence on the crystal morphology, which is generally curled rather than platy as in kaolinite. Common forms are elongated tubes and spheroids. The aim of this investigation is to study the hydration of blended cements with 25% of different calcined clays to evaluate the influence of the content and the morphology of halloysite in the development of the hydration compounds, and compressive strength of mortars. Three clays with different halloysite/kaolinite content, and different morphology were analyzed. The hydrated phases present in pastes at 2, 7, and 28 days were identified by X-ray diffraction (XRD), and the content of CH by differential thermal analysis (DTA/TG). The compressive strength of mortars was tested at 2, 7, and 28 days. The pozzolanic reactivity of the calcined clays was influence by the kaolinite content and morphology of halloysite in natural clays. This results in different crystalline and amorphous aluminic phases obtained at different ages, and that the ensemble results differ, this affects the porosity and the compressive strength.

1 Introduction

One of the main objectives in the cement and concrete industry is promote reduce of CO2 emission and energy consumption, for this purpose the use of supplementary cementing materials (SCMs) in blended cements are a feasible option [1]. The effects of calcined kaolinitic clays as SCMs on the performance of pastes and mortars have been well studied [2–5]. Less attention has been paid to the thermal activation of halloysite than that of kaolinite and its possibility of use as SCMs [6]. Halloysite and kaolinite have identical chemical composition, except that halloysite may have two molecules of H2O, as interlayer water. Additional water in the interlayers of halloysite has a decisive

F. Martirena et al. (eds.), *Calcined Clays for Sustainable Concrete*, RILEM Bookseries 16,
https://doi.org/10.1007/978-94-024-1207-9_73

influence on the crystal morphology, which is generally curled rather than platy as in kaolinite. Common forms are elongated tubes and spheroids [7].

The aim of this investigation is to study the hydration of blended cements with 25% of different calcined clays to evaluate the influence of the content and the morphology of halloysite in the development of the hydration compounds and the compressive strength of mortars. Three clays with different halloysite/kaolinite content, and different morphology were analyzed.

2 Materials and Methods

Three different halloysite/kaolinitic clays from Río Negro Province, Argentine, were select to study as SCMs in blended cements. For this purpose, the clays were calcined at 700 °C, to obtain metakaolinite (MK), an amorphous reactive phase, due to the complete dehydroxylation of clays minerals. Other crystalline phases, present as impurities, such as quartz, cristobalite, trydimite and anatase, remain stable after thermal treatment [8]. The calcined clays were ground and they were labeled as CC1, CC2 and CC3.

The clay minerals present in the natural clays have different morphologies. The halloysite present in natural clays corresponding to CC1 has a tubular morphology, while the halloysite present in natural clay corresponding to CC2 and CC3 has spheroidal morphology, and it coexists with hexagonal plates of kaolinite. The halloysite/kaolinite ratio is 100/0 for the natural clay corresponding to CC1, 75/25 for CC2 and 60/40 for CC3 [8].

To study the hydration of blended cements a normal Portland cement (PC) was used. For this PC, the mineralogical composition of clinker was $C_3S = 63.6\%$, $C_2S = 15.1\%$, $C_3A = 2.8\%$ and $C_4AF = 15\%$; 5.5% gypsum is used as set regulator and limestone is the minor component. Blended cements (bc) were formulated with 25% by mass of PC replacement by halloysite/kaolinitic calcined clays.

The hydration of blended cements (bc) was studied at 2, 7, and 28 days; the hydrated phases were identified on paste and the compressive strength was determined on mortar.

Blended cement pastes were prepared using water to bc ratio (w/bc) of 0.5. At the test-age fragments of paste sample was immersed in acetone during 24 h to stop the hydration, dry overnight in oven at 40 °C and then cooled in a desiccator. The crystalline phases were identified by X-ray diffraction (XRD) using a Bruker D2 PHASER diffractometer. Complementary, thermal analysis (DTA/TG) was performed to analyse the crystalline and amorphous hydrated phases, using a NETZCH STA 409 thermobalance. Pore size distribution was determined at 28 days in fragments of pastes using a mercury intrusion porosimeter (MIP-ThermoFisher Sc PA440), for pore size diameters from 7.3 to 14000 nm.

Compressive strength (CS) was assessed on mortars made with standard sand EN-196-1 (1:3) and w/bc of 0.50. The CS was determined as the average of five specimens using an universal testing machine Instron 4485 at 2, 7, and 28 days.

3 Results

At 2 days, the hydrated crystalline phases detected from PC hydratation were ettringite (Ett) and calcium hydroxide (CH) for PC-pastes and blended cement pastes (Fig. 1a), and unhydrated phases C3S, C2S, and C4AF. For bc-pastes, the intensity of the CH-peaks is lower than that corresponding to PC-paste, showing dilution effect and some CH-consumption to form C-S-H and C-A-H/C-A-S-H phases by the pozzolanic reaction. The hemicarboaluminate (HC, C3A.CH0.5CC0.5.H12) and monocarboaluminate (MC, C3A.CC.H12) phases are formed by the pozzolanic reaction between the reactive alumina from the MK, the CH and calcium carbonate (C) from the limestone filler, a minor component in the PC used. At 2 days, the lowest intensity of the CH-peaks occurs for 25%CC3 paste, showing a high CH-consumption by pozzolanic reaction of this calcined clay obtained from the clay with the highest kaolinite content (halloysite/kaolinite ratio 60/40). The highest intensity of the HC-peaks is for 25%CC1 obtained from the clay with the highest halloysite content and tubular morphology. This allows concluding that the phases obtained as product of the pozzolanic reaction are different when the ratio halloysite/kaolinite is different and the halloysite has different morphology. At later ages (Figs. 1b, c) the intensity of the CH-peaks is low for the three blended cements, and the intensity of HC-peaks and MC-peaks is higher.

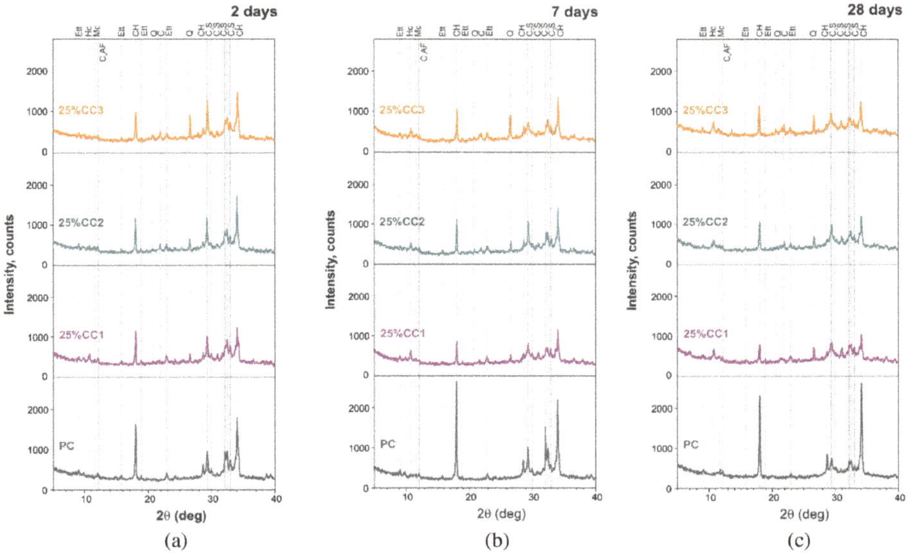

Fig. 1. XRD patterns for hydrated pastes: (a) 2 days, (b) 7 days and (c) 28days

The results obtained by DTA for hydrated pastes are presented in Fig. 2. The first endothermic peaks located at about 110–410 °C characterize the dehydration of C-S-H, C-A-H and C-A-S-H [9]. C-S-H gel is a product from PC hydratation and pozzolanic reaction. The C-A-S-H are cementing compound obtained from pozzolanic reaction where the C-S-H incorporate Al in its structure, like result of pozzolanic reaction too. C-A-H compounds include HC, MC and Ett. Ett (C-A-S-H) starts to loss water at very low temperature (~40 °C) and its complete decomposition into gypsum, hemihydrate (C-A-H) together with an amorphous material occurs at 120 °C [10]. Ett remains stable at 2, 7 and 28 days of hydration. The second endothermic peak located about 410–560 °C characterizes the CH dehydroxylation. Table 1 reports the percentage of CH in mass determined by DTA/TG. For PC-paste, the amount of CH increases with the age. On the other hand, bc-pastes show lower CH-content than the corresponding to PC-paste due to the consumption by the pozzolanic reaction, as determined by XRD (Fig. 1). At 2 days, 25%CC3-paste with calcined clay obtained from the clay with the highest kaolinite content consumes the highest amount of CH, while after 7 days, calcined clays obtained from the clays with high content of halloysite (25%CC2-paste and 25%CC3-paste) consume high amount of CH. At 28 days, 25%CC1-paste with calcined clay obtained from the clay with tubular halloysite consumes the highest amount of CH. The aluminic phases appears as a shoulder of the first endothermic peak in bc-pastes at 2 days (Fig. 2a), and it is more intense for all pastes after 7 days (Figs. 2b, c). At 28 days (Fig. 2c), the area of the first peak is greater for bc-pastes due to the dehydratation of compounds formed during the pozzolanic reaction: C-S-H, C-A-H and C-A-S-H.

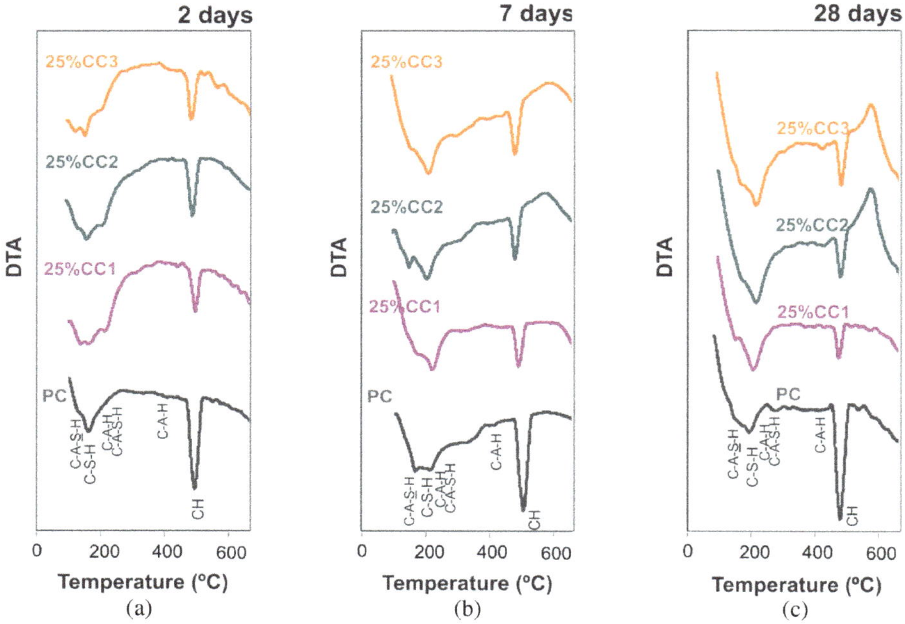

Fig. 2. DTA curves of hydrated pastes: (a) 2 days, (b) 7 days and (c) 28 days.

Table 1. CH content and volume of pores in hydrated pastes. Compressive strength of mortars.

Sample	Amount of CH in hydrated pastes, mass%			Compressive strength of mortars, MPa			Volume of pore at 28 days mm^3/g	
	2 days	7 days	28 days	2 days	7 days	28 days	Diameter 100–14000 nm	Diameter 45–14000 nm
PC	12.75	16.43	16.93	17.5	30.5	37.9	40.7	92.0
25%CC1	6.80	7.09	5.68	12.9	34.8	58.5	13.8	21.2
25%CC2	7.76	7.71	5.96	15.2	35.7	49.1	16.2	19.3
25%CC3	2.33	8.15	6.38	15.9	32.2	48.8	39.4	48.3

For blended cements, the higher compressive strength was developed by the mortar 25%CC3 at 2 days and by the mortar 25%CC1 at 28 days (Table 1). The volume of pores with diameter up to 45 and 100 nm is lower at 28 days in pastes made with blended cements, and it is lowest for bc-pastes with calcined clays obtained from clays with high halloysite content (25%CC1 and 25%CC2), in this pastes the pore refinement is highest.

4 Discussions

The products obtained in pozzolanic reaction at different age were influenced by the disponibility of reactive aluminum: the reactive aluminum present in tubular calcined halloysite at early age produces mainly hemicarboaluminate, which remains stable (Fig. 1), while the calcined kaolinite has more amounts of aluminum available in surface, and then this is incorporated in the C-A-S-H gel (Fig. 2a). This causes that the consumption of CH in pozzolanic reaction to be different; it is higher at early age in blended cements elaborated with high kaolinite calcined clay (Table 1). At later age the CH reacts with the aluminum on the inner surface of the halloysite tubes, has more amount of available and incorporates in C-A-S-H gel (Fig. 2). For this, the volumen of pores that affect the compressive strenght is lowest, and the compressive strenght is highest for blended cements elaborated with tubular halloysite calcined at 28 days (Table 1).

5 Conclusions

Blended cements elaborated with 25% of replacement of calcined halloysite/kaolinitic clays as SCMs are pozzolanic. The hydrated phases obtained are characteristic of the system MK-cement. The CH content is lower than those determined in hydrated pastes elaborated with Portland cement and a higher content of aluminic phases were identified as products of pozzolanic reaction. The pozzolanic reactivity of the calcined clays was influence by the kaolinite content and the morphology of halloysite. This produced that different crystalline and amorphous phases were obtained at different ages, and the assemblage results different affecting the development of the porosity and the compressive strength.

References

1. Damtoft, J.S., Lukasik, J., Herfort, D., Sorrentino, D., Gartner, E.M.: Sustainable development and climate change initiatives. Cem. Concr. Res. **38**, 115–127 (2008). doi: 10.1016/j.cemconres.2007.09.008.10
2. Kakali, G., Perraki, T., Tsivilis, S., Badogiannis, E.: Thermal treatment of kaolin: the effect of mineralogy on the pozzolanic activity. Appl. Clay Sci. **20**, 73–80 (2001). doi:10.1016/S0169-1317(01)00040-0
3. Siddique, R., Klaus, J.: Influence of metakaolin on the properties of mortar and concrete: a review. Appl. Clay Sci. **43**, 392–400 (2009). doi:10.1016/j.clay.2008.11.007
4. Fernandez, R., Martirena, F., Scrivener, K.L.: The origin of the pozzolanic activity of calcined clay minerals: a comparison between kaolinite, illite and montmorillonite. Cem. Concr. Res. **41**, 113–122 (2011). doi:10.1016/j.cemconres.2010.09.013
5. Tironi, A., Trezza, M.A., Scian, A.N., Irassar, E.F.: Kaolinitic calcined clays: factors affecting its performance as pozzolans. Constr. Build. Mater. **28**, 276–281 (2012). doi:10.1016/j.conbuildmat.2011.08.064
6. Rabehi, B., Boumchedda, K., Ghernouti, Y.: Study of calcined halloysite clay as pozzolanic material and its potential use in mortars. Int. J. Phys. Sci. **7**, 5179–5192 (2012). doi:10.5897/IJPS11.184
7. Cravero, F., Fernández, L., Marfíl, S., Sánchez, M., Maiza, P., Martínez, A.: A. Spheroidal halloysites from Patagonia, Argentina: some aspects of their formation and applications. Appl. Clay Sci. **131**, 48–58 (2016). doi:10.1016/j.clay.2016.01.011
8. Tironi, A., Cravero, F., Maggi, J., Scian, A.N., Irassar, E.F.: Utilización de arcillas con elevado contenido de halloysita como puzolanas. In: VII Congreso Internacional de la Asociación Argentina de Tecnología del Hormigón (AATH), T5/67, pp. 545–552 (2016)
9. Morsy, M.S.: Effect of temperature on hydration kinetics and stability of hydration phases of metakaolin-lime sludge-silica fume system. Ceram. Silik. **49**, 225–229 (2005)
10. Zhou, Q., Glasser, F.P.: Thermal stability and decomposition mechanisms of ettringite at <120 °C. Cem. Concr. Res. **31**, 1333–1339 (2001). doi:10.1016/S0008-8846(01)00558-0

Progress of Limestone Calcined Clay Cement in China

Sui Tongbo(✉), Bin Wang, Yuliang Cai, and Shengliang Tang

Sinoma Research Institute, Simona Int'l, CNBM, Beijing, China

Abstract. Waste materials from kaolinite processing was identified with a content of 46% of kaolinite and calcined and used for preparing Limestone Calcined Clay Cement (LC3). It has been found that thermal history of activation in the range of temperature from 750–800 °C does not change much the reactivity of the calcined kaolinitic clay based on strength test. R3 (rapid, relevant and reliable) calorimetry test method was also used for assessing the reactivity of calcined kaolinitic clay, which gives a result in good agreement with mortar strength test. The comparison test reveals that the reduction of flowability due to the addition of calcined clay can be compensated with the combination of limestone, and that the compressive strength of LC3-50 can have equivalent strength to the pure Portland cement, indicating the high reactivity of calcined clay and the synergy when combined with limestone. The design of a flash calciner with a capacity of 600t/d and thermal consumption of 620 kcal/kg in average is also introduced.

1 Background

In addressing the challenge in cement sector on large global CO_2 emission, IEA-CSI Cement Technology Roadmap to 2050 elaborated three categories of best available technologies (BAT), i.e. Energy efficiency, Alternative fuels and Clinker substitutes [1]. Among which the use of clinker substitutes has been proven a solution worldwide that can lead to immediate and remarkable reduction in CO_2 emission of cement.

The development of a new ternary blended cement with combination of limestone-calcined clay-clinker named as LC^3 led by Professor K. Scrivener, EPFL and international teams from Cuba and India set up a good example in developing a relatively inexpensive and widely available SCM source capable of replacing up to 50% of clinker while maintaining similar performance to reference [2–4]. Durability of the LC^3 concrete has also been well proved [5] through laboratory test and field exposure evaluation in terms of chloride ingress, alkali silica reaction, and sulfate attack, etc.

Considering the fact that China is a country with abundant resources of kaolinitic clay (Fig. 1) and more than half of world cement production, the LC^3 solution is expected to lead to a significant contribution to CO_2 mitigation of cement industry.

This article aims to present the latest progress of developing LC^3 using Chinese raw materials and the design of a flash calciner with a capacity of 600tpd and thermal consumption of 620 kcal/kg as an effort to further reducing the energy and emission intensity of LC^3.

F. Martirena et al. (eds.), *Calcined Clays for Sustainable Concrete*, RILEM Bookseries 16,
https://doi.org/10.1007/978-94-024-1207-9_74

A: Raw mining site B: Finished product C: intermediate waste after de-watering

Fig. 1. Mining and further processing of kaolinitic clay

2 Raw Materials and Experiment

2.1 Raw Materials

2.1.1 Origin of Kaolinitic Clay

Waste materials from kaolinite processing was identified for further thermal activation and preparation of LC^3. The typical kaolinite-quartz mineral processing with an annual capacity of 600 thousand tons kaolinite is illustrated below from raw mineral mining to multi-stage separation and the finished product (Fig. 1 A ~ C).

The raw materials used for this study came from the process as shown in Fig. 1C.

2.1.2 Other Raw Materials

Other raw materials used for this study include PI cement, gypsum and limestone with their basic characteristics indicated in Table 1.

Table 1. Basic characteristics of raw materials

Raw materials	Description	Purity	Residue of 80 μm square mesh sieve, %
Reference cement	A pure Portland cement for the test of concrete admixtures	Clinker:gypsum = 95:5	1.23
Gypsum	A natural mineral	90% of $CaSO_4 \cdot 2H_2O$	3.2
Limestone	A natural mineral	91% of $CaCO_3$	1.4

2.2 Experiment

The chemical composition: measured by Bruker XRF analysis. The mineralogical composition: measured via Bruker D8 XRD analysis and Rietveld refinement. Thermal analysis: conducted with Mettler analyzer in a temperature range of 30–1000 °C at a heating rate of 10 °C/min and N_2 atmosphere. Mortar strength test: prismatic mortar specimens with dimensions of 40 × 40 × 160 mm were prepared with w/c ratio of 0.50 and sand/cement ratio of 3.0 and 20 °C standard curing (equal to ISO standard) at targeted ages from 1d, 3d, 7d, 28d to 90d. The reactivity test – mortar strength method: comparison measurement of 28-day age mortar strength for reference cement and targeted samples prepared by reference cement with 30% replacement by calcined clay. The reactivity test – calorimetry method: heat release by isothermal calorimetry for a paste prepared with calcined clay:$Ca(OH)_2$ = 1:3, w/b = 1.2, etc. tested under 40 °C which has been detailed on basis of RILEM TC TRM - Protocol for R^3 model paste preparation and test procedure [7]. Preparation of LC^3-X: the cement proportion = clinker:gypsum:calcined clay:limsestone = X:5: 2(95-X)/3:(95-X)/3. All the components are grounded separately and inter-blended for preparing LC^3-X.

3 The Characterization of the Kaolinitic Clay and Calcined Products

3.1 Mineralogical and Chemical Composition

The moisture of intermediate waste after de-watering ranges from 30–35%. The chemical composition by XRF analysis was presented in Table 2, and the mineralogical composition measured via XRD analysis presented in Fig. 2. Mineral phases were semi-quantitatively analyzed with Rietveld refinement method. Thermal analysis (TGA) was also conducted to identify the kaolinite content through the weight loss between the temperature of 400 °C and 650 °C due to the de-hydroxylation of kaolinite (Fig. 3).

Table 2. Chemical composition of kaolinitic clay

No.	Chemical composition/%												
	SiO_2	Al_2O_3	Fe_2O_3	CaO	MgO	SO_3	K_2O	Na_2O	TiO_2	P_2O_5	MnO	LOI	Σ
G-3	52.72	31.90	1.60	0.09	0.47	0.00	3.47	0.17	0.15	0.026	0.028	9.80	100.42

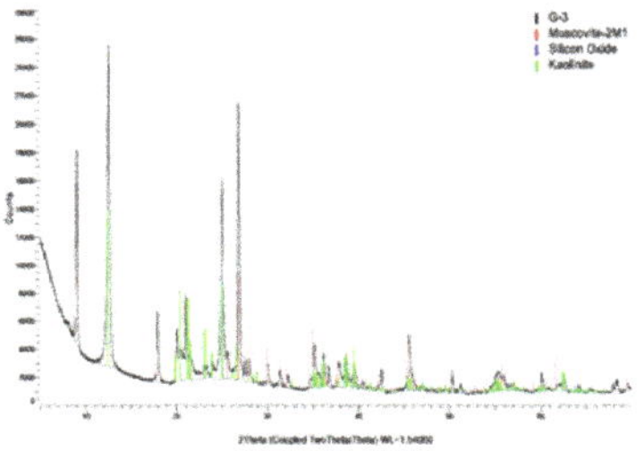

Fig. 2. XRD pattern of kaolinitic clay sample G-3

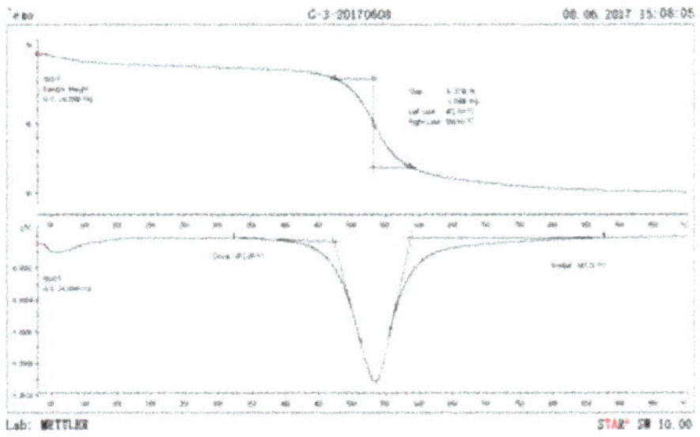

Fig. 3. TGA curve of kaolinitic clay

It has been shown from XRD pattern (Fig. 2) that the raw materials mainly contains muscovite, kaolinite and quartz with kaolinite content 39.40% from Rietveld refinement.

Quantitative analysis of the mineralogical composition of the kaolinitic clay can also be made by TGA through the weight loss 400 °C to 650 °C giving a calculated kaolinite content of 46.64% in the kaolinitic clay waste. This result is a bit higher than that by XRD Rietveld refinement.

3.2 Thermal Activation

The kaolinitic clay was mixed with 5% mass water and molded with pressure into a disk shape green body. The samples, after drying, were calcined in electric furnace under designated thermal activation history (M7506: 750 °C × 60 min, M8006: 800 °C × 60 min and M8018: 800 °C × 180 min). After calcined, the sample were cooled under ambient temperature and ground in a laboratory ball mill for 30 min. Residue of the ground calcined clay after passing through 80 μm square mesh sieve is controlled below 5%. XRD analysis was also conducted for the three calcined clay samples with the results indicated in Fig. 4.

It is found from the pattern in Fig. 4 that the characteristic diffraction peak of kaolinite around $2\theta = 12°$ is remarkably reduced when subject to calcination and completely disappears at 800 °C × 180 min.

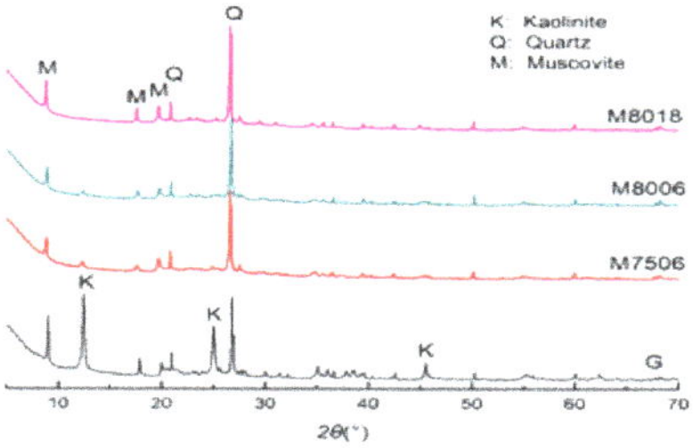

Fig. 4. XRD pattern of calcined clay samples in comparison with unburnt one

3.3 Reactivity Test

Sample M8018 was selected for RILEM TC TRM round robin test with two characterization methods comparing with samples CC1 (Swiss pure meta-kaolin) and CC2 (Indian calcined clay).

3.3.1 Mortar Strength Method

The comparison result on the reactivity of M8018, CC1 and CC2 is shown in Table 3. Certain amount of superplasticizer was also needed for CC1 and CC2 to keep their mortar flow over 160 mm. The results reveal that the reactivity of M8018 lies between that of CC1 and CC2 and that the mortar flowability of the cement with M8018 is higher than CC1 and CC2.

Table 3. Physical property of cement

No.	SCM	28d Compressive Strength, MPa	Reactivity Index, %	Mortar Flow, mm
T-0	Reference	57.1	/	211
T-1	+CC1	60.8	106.5	169
T-2	+CC2	47.5	83.2	170
T-3	+M8018	54.8	96.0	167

3.3.2 Isothermal Calorimetry Method

Heat release by isothermal calorimetry for a paste was measured based on R^3 model with the results indicated in Fig. 5. It can be seen that R^3 calorimetry test can well demonstrate the reactivity of the meta-kaolin/calcined clay with its results in agreement with mortar strength test.

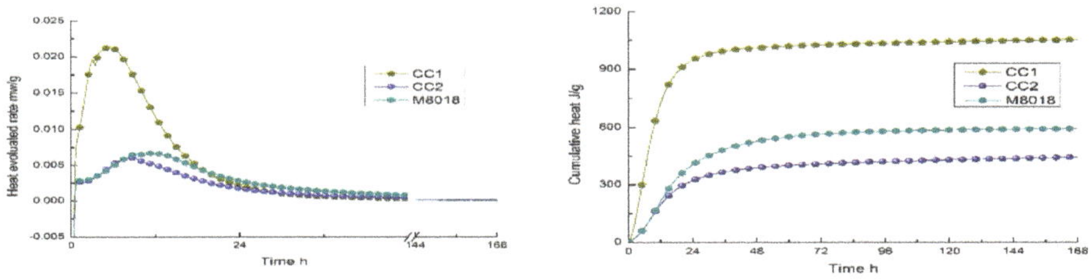

Fig. 5. Isothermal calorimetry test. L: Heat evolution rate curves, R: Cumulative heat curves

4 Performance of LC3 Materials

Mortar strength test was conducted for LC^3-65 and LC^3-50 in comparison with reference cement (Fig. 6). It is found that the early age strength prior to 7-day decreases with the increased addition of blended materials (calcined clay+limestone), while at 28-day and 90-day LC^3-65 and LC^3-50 exhibit similar strength which is very close to the reference cement.

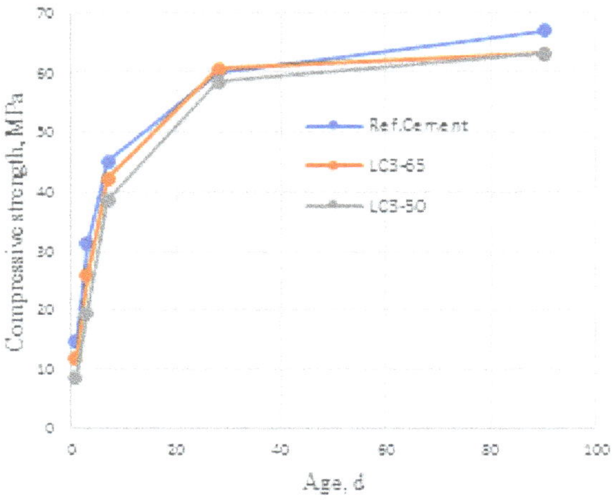

Fig. 6. Mortar strength test of LC3-65 & LC3-50 in comparison with reference cement

Experiment also shows that the addition of calcined clay at the dosage of 20-30% caused considerable reduction in fresh mortar flow rate. This hopefully can be compensated in concrete making when superplasticizers are applied.

5 Energy Efficient Calcining System

An energy efficient 600t/d flash calciner was designed by Sinoma Int'l for the thermal activation of kaolinitic clay. The system is schematically shown in Fig. 7, which includes 5-stage preheating cyclones, decomposing furnace and 2-stage cyclone and fluidized bed cooling system. It should be noticed that this is calcining system only without considering the pre-drying of the raw clay and post-grinding process. Raw clay with 1% of moisture is assumed to feed into the inlet C1 cyclone. Coal as fuel is used for calculating the thermal consumption of the flash calciner, which gives a range of 590–660 kcal/kg and 620 kcal/kg averagely. This indicates a remarkable reduction of energy consumption for calcining system.

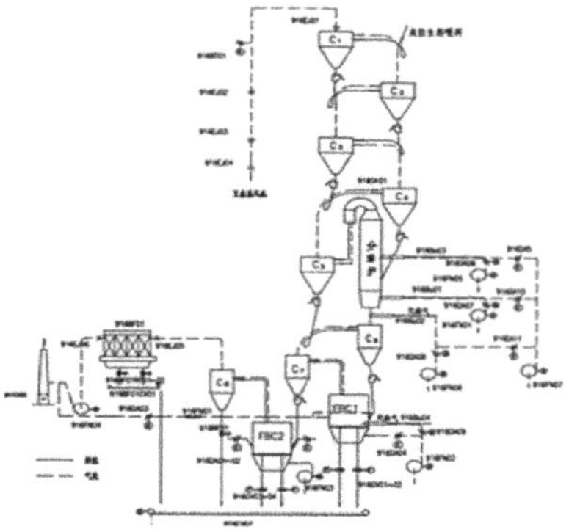

Fig. 7. Process design drawing of flash calciner

6 Conclusion

(1) A kind of Chinese waste material was identified via XRD and TGA for quantitative analysis. The reactivity of the calcined kaolinitic clay was characterized by both mortar strength test and isothermal calorimetry test;

(2) LC^3-65 and LC^3-50 were prepared and proved to give equivalent 28-day compressive strength when compared with reference cement, yet the early age strength decreases with the increase of the blended materials in LC^3;

(3) An energy efficient flash calcining system was designed with a capacity of 600t/d, and thermal consumption is 620 kcal/kg in average using coal as fuel for the calcination.

Acknowledgement. The authors wish to particularly thank the consistent support of LC3 international team, Prof. K. Scrivener, EPFL, Prof. F. Martirena, UCLV, & Asso. Prof. S. Bishnoi, IITD.

References

1. IEA-CSI Cement Technology Roadmap 2009-Carbon Emissions Reductions up to 2050. IEA Publications, Paris (2009)
2. Antoni, M., Rossen, J., Martirena, F., Scrivener, K.: Cement substitution by a combination of metakaolin and limestone. Cem. Concr. Res. **42**, 1579–1589 (2012)
3. Bishnoi, S., Maity, S., Mallik, A., Joseph, S., Krishnan, S.: Pilot scale manufacture of limestone calcined clay cement: The Indian experience. Indian Concr. J. Spec. Issue Future Cem. **88**(7), 22–28 (2014)
4. Hernandez, J.F.M., Scrivener, K.: Development and introduction of a low clinker, low carbon, ternary blend cement in Cuba. In: Scrivener, K., Favier, A. (eds.) Calcined Clays for Sustainable Concrete, pp. 323–329. Springer, Dordrecht (2015)
5. Scrivener, K., Martirena, F., Bishnoi, S.: Presentation of LC3 Workshop, March 2016, Beijing, China, June 2017

6. Chen, Wenrui: Innovation and perspectives of kaolinite of China. Introd. China Non-Met. Miner. Ind. **68**, 3–6 (2008). (in Chinese)
7. Avet, F., Snellings, R., Alujas Diaz, A., Ben Haha, M., Scrivener, K.: Development of a new rapid, relevant and reliable (R3) test method to evaluate the pozzolanic reactivity of calcined kaolinitic clays. Cem. Concr. Res. **85**, 1–11 (2016)

Thermal Activation of Two Complex Clays (Kaolinite-Pyrophillite-Illite) from Tandilia System, Buenos Aires, Argentina

M.A. Trezza(✉), A. Tironi, and E.F. Irassar

Facultad de Ingeniería (UNCPBA), CIFICEN (CONICET-CICPBA- UNCPBA), Av. del Valle 5737, B7400JWI Olavarría, Argentina

Abstract. The aim of this work is focused to identify natural resources with industrial potential to be used as supplementary cementitious materials (SCMs) in Portland blended cements. Two clays obtained from the quarries near to Barker in Tandilia System (Buenos Aires- Argentina) were studied. The geneses of these rocks are by a hydrothermal alteration and include the presence of pyrophillite. Whole-rock were characterized by XRD and FTIR spectroscopy indicating that main clayed minerals are kaolinite ($Si_2Al_2O_5(OH)_4$), illite ($K_{0.66}Si_{3.33}Al_{2.66}O_{10}(OH)_2$) and pyrophillite ($Si_4Al_2O_{10}(OH)_2$) associated with feldspar. The thermal transformation was studied by differential thermal analysis and the phase changes were confirmed by XRD and FTIR. Samples of clays were calcined at different temperature (550 to 1050 °C), the electrical conductivity was measured and the dissolved silica in simulated pore water solution was quantified. The pozzolanic activity was measured by the compressive strength activity index, on blended cement mortars containing 25% by weight of calcined clays.

Results showed pozzolanic activity after 7 days and the compressive strength values exceed the rate of replacement for both clays. The high water demand of sample containing pyrophillite can be attributed to the exfoliation that occurs during the water loss process.

1 Introduction

Calcined clays may be attractive as supplementary cementitious materials (SCMs) since clays are highly abundant and they are a rich source of alumina and silica which can drive the pozzolanic reaction in blended cement [1]. In the natural state, the clay minerals by their crystalline structure do not possess pozzolanic properties. However, they exhibit pozzolanic activity when are calcined at temperatures between 600 and 900 °C and are finely ground [2, 3]. Its use implies lower energy consumption and reduction in CO_2 emission.

Several studies have been carried out on different types of clays to know their thermal behavior and to analyze their potential use as pozzolanic addition in cement industry. The Pyrophillite (P) clays have a structure that collapses above 1100 °C, accompanied by formation of amorphous SiO_2 and poorly ordered mullite. Li et al. [4] report that the amorphous SiO_2 in thermally treated P, is soluble in caustic soda solution. Illite (I) clays

F. Martirena et al. (eds.), *Calcined Clays for Sustainable Concrete*, RILEM Bookseries 16,
https://doi.org/10.1007/978-94-024-1207-9_75

conserve the order of their structure layers despite the thermally treatment, even after dehydroxylation [5]. Thermal activation of kaolinitic (K) clays produce metakaolinite (MK), amorphous phase that reacts with the calcium hydroxide (CH) generated during the hydration of Portland cement (PC) and contributes to its hydration.

Natural clays are complex systems as consequence the aim of this work is focused to identify natural resources with industrial potential to be used as SCMs in Portland blended cements. Two complex clays obtained from the quarries near to Barker in Tandilia System (Buenos Aires- Argentina) were studied. The geneses of these rocks are by a hydrothermal alteration and include the presence of K ($Si_2Al_2O_5(OH)_4$), I ($K_{0.66}Si_{3.33}Al_{2.66}O_{10}(OH)_2$) and P ($Si_4Al_2O_{10}(OH)_2$) associated with feldspars. The thermal transformation of clays was studied at different temperature (550 to 1050 °C) and their potential activity as SCMs were determined in blended cement containing 25% by weight.

2 Materials and Methods

The studied clays were denominated PK1 and PK2. The crystalline phases were identified by X-ray diffraction (XRD) with a Phillips XPW 3710 diffractometer and quantified using the Rietveld method and PANalytical HighScore Plus software. The thermal stability of the phases present in the natural clay was determined by differential thermal analysis combined with thermo-gravimetric analysis (DTA /TG) using NETZSCH STA 409 thermobalance.

The natural clays were placed as a bed in a vessel and calcined in an ORL oven to the set temperature with a ramp of 13 °C min^{-1} and holding this value for 90 min. The temperatures selected, from DTA/TG results, were: 550, 650, 750 and 1050 °C. Calcined samples were grinding until the retained on 45 μm sieve (# 325) was lower than 12%. The transformations of the crystalline phases present in the clays were studied by XRD and FT-IR spectroscopy using a Nicolet Magna 500 spectrophomter.

Pozzolanic activity of calcined clays was proved using electrical conductivity test (EC). The EC was measured using a Jenway 4010 conductivity meter at preestablished time intervals after that the different calcined clays were added to the beaker containing calcium hydroxide saturated solution. The test was performed at 40 ± 1 °C [3]. The alkaline dissolution was carried out in order to quantifying the active silica of different calcined clays in strongly alkaline medium in which they would be found if be used as active additions to the PC. The quantification of dissolved SiO_2 was performed with a colorimetric analytical method.

At the selected calcination temperature, calcined clays were prepared to elaborate blended cements (BC) with 25% w/w of replacement. Mortar (w/cm = 0.5 – sand: cement = 3:1) was elaborated, the flow was determined and the compressive strength (CS) of was measured on standard prisms at 7, 28 and 90 days. The strength activity index (SAI) was calculated as the ratio of the CS of blended cement to the strength of the plain cement at the same age.

3 Results

3.1 Characterization of Clays

Both studied clays (PK1 y PK2) have a similarity chemical composition, with slight differences in the percentages of their main oxides. The crystalline phases were identified by XRD (Fig. 1a). The minerals present quantified by Rietveld method are: PK1: 25%P, 27%I, 40%K, 8%F; and PK2: 13%P, 33%I; 47%K, 7%F.

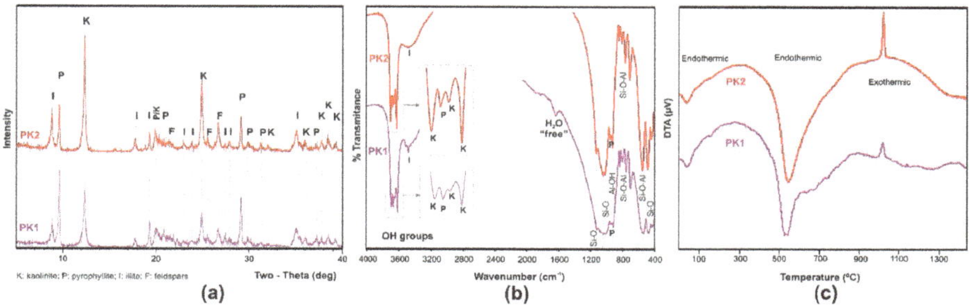

Fig. 1. (a) Diffractograms, (b) IR spectra and (c) DTA/TG of PK1 and PK2 clays

The highest I content quantified for PK2 by DRX-Rietveld is in agreement with the higher content of K_2O (3.65%), and the highest K content is in agreement with the higher content Al_2O_3 (37.00%) in this clay. The FTIR spectra (Fig. 1-b) confirm the assignments made by DRX.

Both clays show similar DTA/TG curves (Fig. 1-c): a wide endothermic peak is observed between 550 and 600 °C, characteristic of the K-dehydroxylation and formation of metakaolinite (MK). In this area, the P-dehydroxylation [6] has also been reported. At approximately 990 °C, an acute exothermic peak corresponding to the formation of the cristaline premullite spinel phase appears. It was reported as typical peak in K-clay [7], as well as in P-clay [6]. The formation of crystalline phase decreases the amount of reactive amorphous phase, necessary to pozzolanic reaction.

3.2 Calcined Clays

Based on DTA/TG results, four calcination temperatures were evaluated: 550, 650, 750 and 1050 °C. Figure 2 show the FTIR characteristic OH-stretching region of calcined clays. As observed by DTA/TG at temperatures between 550 and 600 °C, the K-dehydroxylation was produced to obtained amorphous phase MK. The four bands of OH-stretching corresponding to K disappear at 550 °C. The sharp and narrow band of OH-stretching corresponding to P at 3675 cm^{-1}, decreases greatly with the increase of calcination temperature. It indicates than large amount OH groups of P are removed and disappear over 1000 °C when the dehydroxylation is completed. The mean intensity band centred at 3630 cm^{-1} and the shoulder at 3469 cm^{-1} assigned to the stretching of the OH groups of I, also decreases with the dehydroxylation process, and the formation of amorphous phase.

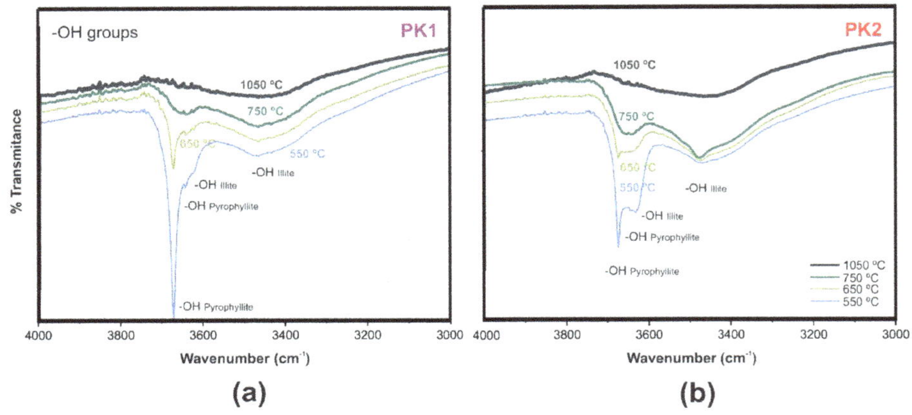

Fig. 2. FT-IR spectra (4000-3000 cm-1) of calcined clays PK1 and PK2.

At temperatures over 550 °C, partial dehydroxylation of the I occurs but this does not affect its crystalline structure and the reduction of the intensity of the basal planes in X-ray diffractograms is seen [5]. It is also observed by XRD that the peak-intensity of I and P decreases with the increase of temperature.

Figure 3a shows the EC as a function of time for the calcined clay–calcium hydroxide solution. During the first 30 min, PK1 calcined at 550, 650 and 750 °C reduce its EC more than calcined PK2. This behaviour indicates the high interaction of this calcined clay with $Ca(OH)_2$. The reactivity of PK1 is similar between 550 and 650 °C, but it decreases at 750 °C. The reactivity of PK2 does not change between 550 to 750 °C. Both

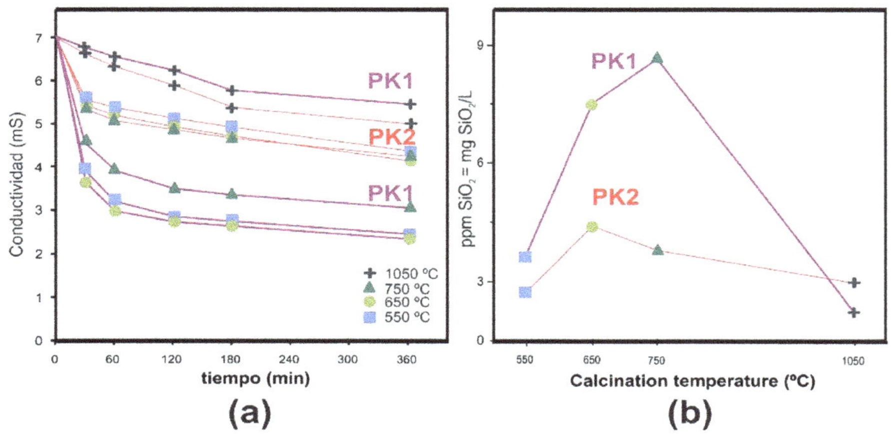

Fig. 3. (a) EC (mS), (b) silica dissolved (mg/l) of calcined clays

calcined clays have low reactivity at 1050 °C, according to the formation of crystalline phase determined by DTA/TG and the consequent decrease of amorphous phase.

The amount of silica dissolved in ppm (mgSiO_2/l solution) in alkaline solution is associated with the amorphization achieved by the heat treatment. Crystalline phases such as quartz and mullite are insoluble in alkaline solution [8]. For PK1 clay, with a higher content of P, where the Si/Al ratio is higher than for K and I, the dissolved SiO_2 content increases significantly with the increase of temperature up to 750 °C (Fig. 3b), where the amount of P decreases. This dissolution decreases if the calcination is done at 1050 °C, at which temperature, as observed by DTA/TG, the premullite [4] has crystallized. The clay PK2 has the same behavior but whit more moderate dissolution.

Silica and alumina reactive contribute to the pozzolanic reactivity in the calcined clays. The higher reactivity of the calcined PK1, with least percentage of K can be attributed to the P and I contribution. PK2 with high illite content shows less pozzolanic activity. At 1050 °C, part of the amorphous phase obtained during the dehydroxylation crystallizes and reduces the amount of amorphous reactive phase.

3.3 Blended Cement

Considering that complete dehydroxylation of the major phases and the biggest silica soluble content (Fig. 3b) occurs between 550 and 650 °C, the selected temperature for calcination was 600 °C to produce the SCMs. The physical characteristics of the calcined and grinding clays were: PK1-600 (Blaine 879 m^2/kg, RoS 45 μm 11.2%, density 2.54), PK2-600 (Blaine 786 m^2/kg, # 45 μm 6.4%, density 2.49). The properties of blended cement mortars with PK1-600 and PK2-600 are reported in Table 1. The mortar flow of blended cement is significantly lower (42 and 63%) than that of the control mortar (122%). This behaviour may be due to the fact that the addition particles that replace the cement particles have large specific surface that requires more water to wetting.

Table 1. Flow, compressive strength (CS) and strength activity index (SAI).

	Flow (%)	7 days		28 days		90 days	
		CS(MPa)	SAI	CS(MPa)	SAI	CS(MPa)	SAI
PC	122	34.5	1.00	46.6	1.00	53.8	1.00
PK1-BC	42	31.4	0.90	48.5	1.04	50.8	0.94
PK2-BC	63	29.4	0.85	48.0	1.03	50.0	0.93

For PK1-600 and PK2-600, the CS at 7 days is lower than the corresponding to plain cement but it overcomes the dilution effect due to stimulation of cement hydration and the pozzolanic reaction. The high reactivity of PK1 produces a quick contribution to CS, but the CS at later ages (28 and 90 days) attains similar values for both calcined clays attributed to the develop of pozzolanic reaction. The SAI attain a value near to 1 for both blended cements at 28 days.

4 Conclusions

The thermal treatment on two complex clays (kaolinite-pyrophillite-illite) from the Tandilia system is characterized by: *dehydration* at low temperature, the *dehydroxylation* of clayed phases from 500 to 600 °C, and the *crystallization* of premullite at 990 °C. The collapse of structure of pyrophillite occurs when the calcination temperature exceeds 750 °C as occurred with the illite phase despite the complete dehydroxylation.

In alkaline solution, complex calcined clays showed increasing percentages of soluble silica when calcination temperature increases from 550 to 750 °C and it was more marked in the PK1. At 1050 °C, the dissolved silica decreases significantly due to the formation of premullite.

For these complex clays, the recommend calcination temperature was 600 °C, considering the high pozzolanic reactivity of these complex calcined clays occurs between 550 and 650°C.

For these clays, mortar flow decreases significantly when 25% by weight of replacement was used and the compressive strength shows a slight reduction at 7 days and a great contribution at later ages reaching to SAI close to 1 at 28 days.

References

1. Ambroise, J., Murat, M., Pera, J.: Hydration reaction and hardening of calcined clays and related minerals. V: extension of the research and general conclusions. Cem. Concr. Res. **15**, 261–268 (1985). doi:10.1016/0008-8846(85)90037-7
2. Tironi, A., Trezza, M.A., Scian, A.N., Irassar, E.F.: Kaolinitic calcined clays: factors affecting its performance as pozzolans. Constr. Build. Mater. **28**, 276–281 (2012). doi:10.1016/j.conbuildmat.2011.08.064
3. Tironi, A., Trezza, M.A., Scian, A.N., Irassar, E.F.: Assessment of pozzolanic activity of different calcined clays. Cem. Concr. Comp. **37**, 319–327 (2013). doi:10.1016/j.cemconcomp.2013.01.002
4. Li, G., Zeng, J., Luo, J., Liu, M., Jing, T., Qiu, G.: Thermal transformation of pyrophyllite and alkali dissolution behavior of silicon. Appl. Clay Sci. **99**, 282–288 (2014). doi:10.1016/j.clay.2014.07.011
5. Fernández, R., Martirena, F., Scrivener, K.L.: The origin of the pozzolanic activity of calcined clay minerals: a comparison between kaolinite, illite and montmorillonite. Cem. Concr. Res. **41**, 113–122 (2011). doi:10.1016/j.cemconres.2010.09.013
6. Mukhopadhyay, T.K., Ghatak, S., Maiti, H.S.: Pyrophyllite as raw material for ceramic applications in the perspective of its pyro-chemical properties. Ceram. Int. **36**, 909–916 (2010). doi:10.1016/j.ceramint.2009.10.026
7. Sanchez Soto, P.J., Justo Erbez, A., de Haro, M.C.J., Pérez Rodriguez, J.L., Raigón Pichardo, M., Pascual Cosp, J.: Caracterización y propiedades cerámicas de una pizarra alumínica que contiene pirofilita. Bol. Soc. Esp. Cerám. Vidrio **33**, 199–205 (1994)
8. Jiang, T., Li, G., Qiu, G., Fan, X., Huang, Z.: Themal activation and alkali dissolution of silicon from illite. Appl. Clay Sci. **40**, 81–89 (2008). doi:10.1016/j.clay.2007.008.002

Assessment of Calcined Clays According to the Main Criterions of Concrete Durability

A. Trümer(✉) and H.-M Ludwig

Bauhaus-University Weimar, Weimar, Germany

Abstract. One main obstacle impeding the implementation of calcined clays into the building material market is the insufficient experience with the long-term behaviour of the concerning concretes. To reduce this insecurity, a comprehensive study was performed which covered most of the relevant durability aspects of concrete made with calcined clay blended cements. For this purpose, different clays based on kaolinite, montmorillonite and illite were regarded which were fired at optimum temperatures according to the strength contribution in cement. After substituting 30% of Portland cement with the respective calcined clay, a number of investigations has been executed to show the performance under different environmental conditions.

While the resistance against chemical working mechanisms like sulphate attack, alkali-silica reaction and chloride induced steel corrosion is improved, exposure to freeze-thaw changes and carbonation seem to limit the application of calcined clays in concrete. From the latter facts, it results the demand for a better understanding of the damaging processes and the material properties suppressing them. Therefore, new criterions appear to which the optimization of calcined clay manufacture has to be adjusted.

1 Introduction

Calcined clays are suitable materials to substitute the Portland cement part in concrete using different synergy effects and giving economical and ecological benefits [1]. However, the experiences made in the last few years showed that the quality of calcined clays varies in a wide range primarily due to the factors clay composition and calcination procedure. This issue prohibits the exact prediction of concrete properties only on the base of former collected data. Moreover, applying composite materials far different from cement clinker always gives a shift in composition of hydration products and quality of microstructure. Both aspects lead to uncertainties concerning the concrete properties, which have to be tested for every single calcined clay.

A lack of knowledge exists, first of all, in the question of influence of calcined clays on the long-term behaviour of concrete. It can be expected that, due to dilution and to the concerning pozzolanic activity, durability properties of concrete will be changed. The smaller amount of information about this issue compared to the high quantity of studies addressing the activation of clays is due to the high effort of performance tests on one hand and, on the other hand, different criterions (determined with different testing methods) for durable concrete in each country.

F. Martirena et al. (eds.), *Calcined Clays for Sustainable Concrete*, RILEM Bookseries 16,
https://doi.org/10.1007/978-94-024-1207-9_76

The presented study describes a first step to evaluate calcined clays with respect to durability demands in Germany. Whereas in former investigations, special issues of sulphate and ASR (alkali-silica reaction) resistance have been addressed, the focus in this work is laid on general requirements demanded from ordinary concretes. These are the material behaviours under exposure to freeze-thaw changes, to chloride ingress and to atmospheric carbonation.

2 Materials and Methods

Three different clays were selected, which can be described as impure, each representing a main clay type common in Germany: kaolin (consisting of 80% kaolinite), illitic clay (45% illite) and bentonite (62% montmorillonite). The clays have been fired in a rotary kiln at optimum temperature with respect to the pozzolanic activity in cement. Here, as in all other tests, the binder was a mixture of 70% ordinary Portland cement (OPC) and 30% of the concerning calcined clay. The following relative strengths at 28 days were reached using the given calcining temperatures:

- calcined kaolin (cK): 700 °C 105%
- calcined illitic clay (cI): 900 °C 91%
- calcined bentonite (cM): 900 °C 106%

Concrete tests were performed on mixtures with 320 kg/m^3 (blended) cement, w/b ratio of 0.50 and natural aggregate of 16 mm maximum size. PCE super plasticizer was used to adjust workable consistency. Slump and compressive strengths were tested according to EN 12350-5 and EN 12390-5, respectively. A ground granulated blast furnace slag (GBFS) was used as reference. To assess durability, the following tests were performed.

The evaluation of the freeze-thaw resistance was done by applying the CIF method (Capillary Suction, Internal Damage and Freeze Thaw Test) described by Auberg [2]. Here, concrete half cubes were casted and stored 1 day in the mould, 6 days under water (20 °C) and 21 days dry (20 °C, 65% rH). One surface of the half cubes is then immersed in water, where it stays unaffected for seven days to allow a presaturation via capillary suction. Afterwards the weathering starts, applying 56 freeze-thaw cycles (FTC) of 12 h each with temperature changes between 20 and −20 °C. During that time, water uptake, mass loss from surface and ultrasound velocity have been measured frequently.

The resistance against chloride ingress was estimated following the instructions of the Federal Waterways Engineering and Research Institute (BAW) [3]. That means drill cores were taken from 56d cured concrete cubes, whose opposite surfaces were immersed in KOH and NaCl solution respectively. After defined application of electrical tension, penetration depth was measured to calculate the chloride migration coefficients.

Carbonation propensity was determined on standard mortars after a 91-day water curing. The prisms produced according to EN 196-1 were exposed to 2.0 vol.-% CO_2

atmosphere to accelerate reaction. Carbonation depths were measured on freshly prepared surfaces applying phenolphthalein indicator. The 56d values were used to calculate carbonation progress under normal atmosphere (0.04 vol.-% CO_2) applying the $\sqrt{t}$-law.

3 Results and Discussion

Former results of the regarded types of calcined clay, which have been presented at LC^3 conference 2015 [4], showed improved durability behaviour of the blended cements compared to pure OPC with respect to sulphate resistance and the risk of damages related to alkali-silica reaction (ASR). To further complete this first overview received by mortar tests, an additional study on concrete was performed.

Figures 1 and 2 show the results of the slump test and the compressive strengths of the produced concretes. Using the same amount of super plasticizer, the slump of the OPC sample is significantly higher compared to concretes with calcined clay. The bad workability of the illite and montmorillonite sample surprises since this behaviour was not observed in mortar. Obviously, the missing affinity of calcined clay particles to the used PCE super plasticizer as well as the changed relation between paste and aggregates have negative effects on the fresh concrete. The latter point seems likewise responsible for a later strength contribution. While, with mortar samples, strength values far beyond dilution have been measured after seven days, the concerning concretes are significantly slower. The relative strengths received before, could not be reached until an age of 28 days. Nevertheless, the order of pozzolanic activity is the same like determined in all mortar tests indicating the metakaolin to be the best material followed by the calcined products of montmorillonitic and illitic clay. However, compared to the latent-hydraulic slag, even the metakaolin reacts slower contradicting again the findings of the mortar tests. Only speculations can be made to explain this lack of correlation between the pozzolanic activity in mortar and concrete. Possibly, the quality of the interfacial transition zone differs in the two systems due to amount and particle size of the respective aggregates. Further investigations are necessary to clarify this phenomenon.

The freeze-thaw resistance of half concrete cubes under CIF testing conditions was assessed in two different ways: first, ultrasound velocity was measured to calculate the dynamic E-modulus giving information about the internal damage, and second, surface damages were measured by weighing the material loss. The results of these methods are presented in Figs. 3 and 4.

Both valuation methods show a clear difference between the performances of the metakaolin sample on the one hand and the concretes with calcined illite and montmorillonite on the other hand. The first fulfil the criterions suggested in [2] for frost-proof concrete safely whereas the two latter fail due to internal defects. One possible reason is a different pore structure resulting in different water uptake during the freeze-thaw weathering (not shown). Regarding only the decrease of dynamic E-modulus, the OPC doesn't perform much better although adsorbing less water. However, the surface resistance of this reference sample is the best under all tested concretes. In comparison to that, the concretes with calcined illitic and montmorillonitic

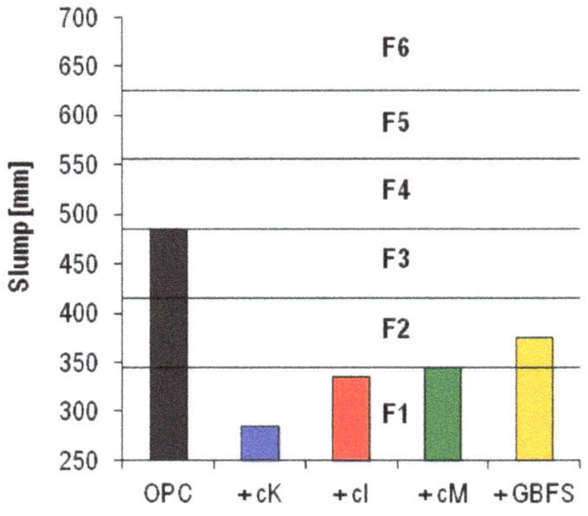

Fig. 1. Slump of liquefied fresh concretes produced with 0.25% PCE super plasticizer

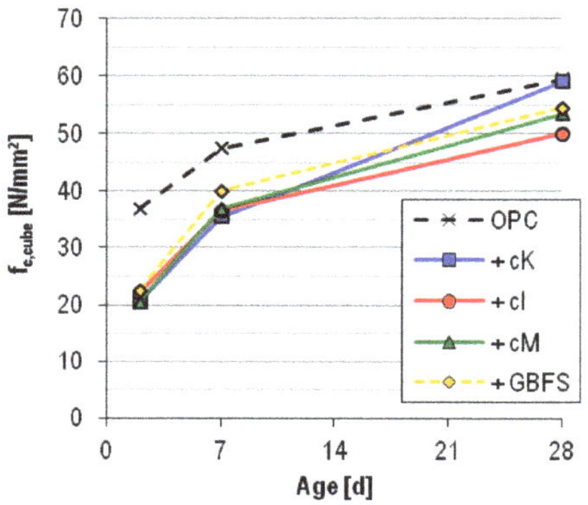

Fig. 2. Compressive strength development of hardened concretes

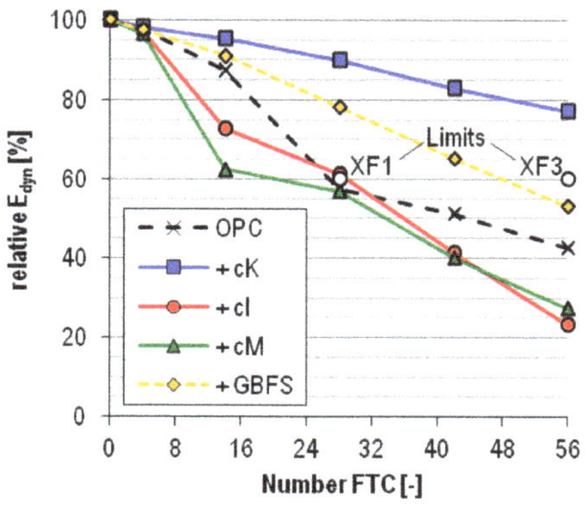

Fig. 3. Relative dynamic E modulus of CIF tested concretes vs. number of freeze-thaw cycles

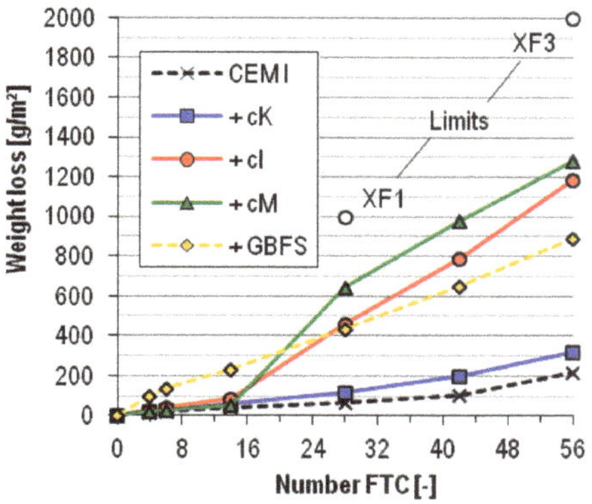

Fig. 4. Compressive strength development of hardened concretes

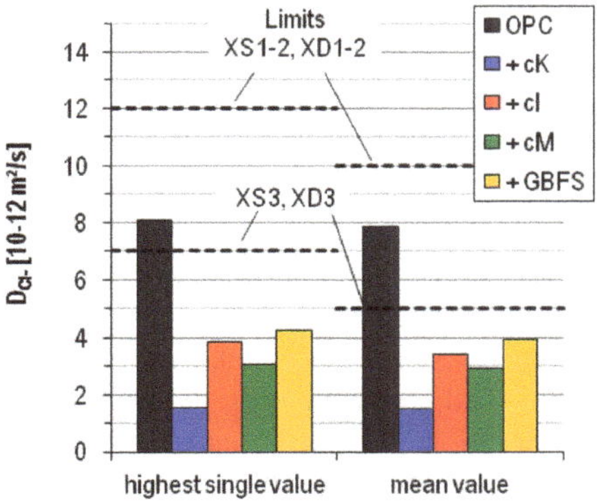

Fig. 5. Chloride migration coefficients of tested concretes

clay show a marked impairment although still satisfying the limits for both XF1 and XF3. In case of the dynamic E-modulus, the material behaviour of these samples is more similar to that of OPC, which, nevertheless, misses the limit values itself. Consequently, it can be stated that calcined clays with sufficient amount of metakaolinite seem to be uncritical in terms of freeze-thaw resistance whereas other materials fail the CIF test. However, due to slow reaction of these calcined clays, their performance is perhaps underestimated under these testing conditions since microstructural densification happens for a big part after 28 days. An extended curing time before starting the weathering could therefore improve the material behaviour.

Chloride penetration was another issue addressed by this study. As shown in Fig. 5, this is a damaging process which is suppressed effectively by using calcined clays as cement substitute. Like before in performance tests concerning sulphate attack and ASR, the improvement can be assigned firstly to inhibited transport mechanisms due to lower capillary porosity. Another positive aspect is the binding of chloride ions by low calcium C-(A)-S-H phases formed via the pozzolanic reaction of the calcined clays.

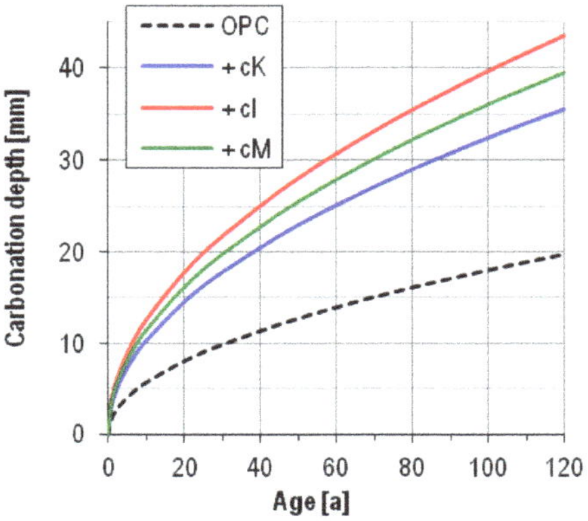

Fig. 6. Carbonation velocity in air extrapolated from accelerated mortar tests using 2.0 vol.-% CO_2

Through these effects, the steel reinforcement is preserved against chloride induced corrosion even under severe conditions (XS3, XD3).

However, steel corrosion can be caused as well by carbonation of cement stone decreasing the pH of pore solution under a critical level of 9.0. Under this condition, the steel is not passivated any longer by a protective layer. Concrete spallings due to high volume of corrosion products are the consequence. Hence, attention has to be paid to the preservation of the reinforcement by controlling the progress of carbonation. Concerning this, it has to be considered that calcined clays, like all other pozzolanic materials, consume Ca(OH)2 and, consequently, reduce the amount of buffer material stabilizing the pH at 12.5. Figure 6 illustrates that the effect on carbonation velocity is similar for all studied calcined clays being twice as fast as the progress in neat OPC mortar. Applying the $\sqrt{t}$-law, this results in a carbonation depth of about 40 mm after 100 years, which is the normally projected life time of concrete constructions. The thickness of concrete layer covering the reinforcement has to be adjusted accordingly.

4 Summary and Conclusions

The study show clearly, that concrete durability depends not only on the quality of the calcined clay but also on the type of exposure. Generally, the before determined order of pozzolanic activity of the three regarded materials could be confirmed (cK > cM > cI). However, the extent of activity is significantly lower in concrete than in mortar systems, which has to initiate further investigations particularly concerning the quality of the interfacial transition zone between cement matrix and aggregates.

Due to this loss of effectiveness, some calcined clays rated as highly active in mortar tests, like the fired bentonite, do not pass every durability test in concrete. Primarily, the freeze-thaw resistance has to be considered critically. Here, only the metakaolin showed satisfying results within in the chosen testing conditions, presumably due to its faster reaction resulting in a denser microstructure of the hardened concrete.

The other tests revealed the expected results known from other pozzolanic materials. In case of chloride ingress, this means significant improvement of resistance of all tested samples. Consequently, chloride induced steel corrosion can be excluded nearly completely. For reinforced concrete, carbonation represents the more dangerous mechanism. Due to consumption of pH stabilizing $Ca(OH)_2$ via pozzolanic reaction, the carbonation front progresses faster into the concrete. As a result, corrosion of the reinforcement has to be prevented by protective measures like increasing the concrete cover.

The results indicate that calcined clays cannot be applied in concrete automatically when pozzolanic activity was proven. Particularly, the freeze-thaw resistance varies dramatically between the different qualities of calcined clays. Consequently, respective tests have to be performed to confirm the suitability of a mixture before applying it in praxis. For research, this aspect opens a wide field of necessary investigations to better understand the mechanisms taking place under freeze-thaw conditions.

References

1. Scrivener, K., Favier, A. (eds.): Calcined Clays for Sustainable Concrete - Proceedings of the 1st International Conference on Calcined Clays for Sustainable Concrete. Springer, Dordrecht (2015)
2. Auberg, R.: Zuverlässige Prüfung des Frost- und Frost-Tausalz-Widerstandes von Beton mit dem CDF- und CIF-Test, Dissertation, Universität GH Essen (1998)
3. Bundesanstalt für Wasserbau, BAW-Merkblatt: Chlorideindringwiderstand von Beton (MCL), Karlsruhe (2012)
4. Trümer, A., Ludwig, H.-M.: Sulphate and ASR resistance of concrete made with calcined clay blended cements. In: Scrivener, K., Favier, A. (eds.) Calcined Clays for Sustainable Concrete - Proceedings of the 1st International Conference on Calcined Clays for Sustainable Concrete, pp. 3–9. Springer, Dordrecht (2015)

Application of Industrially Produced LC^3 to Pavements, AAC Blocks and Other Products

S.K. Wali[1(✉)], S.K. Saxena[1], Mukesh Kumar[1], Soumen Maity[2], and Shashank Bishnoi[3]

[1] JK Lakshmi Cement Ltd, Jhajjar, Haryana, India
[2] Technology and Action for Rural Advancement, New Delhi, India
[3] Indian Institute of Technology Delhi, New Delhi, India

Abstract. LC^3 and LC^2 produced in the plants of JK Lakshmi Cement were used in the construction of several pavements, AAC blocks and other practical applications like plaster and mortars. It was found that in most applications, OPC could be easily replaced by the same weight of LC^3 without a negative impact on performance. Plain, reinforced and fibre-reinforced pavements were cast using the cement and the construction could be carried out using usual construction procedures.

LC^3 and LC^2 could also be used in the production of autoclaved aerated concrete (AAC) blocks, without a significant change to the technology or performance. It was found that a better cohesion and flow could be obtained by the use of LC^3 in place of OPC in mortars and plasters.

For most applications, it was found that a direct replacement of OPC by LC^3 was possible without negatively influencing the performance of the product.

1 Introduction

Concrete is the most widely used building material in the construction of infrastructure such as buildings, bridges, highways, dams, and many other facilities. Concrete usage around the world is second only to water. Ordinary Portland cement (OPC) is the main constituent of the concrete and its production contributes about 5 to 7% of total global greenhouse gas emission [1], which possesses a potential problem for global warming. It is estimated that by the year 2020, emissions will rise above the current levels by approximately 50%.

In order to reduce emission from the cement industry, attempts have been made to prepare blended cement using pozzolanic materials. The major drawback of the blended cement is low early strength. Recently we found that if an appropriate amount of calcined clay is mixed with the clinker at the time of grinding, properties to the concrete are enhanced and emission of greenhouse gases is reduced. Combination of calcined clay, limestone and clinker, also known as Limestone Calcined Clay Cement (LC^3) were used. In this paper properties and applications of LC^3 has been discussed.

F. Martirena et al. (eds.), *Calcined Clays for Sustainable Concrete*, RILEM Bookseries 16,
https://doi.org/10.1007/978-94-024-1207-9_77

2 Industrial Production of LC^3 in India

Indian Government is committed to reduce industrial emission, particularly emission from the cement industries. India is also a partner for signing treaty for climate change. In an attempt JK Lakshmi Cement Ltd., New Delhi, in collaboration with TARA, New Delhi and IIT Delhi developed the process for the industrial manufacture of LC^3 for the first time in India. At the same time, a new environment friendly additive for concrete that makes it easier to produce high performance concrete at a lower cost was produced. Full-scale plant trial production of the cement was conducted at JK Lakshmi Cement Ltd., Jhajjar Unit, Haryana, India on 6th & 7th Oct 2016 under real conditions. Trial applications of the cement have shown that it can reduce as much as 30% CO_2 emissions and 20% energy consumption in cement production. The performance of the concrete produced using this cement exceeds those using cements commercially available today in most aspects.

3 Materials and Experiment

China raw clay (Fig. 1) was purchased from Bhuj, Gujarat mines and the chemical composition is given in Table 1. The calcination of 150 tons of Calcined Clay was done at Shree Ram Minerals in the rotary kiln up to 800° successfully (Fig. 2). Calcined clay is shown in Fig. 3. After the grinding process at JK LAKSHMI CEMENT LTD, a production of 200 MT of LC^3 was obtained. Clinker used was from JK LAKSHMI CEMENT LTD, Sirohi, Rajasthan Plant, mineral gypsum with 80%

Table 1. Chemical composition of Calcined clay and LC^3 (mass%).

Compounds	SiO_2	Al_2O_3	Fe_2O_3	CaO	MgO	K_2O	SO_3	Na_2O
Calcined Clay	61.95	22.97	3.21	11.73	0.00	0.01	0.02	0.11
LC^3	33.28	14.48	4.30	39.65	2.73	0.49	2.13	0.40

Fig. 1. Raw clay mines

Fig. 2. Calcination in rotary kil

sulphate and low grade limestone from the mines of JK LAKSHMI CEMENT LTD, Shirohi, Rajasthan Plant [1] were used. The Raw material proportions are given in Fig. 4. Physical test results of LC3 are given in Table 2.

Fig. 3. Calcined clay

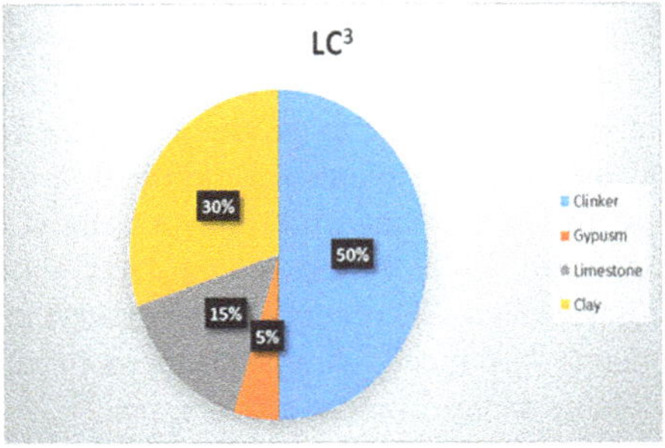

Fig. 4. LC^3 mix proportion

Table 2. Physical test of LC^3

Sl. No	Characteristic	LC^3 Industrial production (JK Lakshmi Plant, Jhajjar)
(i)	Specific gravity	2.88
(ii)	(a) Blaine	637
	(b) Retained on 90μ sieve	6.3%
	(c) Retained on 45μ sieve	22.4
(iii)	Standard Consistency	31.75
(iv)	Initial, min(Setting Time)	115
	Final, min (Setting Time)	155
(v)	1 day, Mpa, Compressive Strength	11.6
	3 days, MPa, Compressive strength	25.40
	7 days, Mpa, Compressive Strength	30.0
	28 Days, MPa, Compressive Strength	39.8
(viii)	Soundness by Le Chatelier Method, MM	0.45 MM
(ix)	Dry shrinkage, Percent of Gauge length	0.01
(x)	Expansion in autoclave	0.02%

4 Industrial Productions of Several Types of Concrete Using LC3 Produced

4.1 Laying of Road Panel at JKLC- Jhajjar with LC3 Cement

To evaluate the performance of LC^3 in concrete, constructing road panels of 64 square meters were cast at JK LAKSHMI CEMENT LTD. plant the data are given in Table 3.

The variation of compressive strength with time in the presence of LC^3 is shown in Fig. 5.

Table 3. Raw materials and Mix design

Used raw material's specifications:

SI. No	Material	Source	Specific gravity
1	LC3	JK Lakshmi LC3 Cement	2.87
2	20 mm aggregate	Kothputli, Rajasthan	2.68
2	10 mm aggregate	Kothputli, Rajasthan	2.66
3	Manufactured sand	Kothputli, Rajasthan	2.56
4	Admixture	PC based (Make: Kunal)	1.08

Mix design for road construction (Per cubic metre):

Grade of concrete	Cement (Kg)	Sand (Kg)	Coarse aggregate 20 mm (Kg)	Coarse aggregate 10 mm (Kg)	Water (Kg)	Chemical admixture (1.1% of Cement)	W/C ratio
M50	547	670	612	408	175	6.56	0.38

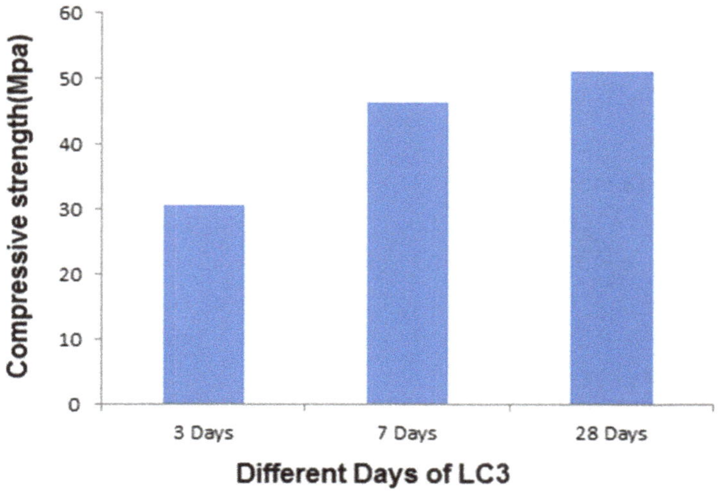

Fig. 5. Compressive strength of LC^3

Fig. 6. Self-compacting concrete with LC^3

Fig. 7. Testing of LC^3 concrete

It was observed that as the LC^3 content increases the slump flow increases. Other features of concrete made from LC^3 are shown in Figs. 6 and 7.

5 AAC BlockProduction at JKLC- Jhajjar with Calcined Clay

Autoclaves aerated concrete (AAC) block production was conducted at JK LAKSHMI CEMENTLTD, Jhajjar Plant with calcined clay with the replacement of OPC grade cement. The composition of the AAC blocks is shown in Fig. 8. The compressive strength is shown in Fig. 9. A 6% increase in the strength of the blocks was observed with the use of calcined clay.

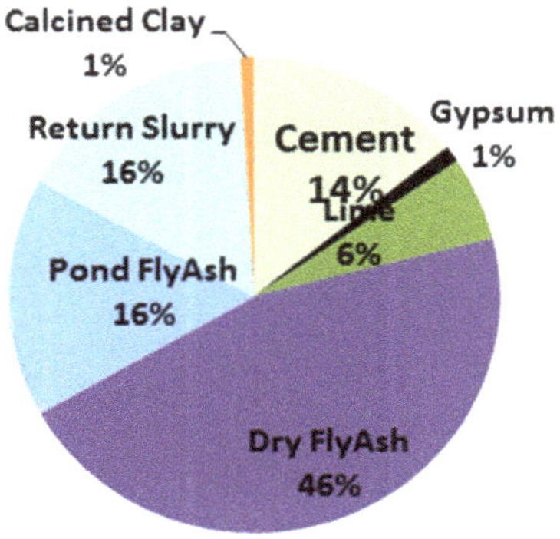

Fig. 8. Composition of AAC blocks

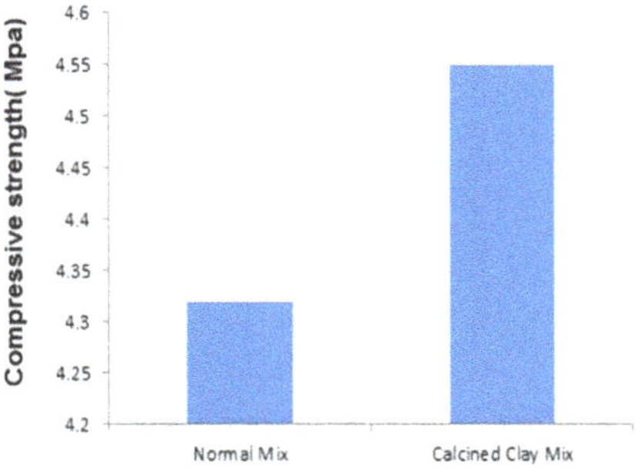

Fig. 9. Compressive strength of AAC blocks

6 Conclusions

This study was carried out to evaluate the fresh and hardened properties of LC^3 containing fiber s and AAC blocks with calcined clay.
The following conclusions can be made from the obtained results:

- It is possible to develop high strength self-compacting LC^3 concretes of strengths ranging from 50–60 MPa,
- The increase in LC^3 content in Concrete Mixes increases the viscosity of the mix.
- Compressive strength increases with adding the 1% of Calcined clay and replacing the OPC cement.

Reference

1. Bishnoi, S., Maity, S., Malik, A., Joseph, S., Krishnan, S.: Pilot scale manufacture of limestone calcined clay cement: the Indian experience. Indian Concr. J. **88**(6), 22–28 (2014)

Machine Learning Approaches to Admixture Design for Clay-Based Cements

N.R. Washburn[1,2(✉)], A. Menon[2], C.M. Childs[1], B. Poczos[3], and K.E. Kurtis[4]

[1] Department of Chemistry, Carnegie Mellon University, Pittsburgh, PA 15217, USA
[2] Department of Materials Science and Engineering, Carnegie Mellon University, Pittsburgh, PA 15217, USA
[3] Department of Machine Learning, Carnegie Mellon University, Pittsburgh, PA 15217, USA
[4] School of Civil and Environmental Engineering, Georgia Institute of Technology, Atlanta, GA 30332, USA

Abstract. Replacement of 30% of ordinary Portland cement (OPC) by metakaolin (MK) reduces the CO2 intensity but negatively impacts the workability. A critical challenge facing adoption of this next-generation infrastructure material is developing admixture systems that impart workability similar to unblended OPC while retaining the advantages in strength and environmental stability conferred by MK. Hierarchical machine learning is a highly-supervised methodology that integrates physical and statistical modelling to understand and optimize complex systems. Here it is applied to designing admixture formulations for OPC-MK blends, providing exceedingly rapid admixture development as well as formulations tailored to specific materials. Elucidating how MK impacts workability of these systems was addressed by screening the effects of superplasticizers, viscosity-modifying admixtures, and water-reducing admixtures on pore solution properties, OPC rheology and the colloidal properties of MK suspensions. Changes in slump spread of 70% OPC/30% MK blends as a function of admixture formulation were fit using regression methods. Increases in slump spread were found to be a strong function of pore solution viscosity, effects of superplasticizer on MK zeta potential and electrosteric interactions, and coupling between pore solution viscosity and osmolality with MK zeta potential and electrosteric interactions, respectively. Work toward designing new admixtures that optimize these interactions will also be pursued.

1 Background

Incorporation of minimally processed metakaolin into cement is an important strategy for reducing the embedded carbon intensity but it negatively impacts workability [1]. Current admixtures designed for Portland cement are inefficient at reducing yield stress, viscosity, and water requirements of clay-based blends. The goals of this research are to use machine learning both to perform molecular design of new admixtures but also

F. Martirena et al. (eds.), *Calcined Clays for Sustainable Concrete*, RILEM Bookseries 16,
https://doi.org/10.1007/978-94-024-1207-9_78

to optimize mix design in tuning the ratios of different classes of additives. The challenges in applying machine learning in this area are the sparse datasets and the time-intensiveness of performing experiments. These make the primary tasks in machine learning – training and testing models – extremely difficult to accomplish.

Machine learning is focused on establishing relationships between system variables (the bottom tier) and complex system responses (the top tier). A purely statistically analysis of n system variables can require of order 2^n or 3^n data points to determine a functional form of the response surface, and the complexity of admixtures and formulations for cement makes it impossible to use traditional approaches where there can be 10–20 molecular or formulation variables. Hierarchical machine learning is a methodology developed to model complex physical systems and represented in Fig. 1. Physical interactions that underlie the system responses are embedded as an intermediate tier that bridges the large parameter space and the complex response space.

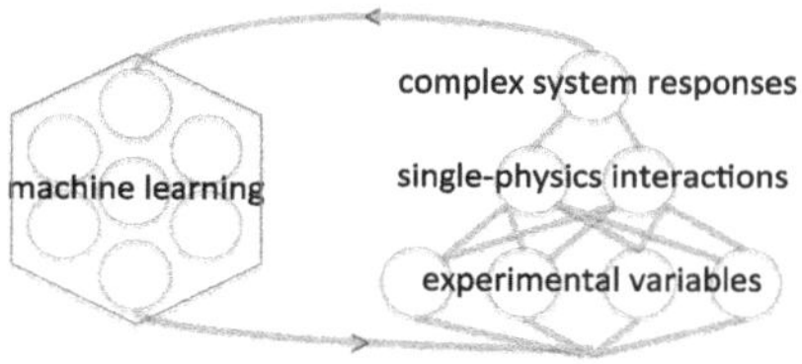

Fig. 1. In hierarchical machine learning, the responses of complex systems (top tier) as a function of changes in experimental variable (bottom tier) are expressed as changes in constituent interactions (middle tier) that are assumed to form the physical basis for the system responses. The top and middle tiers are related through regression analysis of experimental data, and the middle and bottom tiers are related through simple physical models [2]

2 Aim

Here we present the application of hierarchical machine learning to optimizing admixture design for blends of Portland cement and metakaolin. Two aims will be pursued: (1) elucidate molecular design parameters for plasticizers that effectively reduce the yield stress of the pastes, and (2) develop a computational model of formulating the admixture (ADM) classes including superplasticizer (SP), viscosity-modifying agent (VMA), and water-reducing agent (WRA) to optimize the workability of OPC-MK pastes. This approach can be extended to PLCs as well.

3 Method

The adaptation of hierarchical machine learning to cement formulations is shown in Fig. 2. Workability of the paste was system response to be optimize and was primarily considered to be reflected in the slump spread, as measured by mini-slump. Blends of ASTM C150 type I/II Portland cement and metakaolin were considered to be a three-component system, and the single-physics interactions were partitioned into pore solution, Portland cement and metakaolin components. These in turn were considered to be a function of water content and the different admixture classes explored here.

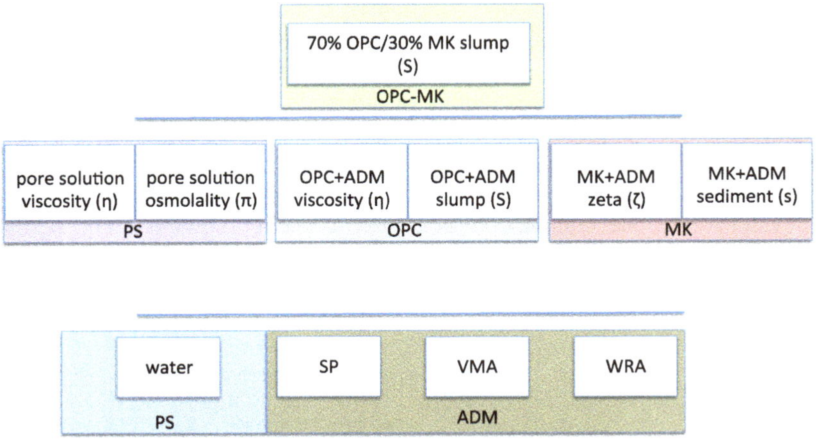

Fig. 2. The HML representation for the 70% OPC/30% MK system for which optimization of the slump spread is the goal. At constant mineral composition, the experimental variables are the water: cementitious ratio and the chemistries and concentrations of the admixtures. Abbreviations used: pore solution (PS), admixture (ADM), superplasticizer (SP), viscosity-modifying admixture (VMA), water-reducing admixture (WRA).

In the initial training set, the effects of the following SP, VMA, and WRA were measured, listed in Table 1.

Table 1. Admixtures tested including superplasticizers (SP), plasticizers (P), water-reducing admixtures (WRA), and viscosity-modifying admixtures (VMA). Members of the training set were polycarboxylate ether (PCE), β-naphthalene sulfonate polycondensates (BNS), lignosulfonate (LS) and polyethylene glycol-grafted (PEGylated) lignosulfonate (LSPEG), kraft lignin (KL) and polyethylene glycol-grafted (PEGylated) kraft lignin (KLPEG), desugarized molasses, and diutan gum.

Admixture	Class
Standard PCE	SP
High-performance PCE	SP
BNS	WRA
LS	WRA
PEGylated LS (LSPEG)	WRA
KL	P
PEGylated KL (KLPEG)	P
Desugarized molasses	WRA
Diutan gum	VMA

The workability of Portland cement/metakaolin blends was assumed to be a function of Portland cement workability, the colloidal properties of metakaolin, and the physicochemical characteristics of the pore solution. The following variables were incorporated in the model of the middle tier, which represents the physical interactions that underlie the rheological behaviour of OPC-MK paste (Table 2):

Table 2. Physical interactions represented in the middle tier. The coefficients will be determined in the regression analysis

Parameter	Symbol	Coefficient
Pore solution viscosity	η_ps	b
Pore solution osmolality	π_ps	c
MK+ADM zeta	ζ_mk	d
MK+ADM sedimentation	s_mk	f
OPC+ADM slump	S_opc	e

4 Results

Slump spread values for 70% portland cement/30% metakaolin blends including some of the plasticizers at various water-to-cementitious materials (w/cm) ratios in the training set are shown in Fig. 3.

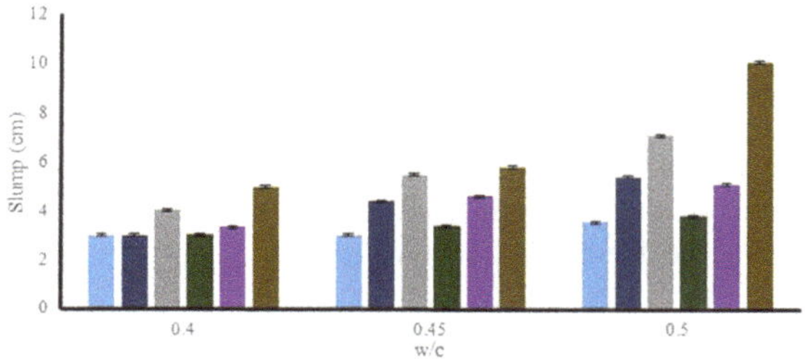

Fig. 3. Slump spread data for OPC-MK blend having different w:cm levels and different admixtures (at 4 mg/mL). Colors represent neat OPC (light blue), LS (dark blue), LSPEG (gray), KL (green), KLPEG (purple), high-performance PCE (brown).

The algorithm seeks to predict these values based on the physical interactions represented in the middle tier of the model. This is accomplished by using standard tools of machine learning, including regularized regression and cross-validation [3].

The first step in the model is developing a model to predict slump spread of the OPC-MK blend as a function of the constituent interactions represented by the middle tier in Fig. 1. The basis set of constituent interactions was the linear terms and quadratic terms based on these. To identify the dominant terms, the least absolute shrinkage and selection operator (LASSO) with leave-one-out cross-validation (LOOCV) were used to find a variable set that would provide optimal fitting of the training set while still permitting fitting to test sets. LASSO is a powerful form of regularized regression using an L1 norm to penalize the inclusion of nonessential variables through the parameter λ:

$$\min_w \left(\|Xw - y\|_2^2 + \lambda \|w\|_1 \right) \tag{1}$$

where X is a given input matrix, y denotes output vector, and w is the vector of model parameters to be optimized. The LASSO trajectory for this dataset is shown in Fig. 4a. Each curve represents the value of a coefficient for the physical variables and their combinations as the value of λ was increased. The degrees of freedom represent the total number of variables in a particular trajectory. The mean square error (MSE) was then calculated for each set of coefficients based on LOOCV, which is shown in Fig. 4b.

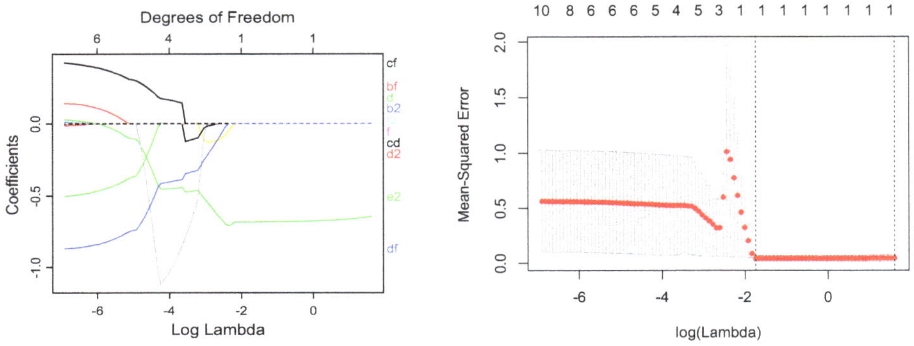

(a) (b)

Fig. 4. (a) LASSO trajectory for the regularized regression predicting the OPC-MK slump as a function of the pore solution, OPC, and MK parameters identified in the middle tier of Fig. 2. (b) MSE plot as a function of the degrees of freedom.

The optimal number of parameters was found to be five, with coefficients:

$$b = -0.12 \quad d = -0.53 \quad be = -0.04 \quad cd = -0.03 \quad df = -0.24$$

The coefficient (d) governing the contribution of the adsorbed admixture to the MK zeta potential was found to be a central factor in determining the slump spread of OPC-MK blends. Since the zeta potential for all polymers was negative, this indicated that a more negative value of admixture zeta potential resulted in an *increase* in the slump spread. The magnitude of the coefficient indicated that this was a leading contribution, although this also depended on the actual values of the zeta potentials. This MK-ADM zeta potential parameter also coupled with the polymer osmotic pressure value (cd) and the estimate of electrosteric interactions obtained via sedimentation (df). The other contributing terms were the viscosity of the pore solution (b), which decreased slump spread as this parameter increased, and the coupling between pore solution viscosity and the measured value of the OPC+ADM slump spread (be). Other parameters, such as osmolality of the pore solution, were found not to influence slump spread of OPC-MK-ADM systems directly.

The predicted form of the expression for OPC-MK-ADM slump spread was:

$$S = -0.12(\eta\_ps) - 0.53(\zeta\_mk) - 0.04(\eta\_ps)(S\_opc) - 0.03(\pi\_ps)(\zeta\_mk) - 0.24(\zeta\_mk)(s\_mk) \quad (2)$$

where the parameters in parentheses represent the values of the pore solution, OPC, or MK characteristics that are tuned by the admixtures.

The next step in the analysis is to parametrize the middle tier of the model by the bottom tier. For the molecular design of admixtures with optimal chemistries, this will involve describing the admixtures by their chemical functional groups and molecular architectures. For formulation optimization, this will involve parametrizing the middle tier by the concentration dependence of the admixture classes. New admixtures or combinations of admixtures will be prepared and their effects tested on the slump spread of OPC-MK blends.

5 Conclusion

Hierarchical machine learning was applied to elucidating the important admixture interactions that determine the slump spread of OPC-MK pastes. Using a small training set based on commercial SP, VMA, and WRA, a model was developed to predict the behavior of these complex materials. The dominant variable was found to be the zeta potential of the MK particles, and development of new admixtures and admixture formulations based on these design principles will be pursued.

References

1. Zaribaf, B.H., Uzal, B., Kurtis, K.: Compatibility of superplasticizers with limestone-metakaolin blended cementitious system. In: Calcined Clays for Sustainable Concrete, pp. 427–434. Springer (2015). doi:10.1007/978-94-017-9939-3_53
2. Menon, A., Gupta, C., Perkins, K.M., DeCost, B.L., Budwal, N., Rios, R.T., Zhang, K., Poczos, B., Washburn, N.R.: Elucidating Multi-Physics Interactions in Suspensions for the Design of Polymeric Dispersants: A Hierarchical Machine Learning Approach (submitted)
3. Tibshirani, R.: Regression shrinkage and selection via the lasso. J. Royal Stat. Soc. B **58**, 267–288 (1996)

Micro-Chemo-Mechanical Characterization of a Limestone-Calcinated-Clay Cement Paste by Statistical Nanoindentation and Quantitative SEM-EDS

William Wilson[1], Luca Sorelli[2](✉), Sreejith Krishnan[3], Shashank Bishnoi[3], and Arezki Tagnit-Hamou[1]

[1] Université de Sherbrooke, Sherbrooke, QC, Canada
[2] CRIB, Université Laval, Québec, QC, Canada
luca.sorelli@gci.ulaval.ca
[3] Department of Civil Engineering, Indian Institute of Technology Delhi, New Delhi, India

Abstract. To address the sustainability concerns associated with Portland cement clinker production, the ternary blend of limestone, calcined clay and cement (which was named LC^3) has recently been demonstrated to be an efficient solution. This work aims to contribute to the development of this promising material by applying latest techniques to characterize the chemo-mechanical properties of the complex heterogeneous microstructure of an LC^3 paste by combining statistical nanoindentation and quantitative SEM-EDS techniques (SNI-QEDS). The results showed that the mechanical properties of LC^3 come from a complex microstructure assemblage of C-A-S-H, Al-rich hydrates and anhydrous grains. Thus, the LC^3 microstructure is composed by large anhydrous grains (clinker and calcined clay) embedded in a cementitious paste made of hydrates incorporating finely graded grains of anhydrous calcined clay, limestone and quartz. Moreover, the partial reaction of the calcined clay, limestone and Portlandite formed C-A-S-H and other Al-rich hydrates (likely including carboaluminates). Notably, the latter exhibited higher mechanical properties than those of C-A-S-H. Finally, the present work provides new knowledge for better understanding the complex LC^3 microstructure towards advanced modelling and mix design optimization.

Keywords: Limestone calcinated clay cement · Microstructure · Nanoindentation · SEM · Quantitative EDS · Hardness · Modulus

1 Background: LC^3 as a Sustainable Alternative to OPC

In the pursuit of sustainable construction materials with low environmental impact for both industrialized and developing countries, the limestone-calcined-clay cements (LC^3) emerge as a promising alternative to conventional Ordinary Portland Cement (OPC). Notably, the synergetic effects between Portland cement, finely ground limestone and calcined clays allow the reduction of the clinker content by up to 50% without impairing the 28-day strengths compared to OPC reference mixtures [1, 2]. As described in

F. Martirena et al. (eds.), *Calcined Clays for Sustainable Concrete*, RILEM Bookseries 16,
https://doi.org/10.1007/978-94-024-1207-9_79

previous studies [1, 3–5], limestone particles not only act as a filler, but they can react with the alumina provided by calcined clays in the presence of calcium hydroxide to form hemi- and mono-carboaluminates [4, 6, 7], which refine the porosity of the cement paste and thus improve its mechanical properties. In the past years, the LC^3 technology has been implemented at the industrial scale in different countries [2, 8, 9] following experimental works covering complementary aspects of the technology, such as the evaluation of the reactivity of calcined kaolinitic clays [10], or the thermodynamic modelling of the hydration of LC^3 mixtures [1]. Nevertheless, the complex LC^3 microstructure is not completely understood and the development of the LC^3 technology would benefit from a better characterization of key microstructure parameters.

2 Aim: Understanding the Mechanical Properties of an LC^3 Microstructure

This study aims to investigate the micromechanical properties of a promising LC^3 cement paste manufactured using locally available Indian OPC, limestone and calcined clay. The use of a "state-of-the-art" technique that couples statistical nanoindentation and quantitative energy-dispersive spectroscopy (SNI-QEDS) provided original insights on the complex assemblage of C-A-S-H, Al-rich hydrates and well-graded anhydrous grains which composes the studied LC^3 system. Notably, the investigated microvolumes including both C-A-S-H and other Al-rich hydrates showed higher mechanical properties than those of the C-A-S-H itself, which suggests that partially hydrated calcined clay and AFm phases (likely including carboaluminates) contribute to increasing the paste's properties.

3 Method: SNI-QEDS on an LC^3 Paste

The experiments were carried on an LC^3 cement paste produced with locally available Indian materials: 55% belite-rich OPC, 30% calcined clay and 15% limestone. The paste was mixed at a water-to-binder ratio of 0.40 and cured for 28 days before stopping the hydration by immersion of ~2 mm thick specimens in isopropanol. The specimens were then levelled with silicon carbide sandpaper and polished using a lapping wheel with 6 μm, 1 μm and 0.25 μm diamond suspensions sprayed on perforated polishing cloths. Ultrasonic cleaning in isopropanol was performed between each step.

As described elsewhere [11], the SNI-QEDS method was employed to statistically investigate the chemo-mechanical properties of the different phases of the hardened LC^3 paste. First, a grid of 30 × 24 nanoindentation points spaced by 10 μm was carried out using a trapezoidal loading with a sufficient holding period to avoid delayed plasticity effects. For each single measurement, the indentation modulus (M) and the indentation hardness (H) were estimated using the Oliver and Pharr method. Then, the location of each probed indentation volume was automatically detected using a Matlab algorithm, the coordinates of the indents were translated into the scanning electron microscope (SEM) and then used as inputs for measurement of the chemical composition at each data point using SEM-EDS quantitative analyses with standards. Furthermore, chemical mappings were performed over the nanoindentation grid. Finally, a statistical deconvolution using the maximum likelihood approach and 5 classification variables (*M, H, Si/Ca, [Al + Fe]/Ca*, and sum of oxides *SOX*) was employed to identify the chemical composition and the micromechanical properties of clusters representing the main phases composing the LC^3 microstructure.

4 Results: Visual Representations of the Clustered Chemo-Mechanical Phases

Figure 1 presents a chemical composite mapping of the investigated Region Of Interest (ROI) on the surface of the LC^3 paste, which consists of the multilayer combination of four chemical mappings (Si in red, Ca in blue, Al in green and Fe in grey). The main phases occurring in the microstructure may be visually identified with colour contrasts: large anhydrous clinker grains (i.e., belite nests) and calcined clay particles are embedded in a complex matrix incorporating both hydrates and finely graded particles of limestone, calcined clay and quartz (occurring as impurity in the raw materials). The grid of nanoindentation datapoints is also displayed on Fig. 1 with markers corresponding to the clusters obtained from the deconvolution of the coupled chemo-mechanical dataset, as described in Fig. 2.

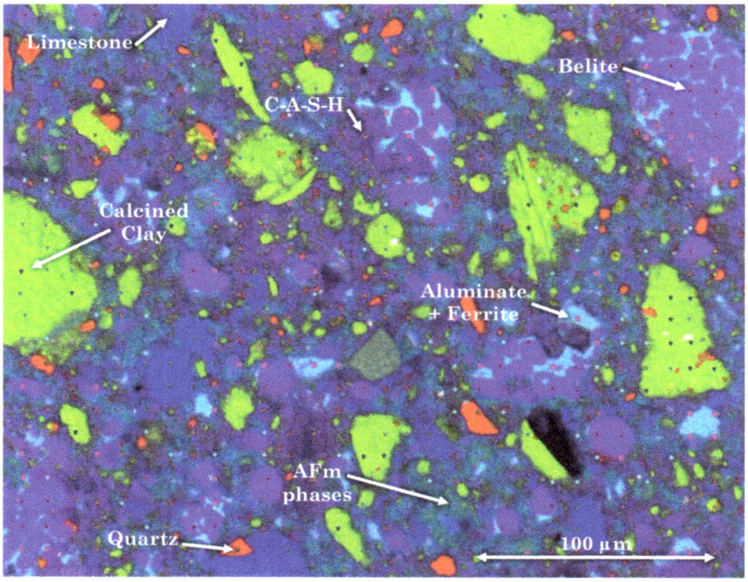

Fig. 1. Chemical composite mapping over the indentation points showing the different anhydrous and hydrous phases along with the clustered data points (see the legend in Fig. 2). The composite mapping consists of a multilayer combination of four elemental mappings (Si in red, Ca in blue, Al in green and Fe in grey).

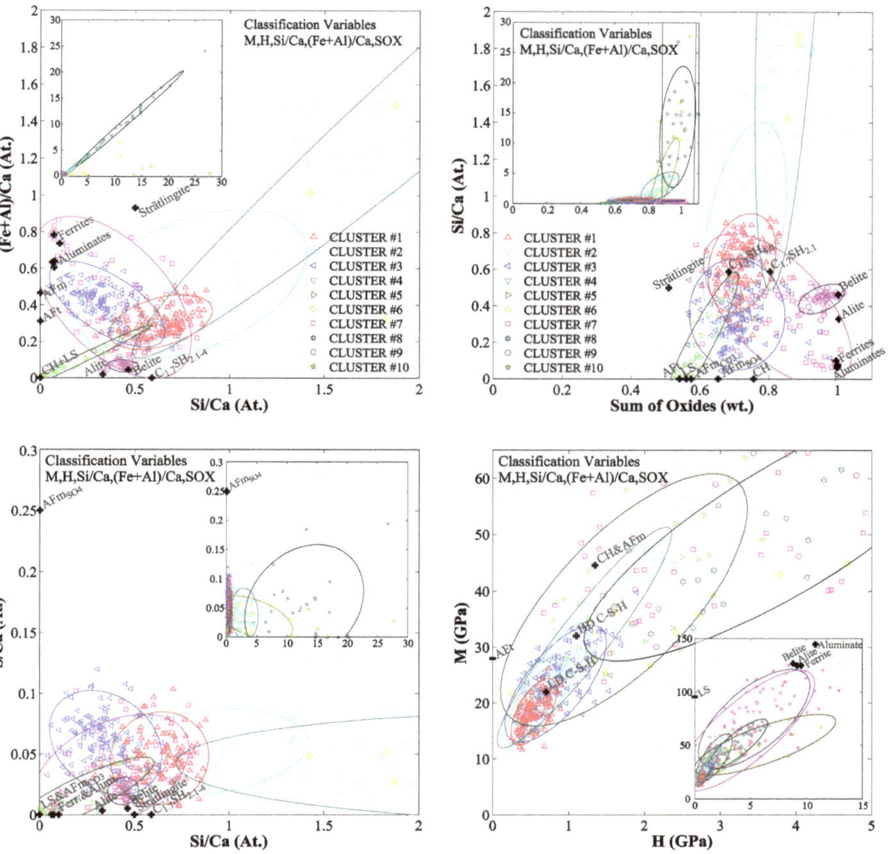

Fig. 2. Representation of the statistical clustering results in the axes: (a) (Fe + Al)/Ca vs. Si/Ca, (b) SOX vs. Si/Ca, (c) S/Ca vs. Si/Ca, and (d) M vs. H. The contour lines represent the 95% confidence interval for each cluster.

As shown in Fig. 2, the SNI-QEDS dataset was described by a combination of 10 clusters, each having preferential (or characteristic) chemo-mechanical properties. Due to the relatively small size of the chemical phases and their possible intermix (at a scale smaller than the probed volume of EDS and nanoindentation, 1–2 μm), those clusters generally include one or two predominant phases sometimes mixed with minor contents of other phases. Figure 2a, b and c present three chemical representations of the results with crosses indicating the theoretical (or typical) composition of the main pure phases [12], which allowed identification of the clusters. The clusters extending towards SOX ≈1 represent anhydrous materials (possibly mixed with hydrates): siliceous clinker phases (○), aluminous clinker phases (□), calcined clay dense inclusions (✡) and quartz grains (☆). Interestingly, the results suggest a variable concentration of calcium in microvolumes associated with calcined clay, although the raw material contains only traces of calcium: this suggests that the clay particles contains both partially hydrated regions (▽) and denser non-reacted clay grains (✲). The range SOX ≈ 0.50–0.80 represents the hydrates, with the central cluster (△) approaching the chemical ratios for C-A-S-H and other clusters representing mixtures of hydrates extending towards the poles of AFm phases (◁), limestone (▷), or calcined clay (, possibly with strätlingite). Notably, the AFm-containing cluster (◁) represents a significant proportion of the hydrates and its relatively low *S/Ca* values compared to the monosulfoaluminate pole (AFm_{SO4}) suggests that the AFm phases include carboaluminates (AFm_{CO3}). Furthermore, ettringite and Portlandite were not observed.

Figure 2d shows the mechanical properties associated with the previously identified clusters for both hydrates and anhydrous grains. The theoretical values are marked with crosses, based on a previous literature review [11]. As expected, the clinker clusters (○ and □) extend towards the highest theoretical values. The dense anhydrous calcined clay cluster (✲) also reach values up to M ≈ 70 GPa and H ≈ 6 GPa, whereas the partially hydrated calcined clay and the limestone containing clusters (▽ and ▷, respectively) exhibits values up to M ≈ 50–60 GPa and H ≈ 2–3 GPa. Interestingly, the mixtures of Al-rich hydrates and C-A-S-H (◁ and ◇) exhibit higher mechanical properties than the cluster associated to C-A-S-H (△), which suggests a reinforcing effect of these phases into the binding matrix of C-A-S-H. Thus, the chemo-mechanical clusters (Fig. 2) and their spatial mapping (Fig. 1) indicates that the calcined clay and the limestone contribute to strength both into products of the AFm family and with residual anhydrous particles, similarly as in the case of high volume natural pozzolan concretes [13]. Further disclosing the microstructure signature of LC^3 will allow advanced modelling and optimization of the microstructure features which govern the material-to-structure optimization.

5 Conclusion: The Macro-Scale Strength of LC^3 Emerges from a Complex Microstructure

Differently from OPC pastes where C-S-H is the main phase responsible for the strength, LC^3 exhibits a complex microstructure assemblage of C-A-S-H, Al-rich hydrates, and anhydrous grains with a wide particle size distribution. Based on a first application of the SNI-QEDS technique, which represents a *state-of-the-art* chemo-mechanical analysis method for a cement-based microstructure, the following preliminary conclusions can be drafted:

1. The chemical microstructure features of LC^3 has been successfully mapped with clusters having preferential chemo-mechanical properties, likely composed by one or more chemical phases;
2. The regions in-between the large dense anhydrous grains (clinker and calcined clay) are filled with hydrates embedding anhydrous or partially hydrated finely graded grains of calcined clay, limestone and quartz;
3. The reaction of the calcined clay provided extra aluminum and silicon which consumed the Portlandite to form C-A-S-H and other Al-rich hydrates (such as carboaluminates);

4. The Al-rich hydrates represent a significant proportion of the hydrates which exhibits mechanical properties higher than those of the C-A-S-H.

This first SNI-QEDS analysis of an LC^3 paste illustrated the complexity of its microstructure and suggested a reinforcing role of the AFm phases in the C-A-S-H matrix. Further investigations will focus on the extent of limestone reaction and its effect on the chemo-mechanical properties of the phases.

References

1. Antoni, M., Rossen, J., Martirena, F., Scrivener, K.: Cement substitution by a combination of metakaolin and limestone. Cem. Concr. Res. **42**, 1579–1589 (2012). doi:10.1016/j.cemconres.2012.09.006
2. Bishnoi, S., Maity, S., Mallik, A., Joseph, S., Krishnan, S.: Pilot scale manufacture of limestone calcined clay cement: the Indian experience. Indian Concr. J. **88**, 22–28 (2014)
3. Matschei, T., Lothenbach, B., Glasser, F.P.: The role of calcium carbonate in cement hydration **37**, 551–558 (2007). doi:10.1016/j.cemconres.2006.10.013
4. Lothenbach, B., Le, G., Gallucci, E., Scrivener, K.: Influence of limestone on the hydration of Portland cements **38**, 848–860 (2008). doi:10.1016/j.cemconres.2008.01.002
5. Damidot, D., Lothenbach, B., Herfort, D., Glasser, F.P.: Thermodynamics and cement science. Cem. Concr. Res. **41**, 679–695 (2011). doi:10.1016/j.cemconres.2011.03.018
6. Ipavec, A., Gabrovšek, R., Vuk, T., Kaučič, V., Maček, J., Meden, A.: Carboaluminate phases formation during the hydration of calcite-containing portland cement. J. Am. Ceram. Soc. **94**, 1238–1242 (2011)
7. Bonavetti, V.L., Rahhal, V.F., Irassar, E.F.: Studies on the carboaluminate formation in limestone filler-blended cements. Cem. Concr. Res. **31**, 853–859 (2001). doi:10.1016/S0008-8846(01)00491-4
8. Emmanuel, A.C., Haldar, P., Maity, S., Bishnoi, S.: Second pilot production of limestone calcined clay cement in India: the experience. Indian Concr. J. **90**, 57–63 (2016). https://www.scopus.com/inward/record.uri?eid=2-s2.0-84967016900&partnerID=40&md5=b55d60c26fbe9c5459d04994b91cb3f9
9. Vizcaino-Andres, L.M., Sanchez-Berriel, S., Damas-Carrera, S., Perez-Hernandez, A., Scrivener, K.L., Martirena-Hernandez, J.F.: Industrial trial to produce a low clinker, low carbon cement. Mater. Constr. **65**, e045 (2015). doi:10.3989/mc.2015.00614
10. Avet, F., Snellings, R., Alujas, A., Ben, M., Scrivener, K.: Development of a new rapid, relevant and reliable (R 3) test method to evaluate the pozzolanic reactivity of calcined kaolinitic clays. Cem. Concr. Res. **85**, 1–11 (2016). doi:10.1016/j.cemconres.2016.02.015
11. Wilson, W., Sorelli, L., Tagnit-Hamou, L.: Automated coupling of statistical nanoindentation and quantitative energy-dispersive spectroscopy (SNI-QEDS): a comprehensive method to disclose the micro-mechanical properties of cement pastes. Cem. Concr. Res. (2017, In Review)
12. Taylor, H.F.W.: Cement Chemistry. Thomas Telford Publishing, London (1997)
13. Wilson, W., Rivera-Torres, J.M., Sorelli, L., Duran-Herrera, A., Tagnit-Hamou, A.: The micromechanical signature of high-volume natural pozzolan concrete by combined statistical nanoindentation and SEM-EDS analyses. Cem. Concr. Res. **91**, 1–12 (2017). doi:10.1016/j.cemconres.2016.10.004

Addressing Key Challenges in MK-PLC Blends at Early Ages: Workability, Slump Retention, and Heat of Hydration

B.H. Zaribaf(✉) and K.E. Kurtis

School of Civil and Environmental Engineering, Georgia Institute of Technology, Atlanta, GA 30332-0355, USA

Abstract. Toward broadening utilization of higher fractions (i.e., >10% by mass) of metakaolin (MK) in concrete construction, several key technical challenges remain. Among these are those associated with differences in early age properties deriving from greater MK additions; these include reduced workability, more rapid slump loss, and increased heat of hydration compared to traditional concrete. Due to competing dilution and acceleration effects associated with the introduction of finely divided limestone (and potentially more fine cement), these effects can be potentially exacerbated or ameliorated when MK is combined with Portland-limestone cement (PLC). This paper examines each of these challenges and offers solutions, in the form of recommended admixtures to improve workability and slump retention and refined heat of hydration models which consider the contributions of MK.

1 Introduction and Background

As with other finely divided supplementary cementitious materials (SCMs) [1], utilization of metakaolin (MK) at greater fractions than 10% cement by mass tends to considerably increase water demand and decrease workability in concrete. In practice, poor workability and rapid slump or flow loss (Fig. 1) can contribute to poor pumpability, increased effort for consolidation, and decreased finishability. Improved workability can be obtained through proper dosing of water-reducing admixtures (WRAs), but guidance is needed regarding the most suitable admixture chemistries and appropriate dosages needed to obtain necessary workability while avoiding incompatibilities (e.g., unpredictable set, prolonged set, segregation).

The fineness of MK also imparts considerable early age reactivity. While supplementary cementitious materials commonly react more slowly and are less exothermic in their reaction than the cement that they partially replace, MK is an important exception to this generalization. When combined with cement, MK has been shown to increase both the rate of cement hydration at early ages and to increase the peak typically associated with aluminate phase reaction, as measured through isothermal calorimetry [2–5]. In mass concrete, undesirable early heat development may lead to thermal cracking or induce delayed ettringite formation, potentially compromising serviceability and durability [6]. In cold-weather or hot-weather concrete construction, knowledge of the rate of hydration can be useful for predicting setting and hardening

F. Martirena et al. (eds.), *Calcined Clays for Sustainable Concrete*, RILEM Bookseries 16,
https://doi.org/10.1007/978-94-024-1207-9_80

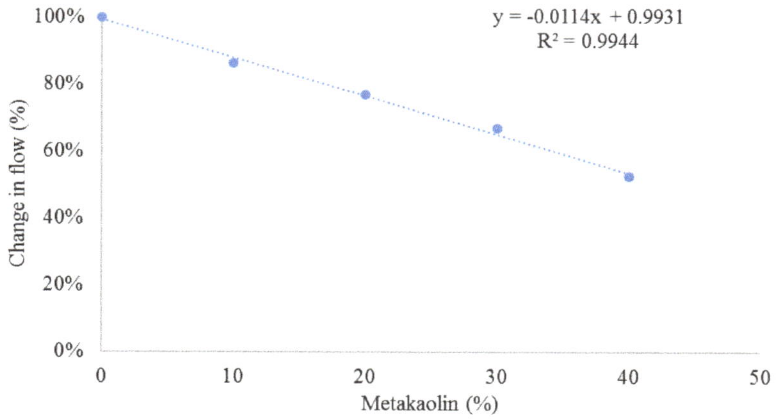

Fig. 1. The change in the flow in cement paste with increase in metakaolin content, as measured by mini-slump.

rates, vital information in construction scheduling. Therefore, an ability to predict heat generation in MK-containing concrete is an important topic of practical importance.

Additional reductions in clinker content can be achieved when MK is combined with Portland limestone cement (PLC). Some PLCs are ground more finely than traditional Portland cements to compensate for potential dilution effects [7]. As a result, the combination of PLC with MK can further influence workability, slump retention and early reaction rates, in ways which are not well-understood or easily predicted due to competing effects of dilution, nucleation, improved particle packing, and enhanced and synergetic reactivity [7, 8].

2 Aims

The goals of this study are to identify field-ready strategies: (1) to overcome reduced workability and to maintain setting behavior and (2) to address the use of MK in heat of hydration models in MK-PLC blends.

3 Materials and Methods

Paste samples were prepared at water-to-binder (w/b) of 0.40, with 0, 10, 20, and 30% by mass MK replacement (denoted MK10, MK20, and MK30), using high-purity metakaolin (Burgess Pigment Company, Sandersville, GA) with Si/Al of 1.15 and a mean particle size of 1.4 μm, as determined by laser particle size diffraction (PSD). To address workability, mini-slump tests were performed to measure flow and slump retention over 60 min, comparing among four commercially available water-reducing admixtures: a mid-range WRA based on lignosulfonate chemistry and three high-range WRAs, with one each of naphthalene formaldehyde condensate (PNS), polymelamine sulfonate (PMS) and polycarboxylate ether (PCE). WRAs were each added first to the mix water and then the water and WRA was added to the cementitious materials. Samples were prepared with no WRA and with dosages of each of four WRA type determined by mini slump to achieve equivalent flow. Later, mini-slump assessments were repeated at 10-minute intervals to assess slump loss over one hour. For flow loss measurements, samples were re-mixed for 10 s before mini slump flow was measured at each time period. Tests were performed on an ASTM C595 Type IL PLC with a mean particle size of 10.2 μm (by laser PSD) and calcium carbonate ($CaCO_3$) content, determined by thermogravitric analysis, of 11.8% by mass.

An empirical model originally developed by Schindler and Folliard [9] was used to estimate the total heat of hydration, based upon cement composition, with the assumption of fully hydrated cement [10]:

$$H_{cem} = 500\rho_{C3S} + 260\rho_{C2S} + 866\rho_{C3A} + 420\rho_{C4AF} + 624\rho_{SO3} + 1186\rho_{FreeCaO} + 850\rho_{MgO} \quad (1)$$

where, H_{cem} is total heat of hydration of the cement (J/g) and ρ_i is weight ratio of i-th compound in terms of total cement content.

The effect of SCMs like slag and fly ash can be accommodated through a modification [10]:

$$H_u = H_{cem}\rho_{cem} + 461\rho_{slag} + 1800\rho_{FA-CAO}\rho_{FA}$$

The degree of hydration at time t, $\alpha_{(t)}$, varies between zero at the initial time and one upon complete hydration:

$$\alpha_{(t)} = \frac{H_{(t)}}{H_u} \quad (2)$$

Where $H_{(t)}$ is the cumulative heat of hydration (J/m^3) released at t. This portion of the study considered two ASTM C150 Type I, one Type V, and one Type II/V cements, as well as two ASTM C595 Type IL PLC sources and then blends of these with 10, 20, and 30% by mass MK. Hydration kinetics and cumulative heat of hydration for each system was measured by isothermal calorimetry performed at 25 °C according to ASTM C1679 [11]. Pastes were mixed externally following ASTM C305 [12] prior to being placed in the calorimeter (TAM AIR). The cumulative heat of hydration of cementitious materials after 7 days was measured, and the degree of hydration was calculated by dividing the measured heat of hydration, by the calculated total heat of hydration for cement that is fully hydrated.

4 Results

4.1 WRAs for Enhanced Workability in MK-PLC Blends

As previously shown in Fig. 1, the flowability as measured by the mini slump test of PLC-MK cement paste decreases by about 11% for each 10% increase in MK content at w/b of 0.40. This behavior demonstrates the need for an efficient WRA to improve the flowability of PLC-MK blends, particularly with at the MK content increases.

However, when examining the dosages required to achieve flow values comparable to the control (0% MK) for each of the four admixture chemistries examined (i.e., LS, PMS, PNS, PCE), the results show that both the LS and PNS require significantly higher dosage than the maximum recommended dosage [106 and thesis]. For the MK10 case, the required dosage for LS far-exceeded the recommended limit and the required dosage of PNS was at the recommended limit. With 20 and 30% MK, the required PNS dosage was a several times the recommended dosage. Use of superplasticizers at such high dosage rates may produce undesirable behavior including excessive set retardation or false set; the latter is primarily found with lignosulfonate [13].

Both the PMS and PCE admixtures produced workability similar to the control samples within the recommended manufacturer dosages, even at 30% MK replacement. Moreover, the relationship between the required PCE admixture dosage and the metakaolin replacement level was found to be linear [4, 13]. This is relationship advantageous for predicting the required dosage of PLC-MK blends at different metakaolin replacement levels, and facilitates practical upscaling of higher fractions of MK in PLC concrete.

4.2 WRAs for Slump Flow Retention in MK-PLC Blends

In addition to improving the workability of MK-PLC mixtures, slump flow retention over time is also important to ensure constructability. Here, the flow loss of PLC and PLC with 10 and 30% MK paste with the two most promising admixture chemistries assessed by mini-slump and calorimetry [13] - PMS and PCE - over a one-hour period (Fig. 2). The highest 1-hour flow losses were observed with the PMS admixture, measuring approximately 60% for the MK10 case and ~80% for MK30. The MK

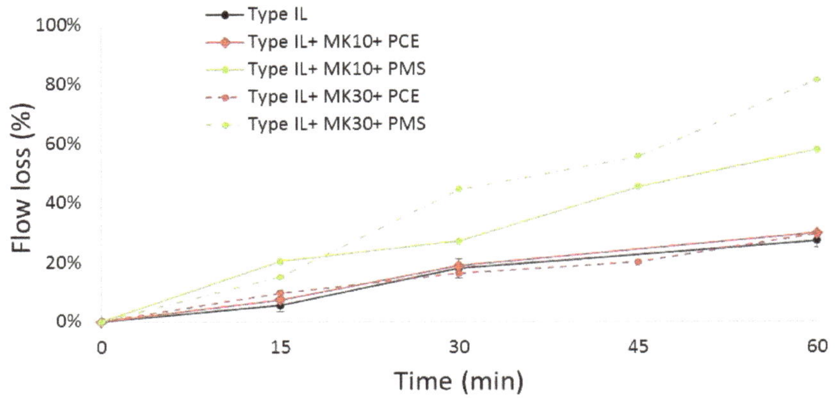

Fig. 2. Flow loss over one hour in Type IL pastes with 10 and 30% MK, as measured by mini-slump. PCE and PMS admixture use in MK-PLC pastes are compared to neat cement control.

paste with PCE showed similar flow loss (25% at 60 min) to the PLC paste. It should be noted that the flow loss was consistently slightly lower for the MK30 paste with the higher PCE content.

4.3 Heat of Hydration Models

For the six cements (A-F) of varying composition considered, along with their blends with 10 and 30% MK, Table 1 shows the calculated ultimate heats of hydration, measured heat of hydration, and degree of hydration at 7 days, as determined through isothermal calorimetry and existing empirically-derived heat of hydration models. Cement A, a Type I Portland cement, has the highest heat of hydration, due to its higher tricalcium silicate content, and the Type V Cement (Cement F) showed the lowest heat of hydration, as expected.

Comparing the measured 7-day heat of hydration of neat cements and MK-blended pastes, a trend of increasing heat with increasing MK content is evident. As a result, predicted degrees of hydration of *greater than 1.0* were calculated for the MK-containing systems (as highlighted in red in Table 1); such results are clearly not realistic. Furthermore, as shown in Fig. 3, the Schindler and Folliard [9] equations tended to underestimate the heat generated at all ages, while equations proposed by Riding et al. [10] (which considered two SCMs, fly ash and slag) tended to

Table 1. Predicted total heat of hydration (Hu), measured total heat of hydration H(t) and degree of hydration (α) at t = 7 days for mixes containing 0, 10, and 30% MK, using equations from [10].

Cement type	Nomenclature	H_u, J/g	$H_{(t)}$, J/g	α at 7 days (unitless)	$\Delta H_{(t)}/\rho_{MK}$, J/g
ASTM C150 Type I and blends	A	485.5	309.0	0.64	
	A+MK10	436.93	355.0	0.81	769
	A+MK30	339.83	415.1	1.22	662
ASTM C595 Type IL and blends	AL	473.19	551.0	1.16	
	AL+MK10	424.36	557.0	1.31	611
	AL+MK30	331.23	544.0	1.64	528
ASTM C150 Type I and blends	C	447.3	367.0	0.82	
	C+MK10	402.6	416.3	1.03	857
	C+MK30	313.1	468.0	1.50	703
ASTM C595 Type IL and blends	DL	462.0	379.0	0.82	
	DL+MK10	415.8	416.0	1.0	749
	DL+MK30	323.4	482.0	1.49	703
ASTM C150 Type V and blends	E	471.5	304.0	0.64	
	E+MK10	424.4	368.0	0.87	944
	E+MK30	330.1	444.0	1.35	770
ASTM C150 Type II/V and blends	F	447.2	318	0.71	
	F+MK10	402.5	370	0.92	838
	F+MK30	326.7	470	1.44	824

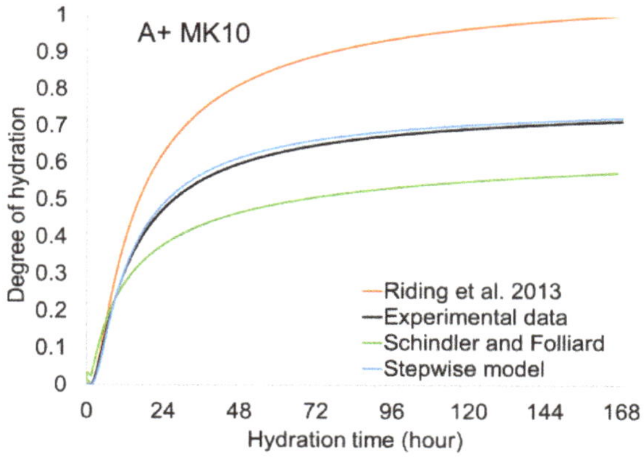

Fig. 3. Comparison of model fits shows the proposed stepwise model gives a better fit for predictions of degree of hydration, based upon heat of hydration data, for MK-PLC systems.

overestimate heat generation. These results demonstrate the need for more accurate predictions of heat generation in the presence of MK.

Analysis of the calorimetry data suggests that heat derived from MK reaction is in the range of 516–944 J/g, with average of 762 J/g of MK. Previous studies show that MK contributes to the heat of hydration even after 7 days. Therefore, the MK heat contribution is multiplied by a factor (1.15) to account for the further heat evolution after 7 days. A modified equation, based upon [10], is proposed which accounts for the MK contribution to the heat of hydration:

$$H_u = H_{cement} . \rho_{cement} + 461 . \rho_{slag} + 1800 . \rho_{FA-CAO} . \rho_{FA} + 876.5 \rho_{MK} \tag{3}$$

From here, a three-parameter equation for the degree of hydration is used to estimate the degree of hydration over time for each cement paste mixture. As shown in an example case for cement A with 10% MK in Fig. 3, the "stepwise" linear regression model developed here shows a good fit to experimental data derived from MK-PLC systems, as well as to modern cements and PLCs. C_3S content, cement fineness, and MK content were found to be significantly related to the ultimate degree of hydration. And, while HRWRAs (needed for workability improvements) can delay cement hydration in the first 6–10 h, their use was not found to affect the ultimate degree of hydration significantly.

5 Conclusion

For maintaining flow at a consistent admixture dosage rate with increasing MK content (up to 30%), polycarboxylate ether WRA chemistry is recommended; this admixture also provided good flow retention properties in MK-PLC systems. Furthermore, an empirically-based equation for heat evolution is proposed which accounts for MK addition and can be used to predict degree of hydration in MK-PLC systems with greater accuracy than existing models.

References

1. Snellings, R.: Assessing understanding and unlocking supplementary cementitious materials. RILEM Tech. Lett. **1**, 50–55 (2016)
2. Justice, J., Kurtis, K.: Influence of metakaolin surface area on properties of cement-based materials. J. Mater. Civ. Eng. **19**, 762–771 (2007)
3. Lagier, F., Kurtis, K.E.: Influence of Portland cement composition on early age reactions with metakaolin. Cem. Concr. Res. **37**, 1411–1417 (2007)
4. Zaribaf, B.H., Uzal, B., Kurtis, K.: Compatibility of superplasticizers with limestone-metakaolin blended cementitious system. In: Scrivener, K., Favier, A. (eds.) Proceedings of Calcined Clays for Sustainable Concrete, pp. 427–434. Springer, Dordrecht (2015)
5. Ramezanianpour, A.A.: Metakaolin, Cement Replacement Materials, pp. 225–255. Springer, Berlin (2014)
6. Branco, F.A., Mendes, P., Mirambell, E.: Heat of hydration effects in concrete structures. ACI Mater. **89**, 139–145 (1992)
7. Nadelman, E.: Ph.D. dissertation: Hydration And Microstructural Development of Portland Limestone Cement-Based Materials, School of Civil and Environmental Engineering, Georgia Institute of Technology (2016)
8. Antoni, M., Rossen, J., Martirena, F., Scrivener, K.: Cement substitution by a combination of metakaolin and limestone. Cem. Concr. Res. **42**, 1579–1589 (2012)
9. Schindler, A.K., Folliard, K.J.: Heat of hydration models for cementitious materials. ACI Mater. **102**, 24–33 (2005)
10. Riding, K.A., Poole, J.L., Folliard, K.J., Juenger, M.C., Schindler, A.K.: Modeling hydration of cementitious systems. ACI Mater. **109**, 225–234 (2012)
11. ASTM C1679-14, Standard Practice for Measuring Hydration Kinetics of Hydraulic Cementitious Mixtures Using Isothermal Calorimetry, ASTM International, West Conshohocken, PA (2014)
12. ASTM C305-14, Standard Practice for Mechanical Mixing of Hydraulic Cement Pastes and Mortars of Plastic Consistency, ASTM International (2014)
13. Zaribaf, B.H.: Ph.D. dissertation: Metakaolin-Portland limestone cements: Evaluating the effects of chemical admixtures on early and late age behavior, School of Civil and Environmental Engineering, Georgia Institute of Technology (2017)

Assessing the Effect of Calcite Impurities in Clay on Optimal Dehydroxylation Parameters for Enhanced Reactivity

F. Zunino(✉) and K. Scrivener

Laboratory of Construction Materials, EPFL STI IMX LMC, Station 12, 1015 Lausanne, Switzerland

Abstract. This study focused on understanding the influence of different amounts of calcite impurities (0 to 10% by mass) on the optimal parameters for clay dehydroxylation and the reactivity of the final calcined clay. A surface response experimental design was used to study the effect of maximum temperature, residence time and calcite content on different experimental responses. Specific surface and particle size distribution were used to characterize the physical properties of the calcined-clays. SEM coupled with EDS was selected to study the particle morphology and the composition of the amorphous phase formed, while XRD was used to assess the mineralogical composition after calcination. TGA analysis was used to measure the remaining amounts of kaolinite and calcite after calcination, while isothermal calorimetry was used to assess the reactivity of the different calcined-clays obtained. It was observed that specific surface area is the most sensitive parameter to variations in calcite content.

1 Introduction

Mineral additions, commonly referred as supplementary cementitious materials (SCMs), are widely used either in blended cements [1, 2]. The use of SCMs, to replace the more energy intensive clinker, leads to a significant reduction of CO_2 emissions per unit volume of concrete [3–5].

Clays are unique among the supplementary cementitious materials because of their worldwide availability, since they are widely distributed throughout in the earth crust [6–8]. The three clay types of major abundance are kaolinite, illite and montmorillonite [9]. Heat treatment of kaolinitic clays between 600 and 800 °C leads to the dehydroxylation of its crystalline structure to give a state of more structural disorder known as metakaolin [10–12].

In some regions, clays are naturally combined with limestone in the quarries. Kaolinite is dehydroxylated between 600 and 800 °C [13]. Usually, clay is calcined at 800 °C to maximize the reactivity of the metakaolin in the calcined clay, higher temperatures lead to reduced reactivity due sintering and then recrystallization [11].

There is lack of systematic studies exploring these interactions at temperatures of 800 °C or lower. This study explores the effect of limestone impurities below 10% by mass on calcined clay reactivity.

F. Martirena et al. (eds.), *Calcined Clays for Sustainable Concrete*, RILEM Bookseries 16,
https://doi.org/10.1007/978-94-024-1207-9_81

2 Materials and Methods

An experimental design aimed to assess the effect of 2 and 8% calcite impurities on calcined clay reactivity was designed. The calcination temperatures were established at 600 °C and 800 °C, above and below the decarbonation threshold of calcite. Furthermore, the effect of residence time was explored at a low (20 min) and high (60 min) level.

Raw clay (kaolinite content of 71% as measured by thermogravimetric analysis (TGA) and Durcal 5 calcium carbonate (calcite) were used. 250 g of material (uncalcined clay and calcite) were weighted in individual 1L plastic containers for each experimental point. Powders were blended together using a Turbula blender for 20 min. Homogeneity of the obtained powder samples was checked with TGA on selected samples. The variability between nominal and experimental calcite content was always lower than 0.1% by mass. Calcination was carried out on 300 mL alumina crucibles. Calcined samples were then stored in sealed containers until analysis.

2.1 X-Ray Diffraction (XRD)

X-Ray Diffraction (XRD) was used to monitor the crystalline phases formed upon hydration. Measurements were carried out on freshly cut cement paste slices at 1, 7 and 28 days. The slices were mounted on a XRD sample holder and measured in Bragg–Brentano mode using a X'Pert PANalytical diffractometer with CuKα source operated at 45 kV and 40 mA. Samples were scanned from 7 to 70° 2θ with a step size of 0.0167 2θ, using a X'Celerator detector, resulting in an equivalent time per step of 30 s.

2.2 Isothermal Calorimetry

Isothermal calorimetry was used to assess the pozzolanic reactivity of calcined clays. For this purposes, the R^3 test proposed by Avet et al. [14] was selected due to its reliability and ease of interpretation for calcined clay benchmarking purposes. The test considering pozzolanic and interaction with calcium carbonate (simulated LC^3-50) was selected. In this procedure, calcined clay is mixed with portlandite, calcium carbonate, potassium sulfate, potassium hydroxide and water at 40 °C, and put into glass ampoules inside the calorimeter under the same temperature conditions.

2.3 Thermogravimetric Analysis (TGA)

TGA was used to assess the amount of calcined kaolinite in the annealed materials, as it has been shown that this magnitude is the overwhelming parameter governing pozzolanic reactivity of calcined clay [14]. This parameter is defined as the difference on kaolinite measured on the raw clay and the calcined clay, allowing to account for differences in reactivity in case of partial dehydroxylation.

3 Results and Discussion

3.1 X-Ray Diffraction

Figure 1 shows the X-ray diffraction pattern for materials calcined at 600 °C, with high initial calcite content (8% by mass), for both low (20 min) and high (60 min) residence time. It can be observed that the amount of calcite remains unchanged despite higher residence time, as confirmed by TGA analysis where 8.02 and 8.11% contents of calcite after calcination were measured for low and high residence time, respectively. The difference in calcined kaolinite content (3.7 and 15.4% for RT = 20 and 60 min respectively, determined by TGA) is attributed directly to the difference in residence time between both samples.

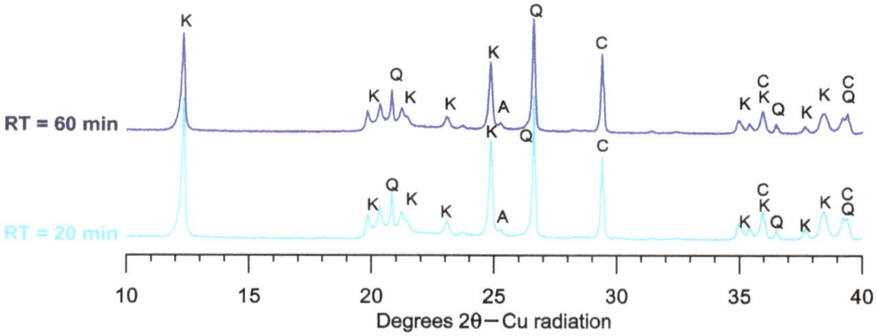

Fig. 1. X-Ray diffraction patterns for materials with 8% initial calcite content and calcined at 600 °C (K = kaolinite; Q = quartz; A = anatase; C = calcite).

Figure 2 shows the X-ray diffraction pattern for materials calcined at 800 °C, with high initial calcite content (8% by mass), for both low (20 min) and high (60 min) residence time.

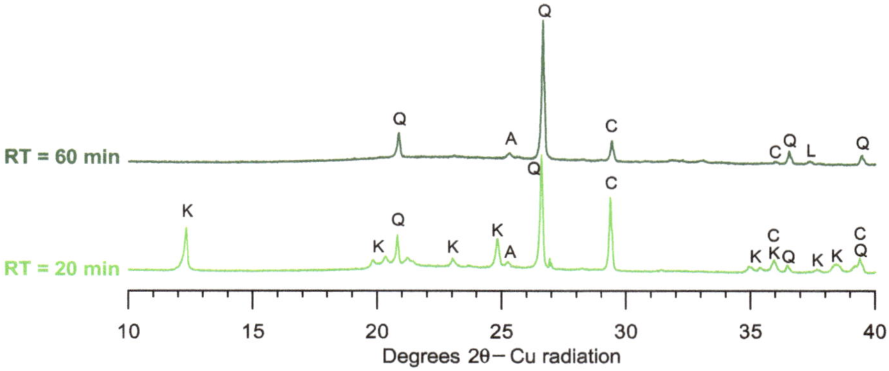

Fig. 2. X-Ray diffraction patterns for materials with 8% initial calcite content and calcined at 800 °C (K = kaolinite; Q = quartz; A = anatase; C = calcite; L = lime).

In this case, most of calcite is decomposed (6.51 and 2.23% remaining calcite content, measured by TGA on RT of 20 min and 60 min, respectively). Despite free lime observed on high residence time sample (0.43%), it doesn't balance with the amount of calcite decomposed. This finding suggests some calcium oxide goes into the amorphous phases. The difference in calcined kaolinite content (41.72 and 55.06% for RT = 20 and 60 min respectively, determined by TGA) is attributed to the difference in residence time between samples. Figure 3 shows the X-ray diffraction pattern for materials calcined at 800 °C for 60 min, with low (2% by mass) and high (8% by mass) initial calcite contents.

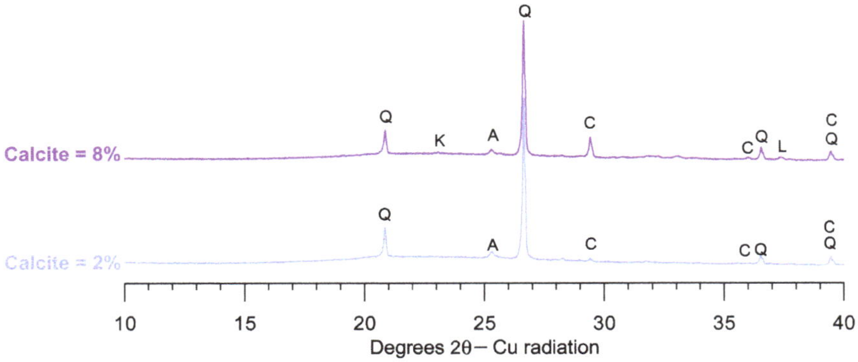

Fig. 3. X-Ray diffraction patterns for materials with 2 and 8% initial calcite content, calcined at 800 °C for 60 min.

As observed in the comparison between samples calcined at 800 °C, calcite is decomposed (0.65 and 2.23% remaining calcite content for 2 and 8% samples respectively, which represents 32.7 and 37.9% of their respective initial contents). When comparing in this case the influence of different amounts of initial calcite under the same calcination conditions, it is observed that calcite seems to negatively influence the dehydroxylation of kaolinite (68.55 and 55.06% calcined kaolinite content for 2 and 8% calcite materials, corresponding to 99 and 84% of their potential, respectively).

3.2 Isothermal Calorimetry – R^3 Test

Isothermal calorimetry tests were performed over the obtained materials using the R^3 method, in order to assess and compare the reactivity of the different calcined clays in a quick and reliable manner. As a general trend, there is a dramatic increase in reactivity observed in all of the materials when increasing the calcination temperature from 600 °C to 800 °C. Figure 4 shows the heat flow (left) and total heat (right) plots for materials calcined at 600 and 800°C, with low (2%) and high (8%) calcite contents, all of them calcined for low residence time (20 min).

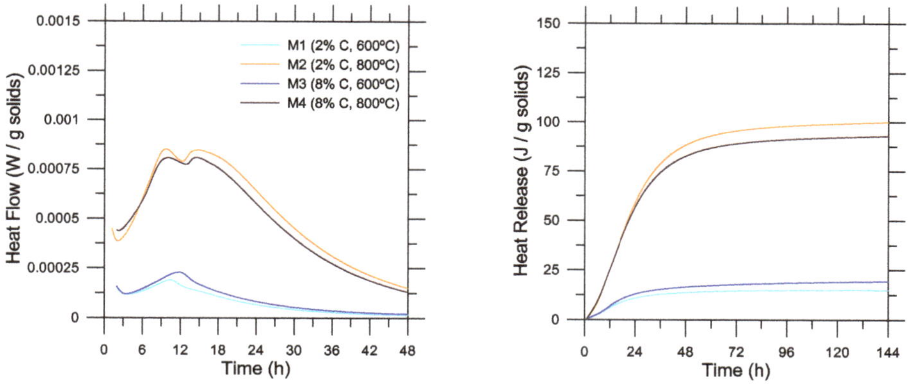

Fig. 4. Isothermal calorimetry results for materials calcined for 20 min, at high and low temperature and with different initial calcite content.

At 600 °C there is little impact of the calcite, and increased contents produces a slight increase in reactivity. At 800 °C, higher contents of calcite seem to have a negative effect on clay reactivity, however, the difference between 2 and 8% initial calcite materials is small (<10 J/g solids at 6 days of hydration).

Figure 5 shows the heat flow (left) and total heat (right) plots for materials calcined at 600 and 800 °C, with low (2%) and high (8%) calcite contents, all of them calcined for high residence time (60 min).

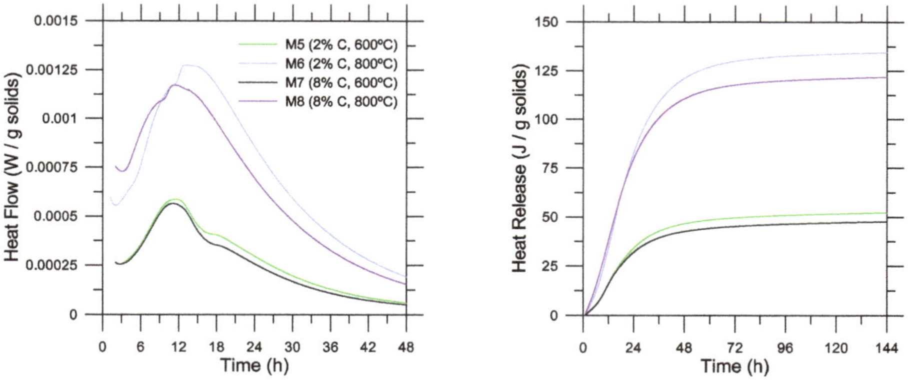

Fig. 5. Isothermal calorimetry results for materials calcined for 60 min, at high and low temperature and with different initial calcite content

At 600 °C and increasing the residence time to 60 min, the effect is similar to the observed at 800 °C and low residence time (20 min). Reaction kinetics don't seem to be significantly affected during the first 24 h of hydration. At 800 °C, the difference in heat evolved at 6 days (144 h) is more significant. Hydration kinetics seems also to been affected (sample M8 exhibits a slight acceleration during the first 12 h), suggesting that the nature of the metakaolin may have changed.

In order to explore additional calcination parameter combinations, further experiments were conducted specially at temperatures below 800 °C. The normalized (corrected by dilution effect) calcined kaolinite content is shown in Fig. 6 for 60 min residence time.

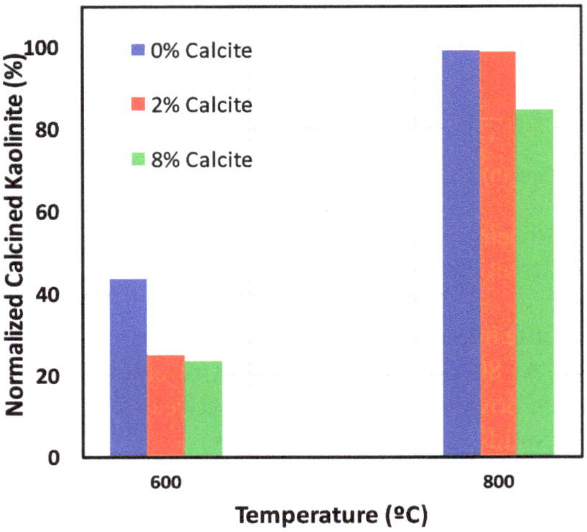

Fig. 6. Normalized calcined kaolinite content for materials annealed between 600 and 800 °C for 20 and 60 min

4 Conclusions

The following preliminary conclusions are derived from the results presented in this report:

1. Calcined kaolinite content appears to still be the overwhelming predictor of calcined clay reactivity on clays with calcite impurities.
2. Calcite has a negative effect on kaolinite dehydroxylation, leading to a reduced reactivity of the material. The effect seems to be amplified with higher residence time, temperature and higher calcite content. The best overall performance for high levels of initial calcite content were obtained with 700 °C and 60 min residence time as calcination parameters.
3. The effect of calcite impurities on clay reactivity seems to be twofold: it migrates into the glassy phase and interferes with the normal dehydroxylation of kaolinite.

Acknowledgements. Authors acknowledge funding provided by the Swiss commission for scholarships for foreign students (FCS), through their Ph.D. scholarship program.

References

1. Lothenbach, B., Scrivener, K., Hooton, R.D.: Supplementary cementitious materials. Cem. Concr. Res. **41**, 1244–1256 (2011). doi:10.1016/j.cemconres.2010.12.001
2. Mehta, P., Monteiro, P.: Concrete: Microstructure, Properties, and Materials, 3rd edn. McGraw-Hill Professional, New York (2005)
3. Zunino, F., Lopez, M.: A methodology for assessing the chemical and physical potential of industrially sourced rice husk ash on strength development and early-age hydration of cement paste. Constr. Build. Mater. **149**, 869–881 (2017)
4. Bentz, D.P., Ferraris, C.F., Jones, S.Z., Lootens, D., Zunino, F.: Limestone and silica powder replacements for cement: early-age performance. Cem. Concr. Compos. **78**, 43–56 (2017). doi:10.1016/j.cemconcomp.2017.01.001
5. Siddique, R.: Waste Materials and By-Products in Concrete. Springer, Berlin (2008)
6. Fernandez Lopez, R.: Calcined Clayey Soils as a Potential Replacement for Cement in Developing Countries, École Polytechnique Fédérale de Lausanne (2009). doi:10.5075/epfl-thesis-4302
7. Tironi, A., Trezza, M.A., Scian, A.N., Irassar, E.F.: Assessment of pozzolanic activity of different calcined clays. Cem. Concr. Compos. **37**, 319–327 (2013). doi:10.1016/j.cemconcomp.2013.01.002
8. Sabir, B., Wild, S., Bai, J.: Metakaolin and calcined clays as pozzolans for concrete: a review. Cem. Concr. Compos. **23**, 441–454 (2001). doi:10.1016/S0958-9465(00)00092-5
9. Fernandez, R., Martirena, F., Scrivener, K.L.: The origin of the pozzolanic activity of calcined clay minerals: a comparison between kaolinite, illite and montmorillonite. Cem. Concr. Res. **41**, 113–122 (2011). doi:10.1016/j.cemconres.2010.09.013
10. Antoni, M.: Investigation of cement substitution by combined addition of calcined clays and limestone, École Polytechnique Fédérale de Lausanne (2011)
11. Alujas, A., Fernández, R., Quintana, R., Scrivener, K.L., Martirena, F.: Pozzolanic reactivity of low grade kaolinitic clays: influence of calcination temperature and impact of calcination products on OPC hydration. Appl. Clay Sci. **108**, 94–101 (2015). doi:10.1016/j.clay.2015.01.028
12. Souri, A., Kazemi-Kamyab, H., Snellings, R., Naghizadeh, R., Golestani-Fard, F., Scrivener, K.: Pozzolanic activity of mechanochemically and thermally activated kaolins in cement. Cem. Concr. Res. **77**, 47–59 (2015). doi:10.1016/j.cemconres.2015.04.017
13. Antoni, M., Rossen, J., Martirena, F., Scrivener, K.: Cement substitution by a combination of metakaolin and limestone. Cem. Concr. Res. **42**, 1579–1589 (2012). doi:10.1016/j.cemconres.2012.09.006
14. Avet, F., Snellings, R., Alujas Diaz, A., Ben Haha, M., Scrivener, K.: Development of a new rapid, relevant and reliable (R3) test method to evaluate the pozzolanic reactivity of calcined kaolinitic clays. Cem. Concr. Res. **85**, 1–11 (2016). doi:10.1016/j.cemconres.2016.02.015

Reactivity and Performance of Limestone Calcined-Clay Cement (LC^3) Cured at Low Temperature

F. Zunino(✉) and K. Scrivener

Laboratory of Construction Materials, EPFL STI IMX LMC, Station 12, 1015 Lausanne, Switzerland

Abstract. Limestone calcined-clay cements (LC^3) take advantage of the synergetic effects of calcium carbonate reaction with the additional aluminium provided by the calcined clay. As temperature decreases, calcium carbonate solubility increase, therefore, the early-age hydration kinetics and the optimal proportioning of the ternary cement system are modified. This study explored the reactivity and mechanical performance of different LC^3 systems cured at 10 and 20 °C. Mixtures containing PC, PC-limestone and a LC^3 blends with 50% clinker factors and 2:1 clay-to-limestone ratio were cast and compared. Hydration kinetics were assessed using isothermal calorimetry at each of the temperatures. The evolution of porosity was studied during hydration by MIP. Compressive strength was measured over time on cement paste cubes. Phase assemblage was monitored using XRD.

1 Introduction

Environmental concerns, such as energy consumption and CO_2 emission reductions, have become of increasing concern in the construction industry during the last decades [1]. Therefore, the use of supplementary cementitious materials (SCMs) as a means to reduce the cement content in concrete mixes [2–5] and enhance the durability of the material to increase the service life of concrete structures [6] have become of increasing interest among researchers.

Mineral additions, commonly referred as supplementary cementitious materials (SCMs), are widely used either in blended cements or added to concrete separately in the mixer. The use of SCMs, leads to a significant reduction of CO_2 emissions per unit volume of concrete, and significant potential for use of wastes and by-products [3, 7].

Some SCMs will react with calcium hydroxide (CH) at ambient temperature to form hydration products such as calcium silicate hydrates (C–S–H) [8]. However, as replacement rate increases, the mechanical properties of concrete are negatively affected, particularly at early age, mainly due to the limited amount of CH available to react with excess SCMs. In addition, some SCMs also negatively affect the setting and early strength gain rate of concrete, imposing restrictions to construction pace [9]. Furthermore, the available amounts of commonly used SCMs, such as fly ash, blast furnace slags and natural pozzolans, are much lower than the worldwide demand of ordinary

F. Martirena et al. (eds.), *Calcined Clays for Sustainable Concrete*, RILEM Bookseries 16,
https://doi.org/10.1007/978-94-024-1207-9_82

Portland cement (OPC). Consequently, research interest has shifted towards alternative and more abundant sources of SCM's such as calcined clays.

Clays are unique among the supplementary cementitious materials because of their worldwide availability, since they are widely distributed throughout in the earth crust. Heat treatment of kaolinitic clays between 600 and 800 °C leads to the dehydroxylation of its crystalline structure to give a state of more structural disorder known as metakaolin [10, 11].

Fine limestone is also commonly used in OPC-based materials. It has been established that limestone additions up to around 5% can react with the aluminate containing phases in OPC, leading to the formation of mono and hemicarbonate (AFm) phases [12–14]. The solubility of limestone is increased at lower temperatures, which is reported to lead to enhanced precipitation of carbonate-AFm phases, leading to a decrease in porosity [15].

The combination of metakaolin and limestone in OPC-based systems can give synergetic benefits from both well-known systems. The additional reactive alumina supplied by metakaolin can enhance limestone reaction and allow higher replacement levels with improved performance [16, 17]. For this reason, so called, LC^3 (limestone calcined clay cements) have become of great interest. This study focus on understanding the effect of lower curing temperature on the hydration and strength development of LC^3 systems. In particular, the effect of the increased solubility of limestone in the presence of aluminates from clay is explored.

2 Materials and Methods

Portland cement classified as CEMI 42.5R was used for the preparation of blended cement pastes. A kaolinitic clay with 60.3% kaolinite content was used. The clay was calcined at 800 °C for 1 h. A limestone with $D_{50} = 5$ μm was used.

Blended LC^3 cement samples were prepared with 2:1 clay-to-limestone ratio (c/l). The gypsum content was adjusted in order to account for the additional aluminates introduced by calcined clay addition. A volumetric water-to-solids ratio (w/s) of 1.234 (equivalent to a w/c ratio 0.4 by weight on the OPC system) was used for all of the mixtures. The detailed mixture design is given in Table 1.

Table 1. Mixture proportioning for LC3 systems.

Mixture ID	Clinker + Gypsum (% wt. binder)	Limestone (% wt. binder)	Calcined Clay (% wt. binder)	Clay/ls ratio	w/b
Control	100	0	0	–	0.40
LS225	77.5	22.5	0	–	0.41
$LC^3$50–2/1	55	15	30	2	0.43

3 Characterization Techniques

Compressive strength was measured over 2 cm cubic samples at 1, 7 and 28 days. Isothermal calorimetry was used to monitor hydration kinetics, along with X-Ray diffraction (XRD) for phase assemblage assessment. Finally, mercury intrusion porosimetry (MIP) was used to assess porosity refinement over time.

4 Results and Discussion

Compressive strength results at 1, 7 and 28 days are shown on Fig. 1. It can be observed that at 1 day, the strength at 10 °C is significantly lower compared to 20 °C curing temperature, mainly due to the slower cement hydration and the significantly reduced rate of reaction of kaolinite at low temperature, reflecting its high apparent activation energy. Therefore, the mechanical behaviour of the systems cured at low temperature seems to be dominated by the Portland cement and limestone fraction. It is also observed that at 7 days, the strength of the system with limestone (LS225) is higher for the samples cured at 10 °C compared to the same mixture cured at 20 °C.

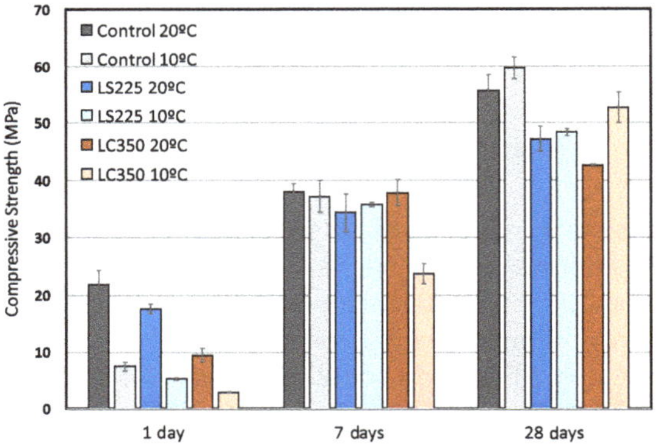

Fig. 1. Compressive strength of cement paste systems cured at 10 °C and 20 °C

At 28 days, the strength of the LC^3 systems cured at 10 °C is higher than the same systems cured at normal temperature (20 °C). On the other hand, the strength of the Control system is also higher for the samples cured at 10 °C at this age. This result may be attributed to the coupled effect of precipitation of C-S-H with lower porosity [18] combined with the reaction of limestone present in the cement (1.7% as measured by TGA). The LC^3-50 system also reaches higher strength when cured at low temperature, suggesting that a similar mechanism as the LS225 system may be occurring and leading to the observed mechanical performance improvement.

When comparing the compressive strength results at 1, 7 and 28 days with the heat release per mL of water, a strong linear correlation is observed (Fig. 2). It is observed that the trend points to higher strength levels for the low curing temperature systems for the same amount of heat release, supporting the supposition that the nature and quality of the hydration products changed compared to normal curing conditions. LC^3 systems seem to deviate from the linear correlation at 20 °C curing condition, probably due to the overlapping of the clay reaction which further contributes to strength with a different heat release footprint associated.

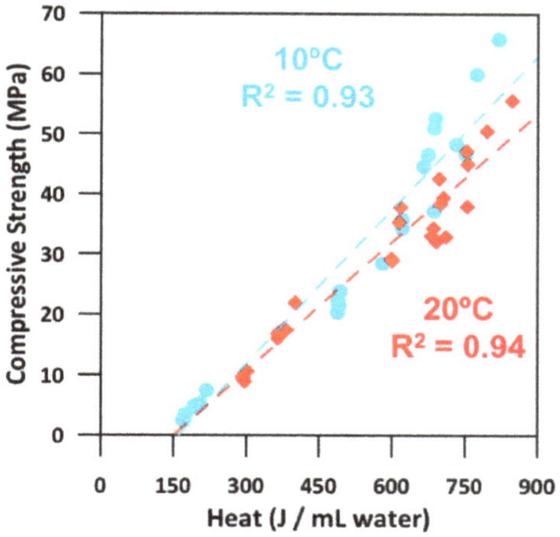

Fig. 2. Heat release versus strength plot for systems cured at 10 °C and 20 °C. Shown are points corresponding to 1, 7 and 28 days strength measurements.

XRD measurements were used with mass balance calculations to obtain the phase assemblage of the systems. This procedure allows computation of the amount of C-A-S-H formed based on the stoichiometry (Ca/Si and Al/Ca ratios) measured on samples of the same material using SEM-EDS. The results for the LC^3 system studied are shown in Fig. 3 for 28-day age at both curing temperatures. In order to account for the expected density difference in C-A-S-H density at normal and low temperature, 2.03 was assumed for 20 °C and 1.91 for 10 °C based on previous findings reported on the literature [18]. It is observed that a higher volume of solids is found in the low temperature sample, which may explain in part the differences in strength observed at this age. This is explained mainly by a slightly higher degree of reaction of clinker and metakaolin on the sample cured at low temperature.

Figure 4 show MIP results for the LC^3-50 2:1. It is observed that at 1 and 7 days, the pore structure is coarser in the low temperature material, and also the total porosity is higher. This is consistent with the observations made in compressive strength, were at this ages the material exhibited lower performance. At 28 days, the pore structure is almost equivalent between the systems.

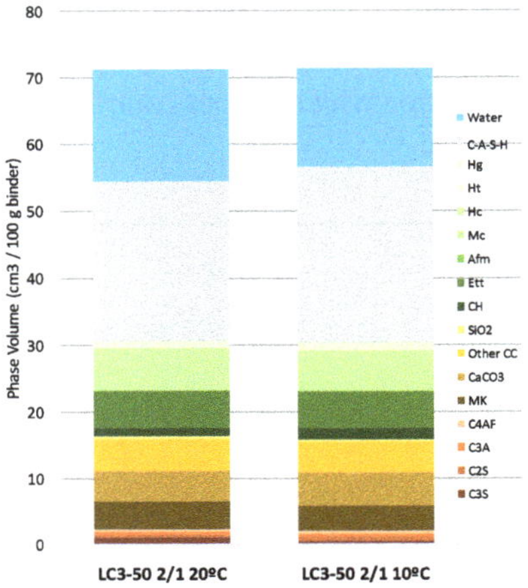

Fig. 3. Phase assemblage for LC3-50 2:1 system at 28 days computed by mass balance, cured at 10 °C and 20 °C.

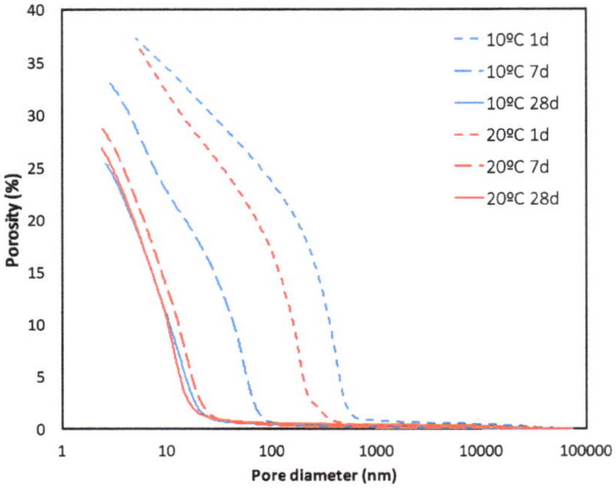

Fig. 4. Cumulative porosity of LC3-50 2:1 system at 1, 7 and 28 days measured by MIP

Figure 5 shows SEM micrographs at 28 days of Control and $LC^3$50 systems, cured at low and normal temperature. In both cases, a coarser pore structure is observed in the 20 °C samples compared to 10 °C. These observations are in good agreement with the compressive strength measurements presented.

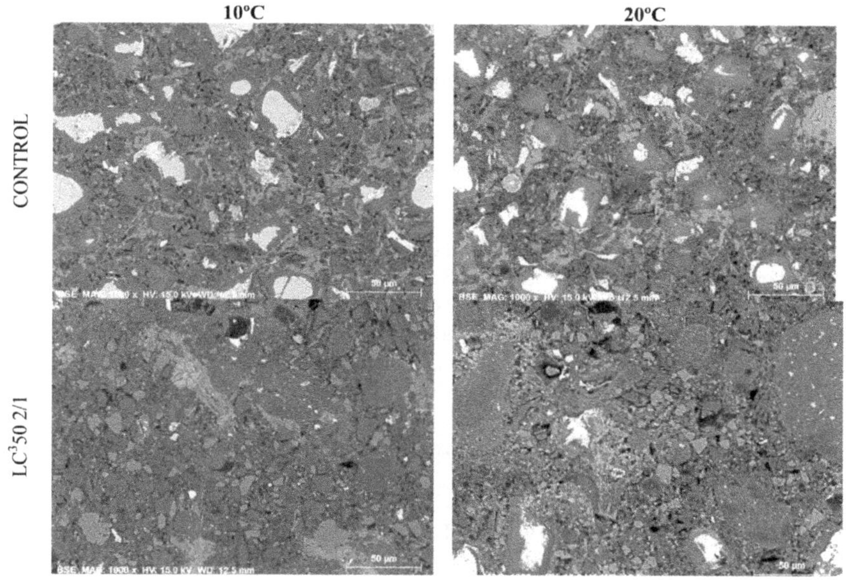

Fig. 5. SEM micrographs (1000x magnification) of control and LC3 systems at 28 days, cured at 10 °C and 20 °C.

5 Conclusions

From the results presented in this study, it is observed that the LC^3 system can take advantage of the enhanced limestone reaction in low temperature curing conditions. However, the reactivity of clay is also significantly reduced. It was observed that at 28 days, the strength of the LC^3 system cured at low temperature is higher compared to their respective counterparts cured at 20 °C. Isothermal calorimetry results also suggest that the nature of the hydration products formed at low temperature is different. Therefore, the observed enhancements are attributed to the coupled effect of low porosity (higher quality) C-S-H formation at low temperature and enhanced AFt phase formation.

Acknowledgements. Authors acknowledge funding provided by the Swiss commission for scholarships for foreign students (FCS), through their Ph.D. scholarship program.

References

1. Schneider, M., Romer, M., Tschudin, M., Bolio, H.: Sustainable cement production—present and future. Cem. Concr. Res. **41**, 642–650 (2011). doi:10.1016/j.cemconres.2011.03.019
2. Lara, R.C., Antoni, M., Díaz, A.A., Scrivener, K., Fernando, J., Hernández, M.: Estudio de la adición de arcillas calcinadas en la durabilidad de hormigones. Rev. Ing. Construcción. **26**, 25–40 (2011)

3. Lothenbach, B., Scrivener, K., Hooton, R.D.: Supplementary cementitious materials. Cem. Concr. Res. **41**, 1244–1256 (2011). doi:10.1016/j.cemconres.2010.12.001
4. Jain, N.: Effect of nonpozzolanic and pozzolanic mineral admixtures on the hydration behavior of ordinary Portland cement. Constr. Build. Mater. **27**, 39–44 (2012). doi:10.1016/j.conbuildmat.2011.08.006
5. Karim, E., El-Hadj, K., Abdelkader, B., Rachid, B.: Analysis of mortar long-term strength with supplementary cementitious materials cured at different temperatures. ACI Mater. J. **107**, 323–331 (2010)
6. Mehta, P.: Durability - critical issues for the future. Concr. Int. **19**, 69–76 (1997)
7. Schneider, M., Romer, M., Tschudin, M., Bolio, H.: Sustainable cement production—present and future. Cem. Concr. Res. **41**, 642–650 (2011). doi:10.1016/j.cemconres.2011.03.019
8. Taylor, H.F.W.: Cement Chemistry, 2nd edn. Thomas Telford, London (1997)
9. Bentz, D.P., Sato, T., De La Varga, I., Weiss, W.J.: Fine limestone additions to regulate setting in high volume fly ash mixtures. Cem. Concr. Compos. **34**, 11–17 (2012). doi:10.1016/j.cemconcomp.2011.09.004
10. Tironi, A., Trezza, M.A., Scian, A.N., Irassar, E.F.: Assessment of pozzolanic activity of different calcined clays. Cem. Concr. Compos. **37**, 319–327 (2013). doi:10.1016/j.cemconcomp.2013.01.002
11. Fernandez, R., Martirena, F., Scrivener, K.L.: The origin of the pozzolanic activity of calcined clay minerals: a comparison between kaolinite, illite and montmorillonite. Cem. Concr. Res. **41**, 113–122 (2011). doi:10.1016/j.cemconres.2010.09.013
12. Bentz, D.P., Ferraris, C.F., Jones, S.Z., Lootens, D., Zunino, F.: Limestone and silica powder replacements for cement: early-age performance. Cem. Concr. Compos. **78**, 43–56 (2017). doi:10.1016/j.cemconcomp.2017.01.001
13. Matschei, T., Lothenbach, B., Glasser, F.P.: The AFm phase in Portland cement. Cem. Concr. Res. **37**, 118–130 (2007). doi:10.1016/j.cemconres.2006.10.010
14. Chowaniec, O.: Limestone Addition in Cement (2012). https://infoscience.epfl.ch/record/174700
15. Bentz, D.P., Stutzman, P.E., Zunino, F.: Low-temperature curing strength enhancement in cement-based materials containing limestone powder. Mater. Struct. **50**, pii:173 (2017)
16. Avet, F., Snellings, R., Alujas Diaz, A., Ben Haha, M., Scrivener, K.: Development of a new rapid, relevant and reliable (R3) test method to evaluate the pozzolanic reactivity of calcined kaolinitic clays. Cem. Concr. Res. **85**, 1–11 (2016). doi:10.1016/j.cemconres.2016.02.015
17. Antoni, M., Rossen, J., Martirena, F., Scrivener, K.: Cement substitution by a combination of metakaolin and limestone. Cem. Concr. Res. **42**, 1579–1589 (2012). doi:10.1016/j.cemconres.2012.09.006
18. Gallucci, E., Zhang, X., Scrivener, K.L.: Effect of temperature on the microstructure of calcium silicate hydrate (C-S-H). Cem. Concr. Res. **53**, 185–195 (2013). doi:10.1016/j.cemconres.2013.06.008

Lightning Source UK Ltd.
Milton Keynes UK
UKHW02n0748210918
329223UK00004B/19/P

9 789402 412062